PERGAMON INTERNATIONAL LIBRARY
of Science, Technology, Engineering and Social Studies
*The 1000-volume original paperback library in aid of education,
industrial training and the enjoyment of leisure*
Publisher: Robert Maxwell, M.C.

COURSE OF THEORETICAL PHYSICS

Volume 6

FLUID MECHANICS

THE PERGAMON TEXTBOOK
INSPECTION COPY SERVICE

An inspection copy of any book published in the Pergamon International Library
will gladly be sent to academic staff without obligation for their consideration for
course adoption or recommendation. Copies may be retained for a period of 60
days from receipt and returned if not suitable. When a particular title is adopted
or recommended for adoption for class use and the recommendation results in a
sale of 12 or more copies, the inspection copy may be retained with our compli-
ments. The Publishers will be pleased to receive suggestions for revised editions
and new titles to be published in this important International Library.

Other Titles in the Series

A Shorter Course of Theoretical Physics

(Based on the Course of Theoretical Physics)

FLUID MECHANICS

by

L. D. LANDAU AND E. M. LIFSHITZ

INSTITUTE OF PHYSICAL PROBLEMS, U.S.S.R. ACADEMY OF SCIENCES

Volume 6 of *Course of Theoretical Physics*

Translated from the Russian by

J. B. SYKES AND W. H. REID

PERGAMON PRESS

Oxford · New York · Toronto · Sydney · Paris · Frankfurt

U.K.	Pergamon Press Ltd., Headington Hill Hall, Oxford OX3 0BW, England
U.S.A.	Pergamon Press Inc., Maxwell House, Fairview Park, Elmsford, New York 10523, U.S.A.
CANADA	Pergamon of Canada, Suite 104, 150 Consumers Road, Willowdale, Ontario M2J 1P9, Canada
AUSTRALIA	Pergamon Press (Aust.) Pty. Ltd., P.O. Box 544, Potts Point, N.S.W. 2011, Australia
FRANCE	Pergamon Press SARL, 24 rue des Ecoles, 75240 Paris, Cedex 05, France
FEDERAL REPUBLIC OF GERMANY	Pergamon Press GmbH, 6242 Kronberg/Taunus, Pferdstrasse 1, Federal Republic of Germany

First edition 1959

Reprinted 1963, 1966, 1975, 1978, 1979

Library of Congress Catalog Card No. 59–10525

Printed in Great Britain by William Clowes & Sons Ltd, Beccles and London

ISBN 0 08 009104 0

CONTENTS

VIII. SOUND

IX. SHOCK WAVES

X. ONE-DIMENSIONAL GAS FLOW

Contents

XVI. DYNAMICS OF SUPERFLUIDS *Page*

XVII. FLUCTUATIONS IN FLUID DYNAMICS

PREFACE TO THE ENGLISH EDITION

THE present book deals with fluid mechanics, i.e. the theory of the motion of liquids and gases.

The nature of the book is largely determined by the fact that it describes fluid mechanics as a branch of theoretical physics, and it is therefore markedly different from other textbooks on the same subject. We have tried to develop as fully as possible all matters of physical interest, and to do so in such a way as to give the clearest possible picture of the phenomena and their interrelation. Accordingly, we discuss neither approximate methods of calculation in fluid mechanics, nor empirical theories devoid of physical significance. On the other hand, accounts are given of some topics not usually found in textbooks on the subject: the theory of heat transfer and diffusion in fluids; acoustics; the theory of combustion; the dynamics of superfluids; and relativistic fluid dynamics.

In a field which has been so extensively studied as fluid mechanics it was inevitable that important new results should have appeared during the several years since the last Russian edition was published. Unfortunately, our preoccupation with other matters has prevented us from including these results in the English edition. We have merely added one further chapter, on the general theory of fluctuations in fluid dynamics.

We should like to express our sincere thanks to Dr Sykes and Dr Reid for their excellent translation of the book, and to Pergamon Press for their ready agreement to our wishes in various matters relating to its publication.

Moscow

L. D. LANDAU
E. M. LIFSHITZ

NOTATION

ρ density

p pressure

T temperature

s entropy per unit mass

ϵ internal energy per unit mass

$w = \epsilon + p/\rho$ heat function (enthalpy)

$\gamma = c_p/c_v$ ratio of specific heats at constant pressure and constant volume

η dynamic viscosity

$\nu = \eta/\rho$ kinematic viscosity

κ thermal conductivity

$\chi = \kappa/\rho c_p$ thermometric conductivity

R   Reynolds number

c velocity of sound

M   ratio of fluid velocity to velocity of sound

CHAPTER I

IDEAL FLUIDS

§1. The equation of continuity

Fluid dynamics concerns itself with the study of the motion of fluids (liquids and gases). Since the phenomena considered in fluid dynamics are macroscopic, a fluid is regarded as a continuous medium. This means that any small volume element in the fluid is always supposed so large that it still contains a very great number of molecules. Accordingly, when we speak of infinitely small elements of volume, we shall always mean those which are "physically" infinitely small, i.e. very small compared with the volume of the body under consideration, but large compared with the distances between the molecules. The expressions *fluid particle* and *point in a fluid* are to be understood in a similar sense. If, for example, we speak of the displacement of some fluid particle, we mean not the displacement of an individual molecule, but that of a volume element containing many molecules, though still regarded as a point.

The mathematical description of the state of a moving fluid is effected by means of functions which give the distribution of the fluid velocity $\mathbf{v} = \mathbf{v}(x, y, z, t)$ and of any two thermodynamic quantities pertaining to the fluid, for instance the pressure $p(x, y, z, t)$ and the density $\rho(x, y, z, t)$. As is well known, all the thermodynamic quantities are determined by the values of any two of them, together with the equation of state; hence, if we are given five quantities, namely the three components of the velocity $\mathbf{v}$, the pressure p and the density ρ, the state of the moving fluid is completely determined.

All these quantities are, in general, functions of the co-ordinates x, y, z and of the time t. We emphasise that $\mathbf{v}(x, y, z, t)$ is the velocity of the fluid at a given point (x, y, z) in space and at a given time t, i.e. it refers to fixed points in space and not to fixed particles of the fluid; in the course of time, the latter move about in space. The same remarks apply to ρ and p.

We shall now derive the fundamental equations of fluid dynamics. Let us begin with the equation which expresses the conservation of matter. We consider some volume V_0 of space. The mass of fluid in this volume is $\int \rho \, dV$, where ρ is the fluid density, and the integration is taken over the volume V_0. The mass of fluid flowing in unit time through an element $\mathbf{df}$ of the surface bounding this volume is $\rho \mathbf{v} \cdot \mathbf{df}$; the magnitude of the vector $\mathbf{df}$ is equal to the area of the surface element, and its direction is along the normal. By convention, we take $\mathbf{df}$ along the outward normal. Then $\rho \mathbf{v} \cdot \mathbf{df}$ is positive if the fluid is flowing out of the volume, and negative if the flow

1

is into the volume. The total mass of fluid flowing out of the volume V_0 in unit time is therefore

$$\oint \rho \mathbf{v} \cdot d\mathbf{f},$$

where the integration is taken over the whole of the closed surface surrounding the volume in question.

Next, the decrease per unit time in the mass of fluid in the volume V_0 can be written

$$-\frac{\partial}{\partial t} \int \rho \, dV.$$

Equating the two expressions, we have

$$\frac{\partial}{\partial t} \int \rho \, dV = -\oint \rho \mathbf{v} \cdot d\mathbf{f}. \tag{1.1}$$

The surface integral can be transformed by Green's formula to a volume integral:

$$\oint \rho \mathbf{v} \cdot d\mathbf{f} = \int \text{div} (\rho \mathbf{v}) \, dV.$$

Thus

$$\int \left[\frac{\partial \rho}{\partial t} + \text{div} (\rho \mathbf{v}) \right] dV = 0.$$

Since this equation must hold for any volume, the integrand must vanish, i.e.

$$\partial \rho / \partial t + \text{div} (\rho \mathbf{v}) = 0. \tag{1.2}$$

This is the *equation of continuity*. Expanding the expression div $(\rho \mathbf{v})$, we can also write (1.2) as

$$\partial \rho / \partial t + \rho \, \text{div} \, \mathbf{v} + \mathbf{v} \cdot \mathbf{grad} \rho = 0. \tag{1.3}$$

The vector

$$\mathbf{j} = \rho \mathbf{v} \tag{1.4}$$

is called the *mass flux density*. Its direction is that of the motion of the fluid, while its magnitude equals the mass of fluid flowing in unit time through unit area perpendicular to the velocity.

§2. Euler's equation

Let us consider some volume in the fluid. The total force acting on this volume is equal to the integral

$$-\oint p \, d\mathbf{f}$$

of the pressure, taken over the surface bounding the volume. Transforming it to a volume integral, we have

$$- \oint p \, d\mathbf{f} = - \int \mathbf{grad}\, p \, dV.$$

Hence we see that the fluid surrounding any volume element dV exerts on that element a force $- dV \,\mathbf{grad}\, p$. In other words, we can say that a force $- \mathbf{grad}\, p$ acts on unit volume of the fluid.

We can now write down the equation of motion of a volume element in the fluid by equating the force $- \mathbf{grad}\, p$ to the product of the mass per unit volume (ρ) and the acceleration $d\mathbf{v}/dt$:

$$\rho \, d\mathbf{v}/dt = - \mathbf{grad}\, p. \tag{2.1}$$

The derivative $d\mathbf{v}/dt$ which appears here denotes not the rate of change of the fluid velocity at a fixed point in space, but the rate of change of the velocity of a given fluid particle as it moves about in space. This derivative has to be expressed in terms of quantities referring to points fixed in space. To do so, we notice that the change $d\mathbf{v}$ in the velocity of the given fluid particle during the time dt is composed of two parts, namely the change during dt in the velocity at a point fixed in space, and the difference between the velocities (at the same instant) at two points $d\mathbf{r}$ apart, where $d\mathbf{r}$ is the distance moved by the given fluid particle during the time dt. The first part is $(\partial \mathbf{v}/\partial t)dt$, where the derivative $\partial \mathbf{v}/\partial t$ is taken for constant x, y, z, i.e. at the given point in space. The second part is

$$dx\frac{\partial \mathbf{v}}{\partial x} + dy\frac{\partial \mathbf{v}}{\partial y} + dz\frac{\partial \mathbf{v}}{\partial z} = (d\mathbf{r}\cdot\mathbf{grad})\mathbf{v}.$$

Thus

$$d\mathbf{v} = (\partial \mathbf{v}/\partial t)dt + (d\mathbf{r}\cdot\mathbf{grad})\mathbf{v},$$

or, dividing both sides by dt,

$$\frac{d\mathbf{v}}{dt} = \frac{\partial \mathbf{v}}{\partial t} + (\mathbf{v}\cdot\mathbf{grad})\mathbf{v}. \tag{2.2}$$

Substituting this in (2.1), we find

$$\frac{\partial \mathbf{v}}{\partial t} + (\mathbf{v}\cdot\mathbf{grad})\mathbf{v} = - \frac{1}{\rho}\mathbf{grad}\, p. \tag{2.3}$$

This is the required equation of motion of the fluid; it was first obtained by L. EULER in 1755. It is called *Euler's equation* and is one of the fundamental equations of fluid dynamics.

If the fluid is in a gravitational field, an additional force $\rho\mathbf{g}$, where $\mathbf{g}$ is the acceleration due to gravity, acts on any unit volume. This force must be added to the right-hand side of equation (2.1), so that equation (2.3) takes the form

$$\frac{\partial \mathbf{v}}{\partial t} + (\mathbf{v} \cdot \mathbf{grad})\mathbf{v} = -\frac{\mathbf{grad}\,p}{\rho} + \mathbf{g}. \tag{2.4}$$

In deriving the equations of motion we have taken no account of processes of energy dissipation, which may occur in a moving fluid in consequence of internal friction (viscosity) in the fluid and heat exchange between different parts of it. The whole of the discussion in this and subsequent sections of this chapter therefore holds good only for motions of fluids in which thermal conductivity and viscosity are unimportant; such fluids are said to be *ideal*.

The absence of heat exchange between different parts of the fluid (and also, of course, between the fluid and bodies adjoining it) means that the motion is adiabatic throughout the fluid. Thus the motion of an ideal fluid must necessarily be supposed adiabatic.

In adiabatic motion the entropy of any particle of fluid remains constant as that particle moves about in space. Denoting by s the entropy per unit mass, we can express the condition for adiabatic motion as

$$\mathrm{d}s/\mathrm{d}t = 0, \tag{2.5}$$

where the total derivative with respect to time denotes, as in (2.1), the rate of change of entropy for a given fluid particle as it moves about. This condition can also be written

$$\partial s/\partial t + \mathbf{v} \cdot \mathbf{grad}\,s = 0. \tag{2.6}$$

This is the general equation describing adiabatic motion of an ideal fluid. Using (1.2), we can write it as an "equation of continuity" for entropy:

$$\partial(\rho s)/\partial t + \mathrm{div}\,(\rho s \mathbf{v}) = 0. \tag{2.7}$$

The product $\rho s \mathbf{v}$ is the "entropy flux density".

It must be borne in mind that the adiabatic equation usually takes a much simpler form. If, as usually happens, the entropy is constant throughout the volume of the fluid at some initial instant, it retains everywhere the same constant value at all times and for any subsequent motion of the fluid. In this case we can write the adiabatic equation simply as

$$s = \mathrm{constant}, \tag{2.8}$$

and we shall usually do so in what follows. Such a motion is said to be *isentropic*.

We may use the fact that the motion is isentropic to put the equation of motion (2.3) in a somewhat different form. To do so, we employ the familiar thermodynamic relation

$$\mathrm{d}w = T\,\mathrm{d}s + V\,\mathrm{d}p,$$

where w is the heat function per unit mass of fluid (enthalpy), $V = 1/\rho$

is the specific volume, and T is the temperature. Since $s =$ constant, we have simply

$$\mathrm{d}w = V\,\mathrm{d}p = \mathrm{d}p/\rho,$$

and so $(\mathbf{grad}\,p)/\rho = \mathbf{grad}\,w$. Equation (2.3) can therefore be written in the form

$$\partial\mathbf{v}/\partial t + (\mathbf{v}\cdot\mathbf{grad})\mathbf{v} = -\mathbf{grad}\,w. \tag{2.9}$$

It is useful to notice one further form of Euler's equation, in which it involves only the velocity. Using a formula well known in vector analysis,

$$\tfrac{1}{2}\mathbf{grad}\,v^2 = \mathbf{v}\times\mathbf{curl}\,\mathbf{v} + (\mathbf{v}\cdot\mathbf{grad})\mathbf{v},$$

we can write (2.9) in the form

$$\partial\mathbf{v}/\partial t + \tfrac{1}{2}\mathbf{grad}\,v^2 - \mathbf{v}\times\mathbf{curl}\,\mathbf{v} = -\mathbf{grad}\,w. \tag{2.10}$$

If we take the curl of both sides of this equation, we obtain

$$\frac{\partial}{\partial t}(\mathbf{curl}\,\mathbf{v}) = \mathbf{curl}(\mathbf{v}\times\mathbf{curl}\,\mathbf{v}), \tag{2.11}$$

which involves only the velocity.

The equations of motion have to be supplemented by the boundary conditions that must be satisfied at the surfaces bounding the fluid. For an ideal fluid, the boundary condition is simply that the fluid cannot penetrate a solid surface. This means that the component of the fluid velocity normal to the bounding surface must vanish if that surface is at rest:

$$v_n = 0. \tag{2.12}$$

In the general case of a moving surface, v_n must be equal to the corresponding component of the velocity of the surface.

At a boundary between two immiscible fluids, the condition is that the pressure and the velocity component normal to the surface of separation must be the same for the two fluids, and each of these velocity components must be equal to the corresponding component of the velocity of the surface.

As has been said at the beginning of §1, the state of a moving fluid is determined by five quantities: the three components of the velocity $\mathbf{v}$ and, for example, the pressure p and the density ρ. Accordingly, a complete system of equations of fluid dynamics should be five in number. For an ideal fluid these are Euler's equations, the equation of continuity, and the adiabatic equation.

PROBLEM

Write down the equations for one-dimensional motion of an ideal fluid in terms of the variables a, t, where a (called a *Lagrangian variable*†) is the x co-ordinate of a fluid particle at some instant $t = t_0$.

---

† Although such variables are usually called Lagrangian, it should be mentioned that the equations of motion in these co-ordinates were first obtained by EULER, at the same time as equations (2.3).

SOLUTION. In these variables the co-ordinate x of any fluid particle at any instant is regarded as a function of t and its co-ordinate a at the initial instant: $x = x(a, t)$. The condition of conservation of mass during the motion of a fluid element (the equation of continuity) is accordingly written $\rho \, dx = \rho_0 \, da$, or

$$\rho \left(\frac{\partial x}{\partial a} \right)_t = \rho_0,$$

where $\rho_0(a)$ is a given initial density distribution. The velocity of a fluid particle is, by definition, $v = (\partial x/\partial t)_a$, and the derivative $(\partial v/\partial t)_a$ gives the rate of change of the velocity of the particle during its motion. Euler's equation becomes

$$\left(\frac{\partial v}{\partial t} \right)_a = -\frac{1}{\rho_0} \left(\frac{\partial p}{\partial a} \right)_t,$$

and the adiabatic equation is

$$(\partial s/\partial t)_a = 0.$$

§3. Hydrostatics

For a fluid at rest in a uniform gravitational field, Euler's equation (2.4) takes the form

$$\mathbf{grad}\, p = \rho \mathbf{g}. \tag{3.1}$$

This equation describes the mechanical equilibrium of the fluid. (If there is no external force, the equation of equilibrium is simply $\mathbf{grad}\, p = 0$, i.e. $p = $ constant; the pressure is the same at every point in the fluid.)

Equation (3.1) can be integrated immediately if the density of the fluid may be supposed constant throughout its volume, i.e. if there is no significant compression of the fluid under the action of the external force. Taking the z-axis vertically upward, we have

$$\partial p/\partial x = \partial p/\partial y = 0, \qquad \partial p/\partial z = -\rho g.$$

Hence

$$p = -\rho g z + \text{constant}.$$

If the fluid at rest has a free surface at height h, to which an external pressure p_0, the same at every point, is applied, this surface must be the horizontal plane $z = h$. From the condition $p = p_0$ for $z = h$, we find that the constant is $p_0 + \rho g h$, so that

$$p = p_0 + \rho g (h - z). \tag{3.2}$$

For large masses of liquid, and for a gas, the density ρ cannot in general be supposed constant; this applies especially to gases (for example, the atmosphere). Let us suppose that the fluid is not only in mechanical equilibrium but also in thermal equilibrium. Then the temperature is the same at every point, and equation (3.1) may be integrated as follows. We use the familiar thermodynamic relation

$$d\Phi = -s\,dT + V\,dp,$$

where Φ is the thermodynamic potential per unit mass. For constant temperature

$$d\Phi = V\,dp = dp/\rho.$$

Hence we see that the expression $(\mathbf{grad}\,p)/\rho$ can be written in this case as $\mathbf{grad}\,\Phi$, so that the equation of equilibrium (3.1) takes the form

$$\mathbf{grad}\,\Phi = \mathbf{g}.$$

For a constant vector $\mathbf{g}$ directed along the negative z-axis we have

$$\mathbf{g} \equiv -\mathbf{grad}\,(gz).$$

Thus

$$\mathbf{grad}\,(\Phi + gz) = 0,$$

whence we find that throughout the fluid

$$\Phi + gz = \text{constant};\qquad(3.3)$$

gz is the potential energy of unit mass of fluid in the gravitational field. The condition (3.3) is known from statistical physics to be the condition for thermodynamic equilibrium of a system in an external field.

We may mention here another simple consequence of equation (3.1). If a fluid (such as the atmosphere) is in mechanical equilibrium in a gravitational field, the pressure in it can be a function only of the altitude z (since, if the pressure were different at different points with the same altitude, motion would result). It then follows from (3.1) that the density

$$\rho = -\frac{1}{g}\frac{dp}{dz}\qquad(3.4)$$

is also a function of z only. The pressure and density together determine the temperature, which is therefore again a function of z only. Thus, in mechanical equilibrium in a gravitational field, the pressure, density and temperature distributions depend only on the altitude. If, for example, the temperature is different at different points with the same altitude, then mechanical equilibrium is impossible.

Finally, let us derive the equation of equilibrium for a very large mass of fluid, whose separate parts are held together by gravitational attraction— a star. Let ϕ be the Newtonian gravitational potential of the field due to the fluid. It satisfies the differential equation

$$\triangle\phi = 4\pi G\rho,\qquad(3.5)$$

where G is the Newtonian constant of gravitation. The gravitational acceleration is $-\mathbf{grad}\,\phi$, and the force on a mass ρ is $-\rho\,\mathbf{grad}\,\phi$. The condition of equilibrium is therefore

$$\mathbf{grad}\,p = -\rho\,\mathbf{grad}\,\phi.$$

Dividing both sides by ρ, taking the divergence of both sides, and using equation (3.5), we obtain

$$\mathrm{div}\left(\frac{1}{\rho}\mathbf{grad}\,p\right) = -4\pi G\rho. \tag{3.6}$$

It must be emphasised that the present discussion concerns only mechanical equilibrium; equation (3.6) does not presuppose the existence of complete thermal equilibrium.

If the body is not rotating, it will be spherical when in equilibrium, and the density and pressure distributions will be spherically symmetrical. Equation (3.6) in spherical co-ordinates then takes the form

$$\frac{1}{r^2}\frac{\mathrm{d}}{\mathrm{d}r}\left(\frac{r^2}{\rho}\frac{\mathrm{d}p}{\mathrm{d}r}\right) = -4\pi G\rho. \tag{3.7}$$

§4. The condition that convection is absent

A fluid can be in mechanical equilibrium (i.e. exhibit no macroscopic motion) without being in thermal equilibrium. Equation (3.1), the condition for mechanical equilibrium, can be satisfied even if the temperature is not constant throughout the fluid. However, the question then arises of the stability of such an equilibrium. It is found that the equilibrium is stable only when a certain condition is fulfilled. Otherwise, the equilibrium is unstable, and this leads to the appearance in the fluid of currents which tend to mix the fluid in such a way as to equalise the temperature. This motion is called *convection*. Thus the condition for a mechanical equilibrium to be stable is the condition that convection is absent. It can be derived as follows.

Let us consider a fluid element at height z, having a specific volume $V(p, s)$, where p and s are the equilibrium pressure and entropy at height z. Suppose that this fluid element undergoes an adiabatic upward displacement through a small interval ξ; its specific volume then becomes $V(p', s)$, where p' is the pressure at height $z + \xi$. For the equilibrium to be stable, it is necessary (though not in general sufficient) that the resulting force on the element should tend to return it to its original position. This means that the element must be heavier than the fluid which it "displaces" in its new position. The specific volume of the latter is $V(p', s')$, where s' is the equilibrium entropy at height $z + \xi$. Thus we have the stability condition

$$V(p', s') - V(p', s) > 0.$$

Expanding this difference in powers of $s' - s = \xi ds/dz$, we obtain

$$\left(\frac{\partial V}{\partial s}\right)_p \frac{ds}{dz} > 0. \tag{4.1}$$

The formulae of thermodynamics give

$$\left(\frac{\partial V}{\partial s}\right)_p = \frac{T}{c_p}\left(\frac{\partial V}{\partial T}\right)_p,$$

where c_p is the specific heat at constant pressure. Both c_p and T are positive, so that we can write (4.1) as

$$\left(\frac{\partial V}{\partial T}\right)_p \frac{ds}{dz} > 0. \tag{4.2}$$

The majority of substances expand on heating, i.e. $(\partial V/\partial T)_p > 0$. The condition that convection is absent then becomes

$$ds/dz > 0, \tag{4.3}$$

i.e. the entropy must increase with height.

From this we easily find the condition that must be satisfied by the temperature gradient dT/dz. Expanding the derivative ds/dz, we have

$$\frac{ds}{dz} = \left(\frac{\partial s}{\partial T}\right)_p \frac{dT}{dz} + \left(\frac{\partial s}{\partial p}\right)_T \frac{dp}{dz} = \frac{c_p}{T}\frac{dT}{dz} - \left(\frac{\partial V}{\partial T}\right)_p \frac{dp}{dz} > 0.$$

Finally, substituting from (3.4) $dp/dz = -g/V$, we obtain

$$\frac{dT}{dz} > -\frac{gT}{c_p V}\left(\frac{\partial V}{\partial T}\right)_p. \tag{4.4}$$

Convection can occur if the temperature falls with increasing height and the magnitude of the temperature gradient exceeds $(gT/c_p V)(\partial V/\partial T)_p$.

If we consider the equilibrium of a column of a perfect gas, then

$$(T/V)(\partial V/\partial T)_p = 1,$$

and the condition for stable equilibrium is simply

$$dT/dz > -g/c_p. \tag{4.5}$$

§5. Bernoulli's equation

The equations of fluid dynamics are much simplified in the case of steady flow. By *steady flow* we mean one in which the velocity is constant in time at any point occupied by fluid. In other words, $\mathbf{v}$ is a function of the co-ordinates only, so that $\partial \mathbf{v}/\partial t = 0$. Equation (2.10) then reduces to

$$\tfrac{1}{2}\mathbf{grad}\,v^2 - \mathbf{v} \times \mathbf{curl}\,\mathbf{v} = -\mathbf{grad}\,w. \tag{5.1}$$

We now introduce the concept of *streamlines*. These are lines such that the tangent to a streamline at any point gives the direction of the velocity at that point; they are determined by the following system of differential equations:

$$\frac{\mathrm{d}x}{v_x} = \frac{\mathrm{d}y}{v_y} = \frac{\mathrm{d}z}{v_z}. \tag{5.2}$$

In steady flow the streamlines do not vary with time, and coincide with the paths of the fluid particles. In non-steady flow this coincidence no longer occurs: the tangents to the streamlines give the directions of the velocities of fluid particles at various points in space at a given instant, whereas the tangents to the paths give the directions of the velocities of given fluid particles at various times.

We form the scalar product of equation (5.1) with the unit vector tangent to the streamline at each point; this unit vector is denoted by **l**. The projection of the gradient on any direction is, as we know, the derivative in that direction. Hence the projection of **grad** w is $\partial w/\partial l$. The vector **v** × **curl v** is perpendicular to **v**, and its projection on the direction of **l** is therefore zero.

Thus we obtain from equation (5.1)

$$\frac{\partial}{\partial l}(\tfrac{1}{2}v^2 + w) = 0.$$

It follows from this that $\tfrac{1}{2}v^2 + w$ is constant along a streamline:

$$\tfrac{1}{2}v^2 + w = \text{constant.} \tag{5.3}$$

In general the constant takes different values for different streamlines. Equation (5.3) is called *Bernoulli's equation*.

If the flow takes place in a gravitational field, the acceleration **g** due to gravity must be added to the right-hand side of equation (5.1). Let us take the direction of gravity as the z-axis, with z increasing upwards. Then the cosine of the angle between the directions of **g** and **l** is equal to the derivative $-\mathrm{d}z/\mathrm{d}l$, so that the projection of **g** on **l** is

$$-g\,\mathrm{d}z/\mathrm{d}l.$$

Accordingly, we now have

$$\frac{\partial}{\partial l}(\tfrac{1}{2}v^2 + w + gz) = 0.$$

Thus Bernoulli's equation states that along a streamline

$$\tfrac{1}{2}v^2 + w + gz = \text{constant.} \tag{5.4}$$

§6. The energy flux

Let us choose some volume element fixed in space, and find how the energy of the fluid contained in this volume element varies with time. The energy of unit volume of fluid is

$$\tfrac{1}{2}\rho v^2 + \rho\epsilon,$$

where the first term is the kinetic energy and the second the internal energy, ϵ being the internal energy per unit mass. The change in this energy is given by the partial derivative

$$\frac{\partial}{\partial t}(\tfrac{1}{2}\rho v^2 + \rho\epsilon).$$

To calculate this quantity, we write

$$\frac{\partial}{\partial t}(\tfrac{1}{2}\rho v^2) = \tfrac{1}{2}v^2\frac{\partial\rho}{\partial t} + \rho\mathbf{v}\cdot\frac{\partial\mathbf{v}}{\partial t},$$

or, using the equation of continuity (1.2) and the equation of motion (2.3),

$$\frac{\partial}{\partial t}(\tfrac{1}{2}\rho v^2) = -\tfrac{1}{2}v^2\operatorname{div}(\rho\mathbf{v}) - \mathbf{v}\cdot\mathbf{grad}\,p - \rho\mathbf{v}\cdot(\mathbf{v}\cdot\mathbf{grad})\mathbf{v}.$$

In the last term we replace $\mathbf{v}\cdot(\mathbf{v}\cdot\mathbf{grad})\mathbf{v}$ by $\tfrac{1}{2}\mathbf{v}\cdot\mathbf{grad}\,v^2$, and $\mathbf{grad}\,p$ by $\rho\,\mathbf{grad}\,w - \rho T\,\mathbf{grad}\,s$ (using the thermodynamic relation $dw = T\,ds + (1/\rho)dp$), obtaining

$$\frac{\partial}{\partial t}(\tfrac{1}{2}\rho v^2) = -\tfrac{1}{2}v^2\operatorname{div}(\rho\mathbf{v}) - \rho\mathbf{v}\cdot\mathbf{grad}\,(\tfrac{1}{2}v^2 + w) + \rho T\mathbf{v}\cdot\mathbf{grad}\,s.$$

In order to transform the derivative $\partial(\rho\epsilon)/\partial t$, we use the thermodynamic relation

$$d\epsilon = T\,ds - p\,dV = T\,ds + (p/\rho^2)d\rho.$$

Since $\epsilon + p/\rho = \epsilon + pV$ is simply the heat function w per unit mass, we find

$$d(\rho\epsilon) = \epsilon\,d\rho + \rho\,d\epsilon = w\,d\rho + \rho T\,ds,$$

and so

$$\frac{\partial(\rho\epsilon)}{\partial t} = w\frac{\partial\rho}{\partial t} + \rho T\frac{\partial s}{\partial t} = -w\operatorname{div}(\rho\mathbf{v}) - \rho T\mathbf{v}\cdot\mathbf{grad}\,s.$$

Here we have also used the general adiabatic equation (2.6).

Combining the above results, we find the change in the energy to be

$$\frac{\partial}{\partial t}(\tfrac{1}{2}\rho v^2 + \rho\epsilon) = -(\tfrac{1}{2}v^2 + w)\operatorname{div}(\rho\mathbf{v}) - \rho\mathbf{v}\cdot\mathbf{grad}(\tfrac{1}{2}v^2 + w),$$

or, finally,

$$\frac{\partial}{\partial t}(\tfrac{1}{2}\rho v^2 + \rho\epsilon) = -\operatorname{div}[\rho\mathbf{v}(\tfrac{1}{2}v^2 + w)]. \tag{6.1}$$

In order to see the meaning of this equation, let us integrate it over some volume:

$$\frac{\partial}{\partial t} \int (\tfrac{1}{2}\rho v^2 + \rho\epsilon)\, \mathrm{d}V = -\int \mathrm{div}\left[\rho\mathbf{v}(\tfrac{1}{2}v^2 + w)\right] \mathrm{d}V,$$

or, converting the volume integral on the right into a surface integral,

$$\frac{\partial}{\partial t} \int (\tfrac{1}{2}\rho v^2 + \rho\epsilon)\, \mathrm{d}V = -\oint \rho\mathbf{v}(\tfrac{1}{2}v^2 + w)\cdot \mathrm{d}\mathbf{f}. \tag{6.2}$$

The left-hand side is the rate of change of the energy of the fluid in some given volume. The right-hand side is therefore the amount of energy flowing out of this volume in unit time. Hence we see that the expression

$$\rho\mathbf{v}(\tfrac{1}{2}v^2 + w) \tag{6.3}$$

may be called the *energy flux density* vector. Its magnitude is the amount of energy passing in unit time through unit area perpendicular to the direction of the velocity.

The expression (6.3) shows that any unit mass of fluid carries with it during its motion an amount of energy $w + \tfrac{1}{2}v^2$. The fact that the heat function w appears here, and not the internal energy ϵ, has a simple physical significance. Putting $w = \epsilon + p/\rho$, we can write the flux of energy through a closed surface in the form

$$-\oint \rho\mathbf{v}(\tfrac{1}{2}v^2 + \epsilon)\cdot \mathrm{d}\mathbf{f} - \oint p\mathbf{v}\cdot \mathrm{d}\mathbf{f}.$$

The first term is the energy (kinetic and internal) transported through the surface in unit time by the mass of fluid. The second term is the work done by pressure forces on the fluid within the surface.

§7. The momentum flux

We shall now give a similar series of arguments for the momentum of the fluid. The momentum of unit volume is $\rho\mathbf{v}$. Let us determine its rate of change, $\partial(\rho\mathbf{v})/\partial t$. We shall use tensor notation.† We have

$$\frac{\partial}{\partial t}(\rho v_i) = \rho\frac{\partial v_i}{\partial t} + \frac{\partial \rho}{\partial t}v_i.$$

---

† The Latin suffixes i, k, ... take the values 1, 2, 3, corresponding to the components of vectors and tensors along the axes x, y, z respectively. We shall write sums of the type $\mathbf{A}\cdot\mathbf{B} = A_1B_1 + A_2B_2 + A_3B_3 = \Sigma A_iB_i$ in the form A_iB_i simply, omitting the summation sign. We shall use a similar procedure in all products involving vectors or tensors: summation over the values 1, 2, 3 is always understood when a Latin suffix appears twice in any term. Such suffixes are sometimes called *dummy suffixes*. In working with dummy suffixes it should be remembered that any pair of such suffixes may be replaced by any other like letters, since the notation used for suffixes that take all possible values obviously does not affect the value of the sum.

Using the equation of continuity (1.2) (with div($\rho\mathbf{v}$) written in the form $\partial(\rho v_k)/\partial x_k$)

$$\frac{\partial \rho}{\partial t} = -\frac{\partial(\rho v_k)}{\partial x_k},$$

and Euler's equation (2.3) in the form

$$\frac{\partial v_i}{\partial t} = -v_k\frac{\partial v_i}{\partial x_k} - \frac{1}{\rho}\frac{\partial p}{\partial x_i},$$

we obtain

$$\frac{\partial}{\partial t}(\rho v_i) = -\rho v_k\frac{\partial v_i}{\partial x_k} - \frac{\partial p}{\partial x_i} - v_i\frac{\partial(\rho v_k)}{\partial x_k}$$

$$= -\frac{\partial p}{\partial x_i} - \frac{\partial}{\partial x_k}(\rho v_i v_k).$$

We write the first term on the right in the form†

$$\frac{\partial p}{\partial x_i} = \delta_{ik}\frac{\partial p}{\partial x_k},$$

and finally obtain

$$\frac{\partial}{\partial t}(\rho v_i) = -\frac{\partial \Pi_{ik}}{\partial x_k}, \tag{7.1}$$

where the tensor Π_{ik} is defined as

$$\Pi_{ik} = p\delta_{ik} + \rho v_i v_k. \tag{7.2}$$

This tensor is clearly symmetrical.

To see the meaning of the tensor Π_{ik}, we integrate equation (7.1) over some volume:

$$\frac{\partial}{\partial t}\int \rho v_i\, dV = -\int \frac{\partial \Pi_{ik}}{\partial x_k}dV.$$

The integral on the right is transformed into a surface integral by Green's formula:‡

$$\frac{\partial}{\partial t}\int \rho v_i\, dV = -\oint \Pi_{ik}\, df_k. \tag{7.3}$$

---

† δ_{ik} denotes the *unit tensor*, i.e. the tensor with components which are unity for $i = k$ and zero for $i \neq k$. It is evident that $\delta_{ik}A_k = A_i$, where A_i is any vector. Similarly, if A_{kl} is a tensor of rank two, we have the relations $\delta_{ik}A_{kl} = A_{il}$, $\delta_{ik}A_{ik} = A_{ii}$, and so on.

‡ The rule for transforming an integral over a closed surface into one over the volume bounded by that surface can be formulated as follows: the surface element df_i must be replaced by the operator $dV \cdot \partial/\partial x_i$, which is to be applied to the whole of the integrand.

The left-hand side is the rate of change of the ith component of the momentum contained in the volume considered. The surface integral on the right is therefore the amount of momentum flowing out through the bounding surface in unit time. Consequently, $\Pi_{ik}df_k$ is the ith component of the momentum flowing through the surface element df. If we write df_k in the form $n_k\,df$, where df is the area of the surface element, and $\mathbf{n}$ is a unit vector along the outward normal, we find that $\Pi_{ik}n_k$ is the flux of the ith component of momentum through unit surface area. We may notice that, according to (7.2), $\Pi_{ik}n_k = pn_i + \rho v_i v_k n_k$. This expression can be written in vector form

$$p\mathbf{n} + \rho\mathbf{v}(\mathbf{v}\cdot\mathbf{n}). \tag{7.4}$$

Thus Π_{ik} is the ith component of the amount of momentum flowing in unit time through unit area perpendicular to the x_k-axis. The tensor Π_{ik} is called the *momentum flux density tensor*. The energy flux is determined by a vector, energy being a scalar; the momentum flux, however, is determined by a tensor of rank two, the momentum itself being a vector.

The vector (7.4) gives the momentum flux in the direction of $\mathbf{n}$, i.e. through a surface perpendicular to $\mathbf{n}$. In particular, taking the unit vector $\mathbf{n}$ to be directed parallel to the fluid velocity, we find that only the longitudinal component of momentum is transported in this direction, and its flux density is $p + \rho v^2$. In a direction perpendicular to the velocity, only the transverse component (relative to $\mathbf{v}$) of momentum is transported, its flux density being just p.

§8. The conservation of circulation

The integral

$$\Gamma = \oint \mathbf{v}\cdot d\mathbf{l},$$

taken along some closed contour, is called the *velocity circulation* round that contour.

Let us consider a closed contour drawn in the fluid at some instant. We suppose it to be a "fluid contour", i.e. composed of the fluid particles that lie on it. In the course of time these particles move about, and the contour moves with them. Let us investigate what happens to the velocity circulation. In other words, let us calculate the time derivative

$$\frac{d}{dt}\oint \mathbf{v}\cdot d\mathbf{l}.$$

We have written here the total derivative with respect to time, since we are seeking the change in the circulation round a "fluid contour" as it moves about, and not round a contour fixed in space.

To avoid confusion, we shall temporarily denote differentiation with respect to the co-ordinates by the symbol δ, retaining the symbol d for differentiation with respect to time. Next, we notice that an element d**l** of the length of the contour can be written as the difference $\delta\mathbf{r}$ between the radius vectors **r** of the points at the ends of the element. Thus we write the velocity circulation as $\oint \mathbf{v} \cdot \delta\mathbf{r}$. In differentiating this integral with respect to time, it must be borne in mind that not only the velocity but also the contour itself (i.e. its shape) changes. Hence, on taking the time differentiation under the integral sign, we must differentiate not only **v** but also $\delta\mathbf{r}$:

$$\frac{d}{dt} \oint \mathbf{v} \cdot \delta\mathbf{r} = \oint \frac{d\mathbf{v}}{dt} \cdot \delta\mathbf{r} + \oint \mathbf{v} \cdot \frac{d\delta\mathbf{r}}{dt}.$$

Since the velocity **v** is just the time derivative of the radius vector **r**, we have

$$\mathbf{v} \cdot \frac{d\delta\mathbf{r}}{dt} = \mathbf{v} \cdot \delta\frac{d\mathbf{r}}{dt} = \mathbf{v} \cdot \delta\mathbf{v} = \delta(\tfrac{1}{2}v^2).$$

The integral of a total differential along a closed contour, however, is zero. The second integral therefore vanishes, leaving

$$\frac{d}{dt} \oint \mathbf{v} \cdot \delta\mathbf{r} = \oint \frac{d\mathbf{v}}{dt} \cdot \delta\mathbf{r}.$$

It now remains to substitute for the acceleration $d\mathbf{v}/dt$ its expression from (2.9):

$$d\mathbf{v}/dt = -\mathbf{grad}\, w.$$

Using Stokes' formula, we then have

$$\oint \frac{d\mathbf{v}}{dt} \cdot \delta\mathbf{r} = \oint \mathbf{curl}\left(\frac{d\mathbf{v}}{dt}\right) \cdot \delta\mathbf{f} = 0,$$

since **curl grad** $w \equiv 0$. Thus, going back to our previous notation, we find[†]

$$\frac{d}{dt} \oint \mathbf{v} \cdot d\mathbf{l} = 0,$$

or

$$\oint \mathbf{v} \cdot d\mathbf{l} = \text{constant.} \tag{8.1}$$

We have therefore reached the conclusion that, in an ideal fluid, the velocity circulation round a closed "fluid" contour is constant in time (*Kelvin's theorem* or the *law of conservation of circulation*).

It should be emphasised that this result has been obtained by using Euler's equation in the form (2.9), and therefore involves the assumption that the

---

[†] This result remains valid in a uniform gravitational field, since in that case **curl g** $\equiv 0$.

flow is isentropic. The theorem does not hold for flows which are not isentropic.†

§9. Potential flow

From the law of conservation of circulation we can derive an important result. Let us at first suppose that the flow is steady, and consider a streamline of which we know that $\boldsymbol{\omega} \equiv \mathbf{curl\ v}$ (the *vorticity*) is zero at some point. We draw an arbitrary infinitely small closed contour to encircle the streamline at that point. By Stokes' theorem, the velocity circulation round any infinitely small contour is equal to $\mathbf{curl\ v} \cdot \mathbf{df}$, where $\mathbf{df}$ is the element of area enclosed by the contour. Since the contour at present under consideration is situated at a point where $\boldsymbol{\omega} \equiv 0$, the velocity circulation round it is zero. In the course of time, this contour moves with the fluid, but always remains infinitely small and always encircles the same streamline. Since the velocity circulation must remain constant, i.e. zero, it follows that $\boldsymbol{\omega}$ must be zero at every point on the streamline.

Thus we reach the conclusion that, if at any point on a streamline $\boldsymbol{\omega} = 0$, the same is true at all other points on that streamline. If the flow is not steady, the same result holds, except that instead of a streamline we must consider the path described in the course of time by some particular fluid particle;‡ we recall that in non-steady flow these paths do not in general coincide with the streamlines.

At first sight it might seem possible to base on this result the following argument. Let us consider steady flow past some body. Let the incident flow be uniform at infinity; its velocity $\mathbf{v}$ is a constant, so that $\boldsymbol{\omega} \equiv 0$ on all streamlines. Hence we conclude that $\boldsymbol{\omega}$ is zero along the whole of every streamline, i.e. in all space.

A flow for which $\boldsymbol{\omega} = 0$ in all space is called a *potential flow* or *irrotational flow*, as opposed to *rotational flow*, in which the vorticity is not everywhere zero. Thus we should conclude that steady flow past any body, with a uniform incident flow at infinity, must be potential flow.

Similarly, from the law of conservation of circulation, we might argue as follows. Let us suppose that at some instant we have potential flow throughout the volume of the fluid. Then the velocity circulation round any closed contour in the fluid is zero.†† By Kelvin's theorem, we could then conclude that this will hold at any future instant, i.e. we should find that, if

---

† Mathematically, it is necessary that there should be a one-to-one relation between p and ρ (which for isentropic flow is $s(p, \rho) = $ constant); then $-(1/\rho)\,\mathbf{grad}\,p$ can be written as the gradient of some function, a result which is needed in deriving Kelvin's theorem.

‡ To avoid misunderstanding, we may mention here that this result has no meaning in turbulent flow (cf. Chapter III). We may also remark that a non-zero vorticity may occur on a streamline after the passage of a shock wave. We shall see that this is because the flow is no longer isentropic, and the law of conservation of circulation cannot then be derived (§106).

†† Here we suppose for simplicity that the fluid occupies a simply-connected region of space. The same final result would be obtained for a multiply-connected region, but restrictions on the choice of contours would have to be made in the derivation.

there is potential flow at some instant, then there is potential flow at all subsequent instants (in particular, any flow for which the fluid is initially at rest must be a potential flow). This is in accordance with the fact that, if $\boldsymbol{\omega} = 0$, equation (2.11) is satisfied identically.

In fact, however, all these conclusions are of only very limited validity. The reason is that the proof given above that $\boldsymbol{\omega} = 0$ all along a streamline is, strictly speaking, invalid for a line which lies in the surface of a solid body past which the flow takes place, since the presence of this surface makes it impossible to draw a closed contour in the fluid encircling such a streamline. The equations of motion of an ideal fluid therefore admit solutions for which *separation* occurs at the surface of the body: the streamlines, having followed the surface for some distance, become separated from it at some point and continue into the fluid. The resulting flow pattern is characterised by the presence of a "surface of tangential discontinuity" proceeding from the body; on this surface the fluid velocity, which is everywhere tangential to the surface, has a discontinuity. In other words, at this surface one layer of fluid "slides" on another. Fig. 1 shows a surface of discontinuity which separates moving fluid from a region of stationary fluid behind the body.

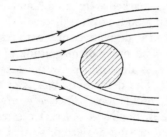

Fig. 1

From a mathematical point of view, the discontinuity in the tangential velocity component corresponds to a surface on which the vorticity is non-zero.

When such discontinuous flows are included, the solution of the equations of motion for an ideal fluid is not unique: besides continuous flow, they admit also an infinite number of solutions possessing surfaces of tangential discontinuity starting from any prescribed line on the surface of the body past which the flow takes place. It should be emphasised, however, that none of these discontinuous solutions is physically significant, since tangential discontinuities are wholly unstable, and therefore the flow would in fact become turbulent (see Chapter III).

The actual physical problem of flow past a given body has, of course, a unique solution. The reason is that ideal fluids do not really exist; any actual fluid has a certain viscosity, however small. This viscosity may have practically no effect on the motion of most of the fluid, but, no matter how small it is, it will be important in a thin layer of fluid adjoining the body.

The properties of the flow in this *boundary layer* decide the choice of one out of the infinity of solutions of the equations of motion for an ideal fluid. It is found that, in the general case of flow past bodies of arbitrary form, solutions with separation must be rejected; separation, if it occurred, would result in turbulence.

In spite of what we have said above, the study of the solutions of the equations of motion for continuous steady potential flow past bodies is in some cases meaningful. Although, in the general case of flow past bodies of arbitrary form, the actual flow pattern bears almost no relation to the pattern of potential flow, for bodies of certain special ("streamlined"—§46) shapes the flow may differ very little from potential flow; more precisely, it will be potential flow except in a thin layer of fluid at the surface of the body and in a relatively narrow "wake" behind the body.

Another important case of potential flow occurs for small oscillations of a body immersed in fluid. It is easy to show that, if the amplitude a of the oscillations is small compared with the linear dimension l of the body ($a \ll l$), the flow past the body will be potential flow. To show this, we estimate the order of magnitude of the various terms in Euler's equation

$$\partial \mathbf{v}/\partial t + (\mathbf{v} \cdot \mathbf{grad})\mathbf{v} = -\mathbf{grad}\, w.$$

The velocity $\mathbf{v}$ changes markedly (by an amount of the same order as the velocity $\mathbf{u}$ of the oscillating body) over a distance of the order of the dimension l of the body. Hence the derivatives of $\mathbf{v}$ with respect to the co-ordinates are of the order of u/l. The order of magnitude of $\mathbf{v}$ itself (at fairly small distances from the body) is determined by the magnitude of $\mathbf{u}$. Thus we have $(\mathbf{v} \cdot \mathbf{grad})\mathbf{v} \sim u^2/l$. The derivative $\partial \mathbf{v}/\partial t$ is of the order of ωu, where ω is the frequency of the oscillations. Since $\omega \sim u/a$, we have $\partial \mathbf{v}/\partial t \sim u^2/a$. It now follows from the inequality $a \ll l$ that the term $(\mathbf{v} \cdot \mathbf{grad})\mathbf{v}$ is small compared with $\partial \mathbf{v}/\partial t$ and can be neglected, so that the equation of motion of the fluid becomes $\partial \mathbf{v}/\partial t = -\mathbf{grad}\, w$. Taking the curl of both sides, we obtain $\partial(\mathbf{curl\ v})/\partial t = 0$, whence $\mathbf{curl\ v} = $ constant. In oscillatory motion, however, the time average of the velocity is zero, and therefore $\mathbf{curl\ v} = $ constant implies that $\mathbf{curl\ v} = 0$. Thus the motion of a fluid executing small oscillations is potential flow to a first approximation.

We shall now obtain some general properties of potential flow. We first recall that the derivation of the law of conservation of circulation, and therefore all its consequences, were based on the assumption that the flow is isentropic. If the flow is not isentropic, the law does not hold, and therefore, even if we have potential flow at some instant, the vorticity will in general be non-zero at subsequent instants. Thus only isentropic flow can in fact be potential flow.

According to Stokes' theorem,

$$\oint \mathbf{v} \cdot d\mathbf{l} = \oint \mathbf{curl\, v} \cdot d\mathbf{f},$$

where the integral on the right is taken over a surface bounded by the contour in question. Hence we see that, in potential flow, the velocity circulation round any closed contour is zero:

$$\oint \mathbf{v} \cdot d\mathbf{l} = 0. \tag{9.1}$$

It follows from this that, in particular, closed streamlines cannot exist in potential flow.† For, since the direction of a streamline is at every point the direction of the velocity, the circulation along such a line can never be zero.

In rotational motion the velocity circulation is not in general zero. In this case there may be closed streamlines, but it must be emphasised that the presence of closed streamlines is not a necessary property of rotational motion.

Like any vector field having zero curl, the velocity in potential flow can be expressed as the gradient of some scalar. This scalar is called the *velocity potential*; we shall denote it by ϕ:

$$\mathbf{v} = \mathbf{grad}\,\phi. \tag{9.2}$$

Writing Euler's equation in the form (2.10)

$$\partial \mathbf{v}/\partial t + \tfrac{1}{2}\mathbf{grad}\,v^2 - \mathbf{v} \times \mathbf{curl}\,\mathbf{v} = -\mathbf{grad}\,w$$

and substituting $\mathbf{v} = \mathbf{grad}\,\phi$, we have

$$\mathbf{grad}\left(\frac{\partial \phi}{\partial t} + \tfrac{1}{2}v^2 + w\right) = 0,$$

whence

$$\partial \phi/\partial t + \tfrac{1}{2}v^2 + w = f(t), \tag{9.3}$$

where $f(t)$ is an arbitrary function of time. This equation is a first integral of the equations of potential flow. The function $f(t)$ in equation (9.3) can be put equal to zero without loss of generality. For, since the velocity is the space derivative of ϕ, we can add to ϕ any function of the time; replacing ϕ by $\phi + \int f(t)dt$, we obtain zero on the right-hand side of (9.3).

For steady flow we have (taking the potential ϕ to be independent of time) $\partial \phi/\partial t = 0$, $f(t) = $ constant, and (9.3) becomes Bernoulli's equation:

$$\tfrac{1}{2}v^2 + w = \text{constant}. \tag{9.4}$$

It must be emphasised here that there is an important difference between the Bernoulli's equation for potential flow and that for other flows. In the general case, the "constant" on the right-hand side is a constant along any given streamline, but is different for different streamlines. In potential flow,

---

† This result, like (9.1), may not be valid for motion in a multiply-connected region of space. In potential flow in such a region, the velocity circulation may be non-zero if the closed contour round which it is taken cannot be contracted to a point without crossing the boundaries of the region.

however, it is constant throughout the fluid. This enhances the importance of Bernoulli's equation in the study of potential flow.

§10. Incompressible fluids

In a great many cases of the flow of liquids (and also of gases), their density may be supposed invariable, i.e. constant throughout the volume of the fluid and throughout its motion. In other words, there is no noticeable compression or expansion of the fluid in such cases. We then speak of *incompressible flow*.

The general equations of fluid dynamics are much simplified for an incompressible fluid. Euler's equation, it is true, is unchanged if we put ρ = constant, except that ρ can be taken under the gradient operator in equation (2.4):

$$\frac{\partial \mathbf{v}}{\partial t} + (\mathbf{v} \cdot \mathbf{grad})\mathbf{v} = -\mathbf{grad}\left(\frac{p}{\rho}\right) + \mathbf{g}. \tag{10.1}$$

The equation of continuity, on the other hand, takes for constant ρ the simple form

$$\operatorname{div} \mathbf{v} = 0. \tag{10.2}$$

Since the density is no longer an unknown function as it was in the general case, the fundamental system of equations in fluid dynamics for an incompressible fluid can be taken to be equations involving the velocity only These may be the equation of continuity (10.2) and equation (2.11):

$$\frac{\partial}{\partial t}(\mathbf{curl\,v}) = \mathbf{curl}\,(\mathbf{v} \times \mathbf{curl\,v}). \tag{10.3}$$

Bernoulli's equation can be written in a simpler form for an incompressible fluid. Equation (10.1) differs from the general Euler's equation (2.9) in that it has $\mathbf{grad}\,(p/\rho)$ in place of $\mathbf{grad}\,w$. Hence we can write down Bernoulli's equation immediately by simply replacing the heat function in (5.4) by p/ρ:

$$\tfrac{1}{2}v^2 + p/\rho + gz = \text{constant}. \tag{10.4}$$

For an incompressible fluid, we can also write p/ρ in place of w in the expression (6.3) for the energy flux, which then becomes

$$\rho\mathbf{v}\left(\tfrac{1}{2}v^2 + \frac{p}{\rho}\right). \tag{10.5}$$

For we have, from a well-known thermodynamic relation, the expression $d\epsilon = T ds - p dV$ for the change in internal energy; for s = constant and $V = 1/\rho$ = constant, $d\epsilon = 0$, i.e. ϵ = constant. Since constant terms in the energy do not matter, we can omit ϵ in $w = \epsilon + p/\rho$.

The equations are particularly simple for potential flow of an incompressible fluid. Equation (10.3) is satisfied identically if **curl v** = 0. Equation (10.2), with the substitution **v** = **grad** ϕ, becomes

$$\triangle\phi = 0, \tag{10.6}$$

i.e. Laplace's equation† for the potential ϕ. This equation must be supplemented by boundary conditions at the surfaces where the fluid meets solid bodies. At fixed solid surfaces, the fluid velocity component v_n normal to the surface must be zero, whilst for moving surfaces it must be equal to the normal component of the velocity of the surface (a given function of time). The velocity v_n, however, is equal to the normal derivative of the potential ϕ: $v_n = \partial\phi/\partial n$. Thus the general boundary conditions are that $\partial\phi/\partial n$ is a given function of co-ordinates and time at the boundaries.

For potential flow, the velocity is related to the pressure by equation (9.3). In an incompressible fluid, we can replace w in this equation by p/ρ:

$$\partial\phi/\partial t + \tfrac{1}{2}v^2 + p/\rho = f(t). \tag{10.7}$$

We may notice here the following important property of potential flow of an incompressible fluid. Suppose that some solid body is moving through the fluid. If the result is potential flow, it depends at any instant only on the velocity of the moving body at that instant, and not, for example, on its acceleration. For equation (10.6) does not explicitly contain the time, which enters the solution only through the boundary conditions, and these contain only the velocity of the moving body.

Fig. 2

From Bernoulli's equation, $\tfrac{1}{2}v^2 + p/\rho$ = constant, we see that, in steady flow of an incompressible fluid (not in a gravitational field), the greatest pressure occurs at points where the velocity is zero. Such a point usually occurs on the surface of a body past which the fluid is moving (at the point O in Fig. 2), and is called a *stagnation point*. If **u** is the velocity of the

---

† The velocity potential was first introduced by EULER, who obtained an equation of the form (10.6) for it; this form later became known as Laplace's equation.

incident current (i.e. the fluid velocity at infinity), and p_0 the pressure at infinity, the pressure at the stagnation point is

$$p_{\max} = p_0 + \tfrac{1}{2}\rho u^2. \tag{10.8}$$

If the velocity distribution in a moving fluid depends on only two co-ordinates (x and y, say), and the velocity is everywhere parallel to the xy-plane, the flow is said to be *two-dimensional* or *plane flow*. To solve problems of two-dimensional flow of an incompressible fluid, it is sometimes convenient to express the velocity in terms of what is called the *stream function*. From the equation of continuity div $\mathbf{v} \equiv \partial v_x/\partial x + \partial v_y/\partial y = 0$ we see that the velocity components can be written as the derivatives

$$v_x = \partial\psi/\partial y, \qquad v_y = -\partial\psi/\partial x \tag{10.9}$$

of some function $\psi(x, y)$, called the stream function. The equation of continuity is then satisfied automatically. The equation that must be satisfied by the stream function is obtained by substituting (10.9) in equation (10.3). We then obtain

$$\frac{\partial}{\partial t}\triangle\psi - \frac{\partial\psi}{\partial x}\frac{\partial}{\partial y}\triangle\psi + \frac{\partial\psi}{\partial y}\frac{\partial}{\partial x}\triangle\psi = 0. \tag{10.10}$$

If we know the stream function we can immediately determine the form of the streamlines for steady flow. For the differential equation of the stream-lines (in two-dimensional flow) is $dx/v_x = dy/v_y$ or $v_y\,dx - v_x\,dy = 0$; it expresses the fact that the direction of the tangent to a streamline is the direction of the velocity. Substituting (10.9), we have

$$\frac{\partial\psi}{\partial x}dx + \frac{\partial\psi}{\partial v}dy = d\psi = 0,$$

whence $\psi =$ constant. Thus the streamlines are the family of curves obtained by putting the stream function $\psi(x, y)$ equal to an arbitrary constant.

If we draw a curve between two points A and B in the xy-plane, the mass flux Q across this curve is given by the difference in the values of the stream function at these two points, regardless of the shape of the curve. For, if v_n is the component of the velocity normal to the curve at any point, we have

$$Q = \rho \oint_A^B v_n\,dl = \rho \oint_A^B (-v_y\,dx + v_x\,dy) = \rho \int_A^B d\psi,$$

or

$$Q = \rho(\psi_B - \psi_A). \tag{10.11}$$

There are powerful methods of solving problems of two-dimensional potential flow of an incompressible fluid past bodies of various profiles, involving the application of the theory of functions of a complex variable.† The basis of these methods is as follows. The potential and the stream function are related to the velocity components by

$$v_x = \partial\phi/\partial x = \partial\psi/\partial y, \qquad v_y = \partial\phi/\partial y = -\partial\psi/\partial x.$$

These relations between the derivatives of ϕ and ψ, however, are the same, mathematically, as the well-known Cauchy–Riemann conditions for a complex expression

$$w = \phi + i\psi \tag{10.12}$$

to be an analytic function of the complex argument $z = x + iy$. This means that the function $w(z)$ has at every point a well-defined derivative

$$\frac{dw}{dz} = \frac{\partial\phi}{\partial x} + i\frac{\partial\psi}{\partial x} = v_x - iv_y. \tag{10.13}$$

The function w is called the *complex potential*, and dw/dz the *complex velocity*. The modulus and argument of the latter give the magnitude v of the velocity and the angle θ between the direction of the velocity and that of the x-axis:

$$dw/dz = ve^{-i\theta}. \tag{10.14}$$

At a solid surface past which the flow takes place, the velocity must be along the tangent. That is, the profile contour of the surface must be a streamline, i.e. $\psi = $ constant along it; the constant may be taken as zero, and then the problem of flow past a given contour reduces to the determination of an analytic function $w(z)$ which takes real values on the contour. The statement of the problem is more involved when the fluid has a free surface; an example is found in Problem 9.

The integral of an analytic function round any closed contour C is well known to be equal to $2\pi i$ times the sum of the residues of the function at its simple poles inside C; hence

$$\oint w' dz = 2\pi i \sum_k A_k,$$

where A_k are the residues of the complex velocity. We also have

$$\oint w' \, dz = \oint (v_x - iv_y)(dx + idy)$$

$$= \oint (v_x \, dx + v_y \, dy) + i \oint (v_x \, dy - v_y \, dx).$$

---

† A more detailed account of these methods and their various applications is given by N. E. KOCHIN, I. A. KIBEL' and N. V. ROZE, *Theoretical Hydromechanics* (*Teoreticheskaya gidromekhanika*), Part 1, 4th ed., Moscow 1948; L. I. SEDOV, *Two-dimensional Problems of Hydrodynamics and Aerodynamics* (*Ploskie zadachi gidrodinamiki i aёrodinamiki*), Moscow 1950.

The real part of this expression is just the velocity circulation Γ round the contour C. The imaginary part, multiplied by ρ, is the mass flux across C; if there are no sources of fluid within the contour, this flux is zero and we then have simply

$$\Gamma = 2\pi i \sum_k A_k; \tag{10.15}$$

all the residues A_k are in this case purely imaginary.

Finally, let us consider the conditions under which the fluid may be regarded as incompressible. When the pressure changes adiabatically by Δp, the density changes by $\Delta\rho = (\partial\rho/\partial p)_s \Delta p$. According to Bernoulli's equation, however, Δp is of the order of ρv^2 in steady flow. Thus $\Delta\rho \sim (\partial\rho/\partial p)_s \rho v^2$. We shall show in §63 that the derivative $(\partial\rho/\partial p)_s$ is the square of the velocity c of sound in the fluid, so that $\Delta\rho \sim \rho v^2/c^2$. The fluid may be regarded as incompressible if $\Delta\rho/\rho \ll 1$. We see that a necessary condition for this is that the fluid velocity should be small compared with that of sound:

$$v \ll c. \tag{10.16}$$

However, this condition is sufficient only in steady flow. In non-steady flow, a further condition must be fulfilled. Let τ and l be a time and a length of the order of the times and distances over which the fluid velocity undergoes significant changes. If the terms $\partial\mathbf{v}/\partial t$ and $(1/\rho)\,\mathbf{grad}\,p$ in Euler's equation are comparable, we find, in order of magnitude, $v/\tau \sim \Delta p/l\rho$ or $\Delta p \sim l\rho v/\tau$, and the corresponding change in ρ is $\Delta\rho \sim l\rho v/\tau c^2$. Now comparing the terms $\partial\rho/\partial t$ and $\rho\,\mathrm{div}\,\mathbf{v}$ in the equation of continuity, we find that the derivative $\partial\rho/\partial t$ may be neglected (i.e. we may suppose ρ constant) if $\Delta\rho/\tau \ll \rho v/l$, or

$$\tau \gg l/c. \tag{10.17}$$

If the conditions (10.16) and (10.17) are both fulfilled, the fluid may be regarded as incompressible. The condition (10.17) has an obvious meaning: the time l/c taken by a sound signal to traverse the distance l must be small compared with the time τ during which the flow changes appreciably, so that the propagation of interactions in the fluid may be regarded as instantaneous.

PROBLEMS

PROBLEM 1. Determine the shape of the surface of an incompressible fluid subject to a gravitational field, contained in a cylindrical vessel which rotates about its (vertical) axis with a constant angular velocity Ω.

SOLUTION. Let us take the axis of the cylinder as the z-axis. Then $v_x = -y\Omega$, $v_y = x\Omega$,

$v_z = 0$. The equation of continuity is satisfied identically, and Euler's equation (10.1) gives

$$x\Omega^2 = \frac{1}{\rho}\frac{\partial p}{\partial x}, \qquad y\Omega^2 = \frac{1}{\rho}\frac{\partial p}{\partial y}, \qquad \frac{1}{\rho}\frac{\partial p}{\partial z} + g = 0.$$

The general integral of these equations is

$$p/\rho = \tfrac{1}{2}\Omega^2(x^2 + y^2) - gz + \text{constant}.$$

At the free surface $p = $ constant, so that the surface is a paraboloid:

$$z = \tfrac{1}{2}\Omega^2(x^2 + y^2)/g,$$

the origin being taken at the lowest point of the surface.

PROBLEM 2. A sphere, of radius R, moves with velocity $\mathbf{u}$ in an incompressible ideal fluid. Determine the potential flow of the fluid past the sphere.

SOLUTION. The fluid velocity must vanish at infinity. The solutions of Laplace's equation $\triangle\phi = 0$ which vanish at infinity are well known to be $1/r$ and the derivatives, of various orders, of $1/r$ with respect to the co-ordinates (the origin is taken at the centre of the sphere). On account of the complete symmetry of the sphere, only one constant vector, the velocity $\mathbf{u}$, can appear in the solution, and, on account of the linearity of both Laplace's equation and the boundary condition, ϕ must involve $\mathbf{u}$ linearly. The only scalar which can be formed from $\mathbf{u}$ and the derivatives of $1/r$ is the scalar product $\mathbf{u}\cdot\mathbf{grad}(1/r)$. We therefore seek ϕ in the form

$$\phi = \mathbf{A}\cdot\mathbf{grad}\,(1/r) = -(\mathbf{A}\cdot\mathbf{n})/r^2,$$

where $\mathbf{n}$ is a unit vector in the direction of $\mathbf{r}$. The constant $\mathbf{A}$ is determined from the condition that the normal components of the velocities $\mathbf{v}$ and $\mathbf{u}$ must be equal at the surface at the sphere, i.e. $\mathbf{v}\cdot\mathbf{n} = \mathbf{u}\cdot\mathbf{n}$ for $r = R$. This condition gives $\mathbf{A} = \tfrac{1}{2}\mathbf{u}R^3$, so that

$$\phi = -\frac{R^3}{2r^2}\mathbf{u}\cdot\mathbf{n}, \qquad \mathbf{v} = \frac{R^3}{2r^3}[3\mathbf{n}(\mathbf{u}\cdot\mathbf{n}) - \mathbf{u}].$$

The pressure distribution is given by equation (10.7):

$$p = p_0 - \tfrac{1}{2}\rho v^2 - \rho\,\partial\phi/\partial t,$$

where p_0 is the pressure at infinity. To calculate the derivative $\partial\phi/\partial t$, we must bear in mind that the origin (which we have taken at the centre of the sphere) moves with velocity $\mathbf{u}$. Hence

$$\partial\phi/\partial t = (\partial\phi/\partial\mathbf{u})\cdot\dot{\mathbf{u}} - \mathbf{u}\cdot\mathbf{grad}\,\phi.$$

The pressure distribution over the surface of the sphere is given by the formula

$$p = p_0 + \tfrac{1}{8}\rho u^2(9\cos^2\theta - 5) + \tfrac{1}{2}\rho R\mathbf{n}\cdot d\mathbf{u}/dt,$$

where θ is the angle between $\mathbf{n}$ and $\mathbf{u}$.

PROBLEM 3. The same as Problem 2, but for an infinite cylinder moving perpendicular to its axis.†

---

† The solution of the more general problems of potential flow past an ellipsoid and an elliptical cylinder may be found in: N. E. KOCHIN, I. A. KIBEL' and N. V. ROZE, *Theoretical Hydromechanics* (*Teoreticheskaya gidromekhanika*), Part 1, 4th ed., pp. 265 and 355, Moscow 1948; H. LAMB, *Hydrodynamics*, 6th ed., §§103–116, Cambridge 1932.

SOLUTION. The flow is independent of the axial co-ordinate, so that we have to solve Laplace's equation in two dimensions. The solutions which vanish at infinity are the first and higher derivatives of $\log r$ with respect to the co-ordinates, where $\mathbf{r}$ is the radius vector perpendicular to the axis of the cylinder. We seek a solution in the form

$$\phi = \mathbf{A} \cdot \mathbf{grad} \log r = \mathbf{A} \cdot \mathbf{n}/r,$$

and from the boundary conditions we obtain $\mathbf{A} = -R^2\mathbf{u}$, so that

$$\phi = -\frac{R^2}{r}\mathbf{u} \cdot \mathbf{n}, \qquad \mathbf{v} = \frac{R^2}{r^2}[2\mathbf{n}(\mathbf{u} \cdot \mathbf{n}) - \mathbf{u}].$$

The pressure at the surface of the cylinder is given by the formula

$$p = p_0 + \tfrac{1}{2}\rho u^2(4\cos^2\theta - 3) + \rho\, R\mathbf{n} \cdot \mathrm{d}\mathbf{u}/\mathrm{d}t.$$

PROBLEM 4. Determine the potential flow of an incompressible ideal fluid in an ellipsoidal vessel rotating about a principal axis with angular velocity Ω, and determine the total angular momentum of the fluid.

SOLUTION. We take Cartesian co-ordinates x, y, z along the axes of the ellipsoid at a given instant, the z-axis being the axis of rotation. The velocity of points in the vessel is

$$\mathbf{u} = \mathbf{\Omega} \times \mathbf{r},$$

so that the boundary condition $v_n = \partial\phi/\partial n = u_n$ is

$$\partial\phi/\partial n = \Omega(xn_y - yn_x),$$

or, using the equation of the ellipsoid $x^2/a^2 + y^2/b^2 + z^2/c^2 = 1$,

$$\frac{x}{a^2}\frac{\partial\phi}{\partial x} + \frac{y}{b^2}\frac{\partial\phi}{\partial y} + \frac{z}{c^2}\frac{\partial\phi}{\partial z} = xy\Omega\left(\frac{1}{b^2} - \frac{1}{a^2}\right).$$

The solution of Laplace's equation which satisfies this boundary condition is

$$\phi = \Omega\frac{a^2 - b^2}{a^2 + b^2}xy. \tag{1}$$

The angular momentum of the fluid in the vessel is

$$M = \rho \int (xv_y - yv_x)\mathrm{d}V.$$

Integrating over the volume V of the ellipsoid, we have

$$M = \frac{\Omega\rho V}{5}\frac{(a^2 - b^2)^2}{a^2 + b^2}.$$

Formula (1) gives the absolute motion of the fluid relative to the instantaneous position of the axes x, y, z which are fixed to the rotating vessel. The motion relative to the vessel (i.e. relative to a rotating system of co-ordinates x, y, z) is found by subtracting the velocity $\mathbf{\Omega} \times \mathbf{r}$ from the absolute velocity; denoting the relative velocity of the fluid by $\mathbf{v}'$, we have

$$v'_x = \frac{\partial\phi}{\partial x} + y\Omega = \frac{2\Omega a^2}{a^2 + b^2}y, \qquad v'_y = -\frac{2\Omega b^2}{a^2 + b^2}x, \qquad v'_z = 0.$$

The paths of the relative motion are found by integrating the equations $\dot{x} = v'_x$, $\dot{y} = v'_y$, and are the ellipses $x^2/a^2 + y^2/b^2 = $ constant, which are similar to the boundary ellipse.

PROBLEM 5. Determine the flow near a stagnation point (Fig. 2).

SOLUTION. A small part of the surface of the body near the stagnation point may be regarded as plane. Let us take it as the xy-plane. Expanding ϕ for x, y, z small, we have as far as the second-order terms

$$\phi = ax + by + cz + Ax^2 + By^2 + Cz^2 + Dxy + Eyz + Fzx;$$

a constant term in ϕ is immaterial. The constant coefficients are determined so that ϕ satisfies the equation $\triangle \phi = 0$ and the boundary conditions $v_z = \partial\phi/\partial z = 0$ for $z = 0$ and all x, y, $\partial\phi/\partial x = \partial\phi/\partial y = 0$ for $x = y = z = 0$ (the stagnation point). This gives $a = b = c = 0$; $C = -A - B$, $E = F = 0$. The term Dxy can always be removed by an appropriate rotation of the x and y axes. We then have

$$\phi = Ax^2 + By^2 - (A + B)z^2. \tag{1}$$

If the flow is axially symmetrical about the z-axis (symmetrical flow past a solid of revolution), we must have $A = B$, so that

$$\phi = A(x^2 + y^2 - 2z^2).$$

The velocity components are $v_x = 2Ax$, $v_y = 2Ay$, $v_z = -4Az$. The streamlines are given by equations (5.2), from which we find $x^2z = c_1$, $y^2z = c_2$, i.e. the streamlines are cubical hyperbolae.

If the flow is uniform in the y-direction (e.g. flow in the z-direction past a cylinder with its axis in the y-direction), we must have $B = 0$ in (1), so that

$$\phi = A(x^2 - z^2).$$

The streamlines are the hyperbolae $xz = $ constant.

PROBLEM 6. Determine the potential flow near an angle formed by two intersecting planes.

SOLUTION. Let us take polar co-ordinates r, θ in the cross-sectional plane (perpendicular to the line of intersection), with the origin at the vertex of the angle; θ is measured from one of the arms of the angle. Let the angle be α radians; for $\alpha < \pi$ the flow takes place within the angle, for $\alpha > \pi$ outside it. The boundary condition that the normal velocity component vanishes means that $\partial\phi/\partial\theta = 0$ for $\theta = 0$ and $\theta = \alpha$. The solution of Laplace's equation satisfying these conditions can be written†

$$\phi = Ar^n \cos n\theta, \qquad n = \pi/\alpha,$$

so that

$$v_r = nAr^{n-1} \cos n\theta, \qquad v_\theta = -nAr^n \sin n\theta.$$

For $n < 1$ (flow outside an angle; Fig. 3), v_r becomes infinite as $1/r^{1-n}$ at the origin. For $n > 1$ (flow inside an angle; Fig. 4), v_r becomes zero for $r = 0$.

The stream function, which gives the form of the streamlines, is $\psi = Ar^n \sin n\theta$. The expressions obtained for ϕ and ψ are the real and imaginary parts of the complex potential $w = Az^n$.

PROBLEM 7. A spherical hole of radius a is suddenly formed in an incompressible fluid filling all space. Determine the time taken for the hole to be filled with fluid (RAYLEIGH 1917).

SOLUTION. The flow after the formation of the hole will be spherically symmetrical, the

---

† We take the solution which involves the lowest positive power of r, since r is small.

velocity at every point being directed to the centre of the hole. For the radial velocity $v_r \equiv v < 0$ we have Euler's equation in spherical polar co-ordinates:

$$\frac{\partial v}{\partial t} + v\frac{\partial v}{\partial r} = -\frac{1}{\rho}\frac{\partial p}{\partial r}. \tag{1}$$

The equation of continuity gives

$$r^2 v = F(t), \tag{2}$$

where $F(t)$ is an arbitrary function of time; this equation expresses the fact that, since the fluid is incompressible, the volume flowing through any spherical surface is independent of the radius of that surface.

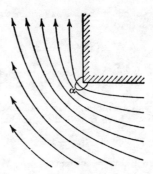

Fig. 3

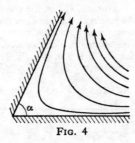

Fig. 4

Substituting v from (2) in (1), we have

$$\frac{F'(t)}{r^2} + v\frac{\partial v}{\partial r} = -\frac{1}{\rho}\frac{\partial p}{\partial r}.$$

Integrating this equation over r from the instantaneous radius $R = R(t) \leqslant a$ of the hole to infinity, we obtain

$$-\frac{F'(t)}{R} + \tfrac{1}{2}V^2 = \frac{p_0}{\rho} \tag{3}$$

where $V = dR(t)/dt$ is the rate of change of the radius of the hole, and p_0 is the pressure at infinity; the fluid velocity at infinity is zero, and so is the pressure at the surface of the hole. From equation (2) for points on the surface of the hole we find

$$F(t) = R^2(t) \, V(t),$$

and, substituting this expression for $F(t)$ in (3), we obtain the equation

$$-\frac{3V^2}{2} - \tfrac{1}{2}R\frac{dV^2}{dR} = \frac{p_0}{\rho}. \tag{4}$$

Integrating with the boundary condition $V = 0$ for $R = a$ (the fluid being initially at rest), we have

$$V \equiv \frac{dR}{dt} = -\sqrt{\left[\frac{2p_0}{3\rho}\left(\frac{a^3}{R^3} - 1\right)\right]}.$$

Hence we have for the required total time for the hole to be filled

$$\tau = \sqrt{\frac{3\rho}{2p_0}} \int_0^a \frac{dR}{\sqrt{[(a/R)^3 - 1]}}.$$

This integral reduces to a beta function, and we have finally

$$\tau = \sqrt{\frac{3a^2\rho\pi}{2p_0}} \frac{\Gamma(5/6)}{\Gamma(1/3)} = 0.915a\sqrt{\frac{\rho}{p_0}}.$$

PROBLEM 8. A sphere immersed in an incompressible fluid expands according to a given law $R = R(t)$. Determine the fluid pressure at the surface of the sphere.

SOLUTION. Let the required pressure be $P(t)$. Calculations exactly similar to those of Problem 7, except that the pressure at $r = R$ is $P(t)$ and not zero, give instead of (3) the equation

$$-\frac{F'(t)}{R} + \tfrac{1}{2}V^2 = \frac{p_0}{\rho} - \frac{P(t)}{\rho}$$

and accordingly instead of (4) the equation

$$\frac{p_0 - P(t)}{\rho} = -\frac{3V^2}{2} - RV\frac{dV}{dR}.$$

Bearing in mind the fact that $V = dR/dt$, we can write the expression for $P(t)$ in the form

$$P(t) = p_0 + \tfrac{1}{2}\rho\left[\frac{d^2(R^2)}{dt^2} + \left(\frac{dR}{dt}\right)^2\right].$$

PROBLEM 9. Determine the form of a jet emerging from an infinitely long slit in a plane wall.

SOLUTION. Let the wall be along the x-axis in the xy-plane, and the aperture be the segment $-\tfrac{1}{2}a \leqslant x \leqslant \tfrac{1}{2}a$ of that axis, the fluid occupying the half-plane $y > 0$. Far from the wall ($y \to \infty$) the fluid velocity is zero, and the pressure is p_0, say.

At the free surface of the jet (BC and $B'C'$ in Fig. 5a) the pressure $p = 0$, while the velocity takes the constant value $v_1 = \surd(2p_0/\rho)$, by Bernoulli's equation. The wall lines are stream-lines, and continue into the free boundary of the jet. Let ψ be zero on the line ABC; then, on the line $A'B'C'$, $\psi = -Q/\rho$, where $Q = \rho a_1 v_1$ is the rate at which the fluid emerges in the jet (a_1, v_1 being the jet width and velocity at infinity). The potential ϕ varies from $-\infty$ to $+\infty$ both along ABC and along $A'B'C'$; let ϕ be zero at B and B'. Then, in the plane of the complex variable w, the region of flow is an infinite strip of width Q/ρ (Fig. 5b). (The points in Fig. 5b, c, d are named to correspond with those in Fig. 5a.)

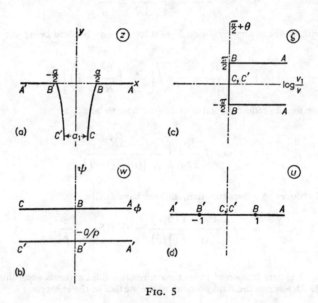

FIG. 5

We introduce a new complex variable, the logarithm of the complex velocity:

$$\zeta = -\log\left[\frac{1}{v\,e^{\frac{1}{2}i\pi}}\frac{dw}{dz}\right] = \log\frac{v_1}{v} + i(\tfrac{1}{2}\pi + \theta); \tag{1}$$

here $v_1 e^{\frac{1}{2}i\pi}$ is the complex velocity of the jet at infinity. On $A'B'$ we have $\theta = 0$; on AB, $\theta = -\pi$; on BC and $B'C'$, $v = v_1$, while at infinity in the jet $\theta = \frac{1}{2}\pi$. In the plane of the complex variable ζ, therefore, the region of flow is a semi-infinite strip of width π in the right half-plane (Fig. 5c). If we can now find a conformal transformation which carries the strip in the w-plane into the half-strip in the ζ-plane (with the points corresponding as in Fig. 5), we shall have determined w as a function of dw/dz, and w can then be found by a simple quadrature.

In order to find the desired transformation, we introduce one further auxiliary complex variable, u, such that the region of flow in the u-plane is the upper half-plane, the points B and B' corresponding to $u = \pm 1$, the points C and C' to $u = 0$, and the infinitely distant points A and A' to $u = \pm\infty$ (Fig. 5d). The dependence of w on this auxiliary variable is given by the conformal transformation which carries the upper half of the u-plane into the strip in the w-plane. With the above correspondence of points, this transformation is

$$w = -\frac{Q}{\rho\pi}\log u. \tag{2}$$

In order to find the dependence of ζ on u, we have to find a conformal transformation of the half-strip in the ζ-plane into the upper half of the u-plane. Regarding this half-strip as a triangle with one vertex at infinity, we can find the desired transformation by means of the well-known Schwarz–Christoffel formula; it is

$$\zeta = -i \sin^{-1} u. \tag{3}$$

Formulae (2) and (3) give the solution of the problem, since they furnish the dependence of dw/dz on w in parametric form.

Let us now determine the form of the jet. On BC we have $w = \phi$, $\zeta = i(\tfrac{1}{2}\pi + \theta)$, while u varies from 1 to 0. From (2) and (3) we obtain

$$\phi = -\frac{Q}{\rho\pi} \log(-\cos\theta), \tag{4}$$

and from (1) we have

$$d\phi/dz = v_1 e^{-i\theta},$$

or

$$dz \equiv dx + i\,dy = \frac{1}{v_1} e^{i\theta}\,d\phi = \frac{a_1}{\pi} e^{i\theta} \tan\theta\,d\theta,$$

whence we find, by integration with the conditions $y = 0$, $x = \tfrac{1}{2}a$ for $\theta = -\pi$, the form of the jet, expressed parametrically. In particular, the compression of the jet is $a_1/a = \pi/(2+\pi)$ = 0·61.

§11. The drag force in potential flow past a body

Let us consider the problem of potential flow of an incompressible ideal fluid past some solid body. This problem is, of course, completely equivalent to that of the motion of a fluid when the same body moves through it. To obtain the latter case from the former, we need only change to a system of co-ordinates in which the fluid is at rest at infinity. We shall, in fact, say in what follows that the body is moving through the fluid.

Let us determine the nature of the fluid velocity distribution at great distances from the moving body. The potential flow of an incompressible fluid satisfies Laplace's equation, $\triangle\phi = 0$. We have to consider solutions of this equation which vanish at infinity, since the fluid is at rest there. We take the origin somewhere inside the moving body; the co-ordinate system moves with the body, but we shall consider the fluid velocity distribution at a particular instant. As we know, Laplace's equation has a solution $1/r$, where r is the distance from the origin. The gradient and higher space derivatives of $1/r$ are also solutions. All these solutions, and any linear combination of them, vanish at infinity. Hence the general form of the required solution of Laplace's equation at great distances from the body is

$$\phi = -\frac{a}{r} + \mathbf{A}\cdot\mathbf{grad}\frac{1}{r} + ...,$$

where a and $\mathbf{A}$ are independent of the co-ordinates; the omitted terms contain higher-order derivatives of $1/r$. It is easy to see that the constant a must be zero. For the potential $\phi = -a/r$ gives a velocity

$$\mathbf{v} = -\mathbf{grad}(a/r) = a\mathbf{r}/r^3.$$

Let us calculate the corresponding mass flux through some closed surface, say a sphere of radius R. On this surface the velocity is constant and equal to a/R^2; the total flux through it is therefore $\rho(a/R^2)4\pi R^2 = 4\pi\rho a$. But the flux of an incompressible fluid through any closed surface must, of course, be zero. Hence we conclude that $a = 0$.

Thus ϕ contains terms of order $1/r^2$ and higher. Since we are seeking the velocity at large distances, the terms of higher order may be neglected, and we have

$$\phi = \mathbf{A}\cdot\mathbf{grad}(1/r) = -\mathbf{A}\cdot\mathbf{n}/r^2, \tag{11.1}$$

and the velocity $\mathbf{v} = \mathbf{grad}\,\phi$ is

$$\mathbf{v} = (\mathbf{A}\cdot\mathbf{grad})\,\mathbf{grad}\frac{1}{r} = \frac{3(\mathbf{A}\cdot\mathbf{n})\mathbf{n}-\mathbf{A}}{r^3}, \tag{11.2}$$

where $\mathbf{n}$ is a unit vector in the direction of $\mathbf{r}$. We see that at large distances the velocity diminishes as $1/r^3$. The vector $\mathbf{A}$ depends on the actual shape and velocity of the body, and can be determined only by solving completely the equation $\triangle\phi = 0$ at all distances, taking into account the appropriate boundary conditions at the surface of the moving body.

The vector $\mathbf{A}$ which appears in (11.2) is related in a definite manner to the total momentum and energy of the fluid in its motion past the body. The total kinetic energy of the fluid (the internal energy of an incompressible fluid is constant) is $E = \frac{1}{2}\int\rho v^2 dV$, where the integration is taken over all space outside the body. We take a region of space V bounded by a sphere of large radius R, whose centre is at the origin, and first integrate only over V, later letting R tend to infinity. We have identically

$$\int v^2\,dV = \int u^2\,dV + \int (\mathbf{v}+\mathbf{u})\cdot(\mathbf{v}-\mathbf{u})\,dV,$$

where $\mathbf{u}$ is the velocity of the body. Since $\mathbf{u}$ is independent of the co-ordinates, the first integral on the right is simply $u^2(V-V_0)$, where V_0 is the volume of the body. In the second integral, we write the sum $\mathbf{v}+\mathbf{u}$ as $\mathbf{grad}\,(\phi+\mathbf{u}\cdot\mathbf{r})$; using the facts that div $\mathbf{v} = 0$ (equation of continuity) and div $\mathbf{u} \equiv 0$, we have

$$\int v^2\,dV = u^2(V-V_0) + \int \text{div}\,[(\phi+\mathbf{u}\cdot\mathbf{r})(\mathbf{v}-\mathbf{u})]dV.$$

The second integral is now transformed into an integral over the surface S of the sphere and the surface S_0 of the body:

$$\int v^2\,dV = u^2(V-V_0) + \oint_{S+S_0} (\phi+\mathbf{u}\cdot\mathbf{r})(\mathbf{v}-\mathbf{u})\cdot d\mathbf{f}.$$

On the surface of the body, the normal components of $\mathbf{v}$ and $\mathbf{u}$ are equal by virtue of the boundary conditions; since the vector $d\mathbf{f}$ is along the normal to the surface, it is clear that the integral over S_0 vanishes identically. On the remote surface S we substitute the expressions (11.1), (11.2) for ϕ and $\mathbf{v}$, and neglect terms which vanish as $R \to \infty$. Writing the surface element on the sphere S in the form $\mathbf{df} = \mathbf{n}R^2\mathrm{do}$, where do is an element of solid angle, we obtain

$$\int v^2 \, \mathrm{d}V = u^2(\tfrac{4}{3}\pi R^3 - V_0) + \int [3(\mathbf{A} \cdot \mathbf{n})(\mathbf{u} \cdot \mathbf{n}) - (\mathbf{u} \cdot \mathbf{n})^2 R^3]\mathrm{do}.$$

Finally, effecting the integration† and multiplying by $\tfrac{1}{2}\rho$, we obtain the following expression for the total energy of the fluid:

$$E = \tfrac{1}{2}\rho(4\pi\mathbf{A} \cdot \mathbf{u} - V_0 u^2). \tag{11.3}$$

As has been mentioned already, the exact calculation of the vector $\mathbf{A}$ requires a complete solution of the equation $\triangle\phi = 0$, taking into account the particular boundary conditions at the surface of the body. However, the general nature of the dependence of $\mathbf{A}$ on the velocity $\mathbf{u}$ of the body can be found directly from the facts that the equation is linear in ϕ, and the boundary conditions are linear in both ϕ and $\mathbf{u}$. It follows from this that $\mathbf{A}$ must be a linear function of the components of $\mathbf{u}$. The energy E given by formula (11.3) is therefore a quadratic function of the components of $\mathbf{u}$, and can be written in the form

$$E = \tfrac{1}{2}m_{ik}u_i u_k, \tag{11.4}$$

where m_{ik} is some constant symmetrical tensor, whose components can be calculated from those of $\mathbf{A}$; it is called the *induced-mass tensor*.

Knowing the energy E, we can obtain an expression for the total momentum $\mathbf{P}$ of the fluid. To do so, we notice that infinitesimal changes in E and $\mathbf{P}$ are related by‡ $\mathrm{d}E = \mathbf{u} \cdot \mathrm{d}\mathbf{P}$; it follows from this that, if E is expressed in

---

† The integration over o is equivalent to averaging the integrand over all directions of the vector $\mathbf{n}$ and multiplying by 4π. To average expressions of the type $(\mathbf{A} \cdot \mathbf{n})(\mathbf{B} \cdot \mathbf{n}) \equiv A_i n_i B_k n_k$, where $\mathbf{A}$, $\mathbf{B}$ are constant vectors, we notice that the mean values $\overline{n_i n_k}$ form a symmetrical tensor, which can be expressed in terms of the unit tensor δ_{ik}: $\overline{n_i n_k} = a\delta_{ik}$. Contracting with respect to the suffixes i and k, and remembering that $n_i n_i = 1$, we find that $a = \tfrac{1}{3}$. Hence

$$\overline{(\mathbf{A} \cdot \mathbf{n})(\mathbf{B} \cdot \mathbf{n})} = \tfrac{1}{3}\delta_{ik}A_i B_k = \tfrac{1}{3}\mathbf{A} \cdot \mathbf{B}.$$

‡ For, let the body be accelerated by some external force $\mathbf{F}$. The momentum of the fluid will thereby be increased; let it increase by $\mathrm{d}\mathbf{P}$ during a time $\mathrm{d}t$. This increase is related to the force by $\mathrm{d}\mathbf{P} = \mathbf{F} \, \mathrm{d}t$, and on scalar multiplication by the velocity $\mathbf{u}$ we have $\mathbf{u} \cdot \mathrm{d}\mathbf{P} = \mathbf{F} \cdot \mathbf{u} \, \mathrm{d}t$, i.e. the work done by the force $\mathbf{F}$ acting through the distance $\mathbf{u} \, \mathrm{d}t$, which in turn must be equal to the increase $\mathrm{d}E$ in the energy of the fluid.

It should be noticed that it would not be possible to calculate the momentum directly as the integral $\rho\mathbf{v} \, \mathrm{d}V$ over the whole volume of the fluid. The reason is that this integral, with the velocity $\mathbf{v}$ distributed in accordance with (11.2), diverges, in the sense that the result of the integration, though finite, depends on how the integral is taken: on effecting the integration over a large region, whose dimensions subsequently tend to infinity, we obtain a value depending on the shape of the region (sphere, cylinder, etc.). The method of calculating the momentum which we use here, starting from the relation $\mathbf{u} \cdot \mathrm{d}\mathbf{P} = \mathrm{d}E$, leads to a completely definite final result, given by formula (11.6), which certainly satisfies the physical relation between the rate of change of the momentum and the forces acting on the body.

the form (11.4), the components of **P** must be

$$P_i = m_{ik}u_k. \tag{11.5}$$

Finally, a comparison of formulae (11.3), (11.4) and (11.5) shows that **P** is given in terms of **A** by

$$\mathbf{P} = 4\pi\rho\mathbf{A} - \rho V_0\mathbf{u}. \tag{11.6}$$

It must be noticed that the total momentum of the fluid is a perfectly definite finite quantity.

The momentum transmitted to the fluid by the body in unit time is $d\mathbf{P}/dt$. With the opposite sign it evidently gives the reaction **F** of the fluid, i.e. the force acting on the body:

$$\mathbf{F} = -d\mathbf{P}/dt. \tag{11.7}$$

The component of **F** parallel to the velocity of the body is called the *drag force*, and the perpendicular component is called the *lift force*.

If it were possible to have potential flow past a body moving uniformly in an ideal fluid, we should have **P** = constant, since **u** = constant, and so **F** = 0. That is, there would be no drag and no lift; the pressure forces exerted on the body by the fluid would balance out (a result known as *d'Alembert's paradox*). The origin of this paradox is most clearly seen by considering the drag. The presence of a drag force in uniform motion of a body would mean that, to maintain the motion, work must be continually done by some external force, this work being either dissipated in the fluid or converted into kinetic energy of the fluid, and the result being a continual flow of energy to infinity in the fluid. There is, however, by definition no dissipation of energy in an ideal fluid, and the velocity of the fluid set in motion by the body diminishes so rapidly with increasing distance from the body that there can be no flow of energy to infinity.

However, it must be emphasised that all these arguments relate only to the motion of a body in an infinite volume of fluid. If, for example, the fluid has a free surface, a body moving uniformly parallel to this surface will experience a drag. The appearance of this force (called *wave drag*) is due to the occurrence of a system of waves propagated on the free surface, which continually remove energy to infinity.

Suppose that a body is executing an oscillatory motion under the action of an external force **f**. When the conditions discussed in §10 are fulfilled, the fluid surrounding the body moves in a potential flow, and we can use the relations previously obtained to derive the equations of motion of the body. The force **f** must be equal to the time derivative of the total momentum of the system, and the total momentum is the sum of the momentum $M\mathbf{u}$ of the body (M being the mass of the body) and the momentum **P** of the fluid:

$$M\,d\mathbf{u}/dt + d\mathbf{P}/dt = \mathbf{f}.$$

Using (11.5), we then obtain

$$M \, du_i/dt + m_{ik} \, du_k/dt = f_i,$$

which can also be written

$$\frac{du_k}{dt}(M\delta_{ik} + m_{ik}) = f_i. \tag{11.8}$$

This is the equation of motion of a body immersed in an ideal fluid.

Let us now consider what is in some ways the converse problem. Suppose that the fluid executes some oscillatory motion on account of some cause external to the body. This motion will set the body in motion also.† We shall derive the equation of motion of the body.

We assume that the velocity of the fluid varies only slightly over distances of the order of the dimension of the body. Let **v** be what the fluid velocity at the position of the body would be if the body were absent; that is, **v** is the velocity of the unperturbed flow. According to the above assumption, **v** may be supposed constant throughout the volume occupied by the body. We denote the velocity of the body by **u** as before.

The force which acts on the body and sets it in motion can be determined as follows. If the body were wholly carried along with the fluid (i.e. if **v** = **u**), the force acting on it would be the same as the force which would act on the liquid in the same volume if the body were absent. The momentum of this volume of fluid is $\rho V_0 \mathbf{v}$, and therefore the force on it is $\rho V_0 \, d\mathbf{v}/dt$. In reality, however, the body is not wholly carried along with the fluid; there is a motion of the body relative to the fluid, in consequence of which the fluid itself acquires some additional motion. The resulting additional momentum of the fluid is $m_{ik}(u_k - v_k)$, since in (11.5) we must now replace **u** by the velocity **u** − **v** of the body relative to the fluid. The change in this momentum with time results in the appearance of an additional reaction force on the body of $-m_{ik} \, d(u_k - v_k)/dt$. Thus the total force on the body is

$$\rho V_0 \frac{dv_i}{dt} - m_{ik}\frac{d}{dt}(u_k - v_k).$$

This force is to be equated to the time derivative of the body momentum. Thus we obtain the following equation of motion:

$$\frac{d}{dt}(Mu_i) = \rho V_0 \frac{dv_i}{dt} - m_{ik}\frac{d}{dt}(u_k - v_k).$$

Integrating both sides with respect to time, we have

$$Mu_i = \rho V_0 v_i - m_{ik}(u_k - v_k),$$

or

$$(M\delta_{ik} + m_{ik})u_k = (m_{ik} + \rho V_0 \delta_{ik})v_k. \tag{11.9}$$

---

† For example, we may be considering the motion of a body in a fluid through which a sound wave is propagated, the wavelength being large compared with the dimension of the body.

We put the constant of integration equal to zero, since the velocity $\mathbf{u}$ of the body in its motion caused by the fluid must vanish when $\mathbf{v}$ vanishes. The relation obtained determines the velocity of the body from that of the fluid. If the density of the body is equal to that of the fluid ($M = \rho V_0$), we have $\mathbf{u} = \mathbf{v}$, as we should expect.

PROBLEMS

PROBLEM 1. Obtain the equation of motion for a sphere executing an oscillatory motion in an ideal fluid, and for a sphere set in motion by an oscillating fluid.

SOLUTION. Comparing (11.1) with the expression for ϕ for flow past a sphere obtained in §10, Problem 2, we see that

$$\mathbf{A} = \tfrac{1}{2}R^3\mathbf{u},$$

where R is the radius of the sphere. The total momentum transmitted to the fluid by the sphere is, according to (11.6), $\mathbf{P} = \tfrac{2}{3}\pi\rho R^3\mathbf{u}$, so that the tensor m_{ik} is

$$m_{ik} = \tfrac{2}{3}\pi\rho R^3\,\delta_{ik}.$$

The drag on the moving sphere is

$$\mathbf{F} = -\tfrac{2}{3}\pi\rho R^3\,\mathrm{d}\mathbf{u}/\mathrm{d}t,$$

and the equation of motion of the sphere oscillating in the fluid is

$$\tfrac{4}{3}\pi R^3(\rho_0 + \tfrac{1}{2}\rho)\frac{\mathrm{d}\mathbf{u}}{\mathrm{d}t} = \mathbf{f},$$

where ρ_0 is the density of the sphere. The coefficient of $\mathrm{d}\mathbf{u}/\mathrm{d}t$ is the *virtual mass* of the sphere; it consists of the actual mass of the sphere and the induced mass, which in this case is half the mass of the fluid displaced by the sphere.

If the sphere is set in motion by the fluid, we have for its velocity, from (11.9),

$$\mathbf{u} = \frac{3\rho}{\rho + 2\rho_0}\mathbf{v}.$$

If the density of the sphere exceeds that of the fluid ($\rho_0 > \rho$), $u < v$, i.e. the sphere "lags behind" the fluid; if $\rho_0 < \rho$, on the other hand, the sphere "goes ahead".

PROBLEM 2. Express the moment of the forces acting on a body moving in a fluid in terms of the vector $\mathbf{A}$.

SOLUTION. As we know from mechanics, the moment $\mathbf{M}$ of the forces acting on a body is determined from its Lagrangian function (in this case, the energy E) by the relation $\delta E = \mathbf{M}\cdot\delta\boldsymbol{\theta}$, where $\delta\boldsymbol{\theta}$ is the vector of an infinitesimal rotation of the body, and δE is the resulting change in E. Instead of rotating the body through an angle $\delta\boldsymbol{\theta}$ (and correspondingly changing the components m_{ik}), we may rotate the fluid through an angle $-\delta\boldsymbol{\theta}$ relative to the body (and correspondingly change the velocity $\mathbf{u}$). We have $\delta\mathbf{u} = -\delta\boldsymbol{\theta}\times\mathbf{u}$, so that

$$\delta E = \mathbf{P}\cdot\delta\mathbf{u} = -\delta\boldsymbol{\theta}\cdot\mathbf{u}\times\mathbf{P}.$$

Using the expression (11.6) for $\mathbf{P}$, we then obtain the required formula:

$$\mathbf{M} = -\mathbf{u}\times\mathbf{P} = 4\pi\rho\mathbf{A}\times\mathbf{u}.$$

§12. Gravity waves

The free surface of a liquid in equilibrium in a gravitational field is a plane. If, under the action of some external perturbation, the surface is moved from its equilibrium position at some point, motion will occur in the liquid. This motion will be propagated over the whole surface in the form of waves, which are called *gravity waves*, since they are due to the action of the gravitational field. Gravity waves appear mainly on the surface of the liquid, they affect the interior also, but less and less at greater and greater depths.

We shall here consider gravity waves in which the velocity of the moving fluid particles is so small that we may neglect the term $(\mathbf{v} \cdot \mathbf{grad})\mathbf{v}$ in comparison with $\partial \mathbf{v}/\partial t$ in Euler's equation. The physical significance of this is easily seen. During a time interval of the order of the period τ of the oscillations of the fluid particles in the wave, these particles travel a distance of the order of the amplitude a of the wave. Their velocity is therefore of the order of a/τ. It varies noticeably over time intervals of the order of τ and distances of the order of λ in the direction of propagation (where λ is the wavelength). Hence the time derivative of the velocity is of the order of v/τ, and the space derivatives are of the order of v/λ. Thus the condition $(\mathbf{v} \cdot \mathbf{grad})\mathbf{v} \ll \partial \mathbf{v}/\partial t$ is equivalent to

$$\frac{1}{\lambda} \left(\frac{a}{\tau}\right)^2 \ll \frac{a}{\tau} \cdot \frac{1}{\tau},$$

or

$$a \ll \lambda, \tag{12.1}$$

i.e. the amplitude of the oscillations in the wave must be small compared with the wavelength. We have seen in §9 that, if the term $(\mathbf{v} \cdot \mathbf{grad})\mathbf{v}$ in the equation of motion may be neglected, we have potential flow. Assuming the fluid incompressible, we can therefore use equations (10.6) and (10.7). The term $\frac{1}{2}v^2$ in the latter equation may be neglected, since it contains the square of the velocity; putting $f(t) = 0$ and including a term $\rho g z$ on account of the gravitational field, we obtain

$$p = -\rho g z - \rho \partial \phi / \partial t. \tag{12.2}$$

We take the z-axis vertically upwards, as usual, and the xy-plane in the equilibrium surface of the fluid.

Let us denote by ζ the z co-ordinate of a point on the surface; ζ is a function of x, y and t. In equilibrium $\zeta = 0$, so that ζ gives the vertical displacement of the surface in its oscillations. Let a constant pressure p_0 (for example, the atmospheric pressure) act on the surface. Then we have at the surface, by (12.2),

$$p_0 = -\rho g \zeta - \rho \partial \phi / \partial t.$$

Instead of the potential ϕ, we can use a potential $\phi' = \phi + (p_0/\rho)t$; this makes no difference, since $\mathbf{v} = \mathbf{grad}\, \phi = \mathbf{grad}\, \phi'$. The term p_0 is removed from the above equation, however, and on dropping the prime we obtain the condition at the surface as

$$g\zeta + (\partial \phi / \partial t)_{z=\zeta} = 0. \tag{12.3}$$

Since the amplitude of the wave oscillations is small, the displacement ζ

is small. Hence we can suppose, to the same degree of approximation, that the vertical component of the velocity of points on the surface is simply the time derivative of ζ:

$$v_z = \partial\zeta/\partial t.$$

But $v_z = \partial\phi/\partial z$, so that

$$(\partial\phi/\partial z)_{z=\zeta} = \partial\zeta/\partial t.$$

Substituting ζ from (12.3) we have

$$\left(\frac{\partial\phi}{\partial z} + \frac{1}{g}\frac{\partial^2\phi}{\partial t^2}\right)_{z=\zeta} = 0.$$

Since the oscillations are small, we can take the value of the parenthesis at $z = 0$ instead of $z = \zeta$. Thus we have finally the following system of equations to determine the motion in a gravitational field:

$$\triangle\phi = 0, \tag{12.4}$$

$$\left(\frac{\partial\phi}{\partial z} + \frac{1}{g}\frac{\partial^2\phi}{\partial t^2}\right)_{z=0} = 0. \tag{12.5}$$

We shall here consider waves on the surface of a fluid whose area is unlimited, and we shall also suppose that the wavelength is small in comparison with the depth of the fluid; we can then regard the fluid as infinitely deep. We shall therefore omit the boundary conditions at the sides and bottom.

Let us consider a gravity wave propagated along the x-axis and uniform in the y-direction; in such a wave, all quantities are independent of y. We shall seek a solution which is a simple periodic function of time and of the co-ordinate x, i.e. we put

$$\phi = f(z)\cos(kx - \omega t).$$

Here ω is what is called the *circular frequency* (we shall say simply the *frequency*) of the wave; $2\pi/\omega$ is the period of the motion at a given point; k is called the *wave number*; $\lambda = 2\pi/k$ is the *wavelength*, i.e. the period of the motion along the x-axis at a given time.

Substituting in the equation

$$\triangle\phi = \frac{\partial^2\phi}{\partial x^2} + \frac{\partial^2\phi}{\partial z^2} = 0,$$

we have

$$d^2f/dz^2 - k^2f = 0.$$

This equation has the solutions e^{kz} and e^{-kz}. We must take the former, since the latter gives an unlimited increase of ϕ as we go into the interior of the fluid (we recall that the fluid occupies the region $z < 0$). Thus we obtain for the velocity potential

$$\phi = Ae^{kz} \cos(kx - \omega t). \tag{12.6}$$

We have also to satisfy the boundary condition (12.5). Substituting (12.6), we obtain

$$k - \omega^2/g = 0,$$

or

$$\omega^2 = kg. \tag{12.7}$$

This gives the relation between the wave number and the frequency of a gravity wave.

The velocity distribution in the moving fluid is found by simply taking the space derivatives of ϕ:

$$v_x = -Ake^{kz} \sin(kx - \omega t), \qquad v_z = Ake^{kz} \cos(kx - \omega t). \tag{12.8}$$

We see that the velocity diminishes exponentially as we go into the fluid. At any given point in space (i.e. for given x, z) the velocity vector rotates uniformly in the xz-plane, its magnitude remaining constant and equal to Ake^{kz}.

Let us also determine the paths of fluid particles in the wave. We temporarily denote by x, z the co-ordinates of a moving fluid particle (and not of a point fixed in space), and by x_0, z_0 the values of x and z at the equilibrium position of the particle. Then $v_x = dx/dt$, $v_z = dz/dt$, and on the right-hand side of (12.8) we may approximate by writing x_0, z_0 in place of x, z, since the oscillations are small. An integration with respect to time then gives

$$x - x_0 = -A\frac{k}{\omega}e^{kz_0} \cos(kx_0 - \omega t),$$

$$z - z_0 = -A\frac{k}{\omega}e^{kz_0} \sin(kx_0 - \omega t). \tag{12.9}$$

Thus the fluid particles describe circles of radius $(Ak/\omega)e^{kz_0}$ about the points (x_0, z_0); this radius diminishes exponentially with increasing depth.

The velocity of propagation U of the wave is, as we shall show in §66, $U = \partial\omega/\partial k$. Substituting here $\omega = \sqrt{(kg)}$, we find that the velocity of propagation of gravity waves on an unbounded surface of infinitely deep fluid is

$$U = \tfrac{1}{2}\sqrt{(g/k)} = \tfrac{1}{2}\sqrt{(g\lambda/2\pi)}. \tag{12.10}$$

It increases with wavelength.

PROBLEMS

PROBLEM 1. Determine the velocity of propagation of gravity waves on an unbounded surface of fluid of depth h.

SOLUTION. At the bottom of the fluid, the normal velocity component must be zero, i.e. $v_z = \partial\phi/\partial z = 0$ for $z = -h$. From this condition we find the ratio of the constants A and B in the general solution

$$\phi = [Ae^{kz} + Be^{-kz}] \cos(kx - \omega t).$$

The result is

$$\phi = A \cos(kx - \omega t) \cosh k(z + h).$$

From the boundary condition (12.5) we find the relation between k and ω to be

$$\omega^2 = gk \tanh kh.$$

The velocity of propagation of the wave is

$$U = \frac{1}{2} \sqrt{\frac{g}{k \tanh kh}} \left[\tanh kh + \frac{kh}{\cosh^2 kh} \right].$$

For $kh \gg 1$ we have the result (12.10), and for $kh \ll 1$ the result (13.10) (see below).

PROBLEM 2. Determine the relation between frequency and wavelength for gravity waves on the surface separating two fluids, the upper fluid being bounded above by a fixed horizontal plane, and the lower fluid being similarly bounded below. The density and depth of the lower fluid are ρ and h, those of the upper fluid are ρ' and h', and $\rho > \rho'$.

SOLUTION. We take the xy-plane as the equilibrium plane of separation of the two fluids. Let us seek a solution having in the two fluids the forms

$$\begin{aligned}
\phi &= A \cosh k(z + h) \cos(kx - \omega t), \\
\phi' &= B \cosh k(z - h') \cos(kx - \omega t),
\end{aligned} \tag{1}$$

so that the conditions at the upper and lower boundaries are satisfied; see the solution to Problem 1. At the surface of separation, the pressure must be continuous; by (12.2), this gives the condition

$$\rho g\zeta + \rho\frac{\partial\phi}{\partial t} = \rho'g\zeta + \rho'\frac{\partial\phi'}{\partial t} \quad \text{for} \quad z = \zeta,$$

or

$$\zeta = \frac{1}{g(\rho - \rho')} \left(\rho'\frac{\partial\phi'}{\partial t} - \rho\frac{\partial\phi}{\partial t} \right). \tag{2}$$

Moreover, the velocity component v_z must be the same for each fluid at the surface of separation. This gives the condition

$$\partial\phi/\partial z = \partial\phi'/\partial z \quad \text{for} \quad z = 0. \tag{3}$$

Now $v_z = \partial\phi/\partial z = \partial\zeta/\partial t$ and, substituting (2), we have

$$g(\rho - \rho')\frac{\partial\phi}{\partial z} = \rho'\frac{\partial^2\phi'}{\partial t^2} - \rho\frac{\partial^2\phi}{\partial t^2}. \tag{4}$$

Substituting (1) in (3) and (4) gives two homogeneous linear equations for A and B, and the condition of compatibility gives

$$\omega^2 = \frac{kg(\rho - \rho')}{\rho \coth kh + \rho' \coth kh'}.$$

For $kh \gg 1$, $kh' \gg 1$ (both fluids very deep),

$$\omega^2 = kg\frac{\rho - \rho'}{\rho + \rho'},$$

while for $kh \ll 1$, $kh' \ll 1$ (long waves),

$$\omega = k\sqrt{\frac{g(\rho - \rho')hh'}{\rho h' + \rho' h}}.$$

PROBLEM 3. Determine the relation between frequency and wavelength for gravity waves propagated simultaneously on the surface of separation and on the upper surface of two fluid layers, the lower (of density ρ) being infinitely deep, and the upper (of density ρ') being of depth h' and having a free upper surface.

SOLUTION. We take the xy-plane as the equilibrium plane of separation of the two fluids. Let us seek a solution having in the two fluids the forms

$$\phi = Ae^{kz}\cos(kx - \omega t),$$
$$\phi' = [Be^{-kz} + Ce^{kz}]\cos(kx - \omega t). \tag{1}$$

At the surface of separation, i.e. for $z = 0$, we have the conditions (see Problem 2)

$$\frac{\partial\phi}{\partial z} = \frac{\partial\phi'}{\partial z}, \qquad g(\rho - \rho')\frac{\partial\phi}{\partial z} = \rho'\frac{\partial^2\phi'}{\partial t^2} - \rho\frac{\partial^2\phi}{\partial t^2}, \tag{2}$$

and at the upper surface, i.e. for $z = h'$, the condition

$$\frac{\partial\phi'}{\partial z} + \frac{1}{g}\frac{\partial^2\phi'}{\partial t^2} = 0. \tag{3}$$

The first equation (2), on substitution in (1), gives $A = C - B$, and the remaining two conditions then give two equations for B and C; from the condition of compatibility we obtain a quadratic equation for ω^2, whose roots are

$$\omega^2 = kg\frac{(\rho - \rho')(1 - e^{-2kh'})}{\rho + \rho' + (\rho - \rho')e^{-2kh'}}, \qquad \omega^2 = kg.$$

For $h' \to \infty$ these roots correspond to waves propagated independently on the surface of separation and on the upper surface.

PROBLEM 4. Determine the possible frequencies of oscillation† (stationary waves) of a fluid of depth h in a rectangular tank of width a and length b.

SOLUTION. We take the x and y axes along two sides of the tank. Let us seek a solution in the form of a stationary wave:

$$\phi = f(x, y)\cosh k(z + h)\cos\omega t.$$

---

† See §68.

We obtain for f the equation

$$\frac{\partial^2 f}{\partial x^2} + \frac{\partial^2 f}{\partial y^2} + k^2 f = 0,$$

and the condition at the free surface gives, as in Problem 1, the relation

$$\omega^2 = gk \tanh kh.$$

We take the solution of the equation for f in the form

$$f = \cos px \cos qy, \qquad p^2 + q^2 = k^2.$$

At the sides of the tank we must have the conditions

$$v_x = \partial\phi/\partial x = 0 \quad \text{for} \quad x = 0, a;$$
$$v_y = \partial\phi/\partial y = 0 \quad \text{for} \quad y = 0, b.$$

Hence we find $p = m\pi/a$, $q = n\pi/b$, where m, n are integers. The possible values of k^2 are therefore

$$k^2 = \pi^2 \left(\frac{m^2}{a^2} + \frac{n^2}{b^2} \right).$$

§13. Long gravity waves

Having considered gravity waves of length small compared with the depth of the fluid, let us now discuss the opposite limiting case of waves of length large compared with the depth. These are called *long* waves.

Let us examine first the propagation of long waves in a channel. The channel is supposed to be along the x-axis, and of infinite length. The cross-section of the channel may have any shape, and may vary along its length. We denote the cross-sectional area of the fluid in the channel by $S = S(x, t)$. The depth and width of the channel are supposed small in comparison with the wavelength.

We shall here consider longitudinal waves, in which the fluid moves along the channel. In such waves the velocity component v_x along the channel is large compared with the components v_y, v_z.

We denote v_x by v simply, and omit small terms. The x-component of Euler's equation can then be written in the form

$$\frac{\partial v}{\partial t} = -\frac{1}{\rho}\frac{\partial p}{\partial x},$$

and the z-component in the form

$$\frac{1}{\rho}\frac{\partial p}{\partial z} = -g;$$

we omit terms quadratic in the velocity, since the amplitude of the wave is again supposed small. From the second equation we have, since the pressure at the free surface ($z = \zeta$) must be p_0,

$$p = p_0 + g\rho(\zeta - z).$$

Substituting this expression in the first equation, we obtain

$$\partial v/\partial t = -g\partial\zeta/\partial x. \tag{13.1}$$

The second equation needed to determine the two unknowns v and ζ can be derived similarly to the equation of continuity; it is essentially the equation of continuity for the case in question. Let us consider a volume of fluid bounded by two plane cross-sections of the channel at a distance dx apart. In unit time a volume $(Sv)_x$ of fluid flows through one plane, and a volume $(Sv)_{x+dx}$ through the other. Hence the volume of fluid between the two planes changes by

$$(Sv)_{x+dx} - (Sv)_x = \frac{\partial(Sv)}{\partial x}dx.$$

Since the fluid is incompressible, however, this change must be due simply to the change in the level of the fluid. The change per unit time in the volume of fluid between the two planes considered is $(\partial S/\partial t)dx$. We can therefore write

$$\frac{\partial S}{\partial t}dx = -\frac{\partial(Sv)}{\partial x}dx,$$

or

$$\frac{\partial S}{\partial t} + \frac{\partial(Sv)}{\partial x} = 0. \tag{13.2}$$

This is the required equation of continuity.

Let S_0 be the equilibrium cross-sectional area of the fluid in the channel. Then $S = S_0 + S'$, where S' is the change in the cross-sectional area caused by the wave. Since the change in the fluid level is small, we can write S' in the form $b\zeta$, where b is the width of the channel at the surface of the fluid. Equation (13.2) then becomes

$$b\frac{\partial\zeta}{\partial t} + \frac{\partial(S_0 v)}{\partial x} = 0. \tag{13.3}$$

Differentiating (13.3) with respect to t and substituting $\partial v/\partial t$ from (13.1), we obtain

$$\frac{\partial^2\zeta}{\partial t^2} - \frac{g}{b}\frac{\partial}{\partial x}\left(S_0\frac{\partial\zeta}{\partial x}\right) = 0. \tag{13.4}$$

If the channel cross-section is the same at all points, then $S_0 = $ constant and

$$\frac{\partial^2\zeta}{\partial t^2} - \frac{gS_0}{b}\frac{\partial^2\zeta}{\partial x^2} = 0. \tag{13.5}$$

This is called a *wave equation*: as we shall show in §63, it corresponds to the propagation of waves with a velocity U which is independent of frequency and is the square root of the coefficient of $\partial^2 \zeta / \partial x^2$. Thus the velocity of propagation of long gravity waves in channels is

$$U = \sqrt{(gS_0/b)}. \tag{13.6}$$

In an entirely similar manner, we can consider long waves in a large tank, which we suppose infinite in two directions (those of x and y). The depth of fluid in the tank is denoted by h. The component v_z of the velocity is now small. Euler's equations take a form similar to (13.1):

$$\frac{\partial v_x}{\partial t} + g \frac{\partial \zeta}{\partial x} = 0, \qquad \frac{\partial v_y}{\partial t} + g \frac{\partial \zeta}{\partial y} = 0. \tag{13.7}$$

The equation of continuity is derived in the same way as (13.2) and is

$$\frac{\partial h}{\partial t} + \frac{\partial (h v_x)}{\partial x} + \frac{\partial (h v_y)}{\partial y} = 0.$$

We write the depth h as $h_0 + \zeta$, where h_0 is the equilibrium depth. Then

$$\frac{\partial \zeta}{\partial t} + \frac{\partial (h_0 v_x)}{\partial x} + \frac{\partial (h_0 v_y)}{\partial y} = 0. \tag{13.8}$$

Let us assume that the tank has a horizontal bottom ($h_0 = $ constant). Differentiating (13.8) with respect to t and substituting (13.7), we obtain

$$\frac{\partial^2 \zeta}{\partial t^2} - gh \left(\frac{\partial^2 \zeta}{\partial x^2} + \frac{\partial^2 \zeta}{\partial y^2} \right) = 0. \tag{13.9}$$

This is again a (two-dimensional) wave equation; it corresponds to waves propagated with a velocity

$$U = \sqrt{(gh)}. \tag{13.10}$$

§14. Waves in an incompressible fluid

There is a kind of gravity wave which can be propagated inside an incompressible fluid. Such waves are due to an inhomogeneity of the fluid caused by the gravitational field. The pressure (and therefore the entropy s) necessarily varies with height; hence any displacement of a fluid particle in height destroys the mechanical equilibrium, and consequently causes an oscillatory motion. For, since the motion is adiabatic, the particle carries with it to its new position its old entropy s, which is not the same as the equilibrium value at the new position.

We shall suppose below that the wavelength is small in comparison with distances over which the gravitational field causes a marked change in density; and we shall regard the fluid itself as incompressible. This means that we can neglect the change in its density caused by the pressure change in the wave. The change in density caused by thermal expansion cannot be neglected, since it is this that causes the phenomenon in question.

Let us write down a system of hydrodynamic equations for this motion. We shall use a suffix 0 to distinguish the values of quantities in mechanical equilibrium, and a prime to mark small deviations from those values. Then the equation of conservation of the entropy $s = s_0 + s'$ can be written, to the first order of smallness,

$$\partial s'/\partial t + \mathbf{v} \cdot \mathbf{grad}\, s_0 = 0, \tag{14.1}$$

where s_0, like the equilibrium values of other quantities, is a given function of the vertical co-ordinate z.

Next, in Euler's equation we again neglect the term $(\mathbf{v} \cdot \mathbf{grad})\mathbf{v}$ (since the oscillations are small); taking into account also the fact that the equilibrium pressure distribution is given by $\mathbf{grad}\, p_0 = \rho_0 \mathbf{g}$, we have to the same accuracy

$$\frac{\partial \mathbf{v}}{\partial t} = -\frac{\mathbf{grad}\, p}{\rho} + \mathbf{g} = -\frac{\mathbf{grad}\, p'}{\rho_0} + \frac{\mathbf{grad}\, p_0}{\rho^2}\rho'.$$

Since, from what has been said above, the change in density is due only to the change in entropy, and not to the change in pressure, we can put

$$\rho' = \left(\frac{\partial \rho_0}{\partial s_0}\right)_p s',$$

and we then obtain Euler's equation in the form

$$\frac{\partial \mathbf{v}}{\partial t} = \frac{\mathbf{g}}{\rho_0}\left(\frac{\partial \rho_0}{\partial s_0}\right)_p s' - \mathbf{grad}\frac{p'}{\rho_0}. \tag{14.2}$$

We can take ρ_0 under the gradient operator, since, as stated above, we always neglect the change in the equilibrium density over distances of the order of a wavelength. The density may likewise be supposed constant in the equation of continuity, which then becomes

$$\text{div } \mathbf{v} = 0. \tag{14.3}$$

We shall seek a solution of equations (14.1)–(14.3) in the form of a plane wave:

$$\mathbf{v} = \text{constant} \times e^{i(\mathbf{k}\cdot\mathbf{r}-\omega t)},$$

and similarly for s' and p'. Substitution in the equation of continuity (14.3) gives

$$\mathbf{v}\cdot\mathbf{k} = 0, \tag{14.4}$$

i.e. the fluid velocity is everywhere perpendicular to the *wave vector* $\mathbf{k}$ (a transverse wave). Equations (14.1) and (14.2) give

$$i\omega s' = \mathbf{v}\cdot\mathbf{grad}\, s_0, \qquad -i\omega\mathbf{v} = \frac{1}{\rho_0}\left(\frac{\partial \rho_0}{\partial s_0}\right)_p s'\mathbf{g} - \frac{i\mathbf{k}}{\rho_0}p'.$$

The condition $\mathbf{v} \cdot \mathbf{k} = 0$ gives with the second of these equations

$$ik^2 p' = \left(\frac{\partial p_0}{\partial s_0}\right)_p s' \, \mathbf{g} \cdot \mathbf{k},$$

and, eliminating $\mathbf{v}$ and s' from the two equations, we obtain the desired relation between the wave vector and the frequency,

$$\omega^2 = -\frac{1}{\rho}\left(\frac{\partial \rho}{\partial s}\right)_p g\frac{ds}{dz}\sin^2\theta. \tag{14.5}$$

Here and henceforward we omit the suffix zero to the equilibrium values of thermodynamic quantities; the z-axis is vertically upwards, and θ is the angle between this axis and the direction of $\mathbf{k}$. If the expression on the right of (14.5) is positive, the condition for the stability of the equilibrium distribution $s(z)$ (the condition that convection is absent—see §4) is fulfilled.

We see that the frequency depends only on the direction of the wave vector, and not on its magnitude. For $\theta = 0$ we have $\omega = 0$; this means that waves of the type considered, with the wave vector vertical, cannot exist.

If the fluid is in both mechanical equilibrium and complete thermodynamic equilibrium, its temperature is constant and we can write

$$\frac{ds}{dz} = \left(\frac{\partial s}{\partial p}\right)_T \frac{dp}{dz} = -\rho g\left(\frac{\partial s}{\partial p}\right)_T.$$

Finally, using the well-known thermodynamic relations

$$\left(\frac{\partial s}{\partial p}\right)_T = \frac{1}{\rho^2}\left(\frac{\partial \rho}{\partial T}\right)_p, \qquad \left(\frac{\partial \rho}{\partial s}\right)_p = \frac{T}{c_p}\left(\frac{\partial \rho}{\partial T}\right)_p,$$

where c_p is the specific heat per unit mass, we find

$$\omega = \sqrt{\frac{T}{c_p}\frac{g}{\rho}\left(\frac{\partial \rho}{\partial T}\right)_p}\,\sin\theta. \tag{14.6}$$

In particular, for a perfect gas,

$$\omega = \frac{g}{\sqrt{(c_p T)}}\sin\theta. \tag{14.7}$$

CHAPTER II

VISCOUS FLUIDS

§15. The equations of motion of a viscous fluid

LET us now study the effect of energy dissipation, occurring during the motion of a fluid, on that motion itself. This process is the result of the thermodynamic irreversibility of the motion. This irreversibility always occurs to some extent, and is due to internal friction (viscosity) and thermal conduction.

In order to obtain the equations describing the motion of a viscous fluid, we have to include some additional terms in the equation of motion of an ideal fluid. The equation of continuity, as we see from its derivation, is equally valid for any fluid, whether viscous or not. Euler's equation, on the other hand, requires modification.

We have seen in §7 that Euler's equation can be written in the form

$$\frac{\partial}{\partial t}(\rho v_i) = -\frac{\partial \Pi_{ik}}{\partial x_k},$$

where Π_{ik} is the momentum flux density tensor. The momentum flux given by formula (7.2) represents a completely reversible transfer of momentum, due simply to the mechanical transport of the different particles of fluid from place to place and to the pressure forces acting in the fluid. The viscosity (internal friction) is due to another, irreversible, transfer of momentum from points where the velocity is large to those where it is small.

The equation of motion of a viscous fluid may therefore be obtained by adding to the "ideal" momentum flux (7.2) a term $-\sigma'_{ik}$ which gives the irreversible "viscous" transfer of momentum in the fluid. Thus we write the momentum flux density tensor in a viscous fluid in the form

$$\Pi_{ik} = p\delta_{ik} + \rho v_i v_k - \sigma'_{ik} = -\sigma_{ik} + \rho v_i v_k. \tag{15.1}$$

The tensor

$$\sigma_{ik} = -p\delta_{ik} + \sigma'_{ik} \tag{15.2}$$

is called the *stress tensor*, and σ'_{ik} the *viscosity stress tensor*. σ_{ik} gives the part of the momentum flux that is not due to the direct transfer of momentum with the mass of moving fluid.†

The general form of the tensor σ'_{ik} can be established as follows. Processes

---

† We shall see below that σ'_{ik} contains a term proportional to δ_{ik}, i.e. of the same form as the term $p\delta_{ik}$. When the momentum flux tensor is put in such a form, therefore, we should specify what is meant by the pressure p; see the end of §49.

of internal friction occur in a fluid only when different fluid particles move with different velocities, so that there is a relative motion between various parts of the fluid. Hence σ'_{ik} must depend on the space derivatives of the velocity. If the velocity gradients are small, we may suppose that the momentum transfer due to viscosity depends only on the first derivatives of the velocity. To the same approximation, σ'_{ik} may be supposed a linear function of the derivatives $\partial v_i/\partial x_k$. There can be no terms in σ'_{ik} independent of $\partial v_i/\partial x_k$, since σ'_{ik} must vanish for $v =$ constant. Next, we notice that σ'_{ik} must also vanish when the whole fluid is in uniform rotation, since it is clear that in such a motion no internal friction occurs in the fluid. In uniform rotation with angular velocity $\boldsymbol{\Omega}$, the velocity $\mathbf{v}$ is equal to the vector product $\boldsymbol{\Omega}\times\mathbf{r}$. The sums

$$\frac{\partial v_i}{\partial x_k} + \frac{\partial v_k}{\partial x_i}$$

are linear combinations of the derivatives $\partial v_i/\partial x_k$, and vanish when $\mathbf{v} = \boldsymbol{\Omega}\times\mathbf{r}$. Hence σ'_{ik} must contain just these symmetrical combinations of the derivatives $\partial v_i/\partial x_k$.

The most general tensor of rank two satisfying the above conditions is

$$\sigma'_{ik} = a\left(\frac{\partial v_i}{\partial x_k} + \frac{\partial v_k}{\partial x_i}\right) + b\frac{\partial v_l}{\partial x_l}\delta_{ik},$$

where a and b are independent of the velocity.† It is convenient, however, to write this expression in a slightly different form, in which a and b are replaced by other constants:

$$\sigma'_{ik} = \eta\left(\frac{\partial v_i}{\partial x_k} + \frac{\partial v_k}{\partial x_i} - \tfrac{2}{3}\delta_{ik}\frac{\partial v_l}{\partial x_l}\right) + \zeta\delta_{ik}\frac{\partial v_l}{\partial x_l}. \tag{15.3}$$

The expression in parentheses has the property of vanishing on contraction with respect to i and k. The constants η and ζ are called *coefficients of viscosity*. As we shall show in §§16 and 49, they are both positive:

$$\eta > 0, \qquad \zeta > 0. \tag{15.4}$$

The equations of motion of a viscous fluid can now be obtained by simply adding the expressions $\partial\sigma'_{ik}/\partial x_k$ to the right-hand side of Euler's equation

$$\rho\left(\frac{\partial v_i}{\partial t} + v_k\frac{\partial v_i}{\partial x_k}\right) = -\frac{\partial p}{\partial x_i}.$$

Thus we have

$$\rho\left(\frac{\partial v_i}{\partial t} + v_k\frac{\partial v_i}{\partial x_k}\right)$$

$$= -\frac{\partial p}{\partial x_i} + \frac{\partial}{\partial x_k}\left\{\eta\left(\frac{\partial v_i}{\partial x_k} + \frac{\partial v_k}{\partial x_i} - \tfrac{2}{3}\delta_{ik}\frac{\partial v_l}{\partial x_l}\right)\right\} + \frac{\partial}{\partial x_i}\left(\zeta\frac{\partial v_l}{\partial x_l}\right). \tag{15.5}$$

---

† In making this statement we use the fact that the fluid is isotropic, as a result of which its properties must be described by scalar quantities only (in this case, a and b).

This is the most general form of the equations of motion of a viscous fluid. The quantities η and ζ are functions of pressure and temperature. In general, p and T, and therefore η and ζ, are not constant throughout the fluid, so that η and ζ cannot be taken outside the gradient operator.

In most cases, however, the viscosity coefficients do not change noticeably in the fluid, and they may be regarded as constant. We then have

$$\frac{\partial \sigma'_{ik}}{\partial x_k} = \eta\left(\frac{\partial^2 v_i}{\partial x_k \partial x_k} + \frac{\partial}{\partial x_i}\frac{\partial v_k}{\partial x_k} - \frac{2}{3}\frac{\partial}{\partial x_i}\frac{\partial v_l}{\partial x_l}\right) + \zeta\frac{\partial}{\partial x_i}\frac{\partial v_l}{\partial x_l}$$

$$= \eta\frac{\partial^2 v_i}{\partial x_k \partial x_k} + (\zeta + \tfrac{1}{3}\eta)\frac{\partial}{\partial x_i}\frac{\partial v_l}{\partial x_l}.$$

But

$$\partial v_l / \partial x_l \equiv \operatorname{div}\mathbf{v}, \qquad \partial^2 v_i / \partial x_k \partial x_k \equiv \triangle v_i.$$

Hence we can write the equation of motion of a viscous fluid, in vector form,

$$\rho\left[\frac{\partial \mathbf{v}}{\partial t} + (\mathbf{v}\cdot\mathbf{grad})\mathbf{v}\right] = -\mathbf{grad}\,p + \eta\triangle\mathbf{v} + (\zeta + \tfrac{1}{3}\eta)\mathbf{grad}\operatorname{div}\mathbf{v}. \quad (15.6)$$

If the fluid may be regarded as incompressible, $\operatorname{div}\mathbf{v} = 0$, and the last term on the right of (15.6) is zero. Thus the equation of motion of an incompressible viscous fluid is

$$\frac{\partial \mathbf{v}}{\partial t} + (\mathbf{v}\cdot\mathbf{grad})\mathbf{v} = -\frac{1}{\rho}\mathbf{grad}\,p + \frac{\eta}{\rho}\triangle\mathbf{v}. \quad (15.7)$$

This is called the *Navier–Stokes equation*. The stress tensor in an incompressible fluid takes the simple form

$$\sigma_{ik} = -p\delta_{ik} + \eta\left(\frac{\partial v_i}{\partial x_k} + \frac{\partial v_k}{\partial x_i}\right). \quad (15.8)$$

We see that the viscosity of an incompressible fluid is determined by only one coefficient. Since most fluids may be regarded as practically incompressible, it is this viscosity coefficient η which is generally of importance. The ratio

$$\nu = \eta/\rho \quad (15.9)$$

is called the *kinematic viscosity* (while η itself is called the *dynamic viscosity*). We give below the values of η and ν for various fluids, at a temperature of 20° C:

	η (g/cm sec)	ν (cm²/sec)
Water	0·010	0·010
Air	0·00018	0·150
Alcohol	0·018	0·022
Glycerine	8·5	6·8
Mercury	0·0156	0·0012

It may be mentioned that the dynamic viscosity of a gas at a given temperature is independent of the pressure. The kinematic viscosity, however, is inversely proportional to the pressure.

The pressure can be eliminated from the Navier–Stokes equation in the same way as from Euler's equation. Taking the curl of both sides of equation (15.7), we obtain, instead of equation (2.11) as for an ideal fluid,

$$\frac{\partial}{\partial t}(\mathbf{curl\,v}) = \mathbf{curl}\,(\mathbf{v}\times\mathbf{curl\,v})+\nu\triangle(\mathbf{curl\,v}). \qquad (15.10)$$

We must also write down the boundary conditions on the equations of motion of a viscous fluid. There are always forces of molecular attraction between a viscous fluid and the surface of a solid body, and these forces have the result that the layer of fluid immediately adjacent to the surface is brought completely to rest, and "adheres" to the surface. Accordingly, the boundary conditions on the equations of motion of a viscous fluid require that the fluid velocity should vanish at fixed solid surfaces:

$$\mathbf{v} = 0. \qquad (15.11)$$

It should be emphasised that both the normal and the tangential velocity component must vanish, whereas for an ideal fluid the boundary conditions require only the vanishing of v_n.†

In the general case of a moving surface, the velocity $\mathbf{v}$ must be equal to the velocity of the surface.

It is easy to write down an expression for the force acting on a solid surface bounding the fluid. The force acting on an element of the surface is just the momentum flux through this element. The momentum flux through the surface element $d\mathbf{f}$ is

$$\Pi_{ik}df_k = (\rho v_i v_k - \sigma_{ik})df_k.$$

Writing df_k in the form $df_k = n_k\,df$, where $\mathbf{n}$ is a unit vector along the normal, and recalling that $\mathbf{v} = 0$ at a solid surface,‡ we find that the force $\mathbf{P}$ acting on unit surface area is

$$P_i = -\sigma_{ik}n_k = pn_i - \sigma'_{ik}n_k. \qquad (15.12)$$

The first term is the ordinary pressure of the fluid, while the second is the force of friction, due to the viscosity, acting on the surface. We must emphasise that $\mathbf{n}$ in (15.12) is a unit vector along the outward normal to the fluid, i.e. along the inward normal to the solid surface.

If we have a surface of separation between two immiscible fluids, the conditions at the surface are that the velocities of the fluids must be equal

---

† We may note that, in general, Euler's equations cannot be satisfied with the boundary condition $\mathbf{v} = 0$.

‡ In determining the force acting on the surface, each surface element must be considered in a frame of reference in which it is at rest. The force is equal to the momentum flux only when the surface is fixed.

and the forces which they exert on each other must be equal and opposite. The latter condition is written

$$n_{1,k}\,\sigma_{1,ik} + n_{2,k}\,\sigma_{2,ik} = 0,$$

where the suffixes 1 and 2 refer to the two fluids. The normal vectors $\mathbf{n_1}$ and $\mathbf{n_2}$ are in opposite directions, i.e. $n_{1,i} = -n_{2,i} = n_i$, so that we can write

$$n_i\,\sigma_{1,ik} = n_i\,\sigma_{2,ik}. \tag{15.13}$$

At a free surface of the fluid the condition

$$\sigma_{ik}\,n_k \equiv \sigma'_{ik}\,n_k - p n_i = 0 \tag{15.14}$$

must hold.

We give below, for reference, expressions for the components of the stress tensor and the Navier–Stokes equation in cylindrical and spherical co-ordinates. In cylindrical co-ordinates r, ϕ, z the components of the stress tensor are

$$\sigma_{rr} = -p + 2\eta \frac{\partial v_r}{\partial r}, \qquad\qquad \sigma_{r\phi} = \eta\left(\frac{1}{r}\frac{\partial v_r}{\partial \phi} + \frac{\partial v_\phi}{\partial r} - \frac{v_\phi}{r}\right),$$

$$\sigma_{\phi\phi} = -p + 2\eta\left(\frac{1}{r}\frac{\partial v_\phi}{\partial \phi} + \frac{v_r}{r}\right), \qquad\qquad \sigma_{\phi z} = \eta\left(\frac{\partial v_\phi}{\partial z} + \frac{1}{r}\frac{\partial v_z}{\partial \phi}\right),$$

$$\sigma_{zz} = -p + 2\eta\frac{\partial v_z}{\partial z}, \qquad\qquad \sigma_{zr} = \eta\left(\frac{\partial v_z}{\partial r} + \frac{\partial v_r}{\partial z}\right). \tag{15.15}$$

The three components of the Navier–Stokes equation and the equation of continuity are

$$\frac{\partial v_r}{\partial t} + v_r \frac{\partial v_r}{\partial r} + \frac{v_\phi}{r}\frac{\partial v_r}{\partial \phi} + v_z \frac{\partial v_r}{\partial z} - \frac{v_\phi^2}{r}$$

$$= -\frac{1}{\rho}\frac{\partial p}{\partial r} + \nu\left(\frac{\partial^2 v_r}{\partial r^2} + \frac{1}{r^2}\frac{\partial^2 v_r}{\partial \phi^2} + \frac{\partial^2 v_r}{\partial z^2} + \frac{1}{r}\frac{\partial v_r}{\partial r} - \frac{2}{r^2}\frac{\partial v_\phi}{\partial \phi} - \frac{v_r}{r^2}\right),$$

$$\frac{\partial v_\phi}{\partial t} + v_r \frac{\partial v_\phi}{\partial r} + \frac{v_\phi}{r}\frac{\partial v_\phi}{\partial \phi} + v_z \frac{\partial v_\phi}{\partial z} + \frac{v_r v_\phi}{r}$$

$$= -\frac{1}{\rho r}\frac{\partial p}{\partial \phi} + \nu\left(\frac{\partial^2 v_\phi}{\partial r^2} + \frac{1}{r^2}\frac{\partial^2 v_\phi}{\partial \phi^2} + \frac{\partial^2 v_\phi}{\partial z^2} + \frac{1}{r}\frac{\partial v_\phi}{\partial r} + \frac{2}{r^2}\frac{\partial v_r}{\partial \phi} - \frac{v_\phi}{r^2}\right),$$

$$\frac{\partial v_z}{\partial t} + v_r \frac{\partial v_z}{\partial r} + \frac{v_\phi}{r}\frac{\partial v_z}{\partial \phi} + v_z \frac{\partial v_z}{\partial z}$$

$$= -\frac{1}{\rho}\frac{\partial p}{\partial z} + \nu\left(\frac{\partial^2 v_z}{\partial r^2} + \frac{1}{r^2}\frac{\partial^2 v_z}{\partial \phi^2} + \frac{\partial^2 v_z}{\partial z^2} + \frac{1}{r}\frac{\partial v_z}{\partial r}\right),$$

$$\frac{\partial v_r}{\partial r} + \frac{1}{r}\frac{\partial v_\phi}{\partial \phi} + \frac{\partial v_z}{\partial z} + \frac{v_r}{r} = 0. \tag{15.16}$$

In spherical co-ordinates r, ϕ, θ we have for the stress tensor

$$\sigma_{rr} = -p + 2\eta \frac{\partial v_r}{\partial r},$$

$$\sigma_{\phi\phi} = -p + 2\eta\left(\frac{1}{r\sin\theta}\frac{\partial v_\phi}{\partial\phi} + \frac{v_r}{r} + \frac{v_\theta\cot\theta}{r}\right),$$

$$\sigma_{\theta\theta} = -p + 2\eta\left(\frac{1}{r}\frac{\partial v_\theta}{\partial\theta} + \frac{v_r}{r}\right),$$

$$\sigma_{r\theta} = \eta\left(\frac{1}{r}\frac{\partial v_r}{\partial\theta} + \frac{\partial v_\theta}{\partial r} - \frac{v_\theta}{r}\right),$$ (15.17)

$$\sigma_{\theta\phi} = \eta\left(\frac{1}{r\sin\theta}\frac{\partial v_\theta}{\partial\phi} + \frac{1}{r}\frac{\partial v_\phi}{\partial\theta} - \frac{v_\phi\cot\theta}{r}\right),$$

$$\sigma_{\phi r} = \eta\left(\frac{\partial v_\phi}{\partial r} + \frac{1}{r\sin\theta}\frac{\partial v_r}{\partial\phi} - \frac{v_\phi}{r}\right),$$

while the equations of motion are

$$\frac{\partial v_r}{\partial t} + v_r\frac{\partial v_r}{\partial r} + \frac{v_\theta}{r}\frac{\partial v_r}{\partial\theta} + \frac{v_\phi}{r\sin\theta}\frac{\partial v_r}{\partial\phi} - \frac{v_\theta^2 + v_\phi^2}{r}$$
$$= -\frac{1}{\rho}\frac{\partial p}{\partial r} + \nu\left[\frac{1}{r}\frac{\partial^2(rv_r)}{\partial r^2} + \frac{1}{r^2}\frac{\partial^2 v_r}{\partial\theta^2} + \frac{1}{r^2\sin^2\theta}\frac{\partial^2 v_r}{\partial\phi^2} + \frac{\cot\theta}{r^2}\frac{\partial v_r}{\partial\theta} - \right.$$
$$\left. - \frac{2}{r^2}\frac{\partial v_\theta}{\partial\theta} - \frac{2}{r^2\sin\theta}\frac{\partial v_\phi}{\partial\phi} - \frac{2v_r}{r^2} - \frac{2\cot\theta}{r^2}v_\theta\right],$$

$$\frac{\partial v_\theta}{\partial t} + v_r\frac{\partial v_\theta}{\partial r} + \frac{v_\theta}{r}\frac{\partial v_\theta}{\partial\theta} + \frac{v_\phi}{r\sin\theta}\frac{\partial v_\theta}{\partial\phi} + \frac{v_r v_\theta}{r} - \frac{v_\phi^2\cot\theta}{r}$$
$$= -\frac{1}{\rho r}\frac{\partial p}{\partial\theta} + \nu\left[\frac{1}{r}\frac{\partial^2(rv_\theta)}{\partial r^2} + \frac{1}{r^2}\frac{\partial^2 v_\theta}{\partial\theta^2} + \frac{1}{r^2\sin^2\theta}\frac{\partial^2 v_\theta}{\partial\phi^2} + \frac{\cot\theta}{r^2}\frac{\partial v_\theta}{\partial\theta} - \right.$$
$$\left. - \frac{2\cos\theta}{r^2\sin^2\theta}\frac{\partial v_\phi}{\partial\phi} + \frac{2}{r^2}\frac{\partial v_r}{\partial\theta} - \frac{v_\theta}{r^2\sin^2\theta}\right],$$

$$\frac{\partial v_\phi}{\partial t} + v_r\frac{\partial v_\phi}{\partial r} + \frac{v_\theta}{r}\frac{\partial v_\phi}{\partial\theta} + \frac{v_\phi}{r\sin\theta}\frac{\partial v_\phi}{\partial\phi} + \frac{v_r v_\phi}{r} + \frac{v_\theta v_\phi\cot\theta}{r}$$
$$= -\frac{1}{\rho r\sin\theta}\frac{\partial p}{\partial\phi} + \nu\left[\frac{1}{r}\frac{\partial^2(rv_\phi)}{\partial r^2} + \frac{1}{r^2}\frac{\partial^2 v_\phi}{\partial\theta^2} + \frac{1}{r^2\sin^2\theta}\frac{\partial^2 v_\phi}{\partial\phi^2} + \right.$$
$$\left. + \frac{\cot\theta}{r^2}\frac{\partial v_\phi}{\partial\theta} + \frac{2}{r^2\sin\theta}\frac{\partial v_r}{\partial\phi} + \frac{2\cos\theta}{r^2\sin^2\theta}\frac{\partial v_\theta}{\partial\phi} - \frac{v_\phi}{r^2\sin^2\theta}\right],$$

$$\frac{\partial v_r}{\partial r} + \frac{1}{r}\frac{\partial v_\theta}{\partial\theta} + \frac{1}{r\sin\theta}\frac{\partial v_\phi}{\partial\phi} + \frac{2v_r}{r} + \frac{v_\theta\cot\theta}{r} = 0. \qquad (15.18)$$

Finally, we give the equation that must be satisfied by the stream function $\psi(x, y)$ in two-dimensional flow of an incompressible viscous fluid. It is obtained by substituting $v_x = \partial\psi/\partial y$, $v_y = -\partial\psi/\partial x$, $v_z = 0$ in equation (15.10):

$$\frac{\partial}{\partial t}(\triangle\psi) - \frac{\partial\psi}{\partial x}\frac{\partial(\triangle\psi)}{\partial y} + \frac{\partial\psi}{\partial y}\frac{\partial(\triangle\psi)}{\partial x} - \nu\triangle^2\psi = 0. \qquad (15.19)$$

§16. Energy dissipation in an incompressible fluid

The presence of viscosity results in the dissipation of energy, which is finally transformed into heat. The calculation of the energy dissipation is especially simple for an incompressible fluid.

The total kinetic energy of an incompressible fluid is

$$E_{\text{kin}} = \tfrac{1}{2}\rho \int v^2 \, dV.$$

We take the time derivative of this energy, writing $\partial(\tfrac{1}{2}\rho v^2)/\partial t = \rho v_i \partial v_i/\partial t$ and substituting for $\partial v_i/\partial t$ the expression for it given by the Navier–Stokes equation:

$$\frac{\partial v_i}{\partial t} = -v_k\frac{\partial v_i}{\partial x_k} - \frac{1}{\rho}\frac{\partial p}{\partial x_i} + \frac{1}{\rho}\frac{\partial \sigma'_{ik}}{\partial x_k}.$$

The result is

$$\frac{\partial}{\partial t}(\tfrac{1}{2}\rho v^2) = -\rho\mathbf{v}\cdot(\mathbf{v}\cdot\mathbf{grad})\mathbf{v} - \mathbf{v}\cdot\mathbf{grad}\,p + v_i\frac{\partial\sigma'_{ik}}{\partial x_k}$$

$$= -\rho(\mathbf{v}\cdot\mathbf{grad})\left(\tfrac{1}{2}v^2 + \frac{p}{\rho}\right) + \operatorname{div}(\mathbf{v}\cdot\boldsymbol{\sigma}') - \sigma'_{ik}\frac{\partial v_i}{\partial x_k}.$$

Here $\mathbf{v}\cdot\boldsymbol{\sigma}'$ denotes the vector whose components are $v_i\sigma'_{ik}$. Since div $\mathbf{v} = 0$ for an incompressible fluid, we can write the first term on the right as a divergence:

$$\frac{\partial}{\partial t}(\tfrac{1}{2}\rho v^2) = -\operatorname{div}\left[\rho\mathbf{v}\left(\tfrac{1}{2}v^2 + \frac{p}{\rho}\right) - \mathbf{v}\cdot\boldsymbol{\sigma}'\right] - \sigma'_{ik}\frac{\partial v_i}{\partial x_k}. \qquad (16.1)$$

The expression in brackets is just the energy flux density in the fluid: the term $\rho\mathbf{v}(\tfrac{1}{2}v^2 + p/\rho)$ is the energy flux due to the actual transfer of fluid mass, and is the same as the energy flux in an ideal fluid (see (10.5)). The second term, $\mathbf{v}\cdot\boldsymbol{\sigma}'$, is the energy flux due to processes of internal friction. For the presence of viscosity results in a momentum flux σ'_{ik}; a transfer of momentum, however, always involves a transfer of energy, and the energy flux is clearly equal to the scalar product of the momentum flux and the velocity.

If we integrate (16.1) over some volume V, we obtain

$$\frac{\partial}{\partial t} \int \tfrac{1}{2}\rho v^2 \, dV = - \oint \left[\rho \mathbf{v}\left(\tfrac{1}{2}v^2 + \frac{p}{\rho}\right) - \mathbf{v}\cdot\boldsymbol{\sigma}' \right]\cdot d\mathbf{f} - \int \sigma'_{ik}\frac{\partial v_i}{\partial x_k} dV. \quad (16.2)$$

The first term on the right gives the rate of change of the kinetic energy of the fluid in V owing to the energy flux through the surface bounding V. The integral in the second term is consequently the decrease per unit time in the kinetic energy owing to dissipation.

If the integration is extended to the whole volume of the fluid, the surface integral vanishes (since the velocity vanishes at infinity†), and we find the energy dissipated per unit time in the whole fluid to be

$$\dot{E}_{\text{kin}} = - \int \sigma'_{ik}\frac{\partial v_i}{\partial x_k}\, dV.$$

In incompressible fluids, the tensor σ'_{ik} is given by (15.8), so that

$$\sigma'_{ik}\frac{\partial v_i}{\partial x_k} = \eta \frac{\partial v_i}{\partial x_k}\left(\frac{\partial v_i}{\partial x_k} + \frac{\partial v_k}{\partial x_i}\right).$$

It is easy to verify that this expression can be written

$$\tfrac{1}{2}\eta\left(\frac{\partial v_i}{\partial x_k} + \frac{\partial v_k}{\partial x_i}\right)^2.$$

Thus we have finally for the energy dissipation in an incompressible fluid

$$\dot{E}_{\text{kin}} = -\tfrac{1}{2}\eta \int \left(\frac{\partial v_i}{\partial x_k} + \frac{\partial v_k}{\partial x_i}\right)^2 dV. \quad (16.3)$$

The dissipation leads to a decrease in the mechanical energy, i.e. we must have $\dot{E}_{\text{kin}} < 0$. The integral in (16.3), however, is always positive. We therefore conclude that the viscosity coefficient η is always positive.

PROBLEM

Transform the integral (16.3) for potential flow into an integral over the surface bounding the region of flow.

SOLUTION. Putting $\partial v_i/\partial x_k = \partial v_k/\partial x_i$ and integrating once by parts, we find

$$\dot{E}_{\text{kin}} = -2\eta \int \left(\frac{\partial v_i}{\partial x_k}\right)^2 dV = -2\eta \int v_i\frac{\partial v_i}{\partial x_k}df_k,$$

or

$$\dot{E}_{\text{kin}} = -\eta \int \mathbf{grad}\, v^2\cdot d\mathbf{f}.$$

---

† We are considering the motion of the fluid in a system of co-ordinates such that the fluid is at rest at infinity. Here, and in similar cases, we speak, for the sake of definiteness, of an infinite volume of fluid, but this implies no loss of generality. For a fluid enclosed in a finite volume, the surface integral again vanishes, because the normal velocity component at the surface vanishes.

§17. **Flow in a pipe**

We shall now consider some simple problems of motion of an incompressible viscous fluid.

Let the fluid be enclosed between two parallel planes moving with a constant relative velocity **u**. We take one of these planes as the xz-plane, with the x-axis in the direction of **u**. It is clear that all quantities depend only on y, and that the fluid velocity is everywhere in the x-direction. We have from (15.7) for steady flow

$$dp/dy = 0, \qquad d^2v/dy^2 = 0.$$

(The equation of continuity is satisfied identically.) Hence $p =$ constant, $v = ay+b$. For $y = 0$ and $y = h$ (h being the distance between the planes) we must have respectively $v = 0$ and $v = u$. Thus

$$v = yu/h. \qquad (17.1)$$

The fluid velocity distribution is therefore linear. The mean fluid velocity, defined as

$$\bar{v} = \frac{1}{h} \int_0^h v \, dy,$$

is

$$\bar{v} = \tfrac{1}{2}u. \qquad (17.2)$$

From (15.12) we find that the normal component of the force on either plane is just p, as it should be, while the tangential friction force on the plane $y = 0$ is

$$\sigma_{xy} = \eta \, dv/dy = \eta u/h; \qquad (17.3)$$

the force on the plane $y = h$ is $-\eta u/h$.

Next, let us consider steady flow between two fixed parallel planes in the presence of a pressure gradient. We choose the co-ordinates as before; the x-axis is in the direction of motion of the fluid. The Navier–Stokes equations give, since the velocity clearly depends only on y,

$$\frac{\partial^2 v}{\partial y^2} = \frac{1}{\eta} \frac{\partial p}{\partial x}, \qquad \frac{\partial p}{\partial y} = 0.$$

The second equation shows that the pressure is independent of y, i.e. it is constant across the depth of the fluid between the planes. The right-hand side of the first equation is therefore a function of x only, while the left-hand side is a function of y only; this can be true only if both sides are constant. Thus $dp/dx =$ constant, i.e. the pressure is a linear function of the co-ordinate x along the direction of flow. For the velocity we now obtain

$$v = \frac{1}{2\eta} \frac{dp}{dx} y^2 + ay + b.$$

The constants a and b are determined from the boundary conditions, $v = 0$ for $y = 0$ and $y = h$. The result is

$$v = -\frac{1}{2\eta}\frac{dp}{dx}[\tfrac{1}{4}h^2 - (y - \tfrac{1}{2}h)^2]. \qquad (17.4)$$

Thus the velocity varies parabolically across the fluid, reaching its maximum value in the middle. The mean fluid velocity (averaged over the depth of the fluid) is again

$$\bar{v} = \frac{1}{h}\int_0^h v\, dy;$$

on calculating this, we find

$$\bar{v} = -\frac{h^2}{12\eta}\frac{dp}{dx}. \qquad (17.5)$$

We may also calculate the frictional force $\sigma_{xy} = \eta(\partial v/\partial y)_{y=0}$ acting on one of the fixed planes. Substitution from (17.4) gives

$$\sigma_{xy} = -\tfrac{1}{2}h\, dp/dx. \qquad (17.6)$$

Finally, let us consider steady flow in a pipe of arbitrary cross-section (the same along the whole length of the pipe, however). We take the axis of the pipe as the x-axis. The fluid velocity is evidently along the x-axis at all points, and is a function of y and z only. The equation of continuity is satisfied identically, while the y and z components of the Navier–Stokes equation again give $\partial p/\partial y = \partial p/\partial z = 0$, i.e. the pressure is constant over the cross-section of the pipe. The x-component of equation (15.7) gives

$$\frac{\partial^2 v}{\partial y^2} + \frac{\partial^2 v}{\partial z^2} = \frac{1}{\eta}\frac{dp}{dx}. \qquad (17.7)$$

Hence we again conclude that $dp/dx = $ constant; the pressure gradient may therefore be written $-\Delta p/l$, where Δp is the pressure difference between the ends of the pipe and l is its length.

Thus the velocity distribution for flow in a pipe is determined by a two-dimensional equation of the form $\triangle v = $ constant. This equation has to be solved with the boundary condition $v = 0$ at the circumference of the cross-section of the pipe. We shall solve the equation for a pipe of circular cross-section. Taking the origin at the centre of the circle and using polar co-ordinates, we have by symmetry $v = v(r)$. Using the expression for the Laplacian in polar co-ordinates, we have

$$\frac{1}{r}\frac{d}{dr}\left(r\frac{dv}{dr}\right) = -\frac{\Delta p}{\eta l}.$$

Integrating, we find

$$v = -\frac{\Delta p}{4\eta l}r^2 + a\log r + b. \tag{17.8}$$

The constant a must be put equal to zero, since the velocity must remain finite at the centre of the pipe. The constant b is determined from the requirement that $v = 0$ for $r = R$, where R is the radius of the pipe. We then find

$$v = \frac{\Delta p}{4\eta l}(R^2 - r^2). \tag{17.9}$$

Thus the velocity distribution across the pipe is parabolic.

It is easy to determine the mass Q of fluid passing each second through any cross-section of the pipe (called the *discharge*). A mass $\rho \cdot 2\pi r v\,dr$ passes each second through an annular element $2\pi r\,dr$ of the cross-sectional area. Hence

$$Q = 2\pi\rho \int_0^R rv\,dr.$$

Using (17.9), we obtain

$$Q = \frac{\pi\Delta p}{8\nu l}R^4. \tag{17.10}$$

The mass of fluid is thus proportional to the fourth power of the radius of the pipe (*Poiseuille's formula*).

PROBLEMS

PROBLEM 1. Determine the flow in a pipe of annular cross-section, the internal and external radii being R_1, R_2.

SOLUTION. Determining the constants a and b in the general solution (17.8) from the conditions that $v = 0$ for $r = R_1$ and $r = R_2$, we find

$$v = \frac{\Delta p}{4\eta l}\left[R_2^2 - r^2 + \frac{R_2^2 - R_1^2}{\log(R_2/R_1)}\log\frac{r}{R_2}\right].$$

The discharge is

$$Q = \frac{\pi\Delta p}{8\nu l}\left[R_2^4 - R_1^4 - \frac{(R_2^2 - R_1^2)^2}{\log(R_2/R_1)}\right].$$

PROBLEM 2. The same as Problem 1, but for a pipe of elliptical cross-section.

SOLUTION. We seek a solution of equation (17.7) in the form $v = Ay^2 + Bz^2 + C$. The constants A, B, C are determined from the requirement that this expression must satisfy the boundary condition $v = 0$ on the circumference of the ellipse (i.e. $Ay^2 + Bz^2 + C = 0$

must be the same as the equation $y^2/a^2+z^2/b^2 = 1$, where a and b are the semi-axes of the ellipse). The result is

$$v = \frac{\Delta p}{2\eta l} \frac{a^2 b^2}{a^2+b^2} \left(1 - \frac{y^2}{a^2} - \frac{z^2}{b^2}\right).$$

The discharge is

$$Q = \frac{\pi \Delta p}{4vl} \frac{a^3 b^3}{a^2+b^2}.$$

PROBLEM 3. The same as Problem 1, but for a pipe whose cross-section is an equilateral triangle of side a.

SOLUTION. The solution of equation (17.7) which vanishes on the bounding triangle is

$$v = \frac{\Delta p}{l} \frac{2}{\sqrt{3}a\eta} h_1 h_2 h_3,$$

where h_1, h_2, h_3 are the lengths of the perpendiculars from a given point in the triangle to its three sides. For each of the expressions Δh_1, Δh_2, Δh_3 (where $\Delta = \partial^2/\partial x^2 + \partial^2/\partial y^2$) is zero; this is seen at once from the fact that each of the perpendiculars h_1, h_2, h_3 may be taken as the axis of y or z, and the result of applying the Laplacian to a co-ordinate is zero. We therefore have

$$\Delta(h_1 h_2 h_3) = 2(h_1 \,\mathbf{grad}\, h_2 \cdot \mathbf{grad}\, h_3 + h_2 \,\mathbf{grad}\, h_3 \cdot \mathbf{grad}\, h_1 +$$
$$+ h_3 \,\mathbf{grad}\, h_1 \cdot \mathbf{grad}\, h_2).$$

But $\mathbf{grad}\, h_1 = \mathbf{n}_1$, $\mathbf{grad}\, h_2 = \mathbf{n}_2$, $\mathbf{grad}\, h_3 = \mathbf{n}_3$, where $\mathbf{n}_1$, $\mathbf{n}_2$, $\mathbf{n}_3$ are unit vectors along the perpendiculars h_1, h_2, h_3. Any two of $\mathbf{n}_1$, $\mathbf{n}_2$, $\mathbf{n}_3$ are at an angle $2\pi/3$, so that $\mathbf{grad}\, h_1 \cdot \mathbf{grad}\, h_2 = \mathbf{n}_1 \cdot \mathbf{n}_2 = \cos(2\pi/3) = -\frac{1}{2}$, and so on. We thus obtain the relation

$$\Delta(h_1 h_2 h_3) = -(h_1 + h_2 + h_3) = -\tfrac{1}{2}\sqrt{3}a,$$

and we see that equation (17.7) is satisfied. The discharge is

$$Q = \frac{\sqrt{3}a^4 \Delta p}{320 vl}.$$

PROBLEM 4. A cylinder of radius R_1 moves with velocity u inside a coaxial cylinder of radius R_2, their axes being parallel. Determine the motion of a fluid occupying the space between the cylinders.

SOLUTION. We take cylindrical co-ordinates, with the z-axis along the axis of the cylinders. The velocity is everywhere along the z-axis and depends only on r (as does the pressure): $v_z = v(r)$. We obtain for v the equation

$$\Delta v = \frac{1}{r} \frac{d}{dr}\left(r \frac{dv}{dr}\right) = 0;$$

the term $(\mathbf{v} \cdot \mathbf{grad})\mathbf{v} = v\, \partial v/\partial z$ vanishes identically. Using the boundary conditions $v = u$ for $r = R_1$ and $v = 0$ for $r = R_2$, we find

$$v = u\frac{\log(r/R_2)}{\log(R_1/R_2)}.$$

The frictional force per unit length of either cylinder is $2\pi\eta u/\log(R_2/R_1)$.

PROBLEM 5. A layer of fluid of thickness h is bounded above by a free surface and below by a fixed plane inclined at an angle α to the horizontal. Determine the flow due to gravity.

SOLUTION. We take the fixed plane as the xy-plane, with the x-axis in the direction of flow (Fig. 6). We seek a solution depending only on z. The Navier–Stokes equations with $v_x = v(z)$ in a gravitational field are

$$\eta\frac{d^2v}{dz^2} + \rho g \sin\alpha = 0, \qquad \frac{dp}{dz} + \rho g \cos\alpha = 0.$$

At the free surface ($z = h$) we must have $\sigma_{xz} = \eta dv/dz = 0$, $\sigma_{zz} = -p = -p_0$ (p_0 being the atmospheric pressure). For $z = 0$ we must have $v = 0$. The solution satisfying these conditions is

$$p = p_0 + \rho g(h - z)\cos\alpha, \qquad v = \frac{\rho g \sin\alpha}{2\eta}z(2h - z).$$

The discharge, per unit length in the y-direction, is

$$Q = \rho\int_0^h v\,dz = \frac{\rho g h^3 \sin\alpha}{3\nu}.$$

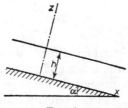

FIG. 6

PROBLEM 6. Determine the way in which the pressure falls along a tube of circular cross-section in which a viscous perfect gas is flowing isothermally (bearing in mind that the dynamic viscosity η of a perfect gas is independent of the pressure).

SOLUTION. Over any short section of the pipe the gas may be supposed incompressible, provided that the pressure gradient is not too great, and we can therefore use formula (17.10), according to which

$$-\frac{dp}{dx} = \frac{8\eta Q}{\pi\rho R^4}.$$

Over greater distances, however, ρ varies, and the pressure is not a linear function of x. According to the equation of state, the gas density $\rho = mp/kT$, where m is the mass of a molecule and k is Boltzmann's constant, so that

$$-\frac{dp}{dx} = \frac{8\eta QkT}{\pi m R^4}\cdot\frac{1}{p}.$$

(The discharge Q of the gas through the tube is obviously the same, whether or not the gas is incompressible.) From this we find

$$p_2{}^2 - p_1{}^2 = \frac{16\eta QkT}{\pi m R^4}l,$$

where p_2, p_1 are the pressures at the ends of a section of the tube of length l.

§18. **Flow between rotating cylinders**

Let us now consider the motion of a fluid between two infinite coaxial cylinders of radii R_1, R_2 ($R_2 > R_1$), rotating about their axis with angular velocities Ω_1, Ω_2. We take cylindrical co-ordinates r, ϕ, z, with the z-axis along the axis of the cylinders. It is evident from symmetry that

$$v_z = v_r = 0, \qquad v_\phi = v(r), \qquad p = p(r).$$

The Navier–Stokes equation in cylindrical co-ordinates gives in this case two equations:

$$\mathrm{d}p/\mathrm{d}r = \rho v^2/r, \tag{18.1}$$

$$\frac{\mathrm{d}^2 v}{\mathrm{d}r^2} + \frac{1}{r}\frac{\mathrm{d}v}{\mathrm{d}r} - \frac{v}{r^2} = 0. \tag{18.2}$$

The latter equation has solutions of the form r^n; substitution gives $n = \pm 1$, so that

$$v = ar + \frac{b}{r}.$$

The constants a and b are found from the boundary conditions, according to which the fluid velocity at the inner and outer cylindrical surfaces must be equal to that of the corresponding cylinder: $v = R_1\Omega_1$ for $r = R_1$, $v = R_2\Omega_2$ for $r = R_2$. As a result we find the velocity distribution to be

$$v = \frac{\Omega_2 R_2^2 - \Omega_1 R_1^2}{R_2^2 - R_1^2} r + \frac{(\Omega_1 - \Omega_2)R_1^2 R_2^2}{R_2^2 - R_1^2}\frac{1}{r}. \tag{18.3}$$

The pressure distribution is then found from (18.1) by straightforward integration.

For $\Omega_1 = \Omega_2 = \Omega$ we have simply $v = \Omega r$, i.e. the fluid rotates rigidly with the cylinders. When the outer cylinder is absent ($\Omega_2 = 0$, $R_2 = \infty$) we have $v = \Omega_1 R_1^2/r$.

Let us also determine the moment of the frictional forces acting on the cylinders. The frictional force acting on unit area of the inner cylinder is along the tangent to the surface and, from (15.12), is equal to the component $\sigma'_{r\phi}$ of the stress tensor. Using formulae (15.15), we find

$$[\sigma'_{r\phi}]_{r=R_1} = \eta\left[\left(\frac{\partial v}{\partial r} - \frac{v}{r}\right)\right]_{r=R_1}$$

$$= -2\eta\frac{(\Omega_1 - \Omega_2)R_2^2}{R_2^2 - R_1^2}.$$

The force acting on unit length of the cylinder is obtained by multiplying by $2\pi R_1$, and the moment M_1 of that force by multiplying the result by R_1. We thus have

$$M_1 = -\frac{4\pi\eta(\Omega_1 - \Omega_2)R_1^2 R_2^2}{R_2^2 - R_1^2}. \tag{18.4}$$

The moment M_2 of the forces acting on the inner cylinder is clearly $-M_1$.[†]

The following general remark may be made concerning the solutions of the equations of motion of a viscous fluid which we have obtained in §§17 and 18. In all these cases the non-linear term $(\mathbf{v} \cdot \mathbf{grad})\mathbf{v}$ in the equations which determine the velocity distribution is identically zero, so that we are actually solving linear equations, a fact which very much simplifies the problem. For this reason all the solutions also satisfy the equations of motion for an incompressible ideal fluid, say in the form (10.2) and (10.3). This is why formulae (17.1) and (18.3) do not contain the viscosity coefficient at all. This coefficient appears only in formulae, such as (17.9), which relate the velocity to the pressure gradient in the fluid, since the presence of a pressure gradient is due to the viscosity; an ideal fluid could flow in a pipe even if there were no pressure gradient.

§19. The law of similarity

In studying the motion of viscous fluids we can obtain a number of important results from simple arguments concerning the dimensions of various physical quantities. Let us consider any particular type of motion, for instance the motion of a body of some definite shape through a fluid. If the body is not a sphere, its direction of motion must also be specified: e.g. the motion of an ellipsoid in the direction of its greatest or least axis. Alternatively, we may be considering flow in a region with boundaries of a definite form (a pipe of given cross-section, etc.).

In such a case we say that bodies of the same shape are *geometrically similar*; they can be obtained from one another by changing all linear dimensions in the same ratio. Hence, if the shape of the body is given, it suffices to specify any one of its linear dimensions (the radius of a sphere or of a cylindrical pipe, one semi-axis of a spheroid of given eccentricity, and so on) in order to determine its dimensions completely.

We shall at present consider steady flow. If, for example, we are discussing flow past a solid body (which case we shall take below, for definiteness), the velocity of the main stream must therefore be constant. We shall suppose the fluid incompressible.

Of the parameters which characterise the fluid itself, only the kinematic

---

[†] The solution of the more complex problem of the motion of a viscous fluid in a narrow space between cylinders whose axes are parallel but not coincident may be found in: N. E. KOCHIN, I. A. KIBEL' and N. V. ROZE, *Theoretical Hydromechanics* (*Teoreticheskaya gidromekhanika*), Part 2, 3rd ed., p. 419, Moscow 1948; A. SOMMERFELD, *Mechanics of Deformable Bodies*, §36, Academic Press, New York 1950.

viscosity $\nu = \eta/\rho$ appears in the equations of hydrodynamics (the Navier–Stokes equations); the unknown functions which have to be determined by solving the equations are the velocity **v** and the ratio p/ρ of the pressure p to the constant density ρ. Moreover, the flow depends, through the boundary conditions, on the shape and dimensions of the body moving through the fluid and on its velocity. Since the shape of the body is supposed given, its geometrical properties are determined by one linear dimension, which we denote by l. Let the velocity of the main stream be u. Then any flow is specified by three parameters, ν, u and l. These quantities have the following dimensions:

$$\nu = \text{cm}^2/\text{sec}, \qquad l = \text{cm}, \qquad u = \text{cm/sec}.$$

It is easy to verify that only one dimensionless quantity can be formed from the above three, namely ul/ν. This combination is called the *Reynolds number* and is denoted by R:

$$R = \rho u l/\eta = ul/\nu. \tag{19.1}$$

Any other dimensionless parameter can be written as a function of R.

We shall now measure lengths in terms of l, and velocities in terms of u, i.e. we introduce the dimensionless quantities $\mathbf{r}/l$, $\mathbf{v}/u$. Since the only dimensionless parameter is the Reynolds number, it is evident that the velocity distribution obtained by solving the equations of incompressible flow is given by a function of the form

$$\mathbf{v} = u\mathbf{f}(\mathbf{r}/l, R). \tag{19.2}$$

It is seen from this expression that, in two different flows of the same type (for example, flow past spheres of different radii by fluids of different viscosities), the velocities $\mathbf{v}/u$ are the same functions of the ratio $\mathbf{r}/l$ if the Reynolds number is the same for each flow. Flows which can be obtained from one another by simply changing the unit of measurement of co-ordinates and velocities are said to be *similar*. Thus flows of the same type with the same Reynolds number are similar. This is called the *law of similarity* (O. REYNOLDS 1883).

A formula similar to (19.2) can be written for the pressure distribution in the fluid. To do so, we must construct from the parameters ν, l, u some quantity with the dimensions of pressure divided by density; this quantity can be u^2, for example. Then we can say that $p/\rho u^2$ is a function of the dimensionless variable $\mathbf{r}/l$ and the dimensionless parameter R. Thus

$$p = \rho u^2 f(\mathbf{r}/l, R). \tag{19.3}$$

Finally, similar considerations can also be applied to quantities which characterise the flow but are not functions of the co-ordinates. Such a quantity is, for instance, the drag force F acting on the body. We can say that the dimensionless ratio of F to some quantity formed from ν, u, l, ρ

and having the dimensions of force must be a function of the Reynolds number alone. Such a combination of ν, u, l, ρ can be $\rho u^2 l^2$, for example. Then

$$F = \rho u^2 l^2 f(R). \tag{19.4}$$

If the force of gravity has an important effect on the flow, then the latter is determined not by three but by four parameters, l, u, ν and the acceleration g due to gravity. From these parameters we can construct not one but two independent dimensionless quantities. These can be, for instance, the Reynolds number and the *Froude number*, which is

$$F = u^2/lg. \tag{19.5}$$

In formulae (19.2)–(19.4) the function f will now depend on not one but two parameters (R and F), and two flows will be similar only if both these numbers have the same values.

Finally, we may say a little regarding non-steady flows. A non-steady flow of a given type is characterised not only by the quantities ν, u, l but also by some time interval τ characteristic of the flow, which determines the rate of change of the flow. For instance, in oscillations, according to a given law, of a solid body, of a given shape, immersed in a fluid, τ may be the period of oscillation. From the four quantities ν, u, l, τ we can again construct two independent dimensionless quantities, which may be the Reynolds number and the number

$$S = u\tau/l, \tag{19.6}$$

sometimes called the *Strouhal number*. Similar motion takes place in these cases only if both these numbers have the same values.

If the oscillations of the fluid occur spontaneously (and not under the action of a given external exciting force), then for motion of a given type S will be a definite function of R:

$$S = f(R).$$

§20. Stokes' formula

The Navier–Stokes equation is considerably simplified in the case of flow at small Reynolds numbers. For steady flow of an incompressible fluid, this equation is

$$(\mathbf{v}\cdot\mathbf{grad})\mathbf{v} = -(1/\rho)\,\mathbf{grad}\,p+(\eta/\rho)\triangle\mathbf{v}.$$

The term $(\mathbf{v}\cdot\mathbf{grad})\mathbf{v}$ is of the order of magnitude of u^2/l, u and l having the same meaning as in §19. The quantity $(\eta/\rho)\triangle\mathbf{v}$ is of the order of magnitude of $\eta u/\rho l^2$. The ratio of the two is just the Reynolds number. Hence the term $(\mathbf{v}\cdot\mathbf{grad})\mathbf{v}$ may be neglected if the Reynolds number is small, and the equation of motion reduces to a linear equation

$$\eta\triangle\mathbf{v}-\mathbf{grad}\,p = 0. \tag{20.1}$$

Together with the equation of continuity

$$\text{div } \mathbf{v} = 0 \tag{20.2}$$

it completely determines the motion. It is useful to note also the equation

$$\triangle \, \mathbf{curl v} = 0, \tag{20.3}$$

which is obtained by taking the curl of equation (20.1).

As an example, let us consider rectilinear and uniform motion of a sphere in a viscous fluid. The problem of the motion of a sphere, it is clear, is exactly equivalent to that of flow past a fixed sphere, the fluid having a given velocity **u** at infinity. The velocity distribution in the first problem is obtained from that in the second problem by simply subtracting the velocity **u**; the fluid is then at rest at infinity, while the sphere moves with velocity $-\mathbf{u}$. If we regard the flow as steady, we must, of course, speak of the flow past a fixed sphere, since, when the sphere moves, the velocity of the fluid at any point in space varies with time.

Thus we must have $\mathbf{v} = \mathbf{u}$ at infinity; we write $\mathbf{v} = \mathbf{v}' + \mathbf{u}$, so that $\mathbf{v}'$ is zero at infinity. Since $\text{div } \mathbf{v} = \text{div } \mathbf{v}' = 0$, $\mathbf{v}'$ can be written as the curl of some vector: $\mathbf{v} = \mathbf{curl A} + \mathbf{u}$. The curl of a polar vector is well known to be an axial vector, and *vice versa*. Since the velocity is an ordinary polar vector, **A** must be an axial vector. Now **v**, and therefore **A**, depend only on the radius vector **r** (we take the origin at the centre of the sphere) and on the parameter **u**; both these vectors are polar. Furthermore, **A** must evidently be a linear function of **u**. The only such axial vector which can be constructed for a completely symmetrical body (the sphere) from two polar vectors is the vector product $\mathbf{r} \times \mathbf{u}$. Hence **A** must be of the form $f'(r)\mathbf{n} \times \mathbf{u}$, where $f'(r)$ is a scalar function of **r**, and **n** is a unit vector in the direction of the radius vector. The product $f'(r)\mathbf{n}$ can be written as the gradient, $\mathbf{grad} f(r)$, of some function $f(r)$, so that the general form of **A** is $\mathbf{grad} f \times \mathbf{u}$. Hence we can write the velocity $\mathbf{v}'$ as

$$\mathbf{v}' = \mathbf{curl} \, [\mathbf{grad} f \times \mathbf{u}].$$

Since **u** is a constant, $\mathbf{grad} f \times \mathbf{u} = \mathbf{curl} \, (f\mathbf{u})$, so that

$$\mathbf{v} = \mathbf{curl} \, \mathbf{curl} \, (f\mathbf{u}) + \mathbf{u}. \tag{20.4}$$

To determine the function f, we use equation (20.3). Since

$$\mathbf{curl v} = \mathbf{curl} \, \mathbf{curl} \, \mathbf{curl} \, (f\mathbf{u}) = (\mathbf{grad} \, \text{div} - \triangle)\mathbf{curl} \, (f\mathbf{u})$$
$$= -\triangle \, \mathbf{curl} \, (f\mathbf{u}),$$

(20.3) takes the form $\triangle^2 \, \mathbf{curl} \, (f\mathbf{u}) = 0$, or, since $\mathbf{u} = $ constant,

$$\triangle^2(\mathbf{grad} f \times \mathbf{u}) = (\triangle^2 \, \mathbf{grad} f) \times \mathbf{u} = 0.$$

It follows from this that

$$\triangle^2 \, \mathbf{grad} f = 0. \tag{20.5}$$

A first integration gives

$$\triangle^2 f = \text{constant.}$$

It is easy to see that the constant must be zero, since the velocity $\mathbf{v}$ must vanish at infinity, and so must its derivatives. The expression $\triangle^2 f$ contains fourth derivatives of f, whilst the velocity is given in terms of the second derivatives of f. Thus we have

$$\triangle^2 f \equiv \frac{1}{r^2}\frac{d}{dr}\left(r^2\frac{d}{dr}\right)\triangle f = 0.$$

Hence

$$\triangle f = 2a/r + A.$$

The constant A must be zero if the velocity is to vanish at infinity. From $\triangle f = 2a/r$ we obtain

$$f = ar + b/r. \tag{20.6}$$

The additive constant is omitted, since it is immaterial (the velocity being given by derivatives of f).

Substituting in (20.4), we have after a simple calculation

$$\mathbf{v} = \mathbf{u} - a\frac{\mathbf{u} + \mathbf{n}(\mathbf{u} \cdot \mathbf{n})}{r} + b\frac{3\mathbf{n}(\mathbf{u} \cdot \mathbf{n}) - \mathbf{u}}{r^3}. \tag{20.7}$$

The constants a and b have to be determined from the boundary conditions: at the surface of the sphere $(r = R)$, $\mathbf{v} = 0$, i.e.

$$\mathbf{u} - a\frac{\mathbf{u} + \mathbf{n}(\mathbf{u} \cdot \mathbf{n})}{R} + b\frac{3\mathbf{n}(\mathbf{u} \cdot \mathbf{n}) - \mathbf{u}}{R^3} = 0.$$

Since this equation must hold for all $\mathbf{n}$, the coefficients of $\mathbf{u}$ and $\mathbf{n}(\mathbf{u} \cdot \mathbf{n})$ must each vanish:

$$\frac{a}{R} + \frac{b}{R^3} - 1 = 0, \qquad -\frac{a}{R} + \frac{3b}{R^3} = 0.$$

Hence $a = \tfrac{3}{4}R$, $b = \tfrac{1}{4}R^3$. Thus we have finally

$$f = \tfrac{3}{4}Rr + \tfrac{1}{4}R^3/r, \tag{20.8}$$

$$\mathbf{v} = -\tfrac{3}{4}R\frac{\mathbf{u} + \mathbf{n}(\mathbf{u} \cdot \mathbf{n})}{r} - \tfrac{1}{4}R^3\frac{\mathbf{u} - 3\mathbf{n}(\mathbf{u} \cdot \mathbf{n})}{r^3} + \mathbf{u}, \tag{20.9}$$

or, in spherical components,

$$v_r = u \cos\theta\left[1 - \frac{3R}{2r} + \frac{R^3}{2r^3}\right],$$

$$v_\theta = -u \sin\theta\left[1 - \frac{3R}{4r} - \frac{R^3}{4r^3}\right]. \tag{20.10}$$

This gives the velocity distribution about the moving sphere. To determine the pressure, we substitute (20.4) in (20.1):

$$\mathbf{grad}\,p = \eta \triangle \mathbf{v} = \eta \triangle \mathbf{curl}\,\mathbf{curl}\,(f\mathbf{u})$$

$$= \eta \triangle (\mathbf{grad}\,\mathrm{div}\,(f\mathbf{u}) - \mathbf{u}\triangle f).$$

But $\triangle^2 f = 0$, and so

$$\mathbf{grad}\,p = \mathbf{grad}[\eta \triangle \mathrm{div}(f\mathbf{u})] = \mathbf{grad}(\eta\mathbf{u}\cdot\mathbf{grad}\,\triangle f).$$

Hence

$$p = \eta\mathbf{u}\cdot\mathbf{grad}\,\triangle f + p_0, \tag{20.11}$$

where p_0 is the fluid pressure at infinity. Substitution for f leads to the final expression

$$p = p_0 - \tfrac{3}{2}\eta\frac{\mathbf{u}\cdot\mathbf{n}}{r^2}R. \tag{20.12}$$

Using the above formulae, we can calculate the force **F** exerted on the sphere by the moving fluid (or, what is the same thing, the drag on the sphere as it moves through the fluid). To do so, we take spherical co-ordinates with the polar axis parallel to **u**; by symmetry, all quantities are functions only of r and of the polar angle θ. The force **F** is evidently parallel to the velocity **u**. The magnitude of this force can be determined from (15.12). Taking from this formula the components, normal and tangential to the surface, of the force on an element of the surface of the sphere, and projecting these components on the direction of **u**, we find

$$F = \oint (-p \cos\theta + \sigma'_{rr} \cos\theta - \sigma'_{r\theta} \sin\theta)df, \tag{20.13}$$

where the integration is taken over the whole surface of the sphere.

Substituting the expressions (20.10) in the formulae

$$\sigma'_{rr} = 2\eta\frac{\partial v_r}{\partial r}, \qquad \sigma'_{r\theta} = \eta\left(\frac{1}{r}\frac{\partial v_r}{\partial\theta} + \frac{\partial v_\theta}{\partial r} - \frac{v_\theta}{r}\right)$$

(see (15.17)), we find that at the surface of the sphere

$$\sigma'_{rr} = 0, \qquad \sigma'_{r\theta} = -(3\eta/2R)u \sin\theta,$$

while the pressure (20.12) is $p = p_0 - (3\eta/2R)u \cos\theta$. Hence the integral (20.13) reduces to $F = (3\eta u/2R)\oint df$, or, finally,[†]

$$F = 6\pi R\eta u. \tag{20.14}$$

This formula (called *Stokes' formula*) gives the drag on a sphere moving

---

[†] With a view to some later applications, we may mention that, if the calculations are done with formula (20.7) for the velocity (the constants a and b being undetermined), we find

$$F = 8\pi a\eta u. \tag{20.14a}$$

slowly in a fluid. We may notice that the drag is proportional to the first powers of the velocity and linear dimension of the body.[†]

This dependence of the drag on the velocity and dimension holds for slowly-moving bodies of other shapes also. The direction of the drag on a body of arbitrary shape is not the same as that of the velocity; the general form of the dependence of **F** on **u** can be written

$$F_i = a_{ik} u_k,$$ (20.15)

where a_{ik} is a tensor of rank two, independent of the velocity. It is important to note that this tensor is symmetrical ($a_{ik} = a_{ki}$), a result which holds in the linear approximation with respect to the velocity, and is a particular case of a general law valid for slow motion accompanied by dissipative processes.[‡]

The solution that we have just obtained for flow past a sphere is not valid at great distances from it, even if the Reynolds number is small. In order to see this, we estimate the magnitude of the term $(\mathbf{v} \cdot \mathbf{grad})\mathbf{v}$, which we neglected in (20.1). At great distances the velocity is **u**. The derivatives of the velocity at these distances are seen from (20.9) to be of the order of uR/r^2. Thus $(\mathbf{v} \cdot \mathbf{grad})\mathbf{v}$ is of the order of u^2R/r^2. The terms retained in equation (20.1), for example $(1/\rho)\,\mathbf{grad}\,p$, are of the order $\eta Ru/\rho r^3$ (cf. (20.12)). The condition

$$u\eta R/\rho r^3 \gg u^2R/r^2$$

holds only at distances $r \ll \nu/u$, where $\nu = \eta/\rho$. At greater distances, the terms we have omitted cannot legitimately be neglected, and the velocity distribution obtained is incorrect.

To obtain the velocity distribution at great distances from the body, we have to take into account the term $(\mathbf{v} \cdot \mathbf{grad})\mathbf{v}$ omitted in (20.1). Since the velocity **v** is nearly equal to **u** at these distances, we can put approximately **u·grad** in place of **v·grad**. We then find for the velocity at great distances the linear equation

$$(\mathbf{u \cdot grad})\mathbf{v} = -(1/\rho)\,\mathbf{grad}\,p + \nu\triangle\mathbf{v}$$ (20.16)

(C. W. OSEEN, 1910).

We shall not pause to give here the solution of this equation for flow

---

[†] The drag can also be calculated for a slowly-moving ellipsoid of any shape. The corresponding formulae are given by H. LAMB, *Hydrodynamics*, 6th ed., §339, Cambridge 1932. We give here the limiting expressions for a plane circular disk of radius R moving perpendicular to its plane:

$$F = 16\eta Ru$$

and for a similar disk moving in its plane:

$$F = 32\eta Ru/3.$$

[‡] See, for instance, *Statistical Physics*, §120, Pergamon Press, London 1958.

past a sphere,† but merely mention that the velocity distribution thus obtained can be used to derive a more accurate formula for the drag on the sphere, which includes the next term in the expansion of the drag in powers of the Reynolds number uR/ν. This formula is‡

$$F = 6\pi\eta uR\left(1 + \frac{3uR}{8\nu}\right).$$ (20.17)

Finally, we may mention that, in solving the problem of flow past an infinite cylinder with the main stream perpendicular to the axis of the cylinder, Oseen's equation has to be used from the start; in this case, equation (20.1) has no solution which satisfies the boundary conditions at the surface of the cylinder and at the same time vanishes at infinity. The drag per unit length of the cylinder is found to be

$$F = \frac{4\pi\eta u}{\tfrac{1}{2} - \gamma - \log(uR/4\nu)},$$ (20.18)

where $\gamma \fallingdotseq 0\cdot577$ is Euler's constant.

PROBLEMS

PROBLEM 1. Determine the motion of a fluid occupying the space between two concentric spheres of radii R_1, R_2 ($R_2 > R_1$), rotating uniformly about different diameters with angular velocities Ω_1, Ω_2; the Reynolds numbers $\Omega_1 R_1^2/\nu$, $\Omega_2 R_2^2/\nu$ are small compared with unity.

SOLUTION. On account of the linearity of the equations, the motion between two rotating spheres may be regarded as a superposition of the two motions obtained when one sphere is at rest and the other rotates. We first put $\Omega_2 = 0$, i.e. only the inner sphere is rotating. It is reasonable to suppose that the fluid velocity at every point is along the tangent to a circle in a plane perpendicular to the axis of rotation with its centre on the axis. On account of the axial symmetry, the pressure gradient in this direction is zero. Hence the equation of motion (20.1) becomes $\triangle \mathbf{v} = 0$. The angular velocity vector Ω_1 is an axial vector. Arguments similar to those given previously show that the velocity can be written as

$$\mathbf{v} = \mathbf{curl}[f(r)\,\Omega_1] = \mathbf{grad}\,f \times \Omega_1.$$

The equation of motion then gives $\mathbf{grad}\,\triangle f \times \Omega_1 = 0$. Since the vector $\mathbf{grad}\,\triangle f$ is parallel to the radius vector, and the vector product $\mathbf{r} \times \Omega_1$ cannot be zero for given Ω_1 and arbitrary $\mathbf{r}$, we must have $\mathbf{grad}\,\triangle f = 0$, so that

$$\triangle f = \text{constant}.$$

---

† A detailed account of the calculations for a sphere and a cylinder is given by N. E. KOCHIN, I. A. KIBEL' and N. V. ROZE, *Theoretical Hydromechanics* (*Teoreticheskaya gidromekhanika*), Part 2, 3rd ed., chapter II, §§25–26, Moscow 1948; H. LAMB, *Hydrodynamics*, 6th ed., §§342–3, Cambridge 1932.

‡ At first sight it might appear that OSEEN'S equation, which does not correctly give the velocity distribution near the sphere, could not be used to calculate the correction to the drag. In fact, however, the contribution to $\mathbf{F}$ due to the motion of the neighbouring fluid (where $u \ll \nu/r$) must be expanded in powers of the vector $\mathbf{u}$. The first non-zero correction term in $\mathbf{F}$ arising from this contribution is then proportional to $u^2\mathbf{u}$, i.e. is of the second order with respect to the Reynolds number; it therefore does not affect the first-order correction in formula (20.17). Further corrections to Stokes' formula cannot be calculated from Oseen's formula.

Integrating, we find

$$f = ar^2 + \frac{b}{r}, \qquad \mathbf{v} = \left(\frac{b}{r^3} - 2a\right)\mathbf{\Omega}_1 \times \mathbf{r}.$$

The constants a and b are found from the conditions that $\mathbf{v} = 0$ for $r = R_2$ and $\mathbf{v} = \mathbf{u}$ for $r = R_1$, where $\mathbf{u} = \mathbf{\Omega}_1 \times \mathbf{r}$ is the velocity of points on the rotating sphere. The result is

$$\mathbf{v} - \frac{R_1^3 R_2^3}{R_2^3 - R_1^3}\left(\frac{1}{r^3} - \frac{1}{R_2^3}\right)\mathbf{\Omega}_1 \times \mathbf{r}.$$

The fluid pressure is constant ($p = p_0$). Similarly, we have for the case where the outer sphere rotates and the inner one is at rest ($\Omega_1 = 0$)

$$\mathbf{v} = \frac{R_1^3 R_2^3}{R_2^3 - R_1^3}\left(\frac{1}{R_1^3} - \frac{1}{r^3}\right)\mathbf{\Omega}_2 \times \mathbf{r}.$$

In the general case where both spheres rotate, we have

$$\mathbf{v} = \frac{R_1^3 R_2^3}{R_2^3 - R_1^3}\left\{\left(\frac{1}{r^3} - \frac{1}{R_2^3}\right)\mathbf{\Omega}_1 \times \mathbf{r} + \left(\frac{1}{R_1^3} - \frac{1}{r^3}\right)\mathbf{\Omega}_2 \times \mathbf{r}\right\}.$$

If the outer sphere is absent ($R_2 = \infty$, $\Omega_2 = 0$), i.e. we have simply a sphere of radius R rotating in an infinite fluid, then

$$\mathbf{v} = (R^3/r^3)\,\mathbf{\Omega} \times \mathbf{r}.$$

Let us calculate the moment of the frictional forces acting on the sphere in this case. If we take spherical co-ordinates with the polar axis parallel to $\mathbf{\Omega}$, we have $v_r = v_\theta = 0$, $v_\phi = v = (R^3\Omega/r^2)\sin\theta$. The frictional force on unit area of the sphere is

$$\sigma'_{r\phi} = \eta\left(\frac{\partial v}{\partial r} - \frac{v}{r}\right)_{r=R} = -3\eta\Omega\sin\theta.$$

The total moment on the sphere is

$$M = \int_0^\pi \sigma'_{r\phi}\, R\sin\theta \cdot 2\pi R^2\sin\theta\, d\theta,$$

whence we find

$$M = -8\pi\eta R^3\Omega.$$

If the inner sphere is absent, $\mathbf{v} = \mathbf{\Omega}_2 \times \mathbf{r}$, i.e. the fluid simply rotates rigidly with the sphere surrounding it.

PROBLEM 2. Determine the velocity of a spherical drop of fluid (of viscosity η') moving under gravity in a fluid of viscosity η (W. Rybczyński 1911).

SOLUTION. We use a system of co-ordinates in which the drop is at rest. For the fluid outside the drop we again seek a solution of equation (20.5) in the form (20.6), so that the velocity has the form (20.7). For the fluid inside the drop, we have to find a solution which does not have a singularity at $r = 0$ (and the second derivatives of f, which determine the velocity, must also remain finite). This solution is

$$f = \tfrac{1}{4}Ar^2 + \tfrac{1}{8}Br^4,$$

and the corresponding velocity is

$$\mathbf{v} = -A\mathbf{u} + Br^2[\mathbf{n}(\mathbf{u} \cdot \mathbf{n}) - 2\mathbf{u}].$$

At the surface of the sphere† the following conditions must be satisfied. The normal velocity components outside ($\mathbf{v}_e$) and inside ($\mathbf{v}_i$) the drop must be zero:

$$v_{i,r} = v_{e,r} = 0.$$

The tangential velocity component must be continuous:

$$v_{i,\theta} = v_{e,\theta},$$

as must be the component $\sigma_{r\theta}$ of the stress tensor:

$$\sigma_{i,r\theta} = \sigma_{e,r\theta}.$$

The condition that the stress tensor components σ_{rr} are equal need not be written down; it would determine the required velocity u, which is more simply found in the manner shown below. From the above four conditions we obtain four equations for the constants a, b, A, B, whose solutions are

$$a = R\frac{2\eta + 3\eta'}{4(\eta + \eta')}, \qquad b = R^3\frac{\eta'}{4(\eta + \eta')}, \qquad A = -BR^2 = \frac{\eta}{2(\eta + \eta')}.$$

By (20.14a), we have for the drag

$$F = 2\pi u\eta R(2\eta + 3\eta')/(\eta + \eta').$$

As $\eta' \to \infty$ (corresponding to a solid sphere) this formula becomes Stokes' formula. In the limit $\eta' \to 0$ (corresponding to a gas bubble) we have $F = 4\pi u\eta R$, i.e. the drag is two-thirds of that on a solid sphere.

Equating F to the force of gravity on the drop, $\frac{4}{3}\pi R^3(\rho - \rho')g$, we find

$$u = \frac{2R^2 g(\rho - \rho')(\eta + \eta')}{3\eta(2\eta + 3\eta')}.$$

PROBLEM 3. Two parallel plane circular disks (of radius R) lie one above the other a small distance apart; the space between them is filled with fluid. The disks approach at a constant velocity u, displacing the fluid. Determine the resistance to their motion (O. REYNOLDS).

SOLUTION. We take cylindrical co-ordinates, with the origin at the centre of the lower disk, which we suppose fixed. The flow is axisymmetric and, since the fluid layer is thin, predominantly radial: $v_z \ll v_r$, and also $\partial v_r/\partial r \ll \partial v_r/\partial z$. Hence the equations of motion become

$$\eta\frac{\partial^2 v_r}{\partial z^2} = \frac{\partial p}{\partial r}, \qquad \frac{\partial p}{\partial z} = 0, \tag{1}$$

$$\frac{1}{r}\frac{\partial(rv_r)}{\partial r} + \frac{\partial v_z}{\partial z} = 0, \tag{2}$$

---

† We may neglect the change of shape of the drop in its motion, since this change is of a higher order of smallness. However, it must be borne in mind that, in order that the moving drop should in fact be spherical, the forces due to surface tension at its boundary must exceed the forces due to pressure differences, which tend to make the drop non-spherical. This means that we must have $\eta u/R \ll \alpha/R$, where α is the surface-tension coefficient, or, substituting $u \sim R^2 g\rho/\eta$,

$$R \ll \sqrt{(\alpha/\rho g)}.$$

with the boundary conditions

$$\text{at } z = 0: \qquad v_r = v_z = 0;$$
$$\text{at } z = h: \qquad v_r = 0, \qquad v_z = -u;$$
$$\text{at } r = R: \qquad p = p_0,$$

where h is the distance between the disks, and p_0 the external pressure. From equations (1) we find

$$v_r = \frac{1}{2\eta} \frac{\partial p}{\partial r} z(z - h).$$

Integrating equation (2) with respect to z, we obtain

$$u = \frac{1}{r} \frac{d}{dr} \int_0^h r v_r \, dz = -\frac{h^3}{12\eta r} \frac{d}{dr} \left(r \frac{dp}{dr} \right),$$

whence

$$p = p_0 + \frac{3\eta u}{h^3} (R^2 - r^2).$$

The total resistance to the moving disk is

$$F = 3\pi\eta u R^4 / 2h^3.$$

§21. The laminar wake

In steady flow of a viscous fluid past a solid body, the flow at great distances behind the body has certain characteristics which can be investigated independently of the particular shape of the body.

Let us denote by $\mathbf{U}$ the constant velocity of the incident current; we take the direction of $\mathbf{U}$ as the x-axis, with the origin somewhere inside the body. The actual fluid velocity at any point may be written $\mathbf{U} + \mathbf{v}$; $\mathbf{v}$ vanishes at infinity.

It is found that, at great distances behind the body, the velocity $\mathbf{v}$ is noticeably different from zero only in a relatively narrow region near the x-axis. This region, called the *laminar wake*,† is reached by fluid particles which move along streamlines passing fairly close to the body. Hence the flow in the wake is essentially rotational. On the other hand, the viscosity has almost no effect at any point on streamlines that do not pass near the body, and the vorticity, which is zero in the incident current, remains practically zero on these streamlines, as it would in an ideal fluid. Thus the flow at great distances from the body may be regarded as potential flow everywhere except in the wake.

We shall now derive formulae relating the properties of the flow in the wake to the forces acting on the body. The total momentum transported by the fluid through any closed surface surrounding the body is equal to the

---

† In contradistinction to the turbulent wake; see §36.

integral of the momentum flux density tensor over that surface, $\oint \Pi_{ik} df_k$. The components of the tensor Π_{ik} are

$$\Pi_{ik} = p\delta_{ik} + \rho(U_i + v_i)(U_k + v_k).$$

We write the pressure in the form $p = p_0 + p'$, where p_0 is the pressure at infinity. The integration of the constant term $p_0\delta_{ik} + \rho U_i U_k$ gives zero, since the vector integral $\oint d\mathbf{f}$ over a closed surface is zero. The integral $U_i \oint \rho v_k df_k$ also vanishes: since the total mass of fluid in the volume considered is constant, the total mass flux $\oint \rho \mathbf{v} \cdot d\mathbf{f}$ through the surface surrounding the volume must be zero. Finally, the velocity $\mathbf{v}$ far from the body is small compared with $\mathbf{U}$. Hence, if the surface in question is sufficiently far from the body, we can neglect the term $\rho v_i v_k$ in Π_{ik} as compared with $\rho U_k v_i$. Thus the total momentum flux is

$$\oint (p'\delta_{ik} + \rho U_k v_i) df_k.$$

Let us now take the fluid volume concerned to be the volume between two infinite planes $x = $ constant, one of them far in front of the body and the other far behind it. The integral over the infinitely distant "lateral" surface vanishes (since $p' = \mathbf{v} = 0$ at infinity), and it is therefore sufficient to integrate only over the two planes. The momentum flux thus obtained is evidently the difference between the total momentum flux entering through the forward plane and that leaving through the backward plane. This difference, however, is just the quantity of momentum transmitted to the body by the fluid per unit time, i.e. the force $\mathbf{F}$ exerted on the body.

Thus the components of the force $\mathbf{F}$ are

$$F_x = \left(\int\int_{x=x_2} - \int\int_{x=x_1} \right)(p' + \rho U v_x) dy\, dz,$$

$$F_y = \left(\int\int_{x=x_2} - \int\int_{x=x_1} \right)\rho U v_y\, dy\, dz,$$

$$F_z = \left(\int\int_{x=x_2} - \int\int_{x=x_1} \right)\rho U v_z\, dy\, dz,$$

where the integration is taken over the infinite planes $x = x_1$ (far behind the body) and $x = x_2$ (far in front of it). Let us first consider the expression for F_x.

Outside the wake we have potential flow, and therefore Bernoulli's equation

$$p + \tfrac{1}{2}\rho(\mathbf{U} + \mathbf{v})^2 = \text{constant} \equiv p_0 + \tfrac{1}{2}\rho U^2$$

holds, or, neglecting the term $\tfrac{1}{2}\rho v^2$ in comparison with $\rho \mathbf{U} \cdot \mathbf{v}$,

$$p' = -\rho U v_x.$$

We see that in this approximation the integrand in F_x vanishes everywhere outside the wake. In other words, the integral over the plane $x = x_2$ (which lies in front of the body and does not intersect the wake) is zero, and the integral over the plane $x = x_1$ need be taken only over the area covered by the cross-section of the wake. Inside the wake, however, the pressure change p' is of the order of ρv^2, i.e. small compared with $\rho U v_x$. Thus we reach the result that the drag on the body is

$$F_x = -\rho U \iint v_x \, dy \, dz, \qquad (21.1)$$

where the integration is taken over the cross-sectional area of the wake far behind the body. The velocity v_x in the wake is, of course, negative: the fluid moves more slowly than it would if the body were absent. Attention is called to the fact that the integral in (21.1) gives the amount by which the discharge through the wake falls short of its value in the absence of the body.

Let us now consider the force (whose components are F_y, F_z) which tends to move the body transversely. This force is called the *lift*. Outside the wake, where we have potential flow, we can write $v_y = \partial\phi/\partial y$, $v_z = \partial\phi/\partial z$; the integral over the plane $x = x_2$, which does not meet the wake, is zero:

$$\iint v_y \, dy \, dz = \iint \frac{\partial\phi}{\partial y} \, dy \, dz = 0, \qquad \iint \frac{\partial\phi}{\partial z} \, dy \, dz = 0,$$

since $\phi = 0$ at infinity. We therefore find for the lift

$$F_y = -\rho U \iint v_y \, dy \, dz, \qquad F_z = -\rho U \iint v_z \, dy \, dz. \qquad (21.2)$$

The integration in these formulae is again taken only over the cross-sectional area of the wake. If the body has an axis of symmetry (not necessarily complete axial symmetry), and the flow is parallel to this axis, then the flow past the body has an axis of symmetry also. In this case the lift is, of course, zero.

Let us return to the flow in the wake. An estimate of the magnitudes of various terms in the Navier–Stokes equation shows that the term $\nu\triangle\mathbf{v}$ can in general be neglected at distances r from the body such that $rU/\nu \gg 1$ (cf. the derivation of the opposite condition at the beginning of §20); these are the distances at which the flow outside the wake may be regarded as potential flow. It is not possible to neglect that term inside the wake even at these distances, however, since the transverse derivatives $\partial^2\mathbf{v}/\partial y^2, \partial^2\mathbf{v}/\partial z^2$ are large compared with $\partial^2\mathbf{v}/\partial x^2$.

The term $(\mathbf{v}\cdot\mathbf{grad})\mathbf{v}$ in the Navier–Stokes equation is of the order of magnitude $(U+v)\,\partial v/\partial x \sim Uv/x$ in the wake. The term $\nu\triangle\mathbf{v}$ is of the order of $\nu\partial^2 v/\partial y^2 \sim \nu v/Y^2$, where Y denotes the width of the wake, i.e. the order of magnitude of the distances from the x-axis at which the velocity $\mathbf{v}$ falls off markedly. If these two magnitudes are comparable, we find

$$Y \sim \sqrt{(\nu x/U)}. \qquad (21.3)$$

This quantity is in fact small compared with x, by the assumed condition $Ux/\nu \gg 1$. Thus the width of the laminar wake increases as the square root of the distance from the body.

In order to determine how the velocity decreases with increasing x in the wake, we return to formula (21.1). The region of integration has an area of the order of Y^2. Hence the integral can be estimated as $F_x \sim \rho U v Y^2$, and by using the relation (21.3) we obtain

$$v \sim F_x/\rho \nu x. \tag{21.4}$$

PROBLEMS

PROBLEM 1. Determine the flow in the laminar wake when there is both drag and lift.

SOLUTION. Writing the velocity in the Navier–Stokes equation in the form $\mathbf{U}+\mathbf{v}$ and omitting terms quadratic in $\mathbf{v}$ (far from the body) we obtain

$$U\frac{\partial \mathbf{v}}{\partial x} = -\mathbf{grad}\left(\frac{p}{\rho}\right) + \nu\left(\frac{\partial^2 \mathbf{v}}{\partial y^2} + \frac{\partial^2 \mathbf{v}}{\partial z^2}\right);$$

we have also neglected the term $\partial^2 \mathbf{v}/\partial x^2$ in $\triangle \mathbf{v}$. We seek a solution in the form $\mathbf{v} = \mathbf{v}_1 + \mathbf{v}_2$, where $\mathbf{v}_1$ satisfies

$$U\frac{\partial \mathbf{v}_1}{\partial x} = \nu\left(\frac{\partial^2 \mathbf{v}_1}{\partial y^2} + \frac{\partial^2 \mathbf{v}_1}{\partial z^2}\right).$$

The term $\mathbf{v}_2$, which appears because of the term $-\mathbf{grad}(p/\rho)$ in the original equation, may be taken as the gradient $\mathbf{grad}\ \Phi$ of some scalar. Since the derivatives with respect to x, far from the body, are small in comparison with those with respect to y and z, we may to the same approximation neglect the term $\partial\Phi/\partial x$ in v_x, i.e. take $v_x = v_{1x}$.

Thus we have for v_x the equation

$$U\frac{\partial v_x}{\partial x} = \nu\left(\frac{\partial^2 v_x}{\partial y^2} + \frac{\partial^2 v_x}{\partial z^2}\right).$$

This equation is formally the same as the two-dimensional equation of heat conduction, with x/U in place of the time, and the viscosity ν in place of the thermometric conductivity. The solution which decreases with increasing y and z (for fixed x) and gives an infinitely narrow wake as $x \to 0$ (in this approximation the dimensions of the body are regarded as small) is (see §51)

$$v_x = -\frac{F_x}{4\pi\rho\nu}\frac{1}{x}e^{-U(y^2+z^2)/4\nu x}. \tag{1}$$

The constant coefficient in this formula is expressed in terms of the drag by means of formula (21.1), in which the integration over y and z may be extended to $\pm\infty$ on account of the rapid decrease of v_x. If we replace the Cartesian co-ordinates by spherical co-ordinates r, θ, ϕ with the polar axis along the x-axis, then the region of the wake ($\sqrt{(y^2+z^2)} \ll x$) corresponds to $\theta \ll 1$. In these co-ordinates formula (1) becomes

$$v_x = -\frac{F_x}{4\pi\rho\nu}\frac{1}{r}e^{-Ur\theta^2/4\nu}. \tag{1'}$$

The term $\partial\Phi/\partial x$ (with Φ given by formula (3) below), which we have omitted, would give a term in v_x which diminishes more rapidly, as $1/r^2$.

v_{1y} and v_{1z} must have the same form as (1). We take the direction of the lift as the y-axis (so that $F_z = 0$). According to (21.2) we have, since $\Phi = 0$ at infinity,

$$\int_{-\infty}^{\infty} \int_{-\infty}^{\infty} v_y \, dy \, dz = \iint \left(v_{1y} + \frac{\partial \Phi}{\partial y} \right) dy \, dz$$

$$= \iint v_{1y} \, dy \, dz = -F_y/\rho U,$$

$$\iint v_{1z} \, dy \, dz = 0.$$

Determining the constants in v_{1y} and v_{1z} from these conditions, we find

$$v_y = -\frac{F_y}{4\pi\rho\nu} \frac{1}{x} e^{-U(y^2+z^2)/4\nu x} + \frac{\partial \Phi}{\partial y}, \qquad v_z = \frac{\partial \Phi}{\partial z}. \tag{2}$$

To determine the function Φ we proceed as follows. By the equation of continuity,

$$\text{div } \mathbf{v} \approx \partial v_y/\partial y + \partial v_z/\partial z = 0;$$

substituting (2), we have

$$\left(\frac{\partial^2}{\partial y^2} + \frac{\partial^2}{\partial z^2} \right) \Phi = -\frac{\partial v_{1y}}{\partial y}.$$

Differentiating this equation with respect to x and using the equation satisfied by v_{1y}, we obtain

$$\left(\frac{\partial^2}{\partial y^2} + \frac{\partial^2}{\partial z^2} \right) \frac{\partial \Phi}{\partial x} = -\frac{\partial}{\partial y} \left(\frac{\partial v_{1y}}{\partial x} \right)$$

$$= -\frac{\nu}{U} \left(\frac{\partial^2}{\partial y^2} + \frac{\partial^2}{\partial z^2} \right) \frac{\partial v_{1y}}{\partial y}.$$

Hence

$$\frac{\partial \Phi}{\partial x} = -\frac{\nu}{U} \frac{\partial v_{1y}}{\partial y}.$$

Finally, substituting the expression for v_{1y} and integrating with respect to x, we have

$$\Phi = -\frac{F_y}{2\pi\rho U} \frac{y}{y^2+z^2} [e^{-U(y^2+z^2)/4\nu x} - 1]. \tag{3}$$

The constant of integration is chosen so that Φ remains finite when $y = z = 0$. In spherical co-ordinates (with the azimuthal angle ϕ measured from the xy-plane)

$$\Phi = -\frac{F_y}{2\pi\rho U} \frac{\cos\phi}{r\theta} [e^{-Ur\theta^2/4\nu} - 1]. \tag{3'}$$

It is seen from (2) and (3) that v_y and v_z, unlike v_x, contain terms which decrease only as $1/\theta^2$ as we move away from the "axis" of the wake, as well as those which decrease exponentially with θ (for a given r).

The qualitative results (21.3) and (21.4) are, as we should expect, in agreement with the above formulae. If there is no lift, the flow in the wake is axially symmetrical.

PROBLEM 2. Determine the flow outside the wake far from the body.

SOLUTION. Outside the wake we assume potential flow. Since we are interested only in the terms in the potential Φ which decrease least rapidly with distance, we seek a solution of Laplace's equation $\triangle \Phi = 0$ as a sum of two terms:

$$\Phi = \frac{a}{r} + \frac{\cos \phi}{r} f(\theta),$$

of which the first is centrally symmetric and belongs to the force F_x, while the second is symmetrical about the xy-plane and belongs to the force F_y.

Using the expression for $\triangle \Phi$ in spherical co-ordinates, we obtain for the function $f(\theta)$ the equation

$$\frac{\mathrm{d}}{\mathrm{d}\theta} \left(\sin \theta \frac{\mathrm{d}f}{\mathrm{d}\theta} \right) - \frac{f}{\sin \theta} = 0.$$

The solution of this equation finite as $\theta \to \pi$ is $f = b \cot \frac{1}{2}\theta$. The coefficient b must be determined so as to give the correct value of F_y. It is simpler, however, to use the fact that in the range $\sqrt{(\nu/Ur)} \ll \theta \ll 1$ this part of Φ must be the same as the expression

$$\Phi = \frac{F_y}{2\pi\rho U} \frac{\cos \phi}{r\theta},$$

obtained from formula (3′), Problem 1, for Φ in the wake. Hence $b = F_y/4\pi\rho U$.

To determine the coefficient a, we notice that the total mass flux through a sphere S of large radius r equals zero, as for any closed surface. The rate of inflow through the part S_0 of S intercepted by the wake is

$$- \iint\limits_{S_0} v_x \, \mathrm{d}y \, \mathrm{d}z = F_x/\rho U.$$

Hence the same quantity must flow out through the rest of the surface of the sphere, i.e. we must have

$$\oint\limits_{S-S_0} \mathbf{v} \cdot \mathbf{df} = F_x/\rho U.$$

Since S_0 is small compared with S, we can put

$$\oint\limits_{S} \mathbf{v} \cdot \mathbf{df} = \oint\limits_{S} \mathbf{grad}\, \Phi \cdot \mathbf{df} = -4\pi a = F_x/\rho U,$$

whence $a = -F_x/4\pi\rho U$.

The complete solution is given by the sum of these two expressions:

$$\Phi = \frac{1}{4\pi\rho Ur} (-F_x + F_y \cos \phi \cot \tfrac{1}{2}\theta), \tag{1}$$

which gives the flow everywhere outside the wake far from the body. The potential diminishes with increasing distance as $1/r$; the velocity $\mathbf{v}$, therefore, diminishes as $1/r^2$. If there is no lift, the flow outside the wake is spherically symmetrical.

§22. The viscosity of suspensions

A fluid in which numerous fine solid particles are suspended (forming a *suspension*) may be regarded as an homogeneous medium if we are concerned with phenomena whose characteristic lengths are large compared with the dimensions of the particles. Such a medium has an effective viscosity η which is different from the viscosity η_0 of the original fluid. The value of η can be calculated for the case where the concentration of the suspended particles is small (i.e. their total volume is small in comparison with that of the fluid). The calculations are relatively simple for the case of spherical particles (A. EINSTEIN, 1906).

It is necessary to consider first the effect of a single solid globule, immersed in a fluid, on flow having a constant velocity gradient. Let the unperturbed flow be described by a linear velocity distribution

$$v_{0i} = \alpha_{ik}x_k, \tag{22.1}$$

where α_{ik} is a constant symmetrical tensor. The fluid pressure is constant:

$$p_0 = \text{constant},$$

and in future we shall take p_0 to be zero, i.e. measure only the deviation from this constant value. If the fluid is incompressible (div $\mathbf{v}_0 = 0$), the sum of the diagonal elements of the tensor α_{ik} must be zero:

$$\alpha_{ii} = 0. \tag{22.2}$$

Now let a small sphere of radius R be placed at the origin. We denote the altered fluid velocity by $\mathbf{v} = \mathbf{v}_0 + \mathbf{v}_1$; $\mathbf{v}_1$ must vanish at infinity, but near the sphere $\mathbf{v}_1$ is not small compared with $\mathbf{v}_0$. It is clear from the symmetry of the flow that the sphere remains at rest, so that the boundary condition is $\mathbf{v} = 0$ for $r = R$.

The required solution of the equations of motion (20.1) to (20.3) may be obtained at once from the solution (20.4), with the function f given by (20.6), if we notice that the space derivatives of this solution are themselves solutions. In the present case we desire a solution depending on the components of the tensor α_{ik} as parameters (and not on the vector $\mathbf{u}$ as in §20). Such a solution is

$$\mathbf{v}_1 = \textbf{curl curl}\,[(\boldsymbol{\alpha}\cdot\textbf{grad})f], \qquad p = \eta_0\alpha_{ik}\partial^2\triangle f/\partial x_i\partial x_k,$$

where $(\boldsymbol{\alpha}\cdot\textbf{grad})f$ denotes a vector whose components are $\alpha_{ik}\partial f/\partial x_k$. Expanding these expressions and determining the constants a and b in the function $f = ar + b/r$ so as to satisfy the boundary conditions at the surface of the sphere, we obtain the following formulae for the velocity and pressure:

$$v_{1i} = \frac{5}{2}\left(\frac{R^5}{r^4} - \frac{R^3}{r^2}\right)\alpha_{kl}n_in_kn_l - \frac{R^5}{r^4}\alpha_{ik}n_k, \tag{22.3}$$

$$p = -5\eta_0\frac{R^5}{r^3}\alpha_{ik}n_in_k, \tag{22.4}$$

where $\mathbf{n}$ is a unit vector in the direction of the radius vector.

Returning now to the problem of determining the effective viscosity of a suspension, we calculate the mean value (over the volume) of the momentum flux density tensor Π_{ik}, which, in the linear approximation with respect to the velocity, is the same as the stress tensor $-\sigma_{ik}$:

$$\bar{\sigma}_{ik} = (1/V) \int \sigma_{ik}\,dV.$$

The integration here may be taken over the volume V of a sphere of large radius, which is then extended to infinity.

First of all, we have the identity

$$\bar{\sigma}_{ik} = \eta_0\left(\overline{\frac{\partial v_i}{\partial x_k}} + \overline{\frac{\partial v_k}{\partial x_i}}\right) - \bar{p}\delta_{ik} +$$

$$+ \frac{1}{V}\int\left\{\sigma_{ik} - \eta_0\left(\frac{\partial v_i}{\partial x_k} + \frac{\partial v_k}{\partial x_i}\right) + p\delta_{ik}\right\}dV. \tag{22.5}$$

The integrand on the right is zero except within the solid spheres; since the concentration of the suspension is supposed small, the integral may be calculated for a single sphere as if the others were absent, and then multiplied by the concentration c of the suspension (the number of spheres per unit volume). The direct calculation of this integral would require an investigation of internal stresses in the spheres. We can circumvent this difficulty, however, by transforming the volume integral into a surface integral over an infinitely distant sphere, which lies entirely in the fluid. To do so, we note that the equation of motion $\partial\sigma_{il}/\partial x_l = 0$ leads to the identity

$$\sigma_{ik} = \partial(\sigma_{il}x_k)/\partial x_l;$$

hence the transformation of the volume integral into a surface integral gives

$$\sigma_{ik} = \eta_0\left(\overline{\frac{\partial v_i}{\partial x_k}} + \overline{\frac{\partial v_k}{\partial x_i}}\right) + c\oint\{\sigma_{il}x_k\,df_l - \eta_0(v_i\,df_k + v_k\,df_i)\}.$$

We have omitted the term in $\bar{p}$, since the mean pressure is necessarily zero; $\bar{p}$ is a scalar, which must be given by a linear combination of the components α_{ik}, and the only such scalar is $\alpha_{ii} = 0$.

In calculating the integral over a sphere of very large radius, only the terms of order $1/r^2$ need be retained in the expression (22.3) for the velocity. A simple calculation gives the value of the integral as

$$c\eta_0 \cdot 20\pi R^3\{5\alpha_{lm}\overline{n_in_kn_ln_m} - \alpha_{il}\overline{n_kn_l}\},$$

where the bar denotes an average with respect to directions of the unit vector

n. Effecting the averaging,† we finally have

$$\bar{\sigma}_{ik} = \eta_0\left(\overline{\frac{\partial v_i}{\partial x_k}} + \overline{\frac{\partial v_k}{\partial x_i}}\right) + 5\eta_0\alpha_{ik}\cdot\tfrac{4}{3}\pi R^3 c. \tag{22.6}$$

The ratio of the second term to the first determines the required relative correction to give the effective viscosity of the suspension. If we are interested only in corrections of the first order of smallness, we can take the first term as $2\eta_0\alpha_{ik}$. We then obtain for the effective viscosity of the suspension

$$\eta = \eta_0(1 + \tfrac{5}{2}\phi), \tag{22.7}$$

where $\phi = \tfrac{4}{3}\pi R^3 c$ is the small ratio of the total volume of the spheres to the total volume of the suspension.

§23. Exact solutions of the equations of motion for a viscous fluid

If the non-linear terms in the equations of motion of a viscous fluid do not vanish identically, the solving of these equations offers great difficulties, and exact solutions can be obtained only in a very small number of cases. Furthermore, it has not yet proved possible to carry out a complete investigation of the steady flow of a viscous fluid in all space round a body in the limit of very large Reynolds numbers. Although, as we shall see, such a flow does not in practice remain steady, the solution of the problem would nevertheless be of great methodological interest.‡

We give below examples of exact solutions of the equations of motion for a viscous fluid.

(1) An infinite plane disk immersed in a viscous fluid rotates uniformly about its axis. Determine the motion of the fluid caused by this motion of the disk (T. VON KÁRMÁN, 1921).

We take cylindrical co-ordinates, with the plane of the disk as the plane $z = 0$. Let the disk rotate about the z-axis with angular velocity Ω. We consider the unbounded volume of fluid on the side $z > 0$. The boundary conditions are

$$v_r = 0, \qquad v_\phi = \Omega r, \qquad v_z = 0 \quad \text{for} \quad z = 0,$$
$$v_r = 0, \qquad v_\phi = 0 \qquad\qquad\qquad \text{for} \quad z = \infty.$$

---

† The required mean values of products of components of the unit vector are symmetrical tensors, which can be formed only from the unit tensor δ_{ik}. We then easily find

$$\overline{n_i n_k} = \delta_{ik},$$

$$\overline{n_i n_k n_l n_m} = \tfrac{1}{15}(\delta_{ik}\delta_{lm} + \delta_{il}\delta_{km} + \delta_{im}\delta_{kl}).$$

‡ The "vanishing viscosity" theory of Oseen is concerned with this problem; it is unsatisfactory, since it is based on an unjustified simplification of the Navier–Stokes equations. Prandtl's boundary-layer theory (see §39) does not solve the problem throughout the volume of the fluid.

The axial velocity v_z does not vanish as $z \to \infty$, but tends to a constant negative value determined by the equations of motion. The reason is that, since the fluid moves radially away from the axis of rotation, especially near the disk, there must be a constant vertical flow from infinity in order to satisfy the equation of continuity. We seek a solution of the equations of motion in the form

$$v_r = r\Omega F(z_1); \quad v_\phi = r\Omega G(z_1); \quad v_z = \sqrt{(\nu\Omega)}H(z_1);$$
$$p = -\rho\nu\Omega P(z_1), \quad \text{where} \quad z_1 = \sqrt{(\Omega/\nu)}z. \tag{23.1}$$

In this velocity distribution, the radial and azimuthal velocities are proportional to the distance from the axis of rotation, while v_z is constant on each horizontal plane.

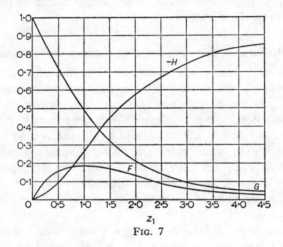

z_1

Fig. 7

Substituting in the Navier–Stokes equation and in the equation of continuity, we obtain the following equations for the functions F, G, H and P:

$$F^2 - G^2 + F'H = F'', \qquad 2FG + G'H = G'',$$
$$HH' = P' + H'', \qquad 2F + H' = 0; \tag{23.2}$$

the prime denotes differentiation with respect to z_1. The boundary conditions are

$$F = 0, \qquad G = 1, \qquad H = 0 \quad \text{for} \quad z_1 = 0.$$
$$F = 0, \qquad G = 0 \qquad\qquad \text{for} \quad z_1 = \infty. \tag{23.3}$$

We have therefore reduced the solution of the problem to the integration of a system of ordinary differential equations in one variable; this can be achieved numerically.† Fig. 7 shows the functions F, G and $-H$ thus obtained.

---

† The numerical integration has also been carried out for another similar problem, in which the fluid rotates uniformly at infinity and the disc is at rest (U.T. BÖDEWADT, *Zeitschrift für angewandte Mathematik und Mechanik* **20**, 241, 1940).

The limiting value of H as $z_1 \to \infty$ is -0.886; in other words, the fluid velocity at infinity is $v_z(\infty) = -0.886\sqrt{(\nu\Omega)}$.

The frictional force acting on unit area of the disk perpendicularly to the radius is $\sigma_{z\phi} = \eta(\partial v_\phi/\partial z)_{z=0}$. Neglecting edge effects, we may write the moment of the frictional forces acting on a disk of large but finite radius R as

$$M = 2 \int_0^R 2\pi r^2 \, \sigma_{z\phi} \, dr = \pi R^4 \rho \sqrt{(\nu\Omega^3)}G'(0).$$

The factor 2 in front of the integral appears because the disk has two sides exposed to the fluid. A numerical calculation of the function G leads to the formula

$$M = -1.94 \, R^4 \rho \sqrt{(\nu\Omega^3)}. \tag{23.4}$$

(2) Determine the steady flow between two plane walls meeting at an angle α (Fig. 8 shows a cross-section of the two planes); the fluid flows out from the line of intersection of the planes (G. HAMEL, 1916).

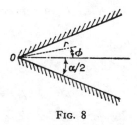

FIG. 8

We take cylindrical co-ordinates r, z, ϕ, with the z-axis along the line of the intersection of the planes (the point O in Fig. 8), and the angle ϕ measured as shown in Fig. 8. The flow is uniform in the z-direction, and we naturally assume it to be entirely radial, i.e.

$$v_\phi = v_z = 0, \qquad v_r = v(r, \phi).$$

The equations (15.16) give

$$v\frac{\partial v}{\partial r} = -\frac{1}{\rho}\frac{\partial p}{\partial r} + \nu\left(\frac{\partial^2 v}{\partial r^2} + \frac{1}{r}\frac{\partial^2 v}{\partial \phi} + \frac{1}{r}\frac{\partial v}{\partial r} - \frac{v}{r^2}\right), \tag{23.5}$$

$$-\frac{1}{\rho r}\frac{\partial p}{\partial \phi} + \frac{2\nu}{r^2}\frac{\partial v}{\partial \phi} = 0, \tag{23.6}$$

$$\partial(rv)/\partial r = 0.$$

It is seen from the last of these that rv is a function of ϕ only. Introducing the function

$$u(\phi) = rv/6\nu, \tag{23.7}$$

we obtain from (23.6)

$$\frac{1}{\rho}\frac{\partial p}{\partial \phi} = \frac{12\nu^2}{r^2}\frac{du}{d\phi},$$

whence

$$\frac{p}{\rho} = \frac{12\nu^2}{r^2}u(\phi) + f(r).$$

Substituting this expression in (23.5), we have

$$\frac{d^2u}{d\phi^2} + 4u + 6u^2 = \frac{1}{6\nu^2}r^3f'(r),$$

from which we see that, since the left-hand side depends only on ϕ and the right-hand side only on r, each must be a constant, which we denote by $2C_1$. Thus $f'(r) = 12\nu^2C_1/r^3$, whence $f(r) = -6\nu^2C_1/r^2 + \text{constant}$, and we have for the pressure

$$\frac{p}{\rho} = \frac{6\nu^2}{r^2}(2u - C_1) + \text{constant}. \tag{23.8}$$

For $u(\phi)$ we have the equation

$$u'' + 4u + 6u^2 = 2C_1,$$

which, on multiplication by u' and one integration, gives

$$\tfrac{1}{2}u'^2 + 2u^2 + 2u^3 - 2C_1u - 2C_2 = 0.$$

Hence we have

$$2\phi = \pm \int \frac{du}{\sqrt{(-u^3 - u^2 + C_1u + C_2)}} + C_3, \tag{23.9}$$

which gives the required dependence of the velocity on ϕ; the function $u(\phi)$ can be expressed in terms of elliptic functions. The three constants C_1, C_2, C_3 are determined from the boundary conditions

$$u(\pm\tfrac{1}{2}\alpha) = 0 \tag{23.10}$$

and from the condition that the same mass Q of fluid passes in unit time through any cross-section $r = \text{constant}$:

$$Q = \rho \int_{-\alpha/2}^{\alpha/2} vr\,d\phi = 6\nu\rho \int_{-\alpha/2}^{\alpha/2} u\,d\phi. \tag{23.11}$$

Q may be either positive or negative. If $Q > 0$, the line of intersection of the planes is a source, i.e. the fluid emerges from the vertex of the angle: this is called *flow in a diverging channel*. If $Q < 0$, the line of intersection is a sink, and we have *flow in a converging channel*. The ratio $|Q|/\nu\rho$ is dimensionless and plays the part of the Reynolds number in the problem considered.

Let us first discuss converging flow ($Q < 0$). To investigate the solution (23.9)–(23.11) we make the assumptions, which will be justified later, that the flow is symmetrical about the plane $\phi = 0$ (i.e. $u(\phi) = u(-\phi)$), and that the function $u(\phi)$ is everywhere negative (i.e. the velocity is everywhere towards the vertex) and decreases monotonically from $u = 0$ at $\phi = \pm\tfrac{1}{2}\alpha$ to $u = -u_0 < 0$ at $\phi = 0$, so that u_0 is the maximum value of $|u|$. Then for $u = -u_0$ we must have $du/d\phi = 0$, whence it follows that $u = -u_0$ is a zero of the cubic expression under the radical in the integrand of (23.9). We can therefore write

$$-u^3 - u^2 + C_1 u + C_2 = (u+u_0)\{-u^2 - (1-u_0)u + q\},$$

where q is another constant. Thus

$$2\phi = \pm \int_{-u_0}^{u} \frac{du}{\sqrt{[(u+u_0)\{-u^2-(1-u_0)u+q\}]}}, \qquad (23.12)$$

the constants u_0 and q being determined from the conditions

$$\alpha = \int_{-u_0}^{0} \frac{du}{\sqrt{[(u+u_0)\{-u^2-(1-u_0)u+q\}]}},$$

$$\tfrac{1}{8}R = \int_{-u_0}^{0} \frac{u\,du}{\sqrt{[(u+u_0)\{-u^2-(1-u_0)u+q\}]}} \qquad (23.13)$$

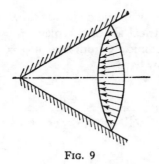

FIG. 9

($R = |Q|/\nu\rho$); the constant q must be positive, since otherwise these integrals would be complex. The two equations just given may be shown to have solutions u_0 and q for any R and $\alpha < \pi$. In other words, convergent symmetrical flow (Fig. 9) is possible for any aperture angle α and any Reynolds number. Let us consider in more detail the flow for very large R. This corresponds to large u_0. Writing (23.12) (for $\phi > 0$) as

$$2(\tfrac{1}{2}\alpha - \phi) = \int_u^0 \frac{du}{\sqrt{[(u+u_0)\{-u^2-(1-u_0)u+q\}]}},$$

we see that the integrand is small throughout the range of integration if $|u|$ is not close to u_0. This means that $|u|$ can differ appreciably from u_0 only for ϕ close to $\pm\tfrac{1}{2}\alpha$, i.e. in the immediate neighbourhood of the walls.† In other words, we have $u \approx \text{constant} = -u_0$ for almost all angles ϕ, and in addition $u_0 = R/6\alpha$, as we see from equations (23.13). The velocity v itself is $|Q|/\rho\alpha r$, giving a non-viscous potential flow with velocity independent of angle and inversely proportional to r. Thus, for large Reynolds numbers, the flow in a converging channel differs very little from potential flow of an ideal fluid. The effect of the viscosity appears only in a very narrow layer near the walls, where the velocity falls rapidly to zero from the value corresponding to the potential flow (Fig. 10).

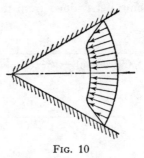

FIG. 10

Now let $Q > 0$, so that we have divergent flow. At first we again suppose that the flow is symmetrical about the plane $\phi = 0$, and that $u(\phi)$ (where now $u > 0$) varies monotonically from zero at $\phi = \pm\tfrac{1}{2}\alpha$ to $u_0 > 0$ at $\phi = 0$. Instead of (23.13) we now have

$$\alpha = \int_0^{u_0} \frac{du}{\sqrt{[(u_0-u)\{u^2+(1+u_0)u+q\}]}},$$

$$\tfrac{1}{6}R = \int_0^{u_0} \frac{u\,du}{\sqrt{[(u_0-u)\{u^2+(1+u_0)u+q\}]}}.$$

(23.14)

---

† The question may be asked how the integral can cease to be small, even if $u \approx -u_0$. The answer is that, for u_0 very large, one of the roots of $-u^2-(1-u_0)u+q = 0$ is close to $-u_0$, so that the radicand has two almost coincident zeros, the whole integral therefore being "almost divergent" at $u = -u_0$.

If we regard u as given, then α increases monotonically as q decreases, and takes its greatest value for $q = 0$:

$$\alpha_{max} = \int_0^{u_0} \frac{du}{\sqrt{[u(u_0 - u)(u + u_0 + 1)]}}.$$

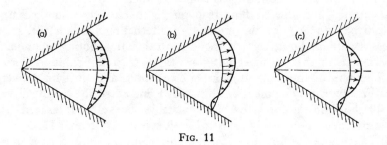

(a)   (b)   (c)

Fig. 11

It is easy to see that for given q, on the other hand, α is a monotonically decreasing function of u_0. Hence it follows that u_0 is a monotonically decreasing function of q for given α, so that its greatest value is for $q = 0$ and is given by the above equation. The maximum $R = R_{max}$ corresponds to the maximum u_0. Using the substitutions $k^2 = u_0/(1 + 2u_0)$, $u = u_0 \cos^2 x$, we can write the dependence of R_{max} on α in the parametric form

$$\alpha = 2\sqrt{(1 - 2k^2)} \int_0^{\pi/2} \frac{dx}{\sqrt{(1 - k^2 \sin^2 x)}},$$

(23.15)

$$R_{max} = -6\alpha \frac{1 - k^2}{1 - 2k^2} + \frac{12}{\sqrt{(1 - 2k^2)}} \int_0^{\pi/2} \sqrt{(1 - k^2 \sin^2 x)} dx.$$

Thus symmetrical flow, everywhere divergent (Fig. 11a), is possible for a given aperture angle only for Reynolds numbers not exceeding a definite value. As $\alpha \to \pi$ ($k \to 0$), $R_{max} \to 0$; as $\alpha \to 0$ ($k \to 1/\sqrt{2}$), R_{max} tends to infinity as $18 \cdot 8/\alpha$.

For $R > R_{max}$ the assumption of symmetrical flow, everywhere divergent, is unjustified, since the conditions (23.14) cannot be satisfied. In the range of angles $-\frac{1}{2}\alpha \leqslant \phi \leqslant \frac{1}{2}\alpha$ the function $u(\phi)$ must now have maxima or minima. The values of $u(\phi)$ corresponding to these extrema must again be zeros of the polynomial under the radical sign. It is therefore clear that the trinomial $u^2 + (1 + u_0)u + q$ (with $u_0 > 0$, $q > 0$) must have two real negative roots in the range mentioned, so that the radicand can be written $(u_0 - u)(u + u_0')(u + u_0'')$, where $u_0 > 0$, $u_0' > 0$, $u_0'' > 0$; we suppose

$u_0' < u_0''$. The function $u(\phi)$ can evidently vary in the range $u_0 \geqslant u \geqslant -u_0'$, $u = u_0$ corresponding to a positive maximum of $u(\phi)$, and $u = -u_0'$ to a negative minimum. Without pausing to make a detailed investigation of the solutions obtained in this way, we may mention that for $R > R_{max}$ a solution appears in which the velocity has one maximum and one minimum, the flow being asymmetric about the plane $\phi = 0$ (Fig. 11b). When R increases further, a symmetrical solution with one minimum and two maxima appears (Fig. 11c), and so on. In all these solutions, therefore, there are regions of both outward and inward flow (though of course the total discharge Q is positive). As $R \to \infty$ the number of alternating minima and maxima increases without limit, so that there is no definite limiting solution. We may emphasise that in divergent flow as $R \to \infty$ the solution does not, therefore, tend to the solution of Euler's equations as it does for convergent flow. Finally, it may be mentioned that, as R increases, the steady divergent flow of the kind described becomes unstable soon after R exceeds R_{max}, and in practice a non-steady or *turbulent* flow occurs (Chapter III).

(3) Determine the flow in a jet emerging from the end of a narrow tube into an infinite space filled with the fluid—the *submerged jet* (L. LANDAU, 1943).

We take spherical co-ordinates r, θ, ϕ, with the polar axis in the direction of the jet at its point of emergence, and with this point as origin. The flow is symmetrical about the polar axis, so that $v_\phi = 0$ and v_θ, v_r are functions of r and θ only. The same total momentum flux (the "momentum of the jet") must pass through any closed surface surrounding the origin (in particular, through an infinitely distant surface). For this to be so, the velocity must be inversely proportional to r, so that

$$v_r = F(\theta)/r, \qquad v_\theta = f(\theta)/r, \tag{23.16}$$

where F and f are some functions of θ only. The equation of continuity is

$$\frac{1}{r^2}\frac{\partial(r^2 v_r)}{\partial r} + \frac{1}{r\sin\theta}\frac{\partial}{\partial\theta}(v_\theta \sin\theta) = 0.$$

Hence we find that

$$F(\theta) = -df/d\theta - f\cot\theta. \tag{23.17}$$

The components $\Pi_{r\phi}$, $\Pi_{\theta\phi}$ of the momentum flux density tensor in the jet vanish identically by symmetry. We assume that the components $\Pi_{\theta\theta}$ and $\Pi_{\phi\phi}$ also vanish; this assumption is justified when we obtain a solution satisfying all the necessary conditions. Using the expressions (15.17) for the components of the tensor σ_{ik}, and formulae (23.16), (23.17), we easily see that the relation

$$\sin^2\theta\, \Pi_{r\theta} = \frac{1}{2}\frac{\partial}{\partial\theta}[\sin^2\theta(\Pi_{\phi\phi} - \Pi_{\theta\theta})]$$

holds between the components of the momentum flux density tensor in the jet. Hence it follows that $\Pi_{r\theta} = 0$. Thus only the component Π_{rr} is non-zero, and it varies as $1/r^2$. It is easy to see that the equations of motion $\partial\Pi_{ik}/\partial x_k = 0$ are automatically satisfied.

Next, we write

$$(\Pi_{\theta\theta} - \Pi_{\phi\phi})/\rho = (f^2 + 2\nu f \cot\theta - 2\nu f')/r^2 = 0,$$

or

$$d(1/f)/d\theta + (1/f)\cot\theta + 1/2\nu = 0.$$

The solution of this equation is

$$f = -2\nu \sin\theta/(A - \cos\theta), \qquad (23.18)$$

and then we have from (23.17)

$$F = 2\nu\left\{\frac{A^2-1}{(A-\cos\theta)^2} - 1\right\}. \qquad (23.19)$$

The pressure distribution is found from the equation

$$\Pi_{\theta\theta}/\rho = p/\rho + f(f + 2\nu\cot\theta)/r^2 = 0,$$

which gives

$$p = \frac{4\rho\nu^2(A\cos\theta - 1)}{r^2(A-\cos\theta)^2}. \qquad (23.20)$$

The constant A can be found in terms of the momentum of the jet, i.e. the total momentum flux in it. This flux is equal to the integral over the surface of a sphere

$$P = \oint \Pi_{rr}\cos\theta\, df = 2\pi \int_0^\pi r^2 \Pi_{rr} \cos\theta \sin\theta\, d\theta.$$

The value of Π_{rr} is given by

$$\frac{1}{\rho}\Pi_{rr} = \frac{4\nu^2}{r^2}\left\{\frac{(A^2-1)^2}{(A-\cos\theta)^4} - \frac{A}{A-\cos\theta}\right\},$$

and a calculation of the integral gives

$$P = 16\pi\nu^2\rho A\left\{1 + \frac{4}{3(A^2-1)} - \tfrac{1}{2}A\log\frac{A+1}{A-1}\right\}. \qquad (23.21)$$

Formulae (23.16)–(23.21) give the solution of the problem.†

---

† The solution here obtained is exact for a jet regarded as emerging from a point source. If the finite dimensions of the tube mouth are taken into account, the solution becomes the first term of an expansion in powers of the ratio of these dimensions to the distance r from the mouth of the tube. This is why, if we calculate from the above solution the total mass flux through a closed surface surrounding the origin, the result is zero. A non-zero total mass flux is obtained when further terms in the above-mentioned expansion are considered; see Yu. B. Rumer, *Prikladnaya matematika i mekhanika* **16**, 255, 1952.

The submerged laminar jet with a non-zero angular momentum has bee ndiscussed by L. G. Loĭtsyanskiĭ (*ibid.* **17**, 3, 1953).

The streamlines are determined by the equation $dr/v_r = rd\theta/v_\theta$, integration of which gives $r\sin^2\theta/(A-\cos\theta) = $ constant. Fig. 12 shows the streamlines in the jet (for $A > 1$).

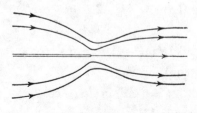

FIG. 12

Let us consider two limiting cases, a weak jet (small momentum P) and a strong jet (large P). As $P \to 0$, the constant A tends to infinity: from (23.21) we have $P = 16\pi\nu^2\rho/A$. For the velocity in this case we have

$$v_\theta = -P\sin\theta/8\pi\nu\rho r, \qquad v_r = P\cos\theta/4\pi\nu\rho r.$$

As $P \to \infty$ (strong jet†), A tends to unity: (23.21) gives $A = 1 + \tfrac{1}{2}\alpha^2$, where $\alpha = 32\pi\nu^2\rho/3P$. For large angles ($\theta \sim 1$), the velocity is given by

$$v_\theta = -(2\nu/r)\cot\tfrac{1}{2}\theta, \qquad v_r = -2\nu/r,$$

but for small angles ($\theta \sim \alpha$) we have

$$v_\theta = -4\nu\theta/(\alpha^2+\theta^2), \qquad v_r = 8\nu\alpha^2/(\alpha^2+\theta^2)^2.$$

§24. Oscillatory motion in a viscous fluid

When a solid body immersed in a viscous fluid oscillates, the flow thereby set up has a number of characteristic properties. In order to study these, it is convenient to begin with a simple but typical example. Let us suppose that an incompressible fluid is bounded by an infinite plane surface which executes a simple harmonic oscillation in its own plane, with frequency ω. We require the resulting motion of the fluid. We take the solid surface as the yz-plane, and the fluid region as $x > 0$; the y-axis is taken in the direction of the oscillation. The velocity u of the oscillating surface is a function of time, of the form $A\cos(\omega t + \alpha)$. It is convenient to write this as the real part of a complex quantity:

$$u = \mathrm{re}(u_0 e^{-i\omega t}),$$

where the constant $u_0 = Ae^{-i\alpha}$ is in general complex, but can always be made real by a proper choice of the origin of time.

---

† However, it must be borne in mind that the flow in a sufficiently strong jet is actually turbulent (§35).

So long as the calculations involve only linear operations on the velocity *u*, we may omit the sign re and proceed as if *u* were complex, taking the real part of the final result. Thus we write

$$u_y = u = u_0\, e^{-i\omega t}. \tag{24.1}$$

The fluid velocity must satisfy the boundary condition $\mathbf{v} = \mathbf{u}$ for $x = 0$, i.e. $v_x = v_z = 0$, $v_y = u$.

It is evident from symmetry that all quantities will depend only on the co-ordinate *x* and the time *t*. From the equation of continuity div $\mathbf{v} = 0$ we therefore have $\partial v_x/\partial x = 0$, whence $v_x = \text{constant} = \text{zero}$, from the boundary condition. Since all quantities are independent of the co-ordinates *y* and *z*, we have $(\mathbf{v}\cdot\mathbf{grad})\mathbf{v} = v_x\, \partial\mathbf{v}/\partial x$, and since v_x is zero it follows that $(\mathbf{v}\cdot\mathbf{grad})\mathbf{v} = 0$ identically. The equation of motion (15.7) becomes

$$\partial\mathbf{v}/\partial t = -(1/\rho)\mathbf{grad}\,p + \nu\triangle\mathbf{v}. \tag{24.2}$$

This is a linear equation. Its *x*-component is $\partial p/\partial x = 0$, i.e. $p = \text{constant}$.

It is further evident from symmetry that the velocity $\mathbf{v}$ is everywhere in the *y*-direction. For $v_y = v$ we have by (24.2)

$$\partial v/\partial t = \nu\partial^2 v/\partial x^2, \tag{24.3}$$

that is, a (one-dimensional) heat conduction equation. We shall look for a solution of this equation which is periodic in *x* and *t*, of the form

$$v = u_0\, e^{i(kx-\omega t)},$$

with a complex amplitude u_0, so that $v = u$ for $x = 0$. Substituting in (24.3), we find $i\omega = \nu k^2$, whence

$$k = \sqrt{(i\omega/\nu)} = \pm(i+1)\sqrt{(\omega/2\nu)},$$

so that the velocity *v* is

$$v = u_0\, e^{-\sqrt{(\omega/2\nu)}x}\, e^{i[\sqrt{(\omega/2\nu)}x - \omega t]}; \tag{24.4}$$

we have taken *k* to have a positive imaginary part, since otherwise the velocity would increase without limit in the interior of the fluid, which is physically impossible.

The solution obtained represents a transverse wave: its velocity $v_y = v$ is perpendicular to the direction of propagation. The most important property of this wave is that it is rapidly damped in the interior of the fluid: the amplitude decreases exponentially as the distance *x* from the solid surface increases.†

Thus transverse waves can occur in a viscous fluid, but they are rapidly damped as we move away from the solid surface whose motion generates the waves.

The distance δ over which the amplitude falls off by a factor of *e* is called the *depth of penetration* of the wave. We see from (24.4) that

$$\delta = \sqrt{(2\nu/\omega)}. \tag{24.5}$$

---

† Over a distance of one wavelength the amplitude diminishes by a factor of $e^{2\pi} \approx 540$.

The depth of penetration therefore diminishes with increasing frequency, but increases with the kinematic viscosity of the fluid.

Let us calculate the frictional force acting on unit area of the plane oscillating in the viscous fluid. This force is evidently in the y-direction, and is equal to the component $\sigma_{xy} = \eta \partial v_y / \partial x$ of the stress tensor; the value of the derivative must be taken at the surface itself, i.e. at $x = 0$. Substituting (24.4), we obtain

$$\sigma_{xy} = \sqrt{(\tfrac{1}{2}\omega\eta\rho)}(i-1)u. \tag{24.6}$$

Supposing u_0 real and taking the real part of (24.6), we have

$$\sigma_{xy} = -\sqrt{(\omega\eta\rho)}u_0 \cos(\omega t + \tfrac{1}{4}\pi).$$

The velocity of the oscillating surface, however, is $u = u_0 \cos \omega t$. There is therefore a phase difference between the velocity and the frictional force.†

It is easy to calculate also the (time) average of the energy dissipation in the above problem. This may be done by means of the general formula (16.3); in this particular case, however, it is simpler to calculate the required dissipation directly as the work done by the frictional forces. The energy dissipated per unit time per unit area of the oscillating plane is equal to the mean value of the product of the force σ_{xy} and the velocity $u_y = u$:

$$-\overline{\sigma_{xy}u} = \tfrac{1}{2}u_0^2\sqrt{(\tfrac{1}{2}\omega\eta\rho)}. \tag{24.7}$$

It is proportional to the square root of the frequency of the oscillations, and to the square root of the viscosity.

An explicit solution can also be given of the problem of a fluid set in motion by a plane surface moving in its plane according to any law $u = u(t)$. We shall not pause to give the corresponding calculations here, since the required solution of equation (24.3) is formally identical with that of an analogous problem in the theory of thermal conduction, which we shall discuss in §52 (the solution is formula (52.15)). In particular, the frictional force on unit area of the surface is given by

$$\sigma_{xy} = \sqrt{\frac{\eta\rho}{\pi}} \int_{-\infty}^{\;} \frac{du(\tau)}{d\tau} \frac{d\tau}{\sqrt{(t-\tau)}}; \tag{24.8}$$

cf. (52.16).

---

† For oscillations of a half-plane (parallel to its edge) there is an additional frictional force due to edge effects. The problem of the motion of a viscous fluid caused by oscillations of a half-plane, and also the more general problem of the oscillations of a wedge of any angle, can be solved by a class of solutions of the equation $\Delta f + k^2 f = 0$, used by A. SOMMERFELD in the theory of diffraction by a wedge; see, for instance, M. VON LAUE, Interferenz und Beugung elektromagnetischer Wellen (Interference and diffraction of electromagnetic waves), *Handbuch der Experimentalphysik* **18**, 333, Akademische Verlagsgesellschaft, Leipzig 1928.

We give here, for reference, only one result: the increase in the frictional force on a half-plane, arising from the edge effect, can be regarded as the result of increasing the area of the half-plane by moving the edge a distance $\tfrac{1}{2}\delta = \sqrt{(\nu/2\omega)}$.

Let us now consider the general case of an oscillating body of arbitrary shape. In the case of an oscillating plane considered above, the term $(\mathbf{v}\cdot\mathbf{grad})\mathbf{v}$ in the equation of motion of the fluid was identically zero. This does not happen, of course, for a surface of arbitrary shape. We shall assume, however, that this term is small in comparison with the other terms, so that it may be neglected. The conditions necessary for this procedure to be valid will be examined below.

We shall therefore begin, as before, from the linear equation (24.2). We take the curl of both sides; the term **curl grad** p vanishes identically, giving

$$\partial(\mathbf{curl\,v})/\partial t = \nu\triangle\mathbf{curl\,v}, \tag{24.9}$$

i.e. **curl v** satisfies a heat conduction equation. We have seen above, however, that such an equation gives an exponential decrease of the quantity which satisfies it. We can therefore say that the vorticity decreases towards the interior of the fluid. In other words, the motion of the fluid caused by the oscillations of the body is rotational in a certain layer round the body, while at larger distances it rapidly changes to potential flow. The depth of penetration of the rotational flow is of the order of $\delta \sim \sqrt{(\nu/\omega)}$.

Two important limiting cases are possible here: the quantity δ may be either large or small compared with the dimension of the oscillating body. Let l be the order of magnitude of this dimension. We first consider the case $\delta \gg l$; this implies that $l^2\omega \ll \nu$. Besides this condition, we shall also suppose that the Reynolds number is small. If a is the amplitude of the oscillations, the velocity of the body is of the order of $a\omega$. The Reynolds number for the motion in question is therefore $\omega al/\nu$. We therefore suppose that

$$l^2\omega \ll \nu, \qquad \omega al/\nu \ll 1. \tag{24.10}$$

This is the case of low frequencies of oscillation, which in turn means that the velocity varies only slowly with time, and therefore that we can neglect the derivative $\partial\mathbf{v}/\partial t$ in the general equation of motion. The term $(\mathbf{v}\cdot\mathbf{grad})\mathbf{v}$, on the other hand, can be neglected because the Reynolds number is small.

The absence of the term $\partial\mathbf{v}/\partial t$ from the equation of motion means that the flow is steady. Thus, for $\delta \gg l$, the flow can be regarded as steady at any given instant. This means that the flow at any given instant is what it would be if the body were moving uniformly with its instantaneous velocity. If, for example, we are considering the oscillations of a sphere immersed in the fluid, with a frequency satisfying the inequalities (24.10) (l being now the radius of the sphere), then we can say that the drag on the sphere will be that given by Stokes' formula (20.14) for uniform motion of the sphere at small Reynolds numbers.

Let us now consider the opposite case, where $l \gg \delta$. In order that the term $(\mathbf{v}\cdot\mathbf{grad})\mathbf{v}$ should again be negligible, it is necessary that the amplitude of the oscillations should be small in comparison with the dimensions of the body:

$$l^2\omega \gg \nu, \qquad a \ll l; \tag{24.11}$$

in this case, it should be noticed, the Reynolds number need not be small. The above inequality is obtained by estimating the magnitude of $(\mathbf{v} \cdot \mathbf{grad})\mathbf{v}$. The operator $(\mathbf{v} \cdot \mathbf{grad})$ denotes differentiation in the direction of the velocity. Near the surface of the body, however, the velocity is nearly tangential. In the tangential direction the velocity changes appreciably only over distances of the order of the dimension of the body. Hence

$$(\mathbf{v} \cdot \mathbf{grad})\mathbf{v} \sim v^2/l \sim a^2\omega^2/l,$$

since the velocity itself is of the order of $a\omega$. The derivative $\partial \mathbf{v}/\partial t$, however, is of the order of $v\omega \sim a\omega^2$. Comparing these, we see that

$$(\mathbf{v} \cdot \mathbf{grad})\mathbf{v} \ll \partial \mathbf{v}/\partial t$$

if $a \ll l$. The terms $\partial \mathbf{v}/\partial t$ and $\nu \triangle \mathbf{v}$ are then easily seen to be of the same order.

We may now discuss the nature of the flow round an oscillating body when the conditions (24.11) hold. In a thin layer near the surface of the body the flow is rotational, but in the rest of the fluid we have potential flow.† Hence the flow everywhere except in the layer adjoining the body is given by the equations

$$\mathbf{curl\,v} = 0, \quad \operatorname{div} \mathbf{v} = 0. \tag{24.12}$$

Hence it follows that $\triangle \mathbf{v} = 0$, and the Navier–Stokes equation reduces to Euler's equation. The flow is therefore ideal everywhere except in the surface layer. Since this layer is thin, in solving equations (24.12) to determine the flow of the rest of the fluid we should take as boundary conditions those which must be satisfied at the surface of the body, i.e. that the fluid velocity is equal to that of the body. The solutions of the equations of motion for an ideal fluid cannot satisfy these conditions, however. We can require only the fulfilment of the corresponding condition for the fluid velocity component normal to the surface.

Although equations (24.12) are inapplicable in the surface layer of fluid, the velocity distribution obtained by solving them satisfies the necessary boundary condition for the normal velocity component, and the actual variation of this component near the surface therefore has no significant properties. The tangential component would be found, by solving the equations (24.12), to have some value different from the corresponding velocity component of the body, whereas these velocity components should be equal also. Hence the tangential velocity component must change rapidly in the surface layer. The nature of this variation is easily determined. Let us consider any portion of the surface of the body, of dimension large compared

---

† For oscillations of a plane surface not only **curl v** but also **v** itself decreases exponentially with characteristic distance δ. This is because the oscillating plane does not displace the fluid, and therefore the fluid remote from it remains at rest. For oscillations of bodies of other shapes the fluid is displaced, and therefore executes a motion where the velocity decreases appreciably only over distances of the order of the dimension of the body.

with δ, but small compared with the dimension of the body. Such a portion
may be regarded as approximately plane, and therefore we can use the results obtained above for a plane surface. Let the x-axis be directed along
the normal to the portion considered, and the y-axis parallel to the tangential
velocity component of the surface there. We denote by v_y the tangential
component of the fluid velocity relative to the body; v_y must vanish on the
surface. Lastly, let $v_0 e^{-i\omega t}$ be the value of v_y found by solving equations
(24.12). From the results obtained at the beginning of this section, we can
say that in the surface layer the quantity v_y will fall off towards the surface
according to the law

$$v_y = v_0 e^{-i\omega t}[1 - e^{-(1-i)x\sqrt{(\omega/2\nu)}}].   \tag{24.13}$$

Finally, the total amount of energy dissipated in unit time will be given by
the integral

$$\bar{E}_{\text{kin}} = -\tfrac{1}{2}\sqrt{(\tfrac{1}{2}\eta\rho\omega)} \oint |v_0|^2 \, df   \tag{24.14}$$

taken over the surface of the oscillating body.

In the Problems at the end of this section we calculate the drag on various
bodies oscillating in a viscous fluid. Here we shall make the following general
remark regarding these forces. Writing the velocity of the body in the complex
form $u = u_0 e^{-i\omega t}$, we obtain a drag F proportional to the velocity u, and also
complex: $F = \beta u$, where $\beta = \beta_1 + i\beta_2$ is a complex constant. This expression
can be written as the sum of two terms with real coefficients:

$$F = (\beta_1 + i\beta_2)u = \beta_1 u - \beta_2 \dot{u}/\omega,   \tag{24.15}$$

one proportional to the velocity u and the other to the acceleration $\dot{u}$.

The (time) average of the energy dissipation is given by the mean product
of the drag and the velocity, where of course we must first take the real
parts of the expressions given above, i.e. $u = \tfrac{1}{2}(u_0 e^{-i\omega t} + u_0{}^* e^{i\omega t})$,
$F = \tfrac{1}{2}(u_0\beta e^{-i\omega t} + u_0{}^*\beta^* e^{i\omega t})$. Noticing that the mean values of $e^{\pm 2i\omega t}$ are
zero, we have

$$\overline{Fu} = \tfrac{1}{4}(\beta + \beta^*)|u_0|^2 = \tfrac{1}{2}\beta_1 |u_0|^2.   \tag{24.16}$$

Thus we see that the energy dissipation arises only from the real part of β;
the corresponding part of the drag (24.15), proportional to the velocity, may
be called the *dissipative part*. The other part of the drag, proportional to
the acceleration and determined by the imaginary part of β, does not involve
the dissipation of energy and may be called the *inertial part*.

Similar considerations hold for the moment of the forces on a body executing rotary oscillations in a viscous fluid.

PROBLEMS

PROBLEM 1. Determine the frictional force on each of two parallel solid planes, between
which is a layer of viscous fluid, when one of the planes oscillates in its own plane.

SOLUTION. We seek a solution of equation (24.3) in the form†

$$v = (A \sin kx + B \cos kx)e^{-i\omega t},$$

and determine A and B from the conditions $v = u = u_0 e^{-i\omega t}$ for $x = 0$ and $v = 0$ for $x = h$, where h is the distance between the planes. The result is

$$v = u\frac{\sin k(h-x)}{\sin kh}.$$

The frictional force per unit area on the moving plane is

$$P_{1x} = \eta(\partial v/\partial x)_{x=0} = -\eta ku \cot kh,$$

while that on the fixed plane is

$$P_{2x} = -\eta(\partial v/\partial x)_{x=h} = \eta ku \operatorname{cosec} kh,$$

the real parts of all quantities being understood.

PROBLEM 2. Determine the frictional force on an oscillating plane covered by a layer of fluid of thickness h, the upper surface being free.

SOLUTION. The boundary condition at the solid plane is $v = u$ for $x = 0$, and that at the free surface is $\sigma_{xy} = \eta \partial v/\partial x = 0$ for $x = h$. We find the velocity

$$v = u\frac{\cos k(h-x)}{\cos kh}.$$

The frictional force is

$$P_x = \eta(\partial v/\partial x)_{x=0} = \eta ku \tan kh.$$

PROBLEM 3. A plane disk of large radius R executes rotary oscillations of small amplitude about its axis, the angle of rotation being $\theta = \theta_0 \cos \omega t$, where $\theta_0 \ll 1$. Determine the moment of the frictional forces acting on the disk.

SOLUTION. For oscillations of small amplitude the term $(\mathbf{v} \cdot \mathbf{grad})\mathbf{v}$ in the equation of motion is always small compared with $\partial \mathbf{v}/\partial t$, whatever the frequency ω. If $R \gg \delta$, the disk may be regarded as infinite in determining the velocity distribution. We take cylindrical co-ordinates, with the z-axis along the axis of rotation, and seek a solution such that $v_r = v_z = 0$, $v_\phi = v = r\Omega(z, t)$. For the angular velocity $\Omega(z, t)$ of the fluid we obtain the equation

$$\partial\Omega/\partial t = \nu\partial^2\Omega/\partial z^2.$$

The solution of this equation which is $-\omega\theta_0 \sin \omega t$ for $z = 0$ and zero for $z = \infty$ is

$$\Omega = -\omega\theta_0 e^{-z/\delta} \sin(\omega t - z/\delta).$$

The moment of the frictional forces on both sides of the disk is

$$M = 2\int_0^R r \cdot 2\pi r\eta (\partial v/\partial z)_{z=0} \, dr = \omega\theta_0\pi\sqrt{(\omega\rho\eta)}R^4 \cos(\omega t - \tfrac{1}{4}\pi).$$

---

† In all the Problems to this section δ denotes the quantity (24.5):

$$\delta = \sqrt{(2\nu/\omega)}, \quad \text{and} \quad k = (1+i)/\delta.$$

PROBLEM 4. Determine the flow between two parallel planes when there is a pressure gradient which varies harmonically with time.

SOLUTION. We take the xz-plane half-way between the two planes, with the x-axis parallel to the pressure gradient, which we write in the form

$$-(1/\rho)\partial p/\partial x = ae^{-i\omega t}.$$

The velocity is everywhere in the x-direction, and is determined by the equation

$$\partial v/\partial t = ae^{-i\omega t} + v\partial^2 v/\partial y^2.$$

The solution of this equation which satisfies the conditions $v = 0$ for $y = \pm\frac{1}{2}h$ is

$$v = \frac{ia}{\omega}e^{-i\omega t}\left[1 - \frac{\cos ky}{\cos\frac{1}{2}kh}\right].$$

The mean value of the velocity over a cross-section is

$$\bar{v} = \frac{ia}{\omega}e^{-i\omega t}\left(1 - \frac{2}{kh}\tan\frac{1}{2}kh\right).$$

For $h/\delta \ll 1$ this becomes

$$\bar{v} \approx ae^{-i\omega t}h^2/12v,$$

in agreement with (17.5), while for $h/\delta \gg 1$ we have

$$\bar{v} \approx (ia/\omega)e^{-i\omega t},$$

in accordance with the fact that in this case the velocity must be almost constant over the cross-section, varying only in a narrow surface layer.

PROBLEM 5. Determine the drag on a sphere of radius R which executes translatory oscillations in a fluid.

SOLUTION. We write the velocity of the sphere in the form $\mathbf{u} = \mathbf{u}_0 e^{-i\omega t}$. As in §20, we seek the fluid velocity in the form $\mathbf{v} = e^{-i\omega t}\,\mathbf{curl\ curl}\,f\mathbf{u}_0$, where f is a function of r only (the origin is taken at the instantaneous position of the centre of the sphere). Substituting in (24.9) and effecting transformations similar to those of §20, we obtain the equation

$$\triangle^2 f + (i\omega/v)\,\triangle f = 0$$

(instead of the equation $\triangle^2 f = 0$ in §20). Hence we have

$$\triangle f = \text{constant} \times e^{ikr}/r,$$

the solution being chosen which decreases exponentially with r. Integrating, we have

$$df/dr = [ae^{ikr}(r - 1/ik) + b]/r^2; \tag{1}$$

the function f itself is not needed, since only the derivatives f' and f'' appear in the velocity. The constants a and b are determined from the condition that $\mathbf{v} = \mathbf{u}$ for $r = R$, and are found to be

$$a = -\frac{3R}{2ik}e^{-ikR}, \qquad b = -\frac{1}{2}R^3\left(1 - \frac{3}{ikR} - \frac{3}{k^2R^2}\right). \tag{2}$$

It may be pointed out that, at large distances $(R \gg \delta)$, $a \to 0$ and $b \to -\frac{1}{2}R^3$, the values for potential flow obtained in §10, Problem 2; this is in accordance with what was said in §24.

The drag is calculated from formula (20.13), in which the integration is over the surface of the sphere. The result is

$$F = 6\pi\eta R\left(1 + \frac{R}{\delta}\right)u + 3\pi R^2\sqrt{(2\eta\rho/\omega)}\left(1 + \frac{2R}{9\delta}\right)\frac{du}{dt}. \qquad (3)$$

For $\omega = 0$ this becomes Stokes' formula, while for large frequencies we have

$$F = \tfrac{2}{3}\pi\rho R^3\frac{du}{dt} + 3\pi R^2\sqrt{(2\eta\rho\omega)}u.$$

The first term in this expression corresponds to the inertial force in potential flow past a sphere (see §11, Problem 1), while the second gives the limit of the dissipative force.

PROBLEM 6. Determine the drag on a sphere moving in an arbitrary manner, the velocity being given by a function $u(t)$.

SOLUTION. We represent $u(t)$ as a Fourier integral:

$$u(t) = \int_{-\infty}^{\infty} u_\omega e^{-i\omega t}d\omega, \qquad u_\omega = \frac{1}{2\pi}\int_{-\infty}^{\infty} u(\tau)e^{i\omega\tau}\,d\tau.$$

Since the equations are linear, the total drag may be written as the integral of the drag forces for velocities which are the separate Fourier components $u_\omega e^{-i\omega t}$; these forces are given by (3) of Problem 5, and are

$$\pi\rho R^3 u_\omega e^{-i\omega t}\left\{\frac{6\nu}{R^2} - \frac{2i\omega}{3} + \frac{3\sqrt{(2\nu)}}{R}(1-i)\sqrt{\omega}\right\}.$$

Noticing that $(du/dt)_\omega = -i\omega u_\omega$, we can rewrite this as

$$\pi\rho R^3 e^{-i\omega t}\left\{\frac{6\nu}{R^2}u_\omega + \tfrac{2}{3}(\dot u)_\omega + \frac{3\sqrt{(2\nu)}}{R}(\dot u)_\omega\frac{1+i}{\sqrt{\omega}}\right\}.$$

On integration over ω, the first and second terms give respectively $u(t)$ and $\dot u(t)$. To integrate the third term, we notice first of all that for negative ω this term must be written in the complex conjugate form, $(1+i)/\sqrt{\omega}$ being replaced by $(1-i)/\sqrt{|\omega|}$; this is because formula (3) of Problem 5 was derived for a velocity $u = u_0 e^{-i\omega t}$ with $\omega > 0$, and for a velocity $u_0 e^{i\omega t}$ we should obtain the complex conjugate. Instead of an integral over ω from $-\infty$ to $+\infty$, we can therefore take twice the real part of the integral from 0 to ∞. We write

$$2\,\mathrm{re}\left\{(1+i)\int_0^\infty \frac{(\dot u)_\omega e^{-i\omega t}}{\sqrt{\omega}}d\omega\right\} = \frac{1}{\pi}\,\mathrm{re}\left\{(1+i)\int_{-\infty}^{\infty}\int_0^\infty \frac{\dot u(\tau)\,e^{i\omega(\tau-t)}}{\sqrt{\omega}}d\omega d\tau\right\}$$

$$= \frac{1}{\pi}\,\mathrm{re}\left\{(1+i)\int_{-\infty}^{\infty}\int_0^\infty \frac{\dot u(\tau)\,e^{-i\omega(t-\tau)}}{\sqrt{\omega}}d\omega d\tau + (1+i)\int_t^{\infty}\int_0^\infty \frac{\dot u(\tau)\,e^{i\omega(\tau-t)}}{\sqrt{\omega}}d\omega d\tau\right\}$$

$$= \sqrt{\frac{2}{\pi}}\,\mathrm{re}\left\{\int_{-\infty}^{t}\frac{\dot u(\tau)}{\sqrt{(t-\tau)}}d\tau + i\int_t^{\infty}\frac{\dot u(\tau)}{\sqrt{(\tau-t)}}d\tau\right\}$$

$$= \sqrt{\frac{2}{\pi}}\int_{-\infty}^{t}\frac{\dot u(\tau)}{\sqrt{(t-\tau)}}d\tau.$$

Thus we have finally for the drag

$$F = 2\pi\rho R^3 \left\{ \frac{1}{3}\frac{du}{dt} + \frac{3\nu u}{R^2} + \frac{3}{R}\sqrt{\frac{\nu}{\pi}}\int_{-\infty}^{t}\frac{du}{d\tau}\frac{d\tau}{\sqrt{(t-\tau)}} \right\}. \tag{1}$$

PROBLEM 7. Determine the drag on a sphere which at time $t = 0$ begins to move with a uniform acceleration, $u = \alpha t$.

SOLUTION. Putting, in formula (1) of Problem 6, $u = 0$ for $t < 0$ and $u = \alpha t$ for $t > 0$ we have for $t > 0$

$$F = 2\pi\rho R^3\alpha \left[\frac{1}{3} + \frac{3\nu t}{R^2} + \frac{6}{R}\sqrt{\frac{t\nu}{\pi}} \right].$$

PROBLEM 8. The same as Problem 7, but for a sphere brought instantaneously into uniform motion.

SOLUTION. We have $u = 0$ for $t < 0$ and $u = u_0$ for $t > 0$. The derivative du/dt is zero except at the instant $t = 0$, when it is infinite, but the time integral of du/dt is finite, and equals u_0. As a result, we have for all $t > 0$

$$F = 6\pi\rho\nu R u_0 \left[1 + \frac{R}{\sqrt{(\pi\nu t)}} \right] + \tfrac{2}{3}\pi\rho R^3 u_0 \delta(t),$$

where $\delta(t)$ is the delta function. For $t \to \infty$ this expression tends asymptotically to the value given by Stokes' formula. The impulsive drag on the sphere at $t = 0$ is obtained by integrating the last term and is $\tfrac{2}{3}\pi\rho R^3 u_0$.

PROBLEM 9. Determine the moment of the forces on a sphere executing rotary oscillations about a diameter in a viscous fluid.

SOLUTION. For the same reasons as in §20, Problem 1, the pressure-gradient term can be omitted from the equation of motion, so that we have $\partial \mathbf{v}/\partial t = \nu \triangle \mathbf{v}$. We seek a solution in the form $\mathbf{v} = \mathbf{curl}\, f\mathbf{\Omega}_0 e^{-i\omega t}$, where $\mathbf{\Omega} = \mathbf{\Omega}_0 e^{-i\omega t}$ is the angular velocity of rotation of the sphere. We then obtain for f, instead of the equation $\triangle f = $ constant,

$$\triangle f + k^2 f = \text{constant.}$$

Omitting an unimportant constant term in the solution of this equation, we find $f = ae^{ikr}/r$ taking the solution which vanishes at infinity. The constant a is determined from the boundary condition that $\mathbf{v} = \mathbf{\Omega} \times \mathbf{r}$ at the surface of the sphere. The result is

$$f = \frac{R^3}{1 - ikR} e^{ik(r-R)}, \qquad \mathbf{v} = (\mathbf{\Omega} \times \mathbf{r})\left(\frac{R}{r}\right)^3 \frac{1 - ikr}{1 - ikR} e^{ik(r-R)},$$

where R is the radius of the sphere. A calculation like that in §20, Problem 1, gives the following expression for the moment of the forces exerted on the sphere by the fluid:

$$M = -\frac{8\pi}{3}\eta R^3\Omega \frac{3 + 6R/\delta + 6(R/\delta)^2 + 2(R/\delta)^3 - 2i(R/\delta)^2(1 + R/\delta)}{1 + 2R/\delta + 2(R/\delta)^2}.$$

For $\omega \to 0$ (i.e. $\delta \to \infty$), we obtain $M = -8\pi\eta R^3\Omega$, corresponding to uniform rotation of the sphere (see §20, Problem 1). In the opposite limiting case $R/\delta \gg 1$, we find

$$M = \frac{4\sqrt{2}}{3}\pi R^4\sqrt{(\eta\rho\omega)}(i-1)\Omega.$$

This expression can also be obtained directly: for $\delta \ll R$ each element of the surface of the sphere may be regarded as plane, and the frictional force acting on it is found by substituting $u = \Omega R \sin \theta$ in formula (24.6).

PROBLEM 10. Determine the moment of the forces on a hollow sphere filled with viscous fluid and executing rotary oscillations about a diameter.

SOLUTION. We seek the velocity in the same form as in Problem 9. For f we take the solution $(a/r) \sin kr$, which is finite everywhere within the sphere, including the centre. Determining a from the boundary condition, we have

$$\mathbf{v} = (\mathbf{\Omega} \times \mathbf{r})\left(\frac{R}{r}\right)^3 \frac{kr \cos kr - \sin kr}{kR \cos kR - \sin kR}.$$

A calculation of the moment of the frictional forces gives the expression

$$M = \tfrac{8}{3}\pi\eta R^3\Omega \frac{k^2 R^2 \sin kR + 3kR \cos kR - 3 \sin kR}{kR \cos kR - \sin kR}.$$

The limiting value for $\delta \gg 1$ is of course the same as in the preceding problem. If $R/\delta \ll 1$ we have

$$M = \tfrac{8}{15}\pi\rho\omega R^5\Omega\left(i - \frac{R^2\omega}{35\nu}\right).$$

The first term corresponds to the inertial forces occurring in the rigid rotation of the whole fluid.

§25. Damping of gravity waves

Arguments similar to those given above can be advanced concerning the velocity distribution near the free surface of a fluid. Let us consider oscillatory motion occurring near the surface (for example, gravity waves). We suppose that the conditions (24.11) hold, the dimension l being now replaced by the wavelength λ:

$$\lambda^2\omega \gg \nu, \qquad a \ll \lambda; \tag{25.1}$$

a is the amplitude of the wave, and ω its frequency. Then we can say that the flow is rotational only in a thin surface layer, while throughout the rest of the fluid we have potential flow, just as we should for an ideal fluid.

The motion of a viscous fluid must satisfy the boundary conditions (15.14) at the free surface; these require that certain combinations of the space derivatives of the velocity should vanish. The flow obtained by solving the equations of ideal-fluid dynamics does not satisfy these conditions, however. As in the discussion of v_y in the previous section, we may conclude that the corresponding velocity derivatives decrease rapidly in a thin surface layer. It is important to notice that this does not imply a large velocity gradient as it does near a solid surface.

Let us calculate the energy dissipation in a gravity wave. Here we must consider the dissipation, not of the kinetic energy alone, but of the mechanical energy E_{mech}, which includes both the kinetic energy and the potential energy in the gravitational field. It is clear, however, that the presence or absence of a gravitational field cannot affect the energy dissipation due to processes of internal friction in the fluid. Hence $\dot{E}_{\text{mech}}$ is given by the same formula (16.3):

$$\dot{E}_{\text{mech}} = -\tfrac{1}{2}\eta \int \left(\frac{\partial v_i}{\partial x_k} + \frac{\partial v_k}{\partial x_i}\right)^2 dV.$$

In calculating this integral for a gravity wave, it is to be noticed that, since the volume of the surface region of rotational flow is small, while the velocity gradient there is not large, the existence of this region may be ignored, unlike what was possible for oscillations of a solid surface. In other words, the integration is to be taken over the whole volume of fluid, which, as we have seen, moves as if it were an ideal fluid.

The flow in a gravity wave for an ideal fluid, however, has already been determined in §12. Since we have potential flow,

$$\partial v_i/\partial x_k = \partial^2\phi/\partial x_k\,\partial x_i = \partial v_k/\partial x_i,$$

so that

$$\dot{E}_{\text{mech}} = -2\eta \int \left(\frac{\partial^2\phi}{\partial x_i\,\partial x_k}\right)^2 dV.$$

The potential ϕ is of the form

$$\phi = \phi_0 \cos(kx - \omega t + \alpha)e^{-kz}.$$

We are interested, of course, not in the instantaneous value of the energy dissipation, but in its mean value $\bar{\dot{E}}_{\text{mech}}$ with respect to time. Noticing that the mean values of the squared sine and cosine are the same, we find

$$\bar{\dot{E}}_{\text{mech}} = -8\eta k^4 \int \overline{\phi^2}\, dV. \tag{25.2}$$

The energy E_{mech} itself may be calculated for a gravity wave by using a theorem of mechanics that, in any system executing small oscillations (of small amplitude, that is), the mean kinetic and potential energies are equal. We can therefore write E_{mech} simply as twice the kinetic energy:

$$E_{\text{mech}} = \rho \int \overline{v^2}\, dV = \rho \int \overline{(\partial\phi/\partial x_i)^2}\, dV,$$

whence

$$E_{\text{mech}} = 2\rho k^2 \int \overline{\phi^2}\, dV. \tag{25.3}$$

The damping of the waves is conveniently characterised by the *damping coefficient* γ, defined as

$$\gamma = |\bar{\dot{E}}_{\text{mech}}|/2E_{\text{mech}}. \tag{25.4}$$

In the course of time, the energy of the wave decreases according to the law $\bar{E}_{mech} = \text{constant} \times e^{-2\gamma t}$; since the energy is proportional to the square of the amplitude, the latter decreases with time as $e^{-\gamma t}$.

Using (25.2), (25.3), we find

$$\gamma = 2\nu k^2. \tag{25.5}$$

Substituting here (12.7), we obtain the damping coefficient for gravity waves in the form

$$\gamma = 2\nu\omega^4/g^2. \tag{25.6}$$

PROBLEMS

PROBLEM 1. Determine the damping coefficient for long gravity waves propagated in a channel of constant cross-section; the frequency is supposed so large that $\sqrt{(\nu/\omega)}$ is small compared with the depth of the fluid in the channel.

SOLUTION. The principal dissipation of energy occurs in the surface layer of fluid, where the velocity changes from zero at the boundary to the value $v = v_0 e^{-i\omega t}$ which it has in the wave. The mean energy dissipation per unit length of the channel is by (24.14) $l|v_0|^2 \sqrt{(\eta\rho\omega/8)}$, where l is the perimeter of the part of the channel cross-section occupied by the fluid. The mean energy of the fluid (again per unit length) is $S\rho\bar{v^2} = \frac{1}{2}S\rho|v_0|^2$, where S is the cross-sectional area of the fluid in the channel. The damping coefficient is $\gamma = l\sqrt{(\nu\omega/8S^2)}$. For a channel of rectangular section, therefore,

$$\gamma = \frac{2h+a}{2\sqrt{2ah}}\sqrt{(\nu\omega)},$$

where a is the width and h the depth of the fluid.

PROBLEM 2. Determine the flow in a gravity wave on a very viscous fluid.

SOLUTION. The calculation of the damping coefficient as shown above is valid only when this coefficient is small, so that the motion may be regarded as that of an ideal fluid to a first approximation. For arbitrary viscosity we seek a solution of the equations of motion

$$\left.\begin{array}{l} \dfrac{\partial v_x}{\partial t} = \nu\left(\dfrac{\partial^2 v_x}{\partial x^2} + \dfrac{\partial^2 v_x}{\partial z^2}\right) - \dfrac{1}{\rho}\dfrac{\partial p}{\partial x}, \\[2ex] \dfrac{\partial v_z}{\partial t} = \nu\left(\dfrac{\partial^2 v_z}{\partial x^2} + \dfrac{\partial^2 v_z}{\partial z^2}\right) - \dfrac{1}{\rho}\dfrac{\partial p}{\partial z} - g, \\[2ex] \dfrac{\partial v_x}{\partial x} + \dfrac{\partial v_z}{\partial z} = 0 \end{array}\right\} \tag{1}$$

which depends on t and x as $e^{-i\omega t + ikx}$, and diminishes in the interior of the fluid ($z < 0$). We find

$$v_x = e^{-i\omega t + ikx}(Ae^{kz} + Be^{mz}), \qquad v_z = e^{-i\omega t + ikx}\left(-iAe^{kz} - \frac{ik}{m}Be^{mz}\right),$$

$$p/\rho = e^{-i\omega t + ikx}\omega Ae^{kz}/k - gz, \qquad \text{where } m = \sqrt{(k^2 - i\omega/\nu)}.$$

The boundary conditions at the fluid surface are

$$\sigma_{zz} = -p + 2\eta\,\partial v_z/\partial z = 0, \qquad \sigma_{xz} = \eta\left(\frac{\partial v_x}{\partial z} + \frac{\partial v_z}{\partial x}\right) = 0 \text{ for } z = \zeta.$$

In the second condition we can immediately put $z = 0$ instead of $z = \zeta$. The first condition, however, should be differentiated with respect to t, after which we replace $g\partial\zeta/\partial t$ by gv_z and then put $z = 0$. The condition that the resulting two homogeneous equations for A and B are compatible gives

$$\left(2 - \frac{i\omega}{\nu k^2}\right)^2 + \frac{g}{\nu^2 k^3} = 4\sqrt{\left(1 - \frac{i\omega}{\nu k^2}\right)}. \tag{2}$$

This equation gives ω as a function of the wave number k; ω is complex, its real part giving the frequency of the oscillations and its imaginary part the damping coefficient. The solutions of equation (2) that have a physical meaning are those whose imaginary parts are negative (corresponding to damping of the wave); only two roots of (2) meet this requirement. If $\nu k^2 \ll \sqrt{(gk)}$ (the condition (25.1)), then the damping coefficient is small, and (2) gives approximately $\omega = \pm\sqrt{(gk)} - i.2\nu k^2$, a result which we already know. In the opposite limiting case $\nu k^2 \gg \sqrt{(gk)}$, equation (2) has two purely imaginary roots, corresponding to damped aperiodic flow. One root is $\omega = -ig/2\nu k$, while the other is much larger (of order νk^2), and therefore of no interest, since the corresponding motion is strongly damped.

CHAPTER III

TURBULENCE

§26. Stability of steady flow

IN SOLVING the equations of steady flow for a viscous fluid, it is often necessary to make certain approximations on account of mathematical difficulties. The validity of these approximate solutions is, of course, restricted, Such, for instance, is the solution of the problem of flow past a sphere given in §20, which is valid only for small Reynolds numbers.

In principle, however, there must be an exact stationary solution of the equations of fluid dynamics for any problem with given steady external conditions; such exact solutions have been considered in §§17, 18 and 23. These solutions formally hold for all Reynolds numbers.

Yet not every solution of the equations of motion, even if it is exact, can actually occur in Nature. The flows that occur in Nature must not only obey the equations of fluid dynamics, but also be stable. For the flow to be stable it is necessary that small perturbations, if they arise, should decrease with time. If, on the contrary, the small perturbations which inevitably occur in the flow tend to increase with time, then the flow is absolutely unstable. Such a flow unstable with respect to infinitely small perturbations cannot exist.

The mathematical investigation of the stability of a given flow with respect to infinitely small perturbations will proceed as follows. On the steady solution concerned (whose velocity distribution is $\mathbf{v}_0\,(x, y, z)$, say), we superpose a non-steady small perturbation $\mathbf{v}_1\,(x, y, z, t)$, which must be such that the resulting velocity $\mathbf{v} = \mathbf{v}_0 + \mathbf{v}_1$ satisfies the equations of motion. The equation for $\mathbf{v}_1$ is obtained by substituting in the equations

$$\frac{\partial \mathbf{v}}{\partial t} + (\mathbf{v}\cdot\mathbf{grad})\mathbf{v} = -\frac{\mathbf{grad}\,p}{\rho} + \nu\triangle\mathbf{v}, \qquad \mathrm{div}\,\mathbf{v} = 0$$

the velocity and pressure $\mathbf{v} = \mathbf{v}_0 + \mathbf{v}_1, p = p_0 + p_1$, where the known functions $\mathbf{v}_0$ and p_0 satisfy the unperturbed equations

$$(\mathbf{v}_0\cdot\mathbf{grad})\mathbf{v}_0 = -\frac{\mathbf{grad}\,p_0}{\rho} + \nu\triangle\mathbf{v}_0, \qquad \mathrm{div}\,\mathbf{v}_0 = 0.$$

Omitting terms above the first order in $\mathbf{v}_1$, we obtain

$$\frac{\partial \mathbf{v}_1}{\partial t} + (\mathbf{v}_0\cdot\mathbf{grad})\mathbf{v}_1 + (\mathbf{v}_1\cdot\mathbf{grad})\mathbf{v}_0$$

$$= -\frac{\mathbf{grad}\,p_1}{\rho} + \nu\triangle\mathbf{v}_1, \qquad \mathrm{div}\,\mathbf{v}_1 = 0. \qquad (26.1)$$

The boundary condition is that $\mathbf{v}_1$ vanishes on fixed solid surfaces.

Thus $\mathbf{v}_1$ satisfies a system of linear differential equations, with coefficients that are functions of the co-ordinates only, and not of the time. The general solution of such equations can be represented as a sum of particular solutions in which $\mathbf{v}_1$ depends on time as $e^{-i\omega t}$. The "frequencies" ω of the perturbations are not arbitrary, but are determined by solving the equations (26.1) with the appropriate boundary conditions. The "frequencies" are in general complex. If there are ω whose imaginary parts are positive, $e^{-i\omega t}$ will increase indefinitely with time. In other words, such perturbations, once having arisen, will increase, i.e. the flow is unstable with respect to such perturbations. For the flow to be stable it is necessary that the imaginary part of any possible "frequency" ω is negative. The perturbations that arise will then decrease exponentially with time.

Such a mathematical investigation of stability is extremely complicated, however. The theoretical problem of the stability of steady flow past bodies of finite dimensions has not yet been solved. It is certain that steady flow is stable for sufficiently small Reynolds numbers. The experimental data seem to indicate that, when R increases, it eventually reaches a value R_{cr} (the *critical Reynolds number*) beyond which the flow is unstable with respect to infinitesimal disturbances. For sufficiently large Reynolds numbers $(R > R_{cr})$, steady flow past solid bodies is therefore impossible. The critical Reynolds number is not, of course, a universal constant, but takes a different value for each type of flow. These values appear to be of the order of 10 to 100; for example, in flow across a cylinder undamped non-steady flow has been observed for $R = ud/\nu = 34$, d being the diameter of the cylinder. Exact measurements of R_{cr}, however, have not been made.

§27. The onset of turbulence

Let us now consider the nature of the non-steady flow which is established as a result of the absolute instability of steady flow at large Reynolds numbers. We begin by examining the properties of this flow at Reynolds numbers only slightly greater than R_{cr}. For $R < R_{cr}$ the imaginary parts of the complex "frequencies" $\omega = \omega_1 + i\gamma_1$ for all possible small velocity perturbations are negative ($\gamma_1 < 0$). For $R = R_{cr}$ there is one frequency whose imaginary part is zero. For $R > R_{cr}$ the imaginary part of this frequency is positive, but, when R is close to R_{cr}, γ_1 is small in comparison with the real part ω_1.[†] The function $\mathbf{v}_1$ corresponding to this frequency is of the form

$$\mathbf{v}_1 = A(t)\mathbf{f}(x, y, z), \tag{27.1}$$

where $\mathbf{f}$ is some complex function of the co-ordinates, and the complex "amplitude" $A(t)$ is[‡]

$$A(t) = \text{constant} \times e^{\gamma_1 t} e^{-i\omega_1 t}. \tag{27.2}$$

[†] It must be borne in mind that the set (or *spectrum*) of all possible frequencies for a given type of flow includes both separate isolated values (the *discrete spectrum*) and the whole of various frequency ranges (the *continuous spectrum*). However, it can be seen that the frequencies with positive imaginary parts in which we are interested occur, in general, only in the discrete spectrum.

[‡] As usual, we understand the real part of (27.2).

This expression for $A(t)$ is actually valid, however, only during a short interval of time after the disruption of the steady flow; the factor $e^{\gamma_1 t}$ increases rapidly with time, whereas the method of determining $\mathbf{v}_1$ given in §26, which leads to expressions like (27.1) and (27.2), applies only when $|\mathbf{v}_1|$ is small. In reality, of course, the modulus $|A|$ of the amplitude of the non-steady flow does not increase without limit, but tends to a finite value. For R close to R_{cr} (we always mean, of course, $R > R_{cr}$), this finite value is small, and can be determined as follows.

Let us find the time derivative of the squared amplitude $|A|^2$. For very small values of t, when (27.2) is still valid, we have $d|A|^2/dt = 2\gamma_1|A|^2$. This expression is really just the first term in an expansion in series of powers of A and A^*. As the modulus $|A|$ increases (still remaining small), subsequent terms in this expansion must be taken into account. The next terms are those of the third order in A. However, we are not interested in the exact value of the derivative $d|A|^2/dt$, but in its time average, taken over times large compared with the period $2\pi/\omega_1$ of the factor $e^{-i\omega_1 t}$; we recall that, since $\omega_1 \gg \gamma_1$, this period is small compared with the time $1/\gamma_1$ required for the amplitude modulus $|A|$ to change appreciably. The third-order terms, however, must contain the periodic factor, and therefore vanish on averaging.† The fourth-order terms include one which is proportional to $A^2 A^{*2} = |A|^4$ and which clearly does not vanish on averaging. Thus we have as far as fourth-order terms

$$\overline{d|A|^2/dt} = 2\gamma_1|A|^2 - \alpha|A|^4. \tag{27.3}$$

where α may be either positive or negative.

Let us suppose that α is positive.‡ We have not put bars above $|A|^2$ and $|A|^4$, since the averaging is only over time intervals short compared with $1/\gamma_1$. For the same reason, in solving the equation we proceed as if the bar were omitted above the derivative also. The solution of equation (27.3) is

$$1/|A|^2 = \alpha/2\gamma_1 + \text{constant} \times e^{-2\gamma_1 t}.$$

Hence it is clear that $|A|^2$ tends asymptotically to a finite limit:

$$|A|^2_{max} = 2\gamma_1/\alpha. \tag{27.4}$$

The quantity γ_1 is some function of the Reynolds number. Near R_{cr} it can be expanded as a series of powers of $R - R_{cr}$. But $\gamma_1(R_{cr}) = 0$, by the definition of the critical Reynolds number. Hence the zero-order term in the expansion is zero, and we have to the first order $\gamma_1 = \text{constant} \times (R - R_{cr})$. Substituting this in (27.4), we see that the modulus $|A|$ of the amplitude is proportional to the square root of $R - R_{cr}$:

$$|A|_{max} \sim \sqrt{(R - R_{cr})}. \tag{27.5}$$

---

† Strictly speaking, the third-order terms give, on averaging, not zero, but fourth-order terms. which we suppose included among the fourth-order terms in the expansion.
‡ This seems to be true for ordinary flow past bodies.

Let us summarise these results. The absolute instability of the flow for $R > R_{cr}$ leads to the appearance of a non-steady periodic flow. For R close to R_{cr} the latter flow can be represented by superposing on the steady flow $v_0(x, y, z)$ a periodic flow $v_1(x, y, z, t)$, with a small but finite amplitude which increases with R proportionally to the square root of $R - R_{cr}$. The velocity distribution in this flow is of the form

$$v_1 = f(x, y, z)e^{-i(\omega_1 t + \beta_1)}, \tag{27.6}$$

where f is a complex function of the co-ordinates, and β_1 is some initial phase. For large $R - R_{cr}$, the separation of the velocity into v_0 and v_1 is no longer meaningful. We then have simply some periodic flow with frequency ω_1. If, instead of the time, we use as an independent variable the phase $\phi_1 \equiv \omega_1 t + \beta_1$, then we can say that the function $v(x, y, z, \phi_1)$ is a periodic function of ϕ_1, with period 2π. This function, however, is no longer a simple trigonometrical function. Its expansion in Fourier series

$$v = \sum_p A_p(x, y, z)e^{-i\phi_1 p} \tag{27.7}$$

(where the summation is over all integers p, positive and negative) includes not only terms with the fundamental frequency ω_1, but also terms whose frequencies are integral multiples of ω_1.

The following important property of this non-steady flow should also be mentioned. Equation (27.3) determines only the modulus of the time factor $A(t)$, and not its phase. The phase $\phi_1 = \omega_1 t + \beta_1$ of the periodic flow remains essentially indeterminate, and depends on the particular initial conditions which happen to occur at the instant when the flow begins. The initial phase β_1 can have any value, depending on these conditions. Thus the periodic flow under consideration is not uniquely determined by the given steady external conditions in which the flow takes place. One quantity—the initial phase of the velocity—remains arbitrary. We may say that the flow has one degree of freedom, whereas steady flow, which is entirely determined by the external conditions, has no degrees of freedom.

Let us now consider the phenomena which occur when the Reynolds number increases further. When this happens, a time finally comes when the periodic flow discussed above in turn becomes unstable. The investigation of this instability would proceed† similarly to the method given above for determining the instability of the original steady flow. The part of the unperturbed flow is now taken by the periodic flow $v_0(x, y, z, t)$ (with frequency ω_1), and in the equations of motion we substitute $v = v_0 + v_2$, where v_2 is a small correction. For v_2 we again obtain a linear equation, but the coefficients are now functions of time as well as of the co-ordinates, being

---

† But has not been carried out even for particular cases, on account of the exceptional mathematical difficulties.

periodic in time with period $2\pi/\omega_1$. The solution of such an equation must be sought in the form $\mathbf{v}_2 = \Pi(x, y, z, t)e^{-i\omega t}$, where $\Pi(x, y, z, t)$ is a periodic function of time, with period $2\pi/\omega_1$. The instability again occurs when a frequency $\omega = \omega_2 + i\gamma_2$ appears such that the imaginary part γ_2 is positive, and the corresponding real part ω_2 then determines the new frequency which appears.

The result, therefore, is that a quasi-periodic flow appears, characterised by two different periods. Just as the flow had one degree of freedom after the appearance of the first periodic flow, so it now involves two arbitrary quantities (phases), i.e. it has two degrees of freedom.

When the Reynolds number increases still further, more and more new periods appear in succession. The range of Reynolds numbers between successive appearances of new frequencies diminishes rapidly in size. The new flows themselves are on a smaller and smaller scale. This means that the order of magnitude of the distances over which the velocity changes appreciably is the smaller, the later the flow in question appears.

For $R > R_{cr}$, therefore, the flow rapidly becomes complicated and confused. Such a flow is said to be *turbulent*; its properties will be investigated in detail in the following sections. In contradistinction to turbulent flow, the regular flow, in which the fluid moves as it were in layers with different velocities, is said to be *laminar*.

We can write down the general form of a function $\mathbf{v}(x, y, z, t)$ whose time dependence is given by some number n of different frequencies ω_j ($j = 1, 2, ..., n$). Instead of one phase $\phi_1 = \omega_1 t + \beta_1$, we now have n different phases $\phi_j = \omega_j t + \beta_j$. The function $\mathbf{v}$ may be regarded as a function of these phases (and of the co-ordinates), and is periodic in each of them, with period 2π. Such a function can be written as a series:

$$\mathbf{v}(x, y, z, t) = \sum_{p_1, p_2, ..., p_n} \mathbf{A}_{p_1 ... p_n}(x, y, z) \exp[-i\sum_{j=1}^{n} p_j\phi_j], \qquad (27.8)$$

the summation being taken over all integrals $p_1, p_2, ..., p_n$. This is a generalisation of formula (27.7). We may notice that the choice of the fundamental frequencies $\omega_1, ..., \omega_n$ is, as we see from (27.8), itself not unique; we could equally well take any n independent linear combinations of ω_j with integral coefficients.†

A flow described by a formula such as (27.8) has n degrees of freedom; it involves n arbitrary initial phases β_j. As the Reynolds number increases, both the number of frequencies and the number of degrees of freedom increase. In the limit as R tends to infinity, the number of degrees of freedom also increases indefinitely.

---

† These linear combinations must be such that from them we can form all possible numbers $\Sigma p_j\phi_j$. It is easy to see that, for this to be so, the determinant of the transformation coefficients relating the old and new frequencies must be unity.

It must be borne in mind that, since the velocity is a periodic function of the phases, with period 2π, the states whose phases differ only by an integral multiple of 2π are physically indistinguishable. In other words, we can say that all the essentially different values of each phase lie in the range $0 \leqslant \phi_j \leqslant 2\pi$. Let us consider any two phases $\phi_1 = \omega_1 t + \beta_1$ and $\phi_2 = \omega_2 t + \beta_2$. Suppose that, at some instant, ϕ_1 has the value α. Then, by what we have just said, ϕ_1 will have values equivalent to α at all instants $t = (\alpha - \beta_1 + 2\pi r)/\omega_1$, where r is any integer. At these instants the phase ϕ_2 will have the values

$$\phi_2 = \omega_2(\alpha - \beta_1)/\omega_1 + \beta_2 + 2\pi r \omega_2/\omega_1.$$

The different frequencies are generally incommensurable, so that ω_2/ω_1 is an irrational number. If we reduce each value of ϕ_2 to the range 0 to 2π by subtracting the appropriate integral multiple of 2π, we therefore obtain, as r goes from 0 to ∞, values for ϕ_2 which are arbitrarily close to any given number in that range. In other words, in the course of a sufficiently long time ϕ_1 and ϕ_2 will simultaneously be arbitrarily close to any given pair of values. The same is obviously true of all the phases. Thus turbulent motion has a certain quasi-periodic property: in the course of a sufficiently long time the fluid passes through states arbitrarily close to any given state, determined by any possible choice of simultaneous values of the phases ϕ_j.

We have introduced the concept of the critical Reynolds number as being the value of R at which instability of steady flow, in the sense described above, first occurs. The critical Reynolds number can, however, be regarded from a somewhat different point of view. For R < R_{cr} there are no stable non-steady solutions of the equations of motion that are not damped in time. After the critical value has been reached, a stable non-steady solution appears, which will actually occur in a moving fluid.

As far as experimental investigations of the flow past ordinary finite bodies are concerned, the two definitions of R_{cr} seem to be the same. Logically, however, this need not be so, and cases could in principle occur where there are two different critical values: one above which non-steady flow can occur without being damped, and another above which steady flow becomes absolutely unstable. The second must obviously be greater than the first. However, since there is at present no indication that such cases of instability actually exist, we shall not pause to investigate them more closely.[†]

§28. Stability of flow between rotating cylinders

To investigate the stability of steady flow between two rotating cylinders (§18) in the limit of very large Reynolds numbers, we can use a simple method like that used in §4 to derive the condition for mechanical stability of a fluid at rest in a gravitational field (RAYLEIGH, 1916). The principle of the method is to consider any small element of the fluid and to suppose that this element

---

[†] We are not here concerned with (e.g.) flow in a pipe, where the loss of stability has unusual properties (see §29).

is displaced from the path which it follows in the flow concerned. As a result of this displacement, forces appear which act on the displaced element. If the original flow is stable, these forces must tend to return the element to its original position.

Each fluid element in the unperturbed flow moves in a circle $r =$ constant about the axis of the cylinders. Let $\mu(r) = mr^2\phi$ be the angular momentum of an element of mass m, ϕ being the angular velocity. The centrifugal force acting on it is μ^2/mr^3; this force is balanced by the radial pressure gradient in the rotating fluid. Let us now suppose that a fluid element at a distance r_0 from the axis is slightly displaced from its path, being moved to a distance $r > r_0$ from the axis. The angular momentum of the element remains equal to its original value $\mu_0 = \mu(r_0)$. The centrifugal force acting on the element in its new position is therefore μ_0^2/mr^3. In order that the element should tend to return to its initial position, this force must be less than the equilibrium value μ^2/mr^3 which is balanced by the pressure gradient at the distance r. Thus the necessary condition for stability is $\mu^2 - \mu_0^2 > 0$. Expanding $\mu(r)$ in powers of the positive difference $r - r_0$, we can write this condition in the form

$$\mu \, d\mu/dr > 0. \tag{28.1}$$

According to formula (18.3), the angular velocity ϕ of the moving fluid particles is

$$\phi = \frac{\Omega_2 R_2^2 - \Omega_1 R_1^2}{R_2^2 - R_1^2} + \frac{(\Omega_1 - \Omega_2)R_1^2 R_2^2}{R_2^2 - R_1^2} \frac{1}{r^2}.$$

Calculating $\mu = mr^2\phi$ and omitting factors which are certainly positive, we can write the condition (28.1) as

$$(\Omega_2 R_2^2 - \Omega_1 R_1^2)\phi > 0. \tag{28.2}$$

The angular velocity ϕ varies monotonically from Ω_1 on the inner cylinder to Ω_2 on the outer cylinder. If the two cylinders rotate in opposite directions, i.e. if Ω_1 and Ω_2 have opposite signs, the function ϕ changes sign between the cylinders, and its product with the constant number $\Omega_2 R_2^2 - \Omega_1 R_1^2$ cannot be everywhere positive. Thus in this case (28.2) does not hold at all points in the fluid, and the flow is unstable.

Now let the two cylinders be rotating in the same direction; taking this direction of rotation as positive, we have $\Omega_1 > 0$, $\Omega_2 > 0$. Then ϕ is everywhere positive, and for the condition (28.2) to be fulfilled it is necessary that

$$\Omega_2 R_2^2 > \Omega_1 R_1^2. \tag{28.3}$$

If $\Omega_2 R_2^2 < \Omega_1 R_1^2$ the flow is unstable. For example, if the outer cylinder is at rest ($\Omega_2 = 0$), while the inner one rotates, then the flow is unstable. If, on the other hand, the inner cylinder is at rest ($\Omega_1 = 0$), the flow is stable.

It must be emphasised that no account has been taken, in the above arguments, of the effect of the viscous forces when the fluid element is displaced.

The method is therefore applicable only for small viscosities, i.e. for large R.

To investigate the stability of the flow for any R, it is necessary to follow the general method, starting from equations (26.1) (G. I. Taylor, 1923). In the present case the unperturbed velocity distribution v_0 depends only on the (cylindrical) radial co-ordinate r, and not on the angle ϕ or the axial co-ordinate z. Thus we have for the perturbation v_1 a system of linear equations with coefficients which contain neither the time nor the co-ordinates ϕ and z. We may seek solutions of these equations in the form

$$v_1 = e^{i(kz-\omega t)} f(r), \qquad (28.4)$$

the direction of the vector f being arbitrary; this solution depends on z through the periodic factor e^{ikz}, and the wave number k determines the periodicity of the perturbation in the z-direction. The possible frequencies ω, obtained by solving the equations with the necessary boundary conditions in a plane perpendicular to the axis ($v_1 = 0$ for $r = R_1$ and $r = R_2$), will then be functions of k, involving R as a parameter: $\omega = \omega(k, R)$. The point where instability appears is determined by the value of R for which the function $\gamma_1 = \text{im } \omega$ first becomes zero for some k. For $R < R_{cr}$, the function $\gamma_1(k, R)$ is always negative, but for $R > R_{cr}$ we have $\gamma_1 > 0$ in some range of k. Let k_{cr} be the value of k for which $\gamma_1 = 0$ when $R = R_{cr}$. The corresponding function (28.4) gives the nature of the flow which occurs (superposed on the original flow) in the fluid at the instant when the original flow ceases to be stable; it is periodic along the axis of the cylinders, with wavelength $2\pi/k_{cr}$.[†]

As well as solutions of the form (28.4), which are independent of the angle ϕ, the system of equations under consideration has also solutions for which v_1 contains a factor $e^{im\phi}$, m being an integer. We are, however, interested only in the solution which corresponds to the first appearance of instability. The solutions with $m \neq 0$ have never been studied in this respect. It is nevertheless natural to suppose that instability occurs first of all with respect to perturbations with $m = 0$, a supposition which is entirely confirmed by experimental results.

It should also be borne in mind that, even for a given k, the solution of the form (28.4) is not unique. In general, a number of solutions with different values of ω correspond to a given k. Again we are interested only in the one which gives the smallest value of R_{cr}.

It is found that a purely imaginary function $\omega(k)$ corresponds to the solution which gives the smallest R_{cr}. Hence, when $k = k_{cr}$, not only im ω but ω itself is zero. This means that the first instability of the steady flow between rotating cylinders leads to the appearance of another flow which is also steady.

[†] For R slightly greater than R_{cr} there is not one value of k, but a whole range, for which im $\omega > 0$. However, it should not be thought that the resulting flow will be a superposition of flows with various periodicities. In reality, for each R a flow of definite periodicity occurs which stabilises the total flow. This periodicity, however, cannot be determined from the linearised equations (26.1).

On account of the great complexity of the calculation,† numerical results have been obtained only for the case where the space between the cylinders is narrow $(R_2 - R_1 \ll R_2)$. Fig. 13 shows an example of the curve separating the regions of unstable (shaded) and stable flow. The right-hand branch of the curve, corresponding to rotation of the two cylinders in the same direction, is asymptotic to the line $\Omega_2 R_2^2 = \Omega_1 R_1^2$. When the Reynolds number increases, for a given type of flow, the two numbers $\Omega_1 R_1^2/\nu$ and $\Omega_2 R_2^2/\nu$ increase by equal factors. In Fig. 13 this corresponds to a movement upwards along a line through the origin having a given slope. In the right-hand part of the diagram (Ω_1 and Ω_2 both positive), such lines for which $\Omega_2 R_2^2/\Omega_1 R_1^2 > 1$ do not meet the curve which bounds the region of instability. If, on the other hand, $\Omega_2 R_2^2/\Omega_1 R_1^2 < 1$, then for sufficiently large Reynolds numbers we enter the region of instability, in accordance with the condition (28.3).

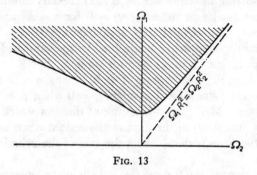

FIG. 13

In the left-hand part of the diagram (Ω_1 and Ω_2 of opposite signs), any line through the origin eventually meets the curve, i.e. the flow can become unstable for any value of the ratio $\Omega_2 R_2^2/\Omega_1 R_1^2$, again in agreement with the results obtained above. For $\Omega_2 = 0$ (when only the inner cylinder rotates), instability sets in when

$$\Omega_1 = 41 \cdot 3 \nu / h \sqrt{(h R_2)}, \tag{28.5}$$

where $h = R_2 - R_1$.

The stability of the flow in the unshaded part of Fig. 13 does not mean, however, that the flow actually remains steady no matter how large R becomes. Experiment shows that there is a limit beyond which stable non-steady flow becomes possible. In this region the steady flow is "metastable": it is stable with respect to small perturbations, but unstable with respect to larger perturbations. If, owing to such perturbations, non-steady flow occurs in some region along the cylinders, it will subsequently "displace" the laminar flow in all space. This non-steady flow has, as soon as it appears, a large number of "degrees of freedom" (in the sense explained in §27), i.e. it is fully developed turbulence.

---

† Further details may be found in the book by C. C. LIN, *The Theory of Hydrodynamic Stability*, Cambridge 1955.

In the shaded part of Fig. 13, the flow again becomes turbulent for sufficiently large R, but there are, it seems, very few data concerning the way in which it appears.

A limiting case of the flow between rotating cylinders, corresponding to large radii and small $h = R_2 - R_1$, is flow between two parallel planes in relative motion (see §17). This flow is stable with respect to infinitely small perturbations for any value of $R = Uh/\nu$, where U is the relative velocity of the planes. Stable turbulent motion becomes possible, however, for values of R greater than about 1500.

§29. Stability of flow in a pipe

The steady flow in a pipe discussed in §17 loses its stability in an unusual manner. Since the flow is uniform in the x-direction (along the pipe), the unperturbed velocity distribution $\mathbf{v}_0$ is independent of x. Similarly to the procedure in §28, we can therefore seek solutions of equations (26.1) in the form

$$\mathbf{v}_1 = e^{i(kx - \omega t)} \mathbf{f}(y, z). \tag{29.1}$$

Here also there is a value $R = R_{cr}$ for which $\gamma_1 = \mathrm{im}\,\omega$ first becomes zero for some value of k. It is of importance, however, that the real part of the function $\omega(k)$ is not now zero.

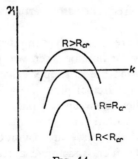

FIG. 14

For values of R only slightly exceeding R_{cr}, the range of values of k for which $\gamma_1(k) > 0$ is small and lies near the point for which $\gamma_1(k)$ is a maximum, i.e. $d\gamma_1/dk = 0$ (as seen from Fig. 14). Let a slight perturbation occur in some part of the flow; it is a wave packet obtained by superposing a series of components of the form (29.1). In the course of time, the components for which $\gamma_1(k) > 0$ will be amplified, while the remainder will be damped. The amplified wave packet thus formed will also be carried downstream with a velocity equal to the group velocity $d\omega/dk$ of the packet; since we are now considering waves whose wave numbers lie in a small range near the point where $d\gamma_1/dk = 0$, the quantity $d\omega/dk \approx d\omega_1/dk$ is real, and is therefore the actual velocity of propagation of the packet.

This downstream displacement of the perturbations is very important, and causes the loss of stability to be totally different from that described in §28.

We have seen that, for flow between rotating cylinders with $R > R_{cr}$ (when there are frequencies with im $\omega > 0$), the original steady flow is no longer possible, since even small perturbations are increased to a finite amplitude. For flow in a pipe, however, the amplification of the perturbation is accompanied by its displacement downstream; if we consider the flow at a given point in the pipe, it is found that the perturbation there is not amplified, but damped. It must also be borne in mind that, since in reality we have pipes of finite length, however great, any perturbation may be carried out of the pipe before it disrupts the laminar flow. Thus, even for $R > R_{cr}$, steady flow in a pipe is effectively stable with respect to small perturbations, and can in principle take place for values of R considerably exceeding R_{cr}.

Since the perturbations increase with the co-ordinate x (downstream), and not with time at a given point, it is reasonable to investigate this type of instability as follows. Let us suppose that, at a given point, a continuously acting perturbation with a given frequency ω is applied to the flow, and examine what will happen to this perturbation as it is carried downstream. Inverting the function $\omega = \omega(k)$, we find what wave number k corresponds to the given (real) frequency ω. If im $k < 0$, the factor e^{ikx} increases with x, i.e. the perturbation is amplified downstream. The curve in the ωR-plane given by the equation im $k(\omega, R) = 0$ defines the region of stability, and separates, for each R, the frequencies of perturbations which are amplified and damped downstream.

The actual calculations are extremely complicated. A complete investigation has been made only for flow between two parallel planes (C. C. LIN, 1946).[†] However, it is reasonable to suppose that the results will be qualitatively the same for flow in a circular pipe.

The limiting curve for flow between two planes is schematically shown in Fig. 15. The shaded area within the curve is the region of instability. As $R \to \infty$, both branches of the curve are asymptotic to the R-axis.[‡] For the smallest value of R at which undamped perturbations are possible we find by calculations $R_{cr} \approx 7700$, R being defined as Uh/ν, with h the distance between the planes and U the fluid velocity averaged across this distance.

Thus, for any frequency between zero and a certain maximum value, there is a finite range of R values for which perturbations with the frequency concerned will be amplified. It is interesting to note that a small but finite viscosity of the fluid has, in a sense, a destabilising effect in comparison with the situation for a strictly ideal fluid. For, when $R \to \infty$, perturbations with any finite frequency are damped, but when a finite viscosity is introduced we eventually reach a region of instability; a further increase in the viscosity (decrease in R) finally brings us out of this region.

---

† A detailed account is given by C. C. LIN, *The Theory of Hydrodynamic Stability*, Cambridge 1955.
‡ The asymptotic equations of the two branches for large R are $\omega h/U = 5 \cdot 0/R^{3/11}$, $\omega h/U = 11 \cdot 2/R^{3/7}$.

These calculations, however, do not answer the question whether, for sufficiently large R, flow in a pipe does not also exhibit true instability with respect to infinitely small perturbations, i.e. instability resulting in the amplification of perturbations with time at a given point. We shall outline the mathematical significance of such an instability. Let us consider some small perturbation which occurs at time $t = 0$ in a finite region. Expanding it as a Fourier integral with respect to x, we can write it as

$$\int\int f(\xi)e^{ik(x-\xi)}d\xi\, dk,$$

where $f(x)$ is a function describing the initial perturbation. In the course of time, each Fourier component of the perturbation will vary as $e^{-i\omega t}$, with a frequency $\omega = \omega(k, R)$, so that the whole perturbation at time t will be given by the integral

$$\int\int f(\xi)e^{ik(x-\xi)-i\omega t}\, d\xi\, dk.$$

Since $f(x)$ is zero except in a finite region, $x - \xi$ has a finite range of values. Hence the behaviour of the integral for large t is essentially determined by the behaviour of the integral

$$\int e^{-i\omega(k)t}\, dk$$

If this integral tends to infinity with t, the flow is in fact absolutely unstable.

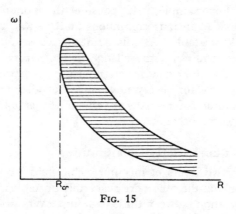

FIG. 15

No such investigation has yet been made, even for a particular case. However, the experimental results concerning flow in pipes give reason to suppose that there is no true instability with respect to arbitrarily small perturbations for any R. This is indicated by the fact that, the more carefully perturbations at the entrance to the pipe are prevented, the larger the Reynolds numbers for which laminar flow can be observed.†

---

† Laminar flow has actually been observed up to $R \approx 50,000$, where $R = Ud/\nu$, d being the diameter of the pipe and U the mean velocity over its cross-section.

However, the experimental results also show that there is another critical Reynolds number (which we denote by R_{cr}'); this determines the limit beyond which stable non-steady flow can exist (cf. the end of §27). If, in any section of the pipe, turbulent flow occurs, then for $R < R_{cr}'$ the turbulent region will be carried downstream and will diminish in size until it disappears altogether; if, on the other hand, $R > R_{cr}'$, the turbulent region will extend in the course of time to include more and more of the flow. If perturbations of the flow occur continually at the entrance to the pipe, then for $R < R_{cr}'$ they will be damped out at some distance down the pipe, no matter how strong they are initially. If, on the other hand, $R > R_{cr}'$, the flow becomes turbulent throughout the pipe, and this can be achieved by perturbations which are the weaker, the greater R.† Thus laminar flow in a pipe with $R > R_{cr}'$ is metastable, being unstable with respect to perturbations of finite intensity; the necessary intensity is the smaller, the greater R.

As has been mentioned at the end of §28, non-steady flow arising by the disruption of metastable laminar flow is already fully-developed turbulence. In this sense the appearance of turbulence in a pipe is essentially different from the appearance of turbulence owing to the absolute instability of steady flow past finite bodies. In the latter case non-steady flow seems to appear in a continuous manner as we pass through R_{cr}, the number of degrees of freedom increasing gradually (as explained in §§26 and 27). For flow in a pipe, however, turbulence appears discontinuously. This difference causes, in particular, the different dependence of the drag on the Reynolds number in the two cases. For example, if we consider the motion of any body in a fluid, the drag force F on it is continuous at $R = R_{cr}$, where steady flow becomes absolutely unstable. At this point the curve $F = F(R)$ can have only a bend corresponding to the change in the nature of the flow. For flow in a pipe, on the other hand, there are essentially two different laws of drag for $R \geqslant R_{cr}$: one for steady flow, and the other for turbulent flow. The drag is discontinuous for whatever value of R marks the transition from one type of flow to the other.

§30. Instability of tangential discontinuities

Flows in which two layers of incompressible fluid move relative to each other, one "sliding" on the other, are absolutely unstable if the fluid is ideal; the surface of separation between these two fluid layers would be a *surface of tangential discontinuity*, on which the fluid velocity tangential to the surface is discontinuous. We shall see below (§35) what is the actual nature of the flow resulting from this instability; here we shall prove the above statement.

If we consider a small portion of the surface of discontinuity and the flow near it, we may regard this portion as plane, and the fluid velocities $\mathbf{v}_1$ and $\mathbf{v}_2$ on each side of it as constants. Without loss of generality we can suppose

---

† For a pipe of circular cross-section R_{cr}' lies between 1600 and 1700. For flow between parallel planes, turbulent flow has been observed from $R = 1400$ upwards.

that one of these velocities is zero; this can always be achieved by a suitable choice of the co-ordinate system. Let $v_2 = 0$, and v_1 be denoted by v simply; we take the direction of v as the x-axis, and the z-axis along the normal to the surface.

Let the surface of discontinuity receive a slight perturbation, in which all quantities—the co-ordinates of points on the surface, the pressure, and the fluid velocity—are periodic functions, proportional to $e^{i(kx-\omega t)}$. We consider the fluid on the side where its velocity is v, and denote by v' the small change in the velocity due to the perturbation. According to the equations (26.1) (with constant $v_0 = v$ and $\nu = 0$), we have the following system of equations for the perturbation v':

$$\operatorname{div} v' = 0, \qquad \frac{\partial v'}{\partial t} + (v \cdot \operatorname{grad})v' = -\frac{\operatorname{grad} p'}{\rho}.$$

Since v is along the x-axis, the second equation can be rewritten

$$\frac{\partial v'}{\partial t} + v\frac{\partial v'}{\partial x} = -\frac{\operatorname{grad} p'}{\rho}. \tag{30.1}$$

If we take the divergence of both sides, then the left-hand side gives zero by virtue of $\operatorname{div} v' = 0$, so that p' must satisfy Laplace's equation:

$$\triangle p' = 0. \tag{30.2}$$

Let $\zeta = \zeta(x, t)$ be the displacement in the z-direction of points on the surface of discontinuity, due to the perturbation. The derivative $\partial\zeta/\partial t$ is the rate of change of the surface co-ordinate ζ for a given value of x. Since the fluid velocity component normal to the surface of discontinuity is equal to the rate of displacement of the surface itself, we have to the necessary approximation

$$\partial\zeta/\partial t = v'_z - v\partial\zeta/\partial x, \tag{30.3}$$

where, of course, the value of v'_z on the surface must be taken.

We seek p' in the form $p' = f(z)\,e^{i(kx-\omega t)}$. Substituting in (30.2), we have for $f(z)$ the equation $\mathrm{d}^2f/\mathrm{d}z^2 - k^2f = 0$, whence $f = \text{constant} \times e^{\pm kz}$. Suppose that the space on the side under consideration (side 1) corresponds to positive values of z. Then we must take $f = \text{constant} \times e^{-kz}$, so that

$$p' = \text{constant} \times e^{i(kx-\omega t)}\,e^{-kz}. \tag{30.4}$$

Substituting this expression in the z-component of equation (30.1), we find†

$$v'_z = kp'_1/i\rho_1(kv - \omega). \tag{30.5}$$

The displacement ζ may also be sought in a form proportional to the same exponential factor $e^{i(kx-\omega t)}$, and we obtain from (30.3) $v'_z = i\zeta(kv - \omega)$.

---

† The case $kv = \omega$, though possible in principle, is not of interest here, since instability can arise only from complex frequencies ω, not from real ω.

This gives, instead of (30.5),

$$p'_1 = -\zeta \rho_1 (kv - \omega)^2 / k. \tag{30.6}$$

The pressure p'_2 on the other side of the surface is given by a similar formula, where now $v = 0$ and the sign is changed (since in this region $z < 0$, and all quantities must be proportional to e^{kz}, not e^{-kz}). Thus

$$p'_2 = \zeta \rho_2 \omega^2 / k. \tag{30.7}$$

We have written different densities ρ_1 and ρ_2 in order to include the case where we have a boundary separating two different immiscible fluids.

Finally, from the condition that the pressures p'_1 and p'_2 are equal on the surface of discontinuity, we obtain $\rho_1 (kv - \omega)^2 = -\rho_2 \omega^2$, from which the desired relation between ω and k is found to be

$$\omega = kv \frac{\rho_1 \pm i \sqrt{(\rho_1 \rho_2)}}{\rho_1 + \rho_2}. \tag{30.8}$$

We see that ω is complex, and there are always ω having a positive imaginary part. Thus tangential discontinuities are unstable, even with respect to infinitely small perturbations. In this form the result is true for very small viscosities, i.e. for very large R. In this case it is meaningless to distinguish instability of the type that is "carried along" from true absolute instability, since, as k increases, the imaginary part of ω increases without limit, and hence the "amplification coefficient" of the perturbation as it is carried along may be as large as we please.

When finite viscosity is taken into account, the tangential discontinuity is no longer sharp; the velocity changes from one value to another across a layer of finite thickness. The problem of the stability of such a flow is mathematically entirely similar to that of the stability of flow in a laminar boundary layer with a point of inflexion in the velocity profile (§41). The experimental results indicate that instability sets in very soon.

§31. Fully developed turbulence

Turbulent flow at fairly large Reynolds numbers is characterised by the presence of an extremely irregular variation of the velocity with time at each point. This is called *fully developed turbulence*. The velocity continually fluctuates about some mean value, and it should be noted that the amplitude of this variation is in general not small in comparison with the magnitude of the velocity itself. A similar irregular variation of the velocity exists between points in the flow at a given instant. The paths of the fluid particles in turbulent flow are extremely complicated, resulting in an extensive mixing of the fluid.

As has been mentioned in the previous section, turbulent flow has a very large number of degrees of freedom. The values of the initial phases β_j corresponding to these degrees of freedom are determined by the initial conditions of the flow. The specification of the exact initial conditions which would determine the value of so many quantities is, however, so unrealistic that even to put the problem in this form is physically meaningless.

The position here is similar to what would happen if we attempted to consider the motion of all the molecules forming a macroscopic body, using the equations of mechanics; here again the problem of specifying the initial conditions which determine the initial values of the co-ordinates and velocities of all the molecules, and then integrating the equations of motion, is physically meaningless. The analogy extends further. A macroscopic body, regarded as composed of individual molecules, has an enormous number of degrees of freedom. An exact microscopic description of the state of the body would involve a determination of the co-ordinates and velocity of every particle composing it. The exact manner in which these quantities vary with time depends on their values at the initial instant. However, owing to the extreme complexity and irregularity of the motion of the molecules, we may suppose that, over a sufficiently long interval of time, the velocities and co-ordinates of the molecules take all possible sets of values, so that the effect of the initial conditions is smoothed out and disappears. This, as is well known, makes possible a statistical discussion of macroscopic bodies.

A similar situation occurs in turbulent flow. For an exact description of the time variation of the velocity distribution in the moving fluid, the values of all the initial phases β_j would have to be given; the values of all the phases $\phi_j = \omega_j t + \beta_j$ at every instant would then be known. We have seen that, whatever the initial phases β_j, over a sufficiently long interval of time the fluid passes through states arbitrarily close to any given state, defined by any possible choice of simultaneous values of the phases ϕ_j. Hence it follows that, in the consideration of turbulent flow, the actual initial conditions cease to have any effect after sufficiently long intervals of time. This shows that the theory of turbulent flow must be a statistical theory. No complete quantitative theory of turbulence has yet been evolved. Nevertheless, several very important qualitative results are known, and the following sections give an account of these.

We introduce the concept of the mean velocity, obtained by averaging over long intervals of time the actual velocity at each point. By such an averaging the irregular variation of the velocity is smoothed out, and the mean velocity varies smoothly from point to point. In what follows we shall denote the mean velocity by $\mathbf{u} = \bar{\mathbf{v}}$. The difference $\mathbf{v}' = \mathbf{v} - \mathbf{u}$ between the true velocity and the mean velocity varies irregularly in the manner characteristic of turbulence; we shall call it the *fluctuating part* of the velocity.

Let us consider in more detail the nature of this irregular motion which is superposed on the mean flow. This motion may in turn be qualitatively regarded as the superposition of *turbulent eddies* of different sizes; by the size of an eddy we mean the order of magnitude of the distances over which the velocity varies appreciably. As the Reynolds number increases, large eddies appear first; the smaller the eddies, the later they appear. For very large

Reynolds numbers, eddies of every size from the largest to the smallest are present. An important part in any turbulent flow is played by the largest eddies, whose size is of the order of the dimensions of the region in which the flow takes place; in what follows we shall denote by l this order of magnitude for any given turbulent flow. These large eddies have the largest amplitudes. The velocity in them is comparable with the variation of the mean velocity over the distance l; we shall denote by Δu the order of magnitude of this variation.† The frequencies corresponding to these eddies are of the order of u/l, the ratio of the mean velocity u (and not its variation Δu) to the dimension l. For the frequency determines the period with which the flow pattern is repeated when observed in some fixed frame of reference. Relative to such a system, however, the whole pattern moves with the fluid at a velocity of the order of u.

The small eddies, on the other hand, which correspond to large frequencies, participate in the turbulent flow with much smaller amplitudes. They may be regarded as a fine detailed structure superposed on the fundamental large turbulent eddies. Only a comparatively small part of the total kinetic energy of the fluid resides in the small eddies.

From the picture of turbulent flow given above, we can draw a conclusion regarding the manner of variation of the fluctuating velocity from point to point at any given instant. Over large distances (comparable with l), the variation of the fluctuating velocity is given by the variation in the velocity of the large eddies, and is therefore comparable with Δu. Over small distances (compared with l), it is determined by the small eddies, and is therefore small (compared with Δu).‡ The same kind of picture is obtained if we observe the variation of the velocity with time at any given point. Over short time intervals (compared with $T \sim l/u$), the velocity does not vary appreciably; over long intervals, it varies by a quantity of the order of Δu.

The length l appears as a characteristic dimension in the Reynolds number R, which determines the properties of a given flow. Besides this Reynolds number, we can introduce the qualitative concept of the Reynolds numbers for turbulent eddies of various sizes. If λ is the order of magnitude of the size of a given eddy, and v_λ the order of magnitude of its velocity, then the corresponding Reynolds number is defined as $R_\lambda \sim v_\lambda \lambda / \nu$. This number is the smaller, the smaller the size of the eddy.

For large Reynolds numbers R, the Reynolds numbers R_λ of the large eddies are also large. Large Reynolds numbers, however, are equivalent to small viscosities. We therefore conclude that, for the large eddies which are the basis of any turbulent flow, the viscosity is unimportant and may be

---

† We are speaking here of the order of magnitude, not of the mean velocity itself, but of its variation (over distances of the order of l), since it is this variation Δu which characterises the velocity of the turbulent flow. The mean velocity itself can have any magnitude, depending on the frame of reference used.

It may also be mentioned that experimental results indicate that the size of the largest eddies is actually somewhat less than l, and their velocity is somewhat less than Δu.

‡ But large compared with the variation of the mean velocity over these small distances.

equated to zero, so that the motion of these eddies obeys Euler's equation. In particular, it follows from this that there is no appreciable dissipation of energy in the large eddies.

The viscosity of the fluid becomes important only for the smallest eddies, whose Reynolds number is comparable with unity. We denote the size of these eddies by λ_0, which we shall determine in the next section. It is in these small eddies, which are unimportant as regards the general pattern of a turbulent flow, that the dissipation of energy occurs.

We thus arrive at the following conception of energy dissipation in turbulent flow. The energy passes from the large eddies to smaller ones, practically no dissipation occurring in this process. We may say that there is a continuous flow of energy from large to small eddies, i.e. from small to large frequencies. This flow of energy is dissipated in the smallest eddies, where the kinetic energy is transformed into heat.†

Since the viscosity of the fluid is important only for the smallest eddies, we may say that none of the quantities pertaining to eddies of sizes $\lambda \gg \lambda_0$ can depend on ν (more exactly, these quantities cannot be changed if ν varies but the other conditions of the motion are unchanged). This circumstance reduces the number of quantities which determine the properties of turbulent flow, and the result is that similarity arguments, involving the dimensions of the available quantities, become very important in the investigation of turbulence.

Let us apply these arguments to determine the order of magnitude of the energy dissipation in turbulent flow. Let ϵ be the mean dissipation of energy per unit time per unit mass of fluid.‡ We have seen that this energy is derived from the large eddies, whence it is gradually transferred to smaller eddies until it is dissipated in eddies of size $\sim \lambda_0$. Hence, although the dissipation is ultimately due to the viscosity, the order of magnitude of ϵ can be determined only by those quantities which characterise the large eddies. These are the fluid density ρ, the dimension l and the velocity Δu. From these three quantities we can form only one having the dimensions of ϵ, namely erg/g sec = cm^2/sec^3. Thus we find

$$\epsilon \sim (\Delta u)^3/l, \qquad (31.1)$$

and this determines the order of magnitude of the energy dissipation in turbulent flow.

In some respects a fluid in turbulent motion may be qualitatively described as having a "turbulent viscosity" ν_{turb} which differs from the true kinematic viscosity ν. Since ν_{turb} characterises the properties of the turbulent flow, its order of magnitude must be determined by ρ, Δu and l. The only quantity that can be formed from these and has the dimensions of kinematic viscosity

---

† For a steady state to be maintained, it is of course necessary that external energy sources should be present which continually supply energy to the large eddies.

‡ In this chapter ϵ denotes the mean dissipation of energy, and not the internal energy of the fluid.

is $l\Delta u$, and therefore

$$\nu_{\text{turb}} \sim l\Delta u. \tag{31.2}$$

The ratio of the turbulent viscosity to the ordinary viscosity is consequently $\nu_{\text{turb}}/\nu \sim R$, i.e. it increases with the Reynolds number.†

The energy dissipation ϵ is expressed in terms of ν_{turb} by

$$\epsilon \sim \nu_{\text{turb}}(\Delta u/l)^2 \tag{31.3}$$

in accordance with the usual definition of viscosity. Whereas ν determines the energy dissipation in terms of the space derivatives of the true velocity, ν_{turb} relates it to the gradient ($\sim \Delta u/l$) of the mean velocity.

We may also apply similarity arguments to determine the order of magnitude Δp of the variation of pressure over the region of turbulent flow. The only quantity having the dimensions of pressure which can be formed from ρ, l and Δu is $\rho(\Delta u)^2$. Hence we must have

$$\Delta p \sim \rho(\Delta u)^2. \tag{31.4}$$

§32. Local turbulence

Let us now consider the properties of the turbulence as regards eddy sizes λ which are small compared with the fundamental eddy size l. We shall refer to these properties as *local* properties of the turbulence. We shall consider fluid that is far from all solid surfaces (more precisely, that is at distances from them large compared with λ).

It is natural to assume that such small-scale turbulence, far from solid bodies, is *isotropic*. This means that, over regions whose dimensions are small compared with l, the properties of the turbulent flow are independent of direction; in particular, they do not depend on the direction of the mean velocity. It must be emphasised that here, and everywhere in the present section, when we speak of the properties of the turbulent flow in a small region of the fluid, we mean the relative motion of the fluid particles in that region, and not the absolute motion of the region as a whole, which is due to the larger eddies.

It is found that several very important results concerning the local properties of turbulence can be obtained immediately from similarity arguments. These results are due to A. N. KOLMOGOROV and to A. M. OBUKHOV (1941). To obtain them, we shall first determine which parameters can be involved in the properties of turbulent flow over regions small compared with l but large compared with the distances λ_0 at which the viscosity of the fluid begins to be important. It is these intermediate distances which we shall discuss below. The

---

† In reality, however, a fairly large numerical coefficient should be included. This is because, as mentioned above, l and Δu may differ quite considerably from the actual scale and velocity of the turbulent flow. The ratio ν_{turb}/ν may be more accurately written $\nu_{\text{turb}}/\nu \sim R/R_{\text{cr}}$, which formula takes into account the fact that ν_{turb} and ν must in reality be comparable in magnitude not for $R \sim 1$, but for $R \sim R_{\text{cr}}$.

parameters in question are the fluid density ρ and another quantity characterising any turbulent flow, the energy ϵ dissipated per unit time per unit mass
of fluid. We have seen that ϵ is the "energy flux" which continually passes
from larger to smaller eddies. Hence, although the energy dissipation is
ultimately due to the viscosity of the fluid and occurs in the smallest eddies,
the quantity ϵ is determined by the properties of larger eddies. It is natural
to suppose that (for given ρ and ϵ) the local properties of the turbulence are
independent of the dimension l and velocity Δu of the flow as a whole. The
fluid viscosity ν also cannot appear in any of the quantities in which we are
at present interested (we recall that we are concerned with distances $\lambda \gg \lambda_0$).

Let us determine the order of magnitude v_λ of the turbulent velocity variation over distances of the order of λ. It must be determined only by ρ, ϵ
and, of course, the distance λ itself. From these three quantities we can
form only one having the dimensions of velocity, namely $(\epsilon\lambda)^{\frac{1}{3}}$. Hence we
can say that the relation

$$v_\lambda \sim (\epsilon\lambda)^{\frac{1}{3}} \tag{32.1}$$

must hold. We thus reach a very important result: the velocity variation over
a small distance is proportional to the cube root of the distance (*Kolmogorov
and Obukhov's law*). The quantity v_λ may also be regarded as the velocity
of turbulent eddies whose size is of the order of λ.†

Let us now put the problem somewhat differently, and determine the order
of magnitude v_τ of the velocity variation at a given point over a time interval
τ which is short compared with the time $T \sim l/u$ characterising the flow as
a whole. To do this, we notice that, since there is a net mean flow, any given
portion of the fluid is displaced, during the interval τ, over a distance of
the order of τu, u being the mean velocity. Hence the portion of fluid which
is at a given point at time τ will have been at a distance τu from that point
at the initial instant. We can therefore obtain the required quantity v_τ by
direct substitution of τu for λ in (32.1):

$$v_\tau \sim (\epsilon\tau u)^{\frac{1}{3}}. \tag{32.2}$$

Thus the velocity variation over a time interval τ is proportional to the cube
root of the interval.

---

† The variation v_λ of the velocity over small distances is fundamentally the variation in the fluctuating part of the velocity; the variation of the mean velocity over small distances is small compared
with the variation of the fluctuating velocity over those distances.

The relation (32.1) may be obtained in another way by expressing a constant quantity, the dis-
sipation ϵ, in terms of quantities characterising the eddies of size λ; ϵ must be proportional to the
squared gradient of the velocity v_λ and to the appropriate turbulent viscosity coefficient

$$\nu_{\text{turb},\lambda} \sim v_\lambda \lambda$$

(cf. (31.2), (31.3)):

$$\epsilon \sim \nu_{\text{turb},\lambda}(v_\lambda/\lambda)^2 \sim v_\lambda^3/\lambda,$$

whence we obtain (32.1).

The quantity v_τ must be distinguished from v_τ', the variation in velocity of a portion of fluid as it moves about. This variation can evidently depend only on ρ and ϵ, which determine the local properties of the turbulence, and of course on τ itself. Forming the only combination of ρ, ϵ and τ that has the dimensions of velocity, we obtain

$$v_\tau' \sim (\epsilon\tau)^{\frac{1}{2}}. \tag{32.3}$$

Unlike the velocity variation at a given point, it is proportional to the square root of τ, not to the cube root. It is easy to see that, for τ small compared with T, v_τ' is always less than v_τ.†

Using the expression (31.1) for ϵ, we can rewrite (32.1) as

$$v_\lambda \sim \Delta u(\lambda/l)^{\frac{1}{3}}. \tag{32.4}$$

Similarly, we can write v_τ as

$$v_\tau \sim \Delta u(\tau/T)^{\frac{1}{3}}, \tag{32.5}$$

where $T \sim l/u$.

Let us now find at what distances the fluid viscosity begins to be important. These distances λ_0 also determine the order of magnitude of the size of the smallest eddies in the turbulent flow (called the "internal scale" of the turbulence, in contradistinction to the "external scale" l). To determine λ_0, we form the Reynolds number $R_\lambda \sim v_\lambda \lambda/\nu$; using (32.4), we obtain

$$R_\lambda \sim \Delta u \cdot \lambda^{4/3}/\nu l^{1/3}.$$

Introducing the Reynolds number $R \sim l\Delta u/\nu$ for the flow as a whole, we can rewrite this as $R_\lambda \sim R(\lambda/l)^{\frac{4}{3}}$. The order of magnitude of λ_0 is that for which $R_{\lambda_0} \sim 1$. Hence we find

$$\lambda_0 \sim l/R^{\frac{3}{4}}. \tag{32.6}$$

The same expression can be obtained by forming from ρ, ϵ and ν the only combination having the dimensions of length, namely $\lambda_0 \sim (\nu^3/\epsilon)^{\frac{1}{4}}$, and expressing ϵ in terms of Δu and l by means of (31.1).

Thus the internal scale of the turbulence is inversely proportional to $R^{\frac{3}{4}}$. For the corresponding velocity we have

$$v_{\lambda_0} \sim \Delta u/R^{\frac{1}{4}}; \tag{32.7}$$

this also decreases when R increases. Finally, the order of magnitude of the frequencies corresponding to eddies of this size is $\omega_0 \sim u/\lambda_0$ or

$$\omega_0 \sim uR^{\frac{3}{4}}/l. \tag{32.8}$$

This gives the order of magnitude of the upper end of the frequency spectrum of the turbulence; the lower end is at frequencies of the order of u/l. Thus the frequency range increases with Reynolds number as $R^{\frac{3}{4}}$.

---

† The inequality $v_\tau' \ll v_\tau$ has in essence been assumed in the derivation of (32.2).

Similar arguments enable us to determine the order of magnitude of the number of degrees of freedom of a turbulent flow. Let us denote by n the number of degrees of freedom per unit volume of the fluid; n has the dimensions $1/cm^3$. This number can depend only on ρ, ϵ and also the viscosity ν, since the latter determines the lower limit of the sizes of the turbulent eddies. From these three quantities we can form only one having the dimensions $1/cm^3$, namely $(\epsilon/\nu^3)^{\frac{3}{4}}$; this is just $1/\lambda_0^3$, a result which might have been expected. Thus we have

$$n \sim 1/\lambda_0^3 \sim R^{9/4}/l^3. \tag{32.9}$$

The total number N of degrees of freedom is obtained by multiplying n by the volume of the region of turbulent flow, which is of the order of l^3:†

$$N \sim R^{9/4}. \tag{32.10}$$

Finally, let us consider the properties of the flow in regions whose dimension λ is small compared with λ_0. In such regions the flow is regular and its velocity varies smoothly. Hence we can expand v_λ in a series of powers of λ and, retaining only the first term, obtain $v_\lambda = \text{constant} \times \lambda$. The order of magnitude of the constant is v_{λ_0}/λ_0, since for $\lambda \sim \lambda_0$ we must have $v_\lambda \sim v_{\lambda_0}$. Substituting (32.6) and (32.7), we find

$$v_\lambda \sim \Delta u \cdot R^{\frac{1}{4}}\lambda/l. \tag{32.11}$$

This formula may also be obtained directly by equating two expressions for the energy dissipation ϵ: the expression $(\Delta u)^3/l$ (31.1), which determines ϵ in terms of quantities characterising the large eddies, and the expression $\nu(v_\lambda/\lambda)^2$, which determines ϵ in terms of the velocity gradient ($\sim v_\lambda/\lambda$) for the eddies in which the energy dissipation actually occurs.

PROBLEM

Two fluid particles are at a small distance λ_1 ($\gg \lambda_0$) apart. Determine the order of magnitude of the time τ required for the particles to move apart to a distance λ_2 ($\lambda_1 \ll \lambda_2 \ll l$).

SOLUTION. If $\lambda \gg \lambda_0$, we have from dimensional considerations $d\lambda/dt \sim (\epsilon\lambda)^{\frac{1}{3}}$. Integrating this and using the fact that $\lambda_2 \gg \lambda_1$, we find $\tau \sim (\lambda_2^2/\epsilon)^{\frac{1}{3}}$.

§33. The velocity correlation

Formula (32.1) determines qualitatively the *correlation of velocities* in local turbulence, i.e. the relation between the velocities at two neighbouring points. Let us now introduce quantities which will serve to characterise this

---

† Formulae (32.6)–(32.10) determine how the corresponding quantities vary with the Reynolds number. Quantitatively, however, it must be borne in mind that a considerable numerical factor may actually appear in all these formulae. The number of degrees of freedom, for example, must be of the order of unity not for $R \sim 1$, but for $R \sim R_{cr}$. Hence we must write the ratio R/R_{cr} in place of R in (32.10):

$$N \sim (R/R_{cr})^{9/4}.$$

correlation quantitatively.† These may be, for instance, the components of the tensor

$$B_{ik} = \overline{(v_{2i} - v_{1i})(v_{2k} - v_{1k})},\qquad(33.1)$$

where $\mathbf{v}_2$ and $\mathbf{v}_1$ are the fluid velocities at two neighbouring points, and the bar denotes an average with respect to time.‡ The radius vector from point 1 to point 2 will be denoted by $\mathbf{r}$; we suppose its magnitude r small compared with l (but not necessarily large compared with the internal scale of turbulence λ_0).

Since local turbulence is isotropic, the tensor B_{ik} cannot depend on any direction in space. The only vector that can appear in the expression for B_{ik} is the radius vector $\mathbf{r}$. In other words, B_{ik} can contain, apart from the absolute magnitude r of $\mathbf{r}$, only the unit tensor δ_{ik} and the unit vector $\mathbf{n}$ in the direction of $\mathbf{r}$. The most general form of such a tensor of rank two is

$$B_{ik} = A(r)\delta_{ik} + B(r)n_i n_k.\qquad(33.2)$$

We take the co-ordinate axes so that one of them is in the direction of $\mathbf{n}$, denoting the velocity component along this axis by v_r and the component perpendicular to $\mathbf{n}$ by v_t. The component B_{rr} is then the mean square relative velocity of two neighbouring fluid particles along the line joining them. Similarly, B_{tt} is the mean square transverse velocity of one particle relative to the other, while B_{rt} is the mean value of the product of these two velocity components. Since $n_r = 1$, $n_t = 0$, we have from (33.2)

$$B_{rr} = A + B,\qquad B_{tt} = A,\qquad B_{rt} = 0\qquad(33.3)$$

Let us now derive a relation between B_{rr} and B_{tt}. To do so, we first notice that the velocity variation over small distances is mainly due to the small eddies. The properties of the local turbulence do not depend on the large eddies that are superposed on it. Hence, to calculate the tensor B_{ik}, it suffices to take the particular case of completely isotropic and homogeneous turbulent flow, in which the mean fluid velocity is zero.†† Expanding the parentheses in (33.1), we have

$$B_{ik} = \overline{v_{1i}v_{1k}} + \overline{v_{2i}v_{2k}} - \overline{v_{1i}v_{2k}} - \overline{v_{1k}v_{2i}}.$$

---

† The results given in this section are due to T. von Kármán and L. Howarth (1938) and to A. N. Kolmogorov (1941). Similar relations for the temperature fluctuations in a non-uniformly heated turbulent flow are given later (see §54, Problems 3 and 4).

‡ If there were no correlation between the velocities at the points 1 and 2, the mean values of the products in (33.1) would reduce to products of the mean value of each factor separately, and would therefore be zero.

†† Such a flow can be imagined as that of a fluid subjected to strong agitation and then left to itself. Of course, the flow will certainly decay with time. The averaging in formula (33.1) must then, strictly speaking, be taken not as an averaging over time but as one over all possible positions of the points 1 and 2 (for a given distance r between them) at a given instant.

Since the flow is completely homogeneous and isotropic, we have $\overline{v_{1i}v_{1k}}$ $= \overline{v_{2i}v_{2k}}$, and $\overline{v_{1i}v_{2k}} = \overline{v_{1k}v_{2i}}$. Thus

$$B_{ik} = 2\overline{v_{1i}v_{1k}} - 2\overline{v_{1i}v_{2k}}. \tag{33.4}$$

We differentiate this expression with respect to the co-ordinates of point 2:

$$\partial B_{ik}/\partial x_{2k} = -2\overline{v_{1i}\partial v_{2k}/\partial x_{2k}}.$$

By the equation of continuity, however, $\partial v_{2k}/\partial x_{2k} = 0$, so that $\partial B_{ik}/\partial x_{2k} = 0$. Since B_{ik} is a function only of the components $x_i = x_{2i} - x_{1i}$ of the vector **r**, differentiation with respect to x_{2k} is equivalent to differentiation with respect to x_k. Substituting (33.2), we have after a simple calculation $A' + B' + 2B/r = 0$, the prime denoting differentiation with respect to r. Substituting (33.3), we can write this as $B'_{rr} + 2(B_{rr} - B_{tt})/r = 0$, whence we have finally the general relation between B_{rr} and B_{tt}:

$$2rB_{tt} = d(r^2 B_{rr})/dr. \tag{33.5}$$

At distances r large compared with λ_0, the velocity difference is proportional to $r^{\frac{1}{3}}$, according to (32.1). The components of the tensor B_{ik} for such r are therefore proportional to $r^{\frac{2}{3}}$. Substituting in (33.5) $B_{rr} = $ constant $\times r^{\frac{2}{3}}$, $B_{tt} = $ constant $\times r^{\frac{2}{3}}$, we obtain the simple relation

$$B_{tt} = \tfrac{4}{3}B_{rr}. \tag{33.6}$$

For distances r small compared with λ_0, the velocity difference is proportional to r, and therefore B_{rr} and B_{tt} are proportional to r^2. Formula (33.5) then gives the relation

$$B_{tt} = 2B_{rr}. \tag{33.7}$$

At these distances ($r \ll \lambda_0$), B_{tt} and B_{rr} can also be separately expressed in terms of the mean energy dissipation ϵ. We write $B_{rr} = ar^2$, where a is constant, and combine (33.2), (33.3), (33.4), obtaining

$$\overline{v_{1i}v_{2k}} = \overline{v_{1i}v_{1k}} - ar^2\delta_{ik} + \tfrac{1}{2}ar^2 n_i n_k.$$

Differentiating this relation, we find

$$\overline{\frac{\partial v_{1i}}{\partial x_{1l}}\frac{\partial v_{2i}}{\partial x_{2l}}} = 15a, \qquad \overline{\frac{\partial v_{1i}}{\partial x_{1l}}\frac{\partial v_{2l}}{\partial x_{2i}}} = 0.$$

Since this holds for arbitrarily small r, we can put $x_{1i} = x_{2i}$, whence

$$\overline{\left(\frac{\partial v_i}{\partial x_l}\right)^2} = 15a, \qquad \overline{\frac{\partial v_i}{\partial x_l}\frac{\partial v_l}{\partial x_i}} = 0.$$

According to the general formula (16.3), however, we have for the mean energy dissipation

$$\epsilon = \tfrac{1}{2}\nu\left(\frac{\partial v_i}{\partial x_l} + \frac{\partial v_l}{\partial x_i}\right)^2 = \nu\left[\left(\frac{\partial v_i}{\partial x_l}\right)^2 + \overline{\frac{\partial v_i}{\partial x_l}\frac{\partial v_l}{\partial x_i}}\right] = 15a\nu,$$

whence $a = \epsilon/15\nu$. We therefore obtain the following relations giving B_{rr} and B_{tt} in terms of the mean energy dissipation:†

$$B_t = \tfrac{2}{15}\epsilon r^2/\nu, \qquad B_{rr} = \tfrac{1}{15}\epsilon r^2/\nu. \tag{33.8}$$

We may also discuss the *triple correlation*

$$B_{ikl} = \overline{(v_{2i}-v_{1i})(v_{2k}-v_{1k})(v_{2l}-v_{1l})}. \tag{33.9}$$

We shall again suppose that the flow is completely homogeneous and isotropic. Let us first consider the auxiliary tensor $\overline{v_{1i}v_{1k}v_{2l}}$. This tensor is symmetrical in the suffixes i and k, and by virtue of the isotropy it must, like B_{ik}, be expressible in terms of n_i and δ_{ik}. The most general form of such a tensor is

$$\overline{v_{1i}v_{1k}v_{2l}} = C(r)\delta_{ik}n_l + D(r)(\delta_{il}n_k + \delta_{kl}n_i) + F(r)n_in_kn_l. \tag{33.10}$$

Differentiating with respect to x_{2l}, we have by the equation of continuity

$$\frac{\partial}{\partial x_{2l}}\overline{(v_{1i}v_{1k}v_{2l})} = \overline{v_{1i}v_{1k}\frac{\partial v_{2l}}{\partial x_{2l}}} = 0.$$

Substituting the expression for $\overline{v_{1i}v_{1k}v_{2l}}$, we have after a simple calculation (here omitted) two equations:

$$d[r^2(3C+2D+F)]/dr = 0,$$

$$C' + 2(C+D)/r = 0.$$

Integration of the former gives $3C+2D+F = \text{constant}/r^2$. For $r = 0$ the functions C, D and F must remain finite. We must therefore put the constant equal to zero, so that $3C+2D+F = 0$. From the two equations thus obtained we find

$$D = -(C+\tfrac{1}{2}rC'), \qquad F = rC'-C. \tag{33.11}$$

We now expand the parentheses in (33.9). It is easy to see that, by virtue of

---

† It might be thought that a possibility exists in principle of obtaining a universal formula, applicable to any turbulent flow, which should give B_{rr} and B_{tt} for all distances r that are small compared with l. In fact, however, there can be no such formula, as we see from the following argument. The instantaneous value of $(v_{2i} - v_{1i})(v_{2k}-v_{1k})$ might in principle be expressed as a universal function of the energy dissipation ϵ at the instant considered. When we average these expressions, however, an important part will be played by the law of variation of ϵ over times of the order of the periods of the large eddies (of size $\sim l$), and this law is different for different flows. The result of the averaging therefore cannot be universal.

the isotropy of the flow, the mean values $\overline{v_{1i}v_{1k}v_{1l}}$ and $\overline{v_{2i}v_{2k}v_{2l}}$ are zero. For all three velocities in these products are taken at the same point; the only tensor in terms of which the tensor $\overline{v_i v_k v_l}$ could be expressed is therefore δ_{ik}. It is, however, impossible to construct a symmetrical tensor of rank three from unit tensors. Such mean values as $\overline{v_{1i}v_{1k}v_{2l}}$ and $\overline{v_{2i}v_{2k}v_{1l}}$, on the other hand, are equal in magnitude and opposite in sign, since the vector n_i in (33.10) changes sign when points 1 and 2 are interchanged. The result is

$$B_{ikl} = 2(\overline{v_{1i}v_{1k}v_{2l}} + \overline{v_{1i}v_{2k}v_{1l}} + \overline{v_{2i}v_{1k}v_{1l}}).$$

Substituting (33.10) and (33.11), we have the expression

$$B_{ikl} = 2(rC' + C)(\delta_{ik}n_l + \delta_{il}n_k + \delta_{kl}n_i) + 6(rC' - C)n_i n_k n_l. \qquad (33.12)$$

Again taking one of the co-ordinate axes parallel to **n**, we obtain the components of the tensor B_{ikl}: $B_{rrr} = -12C$, $B_{rtt} = -2(C + rC')$, $B_{rrt} = B_{ttt} = 0$. Hence we see that the relation

$$6B_{rtt} = \mathrm{d}(rB_{rrr})/\mathrm{d}r \qquad (33.13)$$

holds between the non-zero components B_{rtt} and B_{rrr}.

Finally, it is also possible to find a relation between the components of the tensors B_{ik} and B_{ikl}. To do so, we calculate the derivative $\partial(\overline{v_{1i}v_{2k}})/\partial t$, recalling that a completely homogeneous and isotropic flow necessarily decays with time. Expressing the derivatives $\partial v_{1i}/\partial t$ and $\partial v_{2k}/\partial t$ by means of the Navier–Stokes equation, we obtain

$$\frac{\partial}{\partial t}(\overline{v_{1i}v_{2k}}) = -\frac{\partial}{\partial x_{1l}}(\overline{v_{1i}v_{1l}v_{2k}}) - \frac{\partial}{\partial x_{2l}}(\overline{v_{1i}v_{2k}v_{2l}}) - \frac{\partial}{\partial x_{1i}}\left(\overline{\frac{p_1 v_{2k}}{\rho}}\right) -$$

$$- \frac{\partial}{\partial x_{2k}}\left(\overline{\frac{p_2 v_{1i}}{\rho}}\right) + \nu \triangle_1(\overline{v_{1i}v_{2k}}) + \nu \triangle_2(\overline{v_{1i}v_{2k}}).$$

In using the properties of homogeneity and isotropy, it must be borne in mind that the sign of **r** changes when the points 1 and 2 are interchanged, and therefore the sign of the (first) space derivatives must be changed. The first two terms are therefore equal, and so are the last two terms. The third and fourth terms are zero. For, by virtue of the isotropy, the mean value $\overline{p_1 v_{2k}}$ must be of the form $f(r)n_k$. The divergence $\partial(\overline{p_1 v_{2k}})/\partial x_{2k} = \overline{p_1 \partial v_{2k}/\partial x_{2k}}$ is zero. But the only centrally symmetric vector whose divergence is everywhere zero is a constant times $(1/r^2)n_k$. Such a vector would become infinite for $r = 0$, which is impossible. The constant must therefore be zero. Thus

$$\frac{\partial}{\partial t}(\overline{v_{1i}v_{2k}}) = -2\frac{\partial}{\partial x_{1l}}(\overline{v_{1i}v_{1l}v_{2k}}) + 2\nu \triangle_1 \overline{v_{1i}v_{2k}}. \qquad (33.14)$$

Here we must substitute, in accordance with the formulae derived above,

$$\overline{v_{1i}v_{2k}} = \overline{v_{2i}v_{2k}} - \tfrac{1}{2}B_{ik},$$

$$\overline{v_{1i}v_{1l}v_{2k}} = -\tfrac{1}{12}B_{rrr}\delta_{il}n_k + \tfrac{1}{12}(\tfrac{1}{2}rB_{rrr}' + B_{rrr})(\delta_{ik}n_l + \delta_k n_i) -$$

$$\tag{33.15}$$

$$-\tfrac{1}{12}(rB_{rrr}' - B_{rrr})n_i n_k n_l.$$

In the former expression we replace $\overline{v_{2i}v_{2k}}$ by $\tfrac{1}{3}\overline{v^2}\,\delta_{ik}$, using the complete homogeneity and isotropy of the flow:

$$\overline{v_{1i}v_{2k}} = \tfrac{1}{3}\overline{v^2}\,\delta_{ik} - \tfrac{1}{2}B_{ik}. \tag{33.16}$$

The time derivative of the kinetic energy per unit mass $\tfrac{1}{2}\overline{v^2}$ is just the energy dissipation $-\epsilon$; hence $\partial(\tfrac{1}{3}\overline{v^2})/\partial t = -\tfrac{2}{3}\epsilon$. A simple, though lengthy, calculation gives the equation

$$-\tfrac{2}{3}\epsilon - \frac{1}{2}\frac{\partial B_{rr}}{\partial t} = \frac{1}{6r^4}\frac{\partial (r^4 B_{rrr})}{\partial r} - \frac{\nu}{r^4}\frac{\partial}{\partial r}\left(r^4 \frac{\partial B_{rr}}{\partial r}\right). \tag{33.17}$$

Since r is supposed small, we can with sufficient accuracy put $r = 0$ on the left-hand side, i.e. neglect $\partial B_{rr}/\partial t$ in comparison with ϵ. Multiplying the resulting equation by r^4, integrating over r, and using the fact that the correlation functions vanish for $r = 0$, we obtain the following relation between B_{rr} and B_{rrr}:

$$B_{rrr} = -\tfrac{4}{5}\epsilon r + 6\nu dB_{rr}/dr. \tag{33.18}$$

The relation (33.18), like (33.13), holds for r either greater or less than λ_0. For $r \gg \lambda_0$, the viscosity term is small, and we have simply

$$B_{rrr} = -\tfrac{4}{5}\epsilon r. \tag{33.19}$$

If $r \ll \lambda_0$, we can substitute the expression (33.8) for B_{rr} in (33.18), obtaining $B_{rrr} = 0$; this is because B_{rrr} in this case must be of the third order in r, and so the first-order terms must cancel.†

§34. The turbulent region and the phenomenon of separation

Turbulent flow is in general rotational. However, the distribution of the vorticity $\boldsymbol{\omega}(\equiv \mathbf{curl\ v})$ in the fluid has certain peculiarities in turbulent flow (for very large R): in "steady" turbulent flow past bodies, the whole volume of the fluid can usually be divided into two separate regions. In one of these the flow is rotational, while in the other the vorticity is zero, and we have

---

† The ratio $|B_{rrr}/B_{rr}^{\frac{3}{2}}|$ must have constant values in the ranges $l \gg r \gg \lambda_0$ and $r \ll \lambda_0$. The experimental results show that in fact this quantity is approximately constant for all r, being about $0\cdot 4$.

potential flow. Thus the vorticity is non-zero only in a part of the fluid (though not in general only in a finite part).

That such a limited region of rotational flow can exist is a consequence of the fact that turbulent flow may be regarded as the motion of an ideal fluid, satisfying Euler's equations.† We have seen (§8) that, for the motion of an ideal fluid, the law of conservation of circulation holds. In particular, if at any point on a streamline $\omega = 0$, then the same is true at every point on that streamline. Conversely, if at any point on a streamline $\omega \neq 0$, then ω does not vanish anywhere on the streamline. Hence it is clear that the existence of limited regions of rotational and irrotational flow is compatible with the equations of motion if the region of rotational flow is such that the streamlines within it do not penetrate into the region outside it. Such a distribution of ω will be stable, and the vorticity will remain zero beyond the surface of separation.

One of the properties of the region of rotational turbulent flow is that the exchange of fluid between this region and the surrounding space can occur in only one direction. The fluid can enter this region from the region of potential flow, but can never leave it.

We should emphasise that the arguments given here cannot, of course, be regarded as affording a rigorous proof of the statements made. However, the existence of limited regions of rotational turbulent flow seems to be confirmed by experiment.

The flow is turbulent both in the rotational and in the irrotational region. The nature of the turbulence, however, is totally different in the two regions. To elucidate the reason for this difference, we may point out the following general property of potential flow, which obeys Laplace's equation $\triangle \phi = 0$. Let us suppose that the flow is periodic in the xy-plane, so that ϕ involves x and y through a factor of the form $e^{ik_1 x + ik_2 y}$. Then

$$\partial^2\phi/\partial x^2 + \partial^2\phi/\partial y^2 = -(k_1^2 + k_2^2)\phi = -k^2\phi,$$

and, since the sum of the second derivatives must be zero, the second derivative of ϕ with respect to z must equal ϕ multiplied by a positive coefficient: $\partial^2\phi/\partial z^2 = k^2\phi$. The dependence of ϕ on z is then given by a damping factor of the form e^{-kz} for $z > 0$ (the unlimited increase given by e^{kz} is clearly impossible). Thus, if the potential flow is periodic in some plane, it must be damped in the direction perpendicular to that plane. Moreover, the greater k_1 and k_2 (i.e. the smaller the period of the flow in the xy-plane), the more rapidly the flow is damped along the z-axis. All these arguments remain qualitatively valid in cases where the motion is not strictly periodic, but has only some periodic quality.

From this the following result is immediately obtained. Outside the region of rotational flow, the turbulent eddies must be damped, and must be so

---

† The applicability of these equations to turbulent flow ends at distances of the order of λ_0. The sharp boundary between rotational and irrotational flow is therefore defined only to within such distances.

the more rapidly, the smaller their size. In other words, the small eddies do not penetrate very far into the region of potential flow. Consequently, only the largest eddies are important in this region; they are damped at distances of the order of the (transverse) dimension of the rotational region, which is just the external scale of turbulence in this case. At distances greater than this dimension there is practically no turbulence, and the flow may be regarded as laminar.

We have seen that the energy dissipation in turbulent flow occurs in the smallest eddies; the large eddies do not involve appreciable dissipation, which is why Euler's equation is applicable to them. From what has been said above, we reach the important result that the energy dissipation occurs mainly in the region of rotational turbulent flow, and hardly at all outside that region.

Bearing in mind all these properties of the rotational and irrotational turbulent flow, we shall henceforward, for brevity, call the region of rotational turbulent flow simply the *region of turbulent flow* or the *turbulent region*. In the following sections we shall discuss the form of this region in various cases.

The turbulent region must be bounded in some direction by part of the surface of the body past which the flow takes place. The line bounding this part of the surface is called the *line of separation*. From it begins the surface of separation between the turbulent fluid and the remainder. The formation of a turbulent region in flow past a body is called the *phenomenon of separation*.

The form of the turbulent region is determined by the properties of the flow in the main body of the fluid (i.e. not in the immediate neighbourhood of the surface). A complete theory of turbulence (which does not yet exist) would have to make it possible, in principle, to determine the form of this region by using the equations of motion for an ideal fluid, given the position of the line of separation on the surface of the body. The actual position of the line of separation, however, is determined by the properties of the flow in the immediate neighbourhood of the surface (known as the *boundary layer*), where the viscosity plays a vital part (see §40).

§35. The turbulent jet

The form of the turbulent region, and some other basic properties of it, can be established in certain cases by simple similarity arguments. These cases include, among others, various kinds of free turbulent jet in a space filled with fluid (L. PRANDTL, 1925).

As a first example, let us consider the turbulent region formed when a flow is "separated" at an angle formed by two infinite intersecting planes (shown in cross-section in Fig. 16). For laminar flow (Fig. 3, §10), the flow along one side of the angle (*AO*, say) would turn smoothly and flow along the other side away from the angle (*OB*). In turbulent flow, the pattern is totally different.

The flow along one side of the angle now does not turn on reaching the

vertex, but continues in its former direction. A flow appears along the
other side in the direction BO. The two flows "mix" in the turbulent
region;† the boundaries of this region are shown, dashed, in cross-section
in Fig. 16. The origin of this region can be seen as follows. Let us imagine
a flow in which a uniform stream along AO continues in the same direction,
occupying the whole space above the plane AO and its continuation into the
fluid to the right, while the fluid below this plane is at rest. In other words,
we have a surface of separation (the plane AO produced) between fluid moving
with constant velocity and stationary fluid. Such a surface of discontinuity,
however, is unstable, and cannot exist in practice (see §30). This instability
leads to mixing and the formation of a turbulent region. The flow along
BO arises because fluid must enter the turbulent region from below.

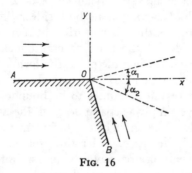

FIG. 16

Let us determine the form of the turbulent region. We take the x-axis
in the direction shown in Fig. 16, the origin being at O. We denote by
Y_1 and Y_2 the distances from the xz-plane to the upper and lower boundaries
of the turbulent region, and require to determine Y_1 and Y_2 as functions of x.
This can easily be done from similarity considerations. Since the planes
are infinite in all directions, there are no constant parameters at our disposal
having the dimensions of length. Hence it follows that Y_1, Y_2 can only be
directly proportional to the distance x:

$$Y_1 = x \tan \alpha_1, \qquad Y_2 = x \tan \alpha_2. \tag{35.1}$$

The proportionality coefficients are simply numerical constants; we write
them as $\tan \alpha_1$, $\tan \alpha_2$, so that α_1 and α_2 are the angles between the two
boundaries of the turbulent region and the x-axis. Thus the turbulent region
is bounded by two planes intersecting along the vertex of the angle.

The values of α_1, α_2 depend only on the size of the angle, and not, for
example, on the velocity of the main stream. They cannot be calculated

---

† We recall that, outside the turbulent region, there is irrotational flow which gradually becomes
laminar as we move away from the boundaries of this region.

theoretically; the experimental results for flow round a right angle are $\alpha_1 = 5°$, $\alpha_2 = 10°$.†

The velocities of the flows along the two sides of the angle are not the same; their ratio is a definite number, again depending only on the size of the angle. When the angle is not close to π, one of the velocities is considerably the greater, namely that of the main stream, which is in the same direction (AO) as the turbulent region. For example, in flow round a right angle, the velocity along the plane AO is thirty times that along BO.

We may also mention that the difference between the fluid pressures on the two sides of the turbulent region is very small. For example, in flow round a right angle it is found that $p_1 - p_2 = 0 \cdot 003 \rho U_1^2$, where U_1 is the velocity of the main stream (along AO), p_1 the pressure in that stream, and p_2 the pressure in the stream along BO.

In the limiting case of flow round an angle of 2π, we have simply the edge of a plate with fluid moving along both sides. The angle $\alpha_1 + \alpha_2$ of the turbulent region is zero, i.e. there is no turbulent region; the velocities of the flows along the two sides of the plate become equal. As the angle AOB increases, a point is reached when the plane BO forms the lower boundary of the turbulent region; the angle AOB is by then obtuse. As the angle increases further, the turbulent region continues to be bounded by the plane BO on one side. Here we have simply a separation, with the line of separation along the vertex of the angle. The angle of the turbulent region remains finite.

As a second example, let us consider the problem of a turbulent jet of fluid issuing from the end of a narrow tube into an infinite space filled with the same fluid. The problem of laminar flow in such a "submerged jet" has been solved in §23. At distances (the only ones we shall consider) large compared with the dimensions of the mouth of the tube, the jet is axially symmetrical, whatever the actual shape of the opening.

Let us determine the form of the turbulent region in the jet. We take the axis of the jet as the x-axis, and denote by R the radius of the turbulent region; we require to determine R as a function of x (which is measured from the end of the tube). As in the previous example, this function is easily determined directly from similarity considerations. At distances large compared with the dimensions of the mouth of the tube, the actual shape and size of the opening cannot affect the form of the jet. Hence we have at our disposal no characteristic parameters of the dimensions of length. It therefore follows as before that R must be proportional to x:

$$R = x \tan \alpha, \tag{35.2}$$

where the numerical constant $\tan \alpha$ is the same for all jets. Thus the turbulent

---

† Here, and elsewhere, we speak of experimental data on the velocity distribution in a transverse cross-section of the turbulent jet, reduced by means of calculations (W. TOLLMIEN 1926) based on the *mixing-length theory* (see the final note to the present section). This theory contains an arbitrary constant, whose value is chosen so as to obtain the best possible agreement with experiment.

region is a cone; the experimental value of the angle 2α is 25 to 30 degrees (Fig. 17).†

The (time average) velocity distribution in a cross-section of the jet has the following properties. The flow is principally along the jet. The longitudinal velocity component falls off rapidly away from the axis of the jet; it becomes $\frac{1}{2}u_0$ (u_0 being the velocity on the axis) at a distance of only $0.35R$ from the axis, and at the boundary of the turbulent region it is of the order of $0.01\,u_0$. The transverse velocity component is approximately uniform in order of magnitude over the cross-section of the turbulent region, and at the boundary of this region it is about $-0.025\,u_0$, being there directed into the jet. This transverse component causes a flow into the turbulent region. The velocity distribution outside the turbulent region (for a given angle α) can be determined theoretically (see Problem 1).

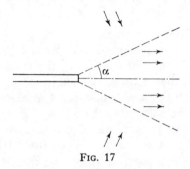

Fig. 17

The velocity in the jet also falls off as we move away from the mouth of the tube. The law of this decrease is easily found. To do so, we use the following method. The total flux of momentum through a spherical surface centred at the tube mouth must be independent of the radius of the surface. The momentum flux density in the jet is of the order of ρu^2, where u is of the order of some mean velocity in the jet; this is the only quantity of the right dimensions that can be formed from the fluid density ρ, the velocity u, and the distance x. The area of the part of the jet cross-section where u is appreciably different from zero is of the order of R^2. Hence the total momentum flux is of the order of $\rho u^2 R^2$. Equating this to a constant and putting $R = \text{constant} \times x$, we obtain

$$u \sim \text{constant}/x, \tag{35.3}$$

i.e. the velocity diminishes inversely as the distance from the mouth of the tube.

---

† Some dependence of the constant α on the initial conditions (velocity profile) in the tube mouth is observed experimentally. It is reasonable to suppose that this dependence is due to the effect of the finite dimensions of the opening, an effect which would disappear at greater distances.

The amount Q of fluid which passes per unit time through a cross-section of the turbulent region of the jet is of the order of the product of its area ($\sim R^2$) and the mean velocity u. Substituting, we find†

$$Q = Bx. \tag{35.4}$$

Thus the discharge through a cross-section of the turbulent region increases with x, i.e. some fluid is, as it were, entrained in the turbulent region.‡ The constant which appears in (35.4) may be determined as follows. At distances of the order of the dimensions of the tube mouth, Q must become the amount Q_0 of fluid emitted from the tube per unit time, which is fixed for any particular jet. Hence we see that $B \sim Q_0/a$, where a gives the transverse dimension of the tube mouth (e.g. the radius, if the opening is circular). Thus we can write

$$B = cQ_0/a, \tag{35.5}$$

where c is a numerical constant which depends only on the form of the opening. If the latter is circular, c is found by experiment to be about $1 \cdot 5$.

The flow in any section of the length of the jet is characterised by the Reynolds number for that section, defined as uR/ν. By virtue of (35.2) and (35.3), however, the product uR is constant along the jet, so that the Reynolds number is the same for all such sections. It can be taken, for instance, as $B/\rho\nu$. The constant B which appears here is the only parameter which determines the flow in the jet. When the "strength" Q_0 of the jet increases (the value of a remaining constant), the Reynolds number $B/\rho\nu$ eventually reaches a critical value, after which the flow simultaneously becomes turbulent along the whole length of the jet.††

---

† If two variable quantities which vary within wide limits are always of the same order of magnitude, then they must be proportional. Hence, in this case (and in similar cases), we can write precisely $Q = \text{constant} \times x$ in place of $Q \sim \text{constant} \times x$.

‡ The total mass flux through any infinite plane across the jet is infinite, i.e. a jet issuing into an infinite space carries with it an infinite amount of fluid.

†† In order to make more detailed calculations for various kinds of turbulent flow, it is customary to employ certain "semi-empirical" theories, based on assumptions concerning the dependence of the turbulent viscosity coefficient on the gradient of the mean velocity. For example, in Prandtl's theory it is assumed that (for plane flow)

$$\nu_{\text{turb}} = l^2 |\partial u_x/\partial y|,$$

where the dependence of l (called the *mixing length*) on the co-ordinates is chosen in accordance with the results of similarity arguments; for instance, in free turbulent jets we put $l = cx$, c being an empirical constant. Such theories usually give good agreement with experiment, and are therefore useful for interpolatory calculations. However, it is not possible to give universal values to the empirical constants which characterise each theory; for example, the value of the ratio of the mixing length l to the transverse dimension of the turbulent region has to be chosen differently in various particular cases. It should also be mentioned that good agreement with experimental results can be obtained with various expressions for the turbulent viscosity.

A more detailed account of these theories is given by L. G. LOĬTSYANSKIĬ, *Aerodynamics of Boundary Layers* (*Aérodinamika pogranichnogo sloya*), Moscow 1941; G. N. ABRAMOVICH, *Free Turbulent Jets of Liquids and Gases* (*Turbulentnye svobodnye strui zhidkostei i gazov*), Moscow 1948; H. SCHLICHTING, *Boundary Layer Theory*, Pergamon Press, London 1955.

PROBLEMS

PROBLEM 1. Determine the mean flow in the jet outside the turbulent region.

SOLUTION. We take spherical co-ordinates r, θ, ϕ, with the polar axis along the axis of the jet, and the origin at its point of emergence. Because the jet is axially symmetrical, the component u_ϕ of the mean velocity is zero, while u_θ and u_r are functions only of r and θ. The same arguments as were used in the problem of the laminar jet (§23) show that u_θ and u_r must be of the forms $u_\theta = f(\theta)/r$, $u_r = F(\theta)/r$. Outside the turbulent region we have potential flow, i.e. **curl u** $= 0$, so that $\partial u_r/\partial\theta - \partial(r u_\theta)/\partial r = 0$. But $r u_\theta$ is independent of r, so that $\partial u_r/\partial\theta = (1/r)\, dF/d\theta = 0$, whence $F = $ constant $= -b$, say, or

$$u_r = -b/r. \tag{1}$$

From the equation of continuity,

$$\frac{1}{r^2}\frac{\partial}{\partial r}(r^2 u_r) + \frac{1}{r\sin\theta}\frac{\partial}{\partial\theta}(u_\theta \sin\theta) = 0,$$

we then obtain

$$f = \frac{\text{constant} - b\cos\theta}{\sin\theta}.$$

The constant of integration must be $-b$ if the velocity is not infinite for $\theta = \pi$ (it does not matter that f is infinite for $\theta = 0$, since the solution in question refers only to the space outside the turbulent region, whereas $\theta = 0$ lies inside that region). Thus

$$u_\theta = -\frac{b(1+\cos\theta)}{r\sin\theta} = -\frac{b}{r}\cot\tfrac{1}{2}\theta. \tag{2}$$

The component of the velocity in the direction of the jet (u_x) and its absolute magnitude are

$$u_x = \frac{b}{r} = \frac{b\cos\theta}{x}, \qquad u = \frac{b}{r\sin\tfrac{1}{2}\theta}. \tag{3}$$

The constant b can be related to the constant B in (35.4). Let us consider a segment of the cone formed by the turbulent region, bounded by two infinitely close cross-sections of the cone. The mass of fluid entering this segment per unit time is $dQ = -2\pi r\rho\sin\alpha \,.\, u_\theta dr = 2\pi b\rho(1+\cos\alpha)dr$, while from formula (35.4) we have $dQ = B\,dx = B\cos\alpha\,dr$. Comparing the two expressions, we obtain

$$b = \frac{B\cos\alpha}{2\pi\rho(1+\cos\alpha)}. \tag{4}$$

At the boundary of the turbulent region, the velocity **u** is directed into this region, making an angle $\tfrac{1}{2}(\pi-\alpha)$ with the positive direction of the x-axis.

Let us compare the mean velocity $\bar{u}_x$ inside the turbulent region (defined as $\bar{u}_x = Q/\pi\rho R^2 = B/\pi\rho x\tan^2\alpha$) with the velocity $(u_x)_{\text{pot}}$ at the boundary of the region. Taking the first equation (3) with $\theta = \alpha$, we find

$$(u_x)_{\text{pot}}/\bar{u}_x = \tfrac{1}{2}(1-\cos\alpha).$$

For $\alpha = 12°$, this ratio is $0\cdot011$, i.e. the velocity at the boundary of the turbulent region is small compared with the mean velocity inside the region.

PROBLEM 2. Determine the law of variation of size and velocity in a submerged turbulent jet issuing from an infinitely long thin slit.

SOLUTION. By the same reasoning as for the axial jet, we conclude that the turbulent region is bounded by two planes intersecting along the slit, i.e. the half-width of the jet is $Y = x \tan \alpha$. The momentum flux in the jet (per unit length of the slit) is of the order of $\rho u^2 Y$. The dependence of the mean velocity u on x is therefore given by $u = \text{constant}/\sqrt{x}$. The discharge through a cross-section of the turbulent region is $Q \sim \rho u Y$, whence $Q = \text{constant} \times \sqrt{x}$. The experimental data give a value of 25° to 33° for the angle 2α of a plane-parallel jet (cf. the third footnote to this section).

§36. The turbulent wake

For Reynolds numbers considerably above the critical value, in flow past a solid body, a long region of turbulent flow is formed behind the body. This is called the *turbulent wake*. At distances large compared with the dimension of the body, simple arguments enable us to determine the form of this wake and the way in which the fluid velocity decreases there (L. PRANDTL, 1926).

As in the investigation of the laminar wake in §21, we denote by **U** the velocity of the incident stream, and take the direction of **U** as the x-axis. The fluid velocity at any point, averaged over the turbulent fluctuations, is written as **U**+**u**. Denoting by a some mean width of the wake, we shall find a as a function of x. If there is no lift, then at large distances from the body the wake is axially symmetrical and circular in cross-section; in this case, a may be the radius of the wake. If a lift force is present, a direction is selected in the yz-plane, and the wake is not axially symmetrical at any distance from the body.

The longitudinal fluid velocity component in the wake is of the order of U, while the transverse component is of the order of some mean value u of the turbulent velocity. The angle between the streamlines and the x-axis is therefore of the order of u/U. The boundary of the wake is, as we know, the boundary beyond which the streamlines of the rotational turbulent motion cannot pass. Hence it follows that the angle between the boundary of the wake and the x-axis is also of the order of u/U. This means that we can write

$$\mathrm{d}a/\mathrm{d}x \sim u/U. \tag{36.1}$$

Next we use formulae (21.1), (21.2), which determine the forces on the body in terms of integrals of the fluid velocity in the wake (the velocity now being interpreted as its mean value). The region of integration in these integrals is of the order of a^2. Hence an estimate of the integral gives $F \sim \rho U u a^2$, where F is of the order of the drag or the lift. Thus

$$u \sim F/\rho U a^2. \tag{36.2}$$

Substituting in (36.1), we find $\mathrm{d}a/\mathrm{d}x \sim F/\rho U^2 a^2$, from which we have by integration

$$a \sim (Fx/\rho U^2)^{\frac{1}{3}}. \tag{36.3}$$

Thus the width of the wake increases as the cube root of the distance from the body. For the velocity u, we have from (36.2) and (36.3)

$$u \sim (FU/\rho x^2)^{\frac{1}{3}}, \tag{36.4}$$

i.e. the mean fluid velocity in the wake is inversely proportional to $x^{\frac{2}{3}}$.

The flow in any cross-section of the wake is characterised by the Reynolds number $R \sim au/\nu$. Substituting (36.2) and (36.3), we obtain

$$R \sim F/\nu\rho Ua \sim (F^2/\rho^2 Ux\nu^3)^{\frac{1}{3}}.$$

We see that this number is not constant along the wake, unlike what we found for the turbulent jet. At sufficiently large distances from the body, R becomes so small that the flow in the wake is no longer turbulent. Beyond this point we have the laminar wake, whose properties have been investigated in §21.

In §21, Problem 2, formulae have been obtained which describe the flow outside the wake and far from the body. These formulae hold for flow outside the turbulent wake as well as outside the laminar wake.

We may mention here some general properties of the velocity distribution round the body. Both inside and outside the turbulent wake, the velocity (by which we always mean **u**) decreases away from the body. However, the longitudinal velocity u_x falls off more rapidly ($\sim 1/x^2$) outside the wake than inside it. Far from the body, therefore, we may suppose u_x to be zero outside the wake. We may say that u_x falls from some maximum value on the axis of the wake to zero at the boundary of the wake. The transverse components u_y, u_z at the boundary are of the same order of magnitude as they are inside the wake, diminishing rapidly as we move away from the wake at a given distance from the body.

§37. Zhukovskiĭ's theorem

The velocity distribution round a body, described at the end of the last section, does not hold for exceptional cases where the thickness of the wake formed behind the body is very small compared with its width. A wake of this kind is formed in flow past bodies whose thickness (in the y-direction) is small compared with their width (in the z-direction); the length (in the direction of flow, the x-direction) may be of any magnitude. That is, we are considering flow past bodies whose cross-section transverse to the flow is very elongated. These bodies include, in particular, *wings*, i.e. bodies whose width, or *span*, is large in comparison with their other dimensions.

It is clear that, in such a case, there is no reason why the velocity component u_y perpendicular to the plane of the turbulent wake should fall off appreciably at distances of the order of the thickness of the wake. On the contrary, this component will now be of the same order of magnitude inside the wake and at considerable distances from it, of the order of the span. Here, of course, we assume that the lift is not zero, since otherwise the transverse velocity practically vanishes.

Let us consider the vertical lift force F_y resulting from such a flow. According to formula (21.2), it it given by the integral

$$F_y = -\rho U \iint u_y \, dy \, dz, \qquad (37.1)$$

where, on account of the nature of the distribution of u_y, the integration must now be taken over the whole transverse plane. Furthermore, since the thickness of the wake (in the y-direction) is small, while the velocity u_y inside the wake is not large compared with its value outside, we can with sufficient accuracy take the integration over y to be over the region outside the wake, writing

$$\int_{-\infty}^{\infty} u_y \, dy \approx \int_{y_1}^{\infty} u_y \, dy + \int_{-\infty}^{y_1} u_y \, dy,$$

where y_1 and y_2 are the co-ordinates of the boundaries of the wake (Fig. 18).

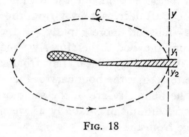

Fɪɢ. 18

Outside the wake, however, we have potential flow, and $u_y = \partial\phi/\partial y$; bearing in mind that $\phi = 0$ at infinity, we therefore obtain

$$\int u_y \, dy = \phi_2 - \phi_1,$$

where ϕ_2 and ϕ_1 are the values of the potential on the two sides of the wake. We may say that $\phi_2 - \phi_1$ is the discontinuity of the potential at the surface of discontinuity which may be substituted for a thin wake. The derivative $u_y = \partial\phi/\partial y$ must remain continuous. A discontinuity in the velocity component normal to the surface of the wake would mean that some quantity of fluid flows into the wake; in the approximation in which the thickness of the wake is neglected, however, this inflow must be zero. Thus we replace the wake by a surface of tangential discontinuity. Next, in the same approximation, the pressure also must be continuous at the wake. Since the variation of the pressure is given in the first approximation, according to Bernoulli's equation, by $\rho U u_x = \rho U \, \partial\phi/\partial x$, it follows that the derivative $\partial\phi/\partial x$ must also be continuous. The derivative $\partial\phi/\partial z$ (the velocity along the wing) is in general discontinuous, however.

Since the derivative $\partial\phi/\partial x$ is continuous, the discontinuity $\phi_2 - \phi_1$ depends only on z, and not on the co-ordinate x along the wake. Thus we have the following formula for the lift:

$$F_y = -\rho U \int (\phi_2 - \phi_1) dz. \tag{37.2}$$

The integration over z may be taken over the width of the wake (of course, $\phi_2 - \phi_1 \equiv 0$ outside the wake).

This formula can be put in a somewhat different form. To do so, we notice that, using well-known properties of an integral of the gradient of a scalar, we can write the difference $\phi_2 - \phi_1$ as a contour integral

$$\oint \mathbf{grad}\,\phi \cdot d\mathbf{l} = \oint (u_y\, dy + u_x\, dx),$$

taken along a contour which starts from the point y_1, encircles the body, and ends at the point y_2, thus passing at every point through the region of potential flow. Since the wake is thin we can, without changing the integral except by quantities of higher order, close this contour by means of the short segment from y_2 to y_1. Denoting by Γ the velocity circulation round the closed contour C enclosing the body (Fig. 18), we have

$$\Gamma = \oint \mathbf{u} \cdot d\mathbf{l} = \phi_2 - \phi_1, \tag{37.3}$$

and for the lift force the formula[†]

$$F_y = -\rho U \int \Gamma\, dz. \tag{37.4}$$

The relation between the lift and the circulation given by this formula constitutes *Zhukovskii's theorem*, first derived by N. E. ZHUKOVSKII in 1906.[‡]

PROBLEMS

PROBLEM 1. Determine the manner of widening of the turbulent wake formed in transverse flow past a cylinder of infinite length.

SOLUTION. The drag f_x per unit length of the cylinder is of the order of $\rho U u Y$. Combining this with the relation (36.1), we find the width Y of the wake to be

$$Y = A\sqrt{(xf_x/\rho U^2)}, \tag{1}$$

where A is a constant. The mean velocity u in the wake falls off in accordance with $u \sim \sqrt{(f_x/\rho x)}$. The Reynolds number $R \sim Yu/\nu \sim f_x/\rho U\nu$ is independent of x, and there is therefore no laminar wake.

---

† The sign of the velocity circulation is always chosen to be that obtained for a counter-clockwise path. The sign in formula (37.3) also depends on the chosen direction of flow. We always suppose that the flow is in the positive direction of the x-axis (from left to right).

‡ Cf. §46 for the application of this theorem to streamlined wings.

We may mention that, according to experimental results, the constant coefficient in (1) is $A = 0.93$ (Y being the half-width of the wake; if Y is taken as the distance at which the velocity u_x falls to half its maximum value (at the centre of the wake), then $A = 0.41$).

PROBLEM 2. Determine the flow outside the wake formed in transverse flow past a body of infinite length.

SOLUTION. Outside the wake we have potential flow; we shall denote the potential by Φ to distinguish it from the angle ϕ in the system of cylindrical co-ordinates which we take, with the z-axis along the length of the body. As in §21, Problem 2, we conclude that we must have

$$\oint \mathbf{u} \cdot \mathbf{df} = \oint \mathbf{grad}\,\Phi \cdot \mathbf{df} = f_x/\rho U,$$

where now the integration is over the surface of a cylinder of large radius and unit length with its axis in the x-direction, and f_x is the drag per unit length of the body. The solution of the two-dimensional Laplace's equation $\triangle \Phi = 0$ that satisfies this condition is $\Phi = (f_x/2\pi\rho U)\log r$. Next, we have for the lift, by formula (37.2), $f_y = \rho U(\Phi_1 - \Phi_2)$. The solution of Laplace's equation that diminishes least rapidly with distance and has a discontinuity on the plane $\phi = 0$ is $\Phi = \text{constant} \times \phi$; since $\phi_2 - \phi_1 = 2\pi$, the constant is $-f_y/2\pi\rho U$. The flow is given by the sum of these two solutions, i.e.

$$\Phi = \frac{1}{2\pi\rho U}(f_x \log r - \phi f_y). \tag{1}$$

The cylindrical components of the velocity $\mathbf{u}$ are

$$u_r = \partial\Phi/\partial r = f_x/2\pi\rho Ur, \qquad u_\phi = (1/r)\partial\Phi/\partial\phi = -f_y/2\pi\rho Ur. \tag{2}$$

The velocity $\mathbf{u}$ is at a constant angle $\tan^{-1}(f_y/f_x)$ to the r-direction.

PROBLEM 3. Determine the manner of bending of the wake behind a body of infinite length when there is a lift force.

SOLUTION. If there is a lift force, the wake (regarded as a surface of discontinuity) is curved in the xy-plane. The function $y = y(x)$ which determines this is given by the equation $dx/(u_x + U) = dy/u_y$. Substituting, by (2) of Problem 2, $u_y \approx -f_y/2\pi\rho Ux$ and neglecting u_x in comparison with U, we obtain

$$dy/dx = -f_y/2\pi\rho U^2 x,$$

whence

$$y = \text{constant} - (f_y/2\pi\rho U^2)\log x.$$

§38. Isotropic turbulence

We have already mentioned in §33 the particular case of turbulent flow that is completely homogeneous and isotropic, the mean velocity being zero throughout the fluid. Such a flow may be imagined as that of a fluid which is vigorously stirred and then left to itself. The motion decays with time, of course.

The further investigation of isotropic turbulence, and in particular the determination of the manner of its decay with time, is based on a conservation law first derived by L. G. LOĬTSYANSKIĬ (1939). This law, which holds only for isotropic turbulence, is a consequence of the general law of conservation of angular momentum, and may be derived as follows.

Let us isolate some fairly large volume in an unbounded fluid, and consider the total fluid angular momentum **M** contained in this volume. **M** has some random value, which is not in general zero. On account of the interaction with the surrounding regions, **M** does not remain strictly constant. However, since the interaction is a surface effect, it is clear that the time T during which **M** varies appreciably must increase with the dimension L of the volume selected. The time T and the dimension L may be arbitrarily large, and in this sense the angular momentum **M** is conserved.

For convenience in what follows, we suppose that the chosen volume of fluid is enclosed in a vessel with fixed solid walls; it is evident that the boundary conditions at the surface of a very large volume cannot have any effect on the volume properties of the flow, in which we are interested.

According to the general definition, the tensor M_{ik}, which is the total angular momentum, is equal to the integral

$$\rho \int (x_i v_k - x_k v_i) \mathrm{d}V$$

taken over the whole volume. We transform this integral as follows:

$$\int x_k v_i \mathrm{d}V = \int \frac{\partial}{\partial x_l}(x_i x_k v_l) \mathrm{d}V - \int x_i x_k \frac{\partial v_l}{\partial x_l} \mathrm{d}V - \int x_i v_k \mathrm{d}V.$$

The first integral on the right-hand side, on being converted into a surface integral, is seen to be zero, since the normal velocity component at the walls bounding the fluid is zero, so that $v_k \, \mathrm{d}f_k = \mathbf{v} \cdot \mathbf{n} \, \mathrm{d}f = 0$. The second integral is zero if the fluid is incompressible (div $\mathbf{v} = 0$). Thus

$$\int x_k v_i \mathrm{d}V = - \int x_i v_k \mathrm{d}V,$$

and we can write

$$M_{ik} = 2\rho \int x_i v_k \mathrm{d}V.$$

The sum of the squared components of M_{ik} is equal to twice the squared absolute magnitude of the angular momentum vector

$$\mathbf{M} = \rho \int \mathbf{r} \times \mathbf{v} \, \mathrm{d}V.$$

We therefore have

$$M^2 = 2\rho^2 [\int x_i v_k \mathrm{d}V]^2.$$

The squared integral can be written as a double integral:

$$M^2 = 2\rho^2 \iint x_i x'_i v_k v'_k \, \mathrm{d}V \, \mathrm{d}V'.$$

Finally, we notice that this expression may be rewritten

$$M^2 = -\rho^2 \iint (x_i - x'_i)^2 v_k v'_k \, \mathrm{d}V \, \mathrm{d}V'; \tag{38.1}$$

the integrals containing the squares x_i^2 and x'^2_i vanish, since

$$\iint x'^2_i v_k v'_k \, \mathrm{d}V \, \mathrm{d}V' = \int x'^2_i v'_k \, \mathrm{d}V' \int v_k \, \mathrm{d}V, \quad \text{and} \quad \int v_k \, \mathrm{d}V = 0$$

because the total linear momentum of an incompressible fluid in a fixed vessel is zero.

The factor $v_k v'_k \equiv \mathbf{v} \cdot \mathbf{v}'$ in the integrand of (38.1 is the scalar product of the velocities at two points having co-ordinates x_k)and x'_k, at a distance $r = \sqrt{[(x_k - x'_k)^2]}$ apart. We average this product over all positions of the points x_k and x'_k (for given r) in the volume concerned: this averaging is the same as the one used in §33 in defining the correlation functions. Since the flow is isotropic, the quantity $\overline{\mathbf{v} \cdot \mathbf{v}'}$ is a function of r only. It falls off rapidly with increasing r, since the velocities of the turbulent flow at two points a great distance apart may be supposed statistically independent: the mean value of their product then reduces to the product of the mean values of the individual velocities, which is zero (the mean velocity being everywhere zero in the flow under consideration).

Effecting this averaging under the integral sign in (38.1), we find

$$M^2 = \rho^2 \int f \, \mathrm{d}V, \quad \text{where} \quad f = -\int \overline{\mathbf{v} \cdot \mathbf{v}'} \, r^2 \, \mathrm{d}V'. \tag{38.2}$$

The integrand in f diminishes rapidly with increasing r, so that the integral converges; this means that, as the dimension L of the region tends to infinity, f tends to a finite limit. Since the flow is homogeneous,[†] the quantity f is constant everywhere in the fluid, and we can write simply $M^2 = \rho^2 f V$. We may point out that the angular momentum is thus found to increase as the square root of the volume of moving fluid, and not proportionally to the volume. This is because the total angular momentum is the sum of a large number of statistically independent components (the angular momenta of various small portions of the fluid) whose mean values are not zero.

Thus we conclude that, for isotropic turbulence, the constancy of M implies the condition

$$\int \overline{\mathbf{v} \cdot \mathbf{v}'} \, r^2 \, \mathrm{d}V' = \text{constant}. \tag{38.3}$$

This is *Loĭtsyanskiĭ's law.*[‡]

[†] Throughout the region, except for a very small part near the surface.

[‡] Doubts have recently been expressed more than once concerning the applicability of the conservation law (38.3), on account of the behaviour of the velocity correlation at very large distances; for example, if this correlation does not decrease sufficiently rapidly, the integral (38.3) may diverge. The whole subject seems to be as yet somewhat unclear.

The integrand in (38.3) is noticeably different from zero in a region whose dimensions are of the order of the scale l of the turbulence (the volume of the region $\sim l^3$), and is there of the order of $v^2 l^2$. Hence we have from (38.3)

$$v^2 l^5 = \text{constant.} \tag{38.4}$$

Using this relation, we can determine the manner of the time decay of isotropic turbulence. To do so, we estimate the time derivative of the kinetic energy of unit volume of the fluid. On the one hand, it may be written as being of the order of $\rho v^2/t$. On the other hand, it must equal the energy dissipated in unit volume per unit time. According to formula (31.1), $\rho\epsilon \sim \rho v^3/l$ (the characteristic velocity here being v). If the two expressions are comparable, we find

$$l \sim vt. \tag{38.5}$$

Substituting (38.5) in (38.4), we see that

$$v = \text{constant}/t^{5/7}. \tag{38.6}$$

Thus the velocity in isotropic turbulence decays with time inversely as $t^{5/7}$. For l we have

$$l = \text{constant} \times t^{2/7}, \tag{38.7}$$

i.e. the external scale of the turbulence increases as $t^{2/7}$ (A. N. Kolmogorov, 1941).

According to formulae (38.6) and (38.7), the Reynolds number $R \sim vl/\nu$ decreases as $t^{-3/7}$, and after a sufficient time it becomes so small that the viscosity begins to be important. The energy dissipation is then determined, on the one hand, by the usual formula (16.3), which gives

$$\epsilon = \tfrac{1}{2}\nu\left(\frac{\partial v_i}{\partial x_k} + \frac{\partial v_k}{\partial x_i}\right)^2 \sim \frac{\nu v^2}{l^2},$$

and, on the other hand, by $\epsilon \sim v^2/t$. Comparing, we obtain

$$l \sim \sqrt{(\nu t)}, \tag{38.8}$$

and then from (38.4) we have

$$v = \text{constant}/t^{5/4}. \tag{38.9}$$

These formulae, which are due to M. D. Millionshchikov (1939), give the manner of decay of isotropic turbulence in the final period, when the effect of the viscosity becomes predominant.

Isotropic turbulent flow can be brought about by passing a stream through a grid having a large number of regularly spaced openings. We denote by $\mathbf{U}$ the velocity of the original flow, taking the x-axis in the direction of $\mathbf{U}$, and the true velocity by $\mathbf{U}+\mathbf{v}$, so that $\mathbf{v}$ is the velocity of the turbulent flow in which we are interested. If we introduce a frame of reference moving with velocity **U**, then relative to this frame the fluid executes a turbulent flow with velocity **v**. As we move away from the grid, the averaged turbulent flow (with velocity $\mathbf{u} = \bar{\mathbf{v}}$) decays faster than the fluctuating flow. This is because the averaged flow has a scale of the order of the dimension a of the grid openings, and these, as we shall see, are small in comparison with the scale of the fluctuating flow. Consequently, at sufficiently large distances x from the grid, the averaged velocity **u** is almost zero, and the turbulent velocity **v** is just the fluctuating velocity. At such distances the turbulence may be regarded as completely isotropic over regions small compared with x (though not necessarily small compared with the external scale of the turbulence). The time decay of the turbulence in the moving frame of reference corresponds to a decay with increasing distance from the grid in the original stationary frame. The manner of this decay is given by the formulae derived above, in which we need only replace t by x/U. Bearing in mind that, at distances from the grid of the order of a (the dimension of the openings), we must have $l \sim a$, we can rewrite formula (38.7) as $l \sim a(x/a)^{2/7}$. For the velocity we have by (38.5), $v \sim lU/x$, whence $v \sim U(a/x)^{5/7}$.

PROBLEM

Using equation (33.17), obtain for isotropic turbulence the quantitative law of decay of the quantities $\overline{v_{r_1} v_{r_2}}$ in the period when viscosity is important (L. G. LOĬTSYANSKIĬ, M. D. MILLIONSHCHIKOV).

SOLUTION. In this case we can neglect the B_{rrr} term in (33.17), as being of a higher order in the (small) velocity. Introducing the quantity

$$b_{rr} \equiv \overline{v_{r_1} v_{r_2}} = \tfrac{1}{3}\overline{v^2} - \tfrac{1}{2}B_{rr}$$

(see (33.16)), we obtain for it the equation

$$\frac{\partial b_{rr}}{\partial t} - \frac{2\nu}{r^4}\frac{\partial}{\partial r}\left(r^4 \frac{\partial b_{rr}}{\partial r}\right) = 0.$$

The solution of this equation that is of interest is

$$b_{rr} = \text{constant} \times e^{-r^2/8\nu t}/t^{5/2};$$

cf. the analogous solution (51.6) of the equation of heat conduction. This gives the asymptotic form of the function b_{rr} for initial conditions such that b_{rr} is any function which decreases sufficiently rapidly with increasing r (just as (51.6) gives the asymptotic law of propagation of heat which at the initial instant is concentrated in a small region of space).

CHAPTER IV

BOUNDARY LAYERS

§39. The laminar boundary layer

WE have several times mentioned the fact that very large Reynolds numbers are equivalent to very small viscosities, and consequently a fluid may be regarded as ideal if R is large. However, this approximation can never be used when the flow in question occurs near solid walls. The boundary conditions for an ideal fluid require only the normal velocity component to vanish; the component tangential to the surface in general remains finite. For a viscous fluid, however, the velocity at a solid wall must vanish entirely.

From this we can conclude that, for large Reynolds numbers, the decrease of the velocity to zero occurs almost exclusively in a thin layer adjoining the wall. This is called the *boundary layer*, and is thus characterised by the presence in it of considerable velocity gradients. The flow in the boundary layer may be either laminar or turbulent. In this section we shall consider the properties of the laminar boundary layer. The boundary of the layer is not, of course, sharp; the transition from the laminar flow in it to the main stream of fluid is continuous.

The rapid decrease of the velocity in the boundary layer is due ultimately to the viscosity, which cannot be neglected even if R is large. Mathematically, this appears in the fact that the velocity gradients in the boundary layer are large, and therefore the viscosity terms in the equations of motion, which contain space derivatives of the velocity, are large even if v is small. The mathematical theory of the boundary layer is due to L. Prandtl.

Let us derive the equations of motion of the fluid in a laminar boundary layer. For simplicity, we consider two-dimensional flow along a plane portion of the surface. This plane is taken as the xz-plane, with the x-axis in the direction of flow. The velocity distribution is independent of z, and the velocity has no z-component.

The exact Navier–Stokes equations and the equation of continuity are then

$$v_x \frac{\partial v_x}{\partial x} + v_y \frac{\partial v_x}{\partial y} = -\frac{1}{\rho}\frac{\partial p}{\partial x} + v\left(\frac{\partial^2 v_x}{\partial x^2} + \frac{\partial^2 v_x}{\partial y^2}\right), \tag{39.1}$$

$$v_x \frac{\partial v_y}{\partial x} + v_y \frac{\partial v_y}{\partial y} = -\frac{1}{\rho}\frac{\partial p}{\partial y} + v\left(\frac{\partial^2 v_y}{\partial x^2} + \frac{\partial^2 v_y}{\partial y^2}\right), \tag{39.2}$$

$$\frac{\partial v_x}{\partial x} + \frac{\partial v_y}{\partial y} = 0. \tag{39.3}$$

The flow is supposed steady, and the time derivatives are therefore omitted.

Since the boundary layer is thin, it is clear that the flow in it takes place mainly parallel to the surface, i.e. the velocity v_y is small compared with v_x (as is seen immediately from the equation of continuity).

The velocity varies rapidly along the y-axis, an appreciable change in it occurring at distances of the order of the thickness δ of the boundary layer. Along the x-axis, on the other hand, the velocity varies slowly, an appreciable change in it occurring only over distances of the order of a length l characteristic of the problem (the dimension of the body, say). Hence the y-derivatives of the velocity are large in comparison with the x-derivatives. It follows that, in equation (39.1), the derivative $\partial^2 v_x/\partial x^2$ may be neglected in comparison with $\partial^2 v_x/\partial y^2$; comparing (39.1) with (39.2), we see that the derivative $\partial p/\partial y$ is small in comparison with $\partial p/\partial x$ (the ratio being of the same order as v_y/v_x). In the approximation considered we can put simply

$$\partial p/\partial y = 0, \qquad (39.4)$$

i.e. suppose that there is no transverse pressure gradient in the boundary layer. In other words the pressure in the boundary layer is equal to the pressure $p(x)$ in the main stream, and is a given function of x for the purpose of solving the boundary-layer problem. In equation (39.1) we can now write, instead of $\partial p/\partial x$, the total derivative $dp(x)/dx$; this derivative can be expressed in terms of the velocity $U(x)$ of the main stream. Since we have potential flow outside the boundary layer, Bernoulli's equation, $p + \frac{1}{2}\rho U^2$ = constant, holds, whence $(1/\rho)dp/dx = -U dU/dx$.

Thus we obtain the equations of motion in the laminar boundary layer in the form

$$v_x \frac{\partial v_x}{\partial x} + v_y \frac{\partial v_x}{\partial y} - \nu \frac{\partial^2 v_x}{\partial y^2} = -\frac{1}{\rho}\frac{dp}{dx}$$

$$= U\frac{dU}{dx}, \qquad (39.5)$$

$$\frac{\partial v_x}{\partial x} + \frac{\partial v_y}{\partial y} = 0.$$

It can easily be shown that these equations, though derived for flow along a plane wall, remain valid in the more general case of any two-dimensional flow (transverse flow past a cylinder of infinite length and arbitrary cross-section). Here x is the distance measured along the circumference of the cross-section from some point on it, and y is the distance from the surface.

Let U_0 be a velocity characteristic of the problem (for example, the velocity of the main stream at infinity). Instead of the co-ordinates x, y and the velocities v_x, v_y, we introduce the dimensionless variables x', y', v'_x, v'_y:

$$x = lx', \qquad y = ly'/\sqrt{R}, \qquad v_x = U_0 v'_x, \qquad v_y = U_0 v'_y/\sqrt{R} \qquad (39.6)$$

(and correspondingly $U = U_0 U'$), where $R = U_0 l/\nu$. Then the equations (39.5) take the form

$$v'_x \frac{\partial v'_x}{\partial x'} + v'_y \frac{\partial v'_y}{\partial y'} - \frac{\partial^2 v'_x}{\partial y'^2} = U' \frac{dU'}{dx'},$$

$$\frac{\partial v'_x}{\partial x'} + \frac{\partial v'_y}{\partial y'} = 0.$$

(39.7)

These equations (and the boundary conditions on them) do not involve the viscosity. This means that their solutions are independent of the Reynolds number. Thus we reach the important result that, when the Reynolds number is changed, the whole flow pattern in the boundary layer simply undergoes a similarity transformation, longitudinal distances and velocities remaining unchanged, while transverse distances and velocities vary as $1/\sqrt{R}$.

Next, we can say that the dimensionless velocities v'_x, v'_y obtained by solving equations (39.7) must be of the order of unity, since they do not depend on R. The same is true of the thickness δ of the boundary layer in terms of the co-ordinates x', y'. From formulae (39.6) we can therefore conclude that

$$v_y \sim U_0/\sqrt{R},$$

(39.8)

i.e. the ratio of the transverse and longitudinal velocities is inversely proportional to $\sqrt{R}$, and that

$$\delta \sim l/\sqrt{R},$$

(39.9)

i.e. the thickness of the boundary layer diminishes with increasing Reynolds number as $1/\sqrt{R}$.

Let us apply the equations for the boundary layer to the case of plane-parallel flow along a flat plate. Let the plane of the plate be the xz half-plane with $x > 0$ (the leading edge of the plate thus being the line $x = 0$). We suppose the plate to extend indefinitely in the positive x-direction. The velocity of the main stream in this case is evidently constant ($U = $ constant). The equations (39.5) become

$$v_x \frac{\partial v_x}{\partial x} + v_y \frac{\partial v_x}{\partial y} = \nu \frac{\partial^2 v_x}{\partial y^2}, \qquad \frac{\partial v_x}{\partial x} + \frac{\partial v_y}{\partial y} = 0.$$

(39.10)

The boundary conditions at the surface of the plate are that both velocity components should vanish: $v_x = v_y = 0$ for $y = 0$, $x \geqslant 0$. As we move away from the plate, the velocity must approach asymptotically the velocity U of the incident flow, i.e. $v_x = U$ for $y \to \pm \infty$. In the solution of the equations for the boundary layer, as we have seen, v_x/U and $v_y\sqrt{(l/U\nu)}$ can be functions only of $x' = x/l$ and $y' = y\sqrt{(U/l\nu)}$. In the problem under consideration, however, the plate is infinite in extent and there are no characteristic lengths l. Hence v_x/U can depend only on a combination of x' and y'

which does not involve l, namely $y'/\sqrt{x'} = y\sqrt{(U/vx)}$. Similarly, the product $v'_y\sqrt{x'}$ must be a function of $y'/\sqrt{x'}$. Thus we can seek a solution in the form

$$v_x = Uf[y\sqrt{(U/vx)}], \qquad v_y = \sqrt{(Uv/x)}f_1[y\sqrt{(U/vx)}], \qquad (39.11)$$

where f and f_1 are some dimensionless functions. Using the second equation (39.10), we can express f_1 in terms of f. The problem thus reduces to the determination of a single function f of a single variable $\xi = y\sqrt{(U/vx)}$.†

In what follows we shall be interested only in the distribution of the longitudinal velocity v_x (since v_y is small). We can draw an important conclusion from formula (39.11) without even determining the function f. The velocity v_x increases from zero at the surface of the plate to a definite fraction of U for a given value of the argument of f, i.e. for $y\sqrt{(U/vx)} =$ any given constant. Hence we can conclude that the thickness of the boundary layer in flow along a plate is given in order of magnitude by

$$\delta \sim \sqrt{(vx/U)}. \qquad (39.12)$$

Thus, as we move away from the edge of the plate, δ increases as the square root of the distance from the edge.

The function f can be determined by numerical integration. A graph of this function is shown in Fig. 19. We see that f tends very rapidly to its limiting value of unity.‡

The frictional force on unit area of the surface of the plate is

$$\sigma_{xy} = \eta(\partial v_x/\partial y)_{y=0}.$$

A numerical calculation gives

$$\sigma_{xy} = 0{\cdot}332\sqrt{(\eta\rho U^3/x)}. \qquad (39.13)$$

If the plate is of length l (in the x-direction), then the total frictional force on it per unit length in the z-direction is

$$F = 2\int_0^l \sigma_{xy}\,dx.$$

The factor 2 is due to the fact that the plate has two sides exposed to the

---

† It is easily shown that, if the function $\phi(\xi)$ is such that $f(\xi) = \phi'(\xi)$, then $f_1(\xi) = \frac{1}{2}(\xi\phi' - \phi)$, while ϕ satisfies the equation $\phi\phi'' + 2\phi''' = 0$, with the boundary conditions $\phi = \phi' = 0$ for $\xi = 0$, $\phi' = 1$ for $\xi = \infty$.

‡ The "displacement thickness" δ^*, sometimes used to characterise the thickness of the boundary layer, is defined by

$$\int_0^\infty (U - v_x)dy = U\delta^*.$$

It is equal to $1{\cdot}72\sqrt{(vx/U)}$.

fluid. Substituting (39.13), we have

$$F = 1 \cdot 328 \sqrt{(\eta \rho l U^3)} \tag{39.14}$$

(H. BLASIUS, 1908). We may point out that the frictional force is proportional to the $\frac{3}{2}$ power of the velocity of the main stream. Formula (39.14) can be applied only to fairly long plates, for which the Reynolds number Ul/ν is fairly large. The force is customarily expressed in terms of the *drag coefficient*, defined as the dimensionless ratio

$$C = F/\tfrac{1}{2}\rho \, U^2 \cdot 2l. \tag{39.15}$$

By (39.14), this quantity, for laminar flow along a plate, is inversely proportional to the square root of the Reynolds number:

$$C = 1 \cdot 328/\sqrt{R}. \tag{39.16}$$

The quantitative formulae obtained above relate, of course, only to flow along a flat plate. The qualitative results, however, such as (39.8) and (39.9), hold for flow past bodies of any shape; in such cases l is the dimension of the body in the direction of flow.

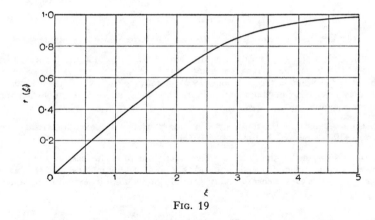

FIG. 19

We may make special mention of two cases of the boundary layer. If we have a plane disk, of large radius, rotating in the fluid about an axis perpendicular to its plane, then to estimate the thickness of the boundary layer we must replace U in (39.12) by Ωx, where Ω is the angular velocity of rotation. We then find

$$\delta \sim \sqrt{(\nu/\Omega)}. \tag{39.17}$$

We see that the thickness of the boundary layer may be regarded as a constant over the surface of the disk, in accordance with the exact solution of this problem obtained in §23. The magnitude of the frictional forces on the disk, as obtained from the equations for the boundary layer, is of course (23.4), since this formula is exact and therefore holds for laminar flow with any value of R.

Finally, let us consider the laminar boundary layer formed at the walls of a pipe near the point of entry of fluid. The fluid usually enters the pipe with a velocity distribution which is almost constant over the cross-section, and the velocity falls to zero entirely within the boundary layer. As we move away from the entrance to the pipe, the fluid layers nearer the axis are retarded. Since the mass of fluid that passes each cross-section is the same, the inner part of the stream, where the velocity is still uniform, must be accelerated as its diameter is reduced. This continues until a Poiseuille velocity distribution is asymptotically reached; this distribution is thus found only at some distance from the entrance to the pipe. It is easy to determine the order of magnitude of the length l of the "inlet section". It is given by the fact that, at a distance l from the entrance, the thickness of the boundary layer is of the same order of magnitude as the radius a of the pipe, so that the boundary layer fills almost the whole cross-section. Putting in (39.12) $x \sim l$ and $\delta \sim a$, we obtain

$$l \sim a^2 U/\nu \sim a\mathrm{R}. \tag{39.18}$$

Thus the length of the inlet section is proportional to the Reynolds number.†

PROBLEMS

PROBLEM 1. Determine the thickness of the boundary layer near a stagnation point (see §10).

SOLUTION. Near the stagnation point the fluid velocity (outside the boundary layer) is proportional to the distance x from that point, so that we can put $U = cx$. By estimating the magnitudes of the terms in the equations (39.5) we find $\delta \sim \sqrt{(\nu/c)}$. Thus the thickness of the boundary layer near the stagnation point is finite (and, in particular, does not vanish at the stagnation point itself).

PROBLEM 2. Determine the flow in the boundary layer in a converging channel between two non-parallel planes (K. POHLHAUSEN, 1921).

SOLUTION. Considering the boundary layer along one of the planes, we measure the coordinate x along that plane from the point O (Fig. 8, §23). For an ideal fluid we should have the velocity $U = Q/\alpha x \rho$; the corresponding pressure gradient is, by Bernoulli's equation, given by

$$\frac{1}{\rho}\frac{\mathrm{d}p}{\mathrm{d}x} = -\frac{\mathrm{d}}{\mathrm{d}x}(\tfrac{1}{2}U^2) = \frac{Q^2}{\alpha^2 x^3 \rho^2}.$$

It is easy to see that v_x and v_y must be sought in the form

$$v_x = (Q/\rho\alpha x)f(y/x), \qquad v_y = (Q/\rho\alpha x)f_1(y/x).$$

From the equation of continuity we obtain $f_1 = (y/x)f$, and the first equation (39.5) then gives for the function f

$$(\rho\nu\alpha/Q)f'' = 1-f^2,$$

where the prime denotes the differentiation of f with respect to its argument $\xi = y/x$. The

---

† We shall not discuss the theory of the boundary layer for a compressible fluid, which is, of course, considerably more complicated than that for an incompressible fluid. An account of this theory may be found in: N. E. KOCHIN, I. A. KIBEL' and N. V. ROZE, *Theoretical Hydromechanics* (*Teoreticheskaya gidromekhanika*), Part 2, 3rd ed., Chapter II, §§35, 36, Moscow 1948; H. SCHLICHTING, *Boundary Layer Theory*, Pergamon Press, London 1955; L. HOWARTH ed., *Modern Developments in Fluid Dynamics: High Speed Flow*, vol. 1, Oxford 1953.

boundary conditions are $f(0) = 0$, $f(\infty) = 1$ (since we must have $(v_x)_{y=0} = 0$, $(v_x)_{y=\infty} = Q/\rho\alpha x$). A first integral of the equation is

$$(\nu\alpha\rho/2Q)f'^2 = f - \tfrac{1}{3}f^3 + \text{constant.}$$

Since f tends to unity as $y \to \infty$, we see that f' tends to a definite limit, which can only be zero. The constant being thereby determined, we find

$$(\nu\alpha\rho/2Q)f'^2 = -\tfrac{1}{3}(f-1)^2(f+2).$$

Since the right-hand side is always negative for $0 \leqslant f \leqslant 1$, we must have $Q < 0$. That is, a boundary layer of the type in question is formed only by flow in a converging channel (and only at large Reynolds numbers $R = |Q|/\nu\rho$), and not by flow in a diverging channel, in accordance with the results of §23. Integrating again, we have finally

$$f = 3 \tanh^2[\log(\sqrt{2} + \sqrt{3}) + \xi\sqrt{(R/2\alpha)}] - 2.$$

§40. Flow near the line of separation

In describing the line of separation (§34) we have already mentioned that the actual position of this line on the surface of the body is determined by the properties of the flow in the boundary layer. We shall see below that, from a mathematical point of view, the line of separation is a line whose points are singular points of the solutions of (Prandtl's) equations of motion in the boundary layers. The problem is to determine the properties of these solutions near such a line of singularities.†

We know already that, from the line of separation, there begins a surface which extends into the fluid and marks off the region of turbulent flow. The flow is rotational throughout the turbulent region, whereas in the absence of separation it would be rotational only in the boundary layer, where the viscosity is important; the vorticity would be zero in the main stream. Hence we can say that separation causes the vorticity to "penetrate" from the boundary layer into the fluid. By the conservation of circulation, however, this "penetration" can occur only by the direct mixing of fluid moving near the surface (in the boundary layer) with the main stream. In other words, the flow in the boundary layer must be separated from the surface of the body, the streamlines consequently leaving the surface layer and entering the interior of the fluid. This phenomenon is therefore called *separation* or *separation of the boundary layer*.

The equations of motion in the boundary layer lead, as we have seen, to the result that the tangential velocity component (v_x) in the boundary layer is large compared with the component (v_y) normal to the surface of the body. This relation between v_x and v_y derives from our basic assumptions regarding the nature of the flow in the boundary layer, and must necessarily be found wherever Prandtl's equations have physically meaningful solutions. Mathematically, it is found at all points not lying in the immediate neighbourhood of singular points. But if $v_y \ll v_x$ it follows that the fluid moves along the

---

† The treatment of the problem given here, due to L. D. Landau, is somewhat different from that usually given.

surface of the body, and moves away from the surface only very slightly, so that there can be no separation. We therefore reach the conclusion that separation can occur only on a line whose points are singularities of the solution of Prandtl's equations.

The nature of these singularities also follows immediately. For, as we approach the line of separation, the flow deviates from the boundary layer towards the interior of the fluid. In other words, the normal velocity component ceases to be small compared with the tangential component, and is now of at least the same order of magnitude. We have seen (cf. (39.8)) that the ratio v_y/v_x is of the order of $1/\sqrt{R}$, so that an increase of v_y to the point where $v_y \sim v_x$ means an increase by a factor of $\sqrt{R}$. Hence, for sufficiently large Reynolds numbers (which, of course, we are considering) we may suppose that v_y increases by an infinite factor. If we use Prandtl's equations in dimensionless form (see (39.7)), the situation just described is formally equivalent to an infinite value of the dimensionless velocity v'_y on the line of separation.

In order to simplify the subsequent discussion a little, we shall consider the two-dimensional problem of transverse flow past a body of infinite length. As usual, x is the co-ordinate along the surface in the direction of flow, while y is the distance from the surface of the body. Instead of a line of separation, we now have a point of separation, namely the intersection of the line of separation with the xy-plane; in the co-ordinates used, this is the point $x = $ constant $\equiv x_0$, $y = 0$. Let $x < x_0$ be the region in front of the point of separation.

According to the above results, we have for all† y

$$v_y(x_0, y) = \infty. \tag{40.1}$$

In Prandtl's equations, however, v_y is a kind of parameter, which is usually of no interest (on account of its smallness) in investigating the flow in the boundary layer. Hence it is necessary to ascertain the properties of the function v_y near the line of separation.

It is clear from (40.1) that, for $x = x_0$, the derivative $\partial v_y/\partial y$ also becomes infinite. From the equation of continuity, $\partial v_x/\partial x + \partial v_y/\partial y = 0$, it then follows that $(\partial v_x/\partial x)_{x=x_0}$ is infinite, or $\partial x/\partial v_x = 0$, where x is regarded as a function of v_x and y. We denote by $v_0(y)$ the value of the function $v_x(x, y)$ for $x = x_0$: $v_0(y) = v_x(x_0, y)$. Near the point of separation, the differences $v_x - v_0$ and $x_0 - x$ are small, and we can expand $x_0 - x$ in powers of $v_x - v_0$ (for a given y). Since $(\partial x/\partial v_x)_{v=v_0} = 0$, the first-order term in this expansion must vanish identically, and we have as far as terms of the second order $x_0 - x = f(y)(v_x - v_0)^2$, or

$$v_x = v_0(y) + \alpha(y)\sqrt{(x_0 - x)}, \tag{40.2}$$

---

† Except $y = 0$, where we must always have $v_y = 0$ in accordance with the boundary conditions at the surface of the body.

where $\alpha = 1/\sqrt{f}$ is some function of y alone. Putting now

$$\frac{\partial v_y}{\partial y} = -\frac{\partial v_x}{\partial x} = \frac{\alpha(y)}{2\sqrt{(x_0-x)}}$$

and integrating, we have for v_y

$$v_y = \beta(y)/\sqrt{(x_0-x)}, \tag{40.3}$$

where

$$\beta(y) = \tfrac{1}{2} \int \alpha(y)\,dy$$

is another function of y.

Next, we use the first equation (39.5):

$$v_x \frac{\partial v_x}{\partial x} + v_y \frac{\partial v_x}{\partial y} = \nu \frac{\partial^2 v_x}{\partial y^2} - \frac{1}{\rho}\frac{dp}{dx}. \tag{40.4}$$

The derivative $\partial^2 v_x/\partial y^2$ does not become infinite for $x = x_0$, as we see from (40.2). The same is true of dp/dx, which is determined by the flow outside the boundary layer. Both terms on the left-hand side of equation (40.4) become infinite, however. In the first approximation we can therefore write for the region near the point of separation $v_x \partial v_x/\partial x + v_y \partial v_x/\partial y = 0$. Substituting $\partial v_x/\partial x = -\partial v_y/\partial y$, we can rewrite this as

$$v_x \frac{\partial v_y}{\partial y} - v_y \frac{\partial v_x}{\partial y} = v_x^2 \frac{\partial}{\partial y}\left(\frac{v_y}{v_x}\right) = 0.$$

Since the velocity v_x does not in general vanish for $x = x_0$, it follows that $\partial(v_y/v_x)/\partial y = 0$, i.e. the ratio v_y/v_x is independent of y. From (40.2) and (40.3), we have to within terms of higher order

$$\frac{v_y}{v_x} = \frac{\beta(y)}{v_0(y)\sqrt{(x_0-x)}}.$$

If this is a function of x alone, we must have $\beta(y) = \tfrac{1}{2}Av_0(y)$, where A is a numerical constant. Thus

$$v_y = \frac{Av_0(y)}{2\sqrt{(x_0-x)}}. \tag{40.5}$$

Finally, noticing that α and β in (40.2) and (40.3) obey the relation $\alpha = 2\beta'$, we obtain $\alpha = A\,dv_0/dy$, so that

$$v_x = v_0(y) + A(dv_0/dy)\sqrt{(x_0-x)}. \tag{40.6}$$

Formulae (40.5) and (40.6) determine v_x and v_y as functions of x near the point of separation. We see that each can be expanded in this region in powers of $\sqrt{(x_0-x)}$, the expansion of v_y beginning with the -1 power, so that v_y becomes infinite as $(x_0-x)^{-\frac{1}{2}}$ for $x \to x_0$. For $x > x_0$, i.e. beyond the point of separation, the expansions (40.5) and (40.6) are physically meaningless, since the square roots become imaginary; this means that the solutions of Prandtl's equations which give the flow up to the point of separation cannot meaningfully be continued beyond that point.

From the boundary conditions at the surface of the body, we must always have $v_x = v_y = 0$ for $y = 0$. We therefore conclude from (40.5) and (40.6) that

$$v_0(0) = 0, \qquad (dv_0/dy)_{y=0} = 0. \tag{40.7}$$

Thus we have the important result (due to Prandtl) that, at the point of separation itself ($x = x_0$, $y = 0$), not only the velocity v_x but also its first derivative with respect to y is zero.

It must be emphasised that the equation $\partial v_x/\partial y = 0$ on the line of separation holds only when v_y becomes infinite for that value of x. If the constant A in (40.5) happens to be zero, so that $v_y(x_0, y) \neq \infty$, then the point $x = x_0$, $y = 0$ at which the derivative $\partial v_x/\partial y$ vanishes would have no other particular properties, and would not be a point of separation. A can vanish, however, only by chance, and such an event is therefore unlikely. In practice a point on the surface of the body at which $\partial v_x/\partial y = 0$ is always a point of separation.

If there is no separation at the point $x = x_0$ (i.e. if $A = 0$), then for $x > x_0$ we have $(\partial v_x/\partial y)_{y=0} < 0$, i.e. v_x becomes negative (of increasing absolute magnitude) as we move away from the surface, y being still small. That is, the fluid beyond the point $x = x_0$ moves, in the lower parts of the boundary layer, in the direction opposite to that of the main stream; there is a "back-flow" of fluid at this point. It must be emphasised that from such arguments we cannot conclude that there is necessarily a point of separation where $\partial v_x/\partial y = 0$; the whole flow pattern with the "back-flow" might lie (as it does for $A = 0$) entirely within the boundary layer and not enter the main stream, whereas it is characteristic of separation that the flow enters the main body of the fluid.

It has been shown in the previous section that the flow pattern in the boundary layer is similar for different Reynolds numbers, and, in particular the scale in the x-direction remains unchanged. It follows from this that the value x_0 of the co-ordinate x for which the derivative $(\partial v_x/\partial y)_{y=0}$ is zero is the same for all R. Thus we have the important result that the position of the point of separation on the surface of the body is independent of the Reynolds number (so long as the boundary layer remains laminar, of course; see §45).

Let us also ascertain the properties of the pressure distribution $p(x)$ near the point of separation. For $y = 0$ the left-hand side of equation (40.4) is zero together with v_x and v_y, and there remains

$$\nu(\partial^2 v_x/\partial y^2)_{y=0} = (1/\rho)dp/dx. \tag{40.8}$$

It is clear from this that the sign of dp/dx is the same as that of $(\partial^2 v_x/\partial y^2)_{y=0}$.

When $(\partial v_x/\partial y)_{y=0} > 0$ we can say nothing regarding the sign of the second derivative. However, since v_x is positive and increases away from the wall (in front of the point of separation), we must always have $(\partial^2 v_x/\partial y^2)_{y=0} > 0$ at $x = x_0$ itself, where $\partial v_x/\partial y = 0$. Hence we conclude that

$$(dp/dx)_{x=x_0} > 0, \tag{40.9}$$

i.e. the fluid near the point of separation moves from the lower pressure to the higher pressure. The pressure gradient is related to the gradient of the velocity $U(x)$ outside the boundary layer by $(1/\rho)dp/dx = -U\,dU/dx$. Since the positive direction of the axis is the same as the direction of the main stream, $U > 0$, and therefore

$$(dU/dx)_{x=x_0} < 0, \tag{40.10}$$

i.e. the velocity U decreases in the direction of flow near the point of separation.

From the results obtained above we can deduce that there must be separation somewhere on the surface of the body. For there is on both the front and the back of the body a point (the stagnation point) at which the fluid velocity is zero for potential flow of an ideal fluid. Consequently, for some value of x, the velocity $U(x)$ must begin to decrease, and finally it becomes zero. It is clear, however, that the fluid moving over the surface of the body is retarded the more strongly, the closer it is to the surface (i.e. the smaller is y). Hence, before the velocity $U(x)$ is zero at the outer limit of the boundary layer, the velocity in the immediate neighbourhood of the surface must be zero. Mathematically, this evidently means that the derivative $\partial v_x/\partial y$ must always vanish (and therefore there must be separation) for some x less than the value for which $U(x) = 0$.

In flow past bodies of any form the calculations can be carried out in an entirely similar manner, and they lead to the result that the derivatives $\partial v_x/\partial y$, $\partial v_z/\partial y$ of the two velocity components v_x and v_z tangential to the surface of the body vanish on the line of separation (the y-axis, as before, is along the normal to the portion of the surface considered).

We may give a simple argument which demonstrates the necessity of separation in cases where the fluid would otherwise have a rapid increase of pressure (and therefore a rapid decrease in the velocity U) in the direction of its flow past the body. Over a small distance $\Delta x = x_2 - x_1$, let the pressure p increase rapidly from p_1 to p_2 ($p_2 \gg p_1$). Over the same distance Δx, the fluid velocity U outside the boundary layer falls from its initial value U_1 to a considerably smaller value U_2 determined by Bernoulli's equation:

$$\tfrac{1}{2}(U_1^2 - U_2^2) = (p_2 - p_1)/\rho.$$

Since p is independent of y, the pressure increase $p_2 - p_1$ is the same at all distances from the surface. If the pressure gradient $dp/dx \sim (p_2 - p_1)/\Delta x$ is sufficiently high, the term $\nu \partial^2 v_x/\partial y^2$ involving the viscosity may be omitted from the equation of motion (40.4) (if, of course, y is not small). Then, to estimate the change in the velocity v in the boundary layer, we can use Bernoulli's equation, putting $\frac{1}{2}(v_2^2 - v_1^2) = -(p_2 - p_1)/\rho$, or, from the equation previously obtained, $v_2^2 = v_1^2 - (U_1^2 - U_2^2)$. The velocity v_1 in the boundary layer is less than that of the main stream, and we can select a value of y for which $v_1^2 < U_1^2 - U_2^2$. The velocity v_2 is then imaginary, showing that Prandtl's equations have no physically significant solutions. In fact, there must be separation in the distance Δx, as a result of which the pressure gradient is reduced.

An interesting case of the appearance of separation is given by flow at an angle formed by two intersecting solid surfaces. For laminar potential flow outside an angle (Fig. 3), the fluid velocity at the vertex of the angle would become infinite (see §10, Problem 6), increasing in the stream approaching the vertex and diminishing in the stream leaving the vertex. In reality, the rapid decrease in velocity (and corresponding increase in pressure) beyond the vertex would lead to separation, the line of separation being the line of intersection of the surfaces. The resulting flow pattern is that discussed in §35.

In laminar flow inside an angle (Fig. 4), the fluid velocity is zero at the vertex. In this case the velocity diminishes (and the pressure increases) in the flow approaching the vertex. The result is in general the appearance of separation, the line of separation being upstream from the vertex of the angle.

PROBLEM

Determine the least possible increase Δp in the pressure which can occur (in the main stream) over a distance Δx and cause separation.

SOLUTION. Let y be a distance from the surface of the body at which, firstly, Bernoulli's equation can be applied and, secondly, the squared velocity $v^2(y)$ in the boundary layer is less than the change $|\Delta U^2|$ in the squared velocity outside that layer. For $v(y)$ we can write, in order of magnitude, $v(y) \approx y \, dv/dy \sim Uy/\delta$, where $\delta \sim \sqrt{(\nu l/U)}$ is the width of the boundary layer and l the dimension of the body. Equating, in order of magnitude, the two terms on the right-hand side of equation (40.4), we find

$$(1/\rho)\Delta p/\Delta x \sim \nu v(y)/y^2 \sim \nu U/\delta y.$$

From the condition

$$v^2 = |\Delta U^2| = (2/\rho)\Delta p \text{ we have } U^2 y^2/\delta^2 \sim \Delta p/\rho.$$

Eliminating y, we finally obtain

$$\Delta p \sim \rho U^2 (\Delta x/l)^{\frac{2}{3}}.$$

§41. Stability of flow in the laminar boundary layer

Laminar flow in the boundary layer, like any other laminar flow, becomes to some extent unstable at sufficiently large Reynolds numbers. The manner of the loss of stability in the boundary layer is similar to that which occurs for flow in a pipe (§29).

The Reynolds number for flow in the boundary layer varies over the surface of the body. For example, in flow along a plate we could define the Reynolds number as $R_x = Ux/\nu$, where x is the distance from the leading edge of the plate, and U the fluid velocity outside the boundary layer. A more suitable definition for the boundary layer, however, is one in which the length parameter directly characterises the thickness of the layer; such, for instance, is the "displacement thickness" δ^* (see the second footnote to §39). We then have $R_{\delta^*} = U\delta^*/\nu$. Since the dependence of the boundary-layer thickness on the distance x is given by formula (39.12), it is clear that $R_{\delta^*} \sim \sqrt{R_x}$.†

Because the change of the layer thickness with distance is comparatively slow, it may be neglected in investigating the stability of flow in a small portion of the layer, and we may consider a rectilinear two-dimensional flow, with a velocity profile which does not vary along the x-axis.‡ Then, from a mathematical point of view, the problem is entirely analogous to that of the stability of flow between two parallel planes, discussed in §29. The only difference is in the form of the velocity profile; instead of a symmetrical profile with $v = 0$ on both sides, we now have an unsymmetrical profile in which the velocity changes from zero at the surface of the body to some given value U, the velocity of the flow outside the boundary layer. The investigation leads to the following results (LIN, 1945; see C. C. LIN, *The Theory of Hydrodynamic Stability*, Cambridge 1955).

The form of the limiting curve of stability in the ωR-plane (see §29) depends on the form of the velocity profile in the boundary layer. If the velocity profile has no point of inflexion, and the velocity v_x increases monotonically with the curve $v_x = v_x(y)$ everywhere convex upwards (Fig. 20a), then the boundary of the stable region is completely similar in form to that which is obtained for flow in a pipe: there is a minimum value $R = R_{cr}$ at which amplified perturbations first appear, and for $R \to \infty$ both branches of the curve are asymptotic to the axis of abscissae (Fig. 21a). For the velocity profile which occurs in the boundary layer on a flat plate, the critical Reynolds number is found by calculation to be $R_{\delta^*, cr} \approx 420$.††

A velocity profile of the kind shown in Fig. 20a cannot occur if the fluid velocity outside the boundary layer decreases downstream. In this case the velocity profile must have a point of inflexion. For, let us consider a small portion of the surface, which we may regard as plane, and let x be again the co-ordinate in the direction of flow, and y the distance from the wall. From (40.8) we have

$$\nu(\partial^2 v_x/\partial y^2)_{y=0} = (1/\rho)dp/dx = -U\,\partial U/\partial x,$$

whence we see that, if U decreases downstream ($\partial U/\partial x < 0$), we must have $\partial^2 v_x/\partial y^2 > 0$ near the surface, i.e. the curve $v_x = v_x(y)$ is concave upwards. As y increases, the velocity v_x must tend asymptotically to the finite limit U.

---

† For example, in a laminar boundary layer on a flat plate $R_{\delta^*} = 1.72\sqrt{R_x}$.

‡ In doing so, of course, we pass over the question of the effect which the curvature of the surface may have on the stability of the boundary layer.

†† For $R_{\delta^*} \to \infty$, ω tends to zero, on the two branches I and II of the limiting curve, as $R_{\delta^*}^{-\frac{1}{3}}$ and $R_{\delta^*}^{-1/5}$ respectively.

It is then clear from geometrical considerations that the curve must become convex upwards, and therefore must have a point of inflexion (Fig. 20b). In this case the form of the curve defining the stable region is slightly changed: the two branches have different asymptotes for $R \to \infty$, one tending to the axis of abscissae and the other to a finite non-zero value of ω (Fig. 21b). The presence of a point of inflexion also reduces considerably the value of R_{cr}.

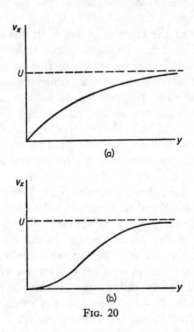

FIG. 20

The fact that the Reynolds number increases along the boundary layer makes the behaviour of the perturbations as they are carried downstream somewhat unusual. Let us consider flow along a flat plate, and suppose that a perturbation of given frequency ω occurs at some point in the boundary layer. Its propagation downstream corresponds to a movement in Fig. 21a to the right along a horizontal line $\omega =$ constant. The perturbation is at first damped: then, on reaching branch I of the stability curve, it begins to be amplified. This continues until branch II is reached, whereupon the perturbation is again damped. The total "amplification coefficient" for the perturbation during its passage through the region of instability increases very rapidly as this region moves towards large R (i.e. as the corresponding horizontal segment between branches I and II moves downwards).

These results, however, do not answer the question whether true absolute instability occurs in the laminar boundary layer for sufficiently large R— that is, instability due to the amplification in time of perturbations at a given point (see §29). As with flow in a pipe, no such investigation has yet been made.

The experimental results for flow along a flat plate show that the point where turbulence appears in the boundary layer[†] depends to a considerable extent on the intensity of the perturbations in the main stream. For marked perturbations, the boundary layer was observed to become turbulent for $R_{\delta^*} \approx 560$. As the intensity of the perturbations diminishes, the onset of turbulence is postponed to higher values of R_{δ^*}, which seem to tend to a finite limit of about 3000.

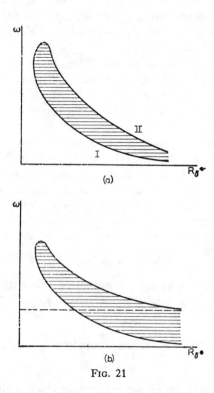

Fig. 21

It is possible that the existence of the limit indicates the presence of true absolute instability for sufficiently high values of R. On the other hand, it may be that, because of the extremely rapid increase of the "amplification coefficient" with R, the "displacement" instability of the kind described above may give the appearance of true instability.

§42. The logarithmic velocity profile

Let us consider plane-parallel turbulent flow along an unbounded plane

---

[†] Because the Reynolds number varies along the plate, the whole boundary layer does not become turbulent immediately, but only the part where R_{δ^*} exceeds a certain value. For a given incident velocity, this means that turbulence begins at a definite distance from the leading edge; as the velocity increases, this distance approaches zero.

surface; the term "plane-parallel" applies, of course, to the time average of the flow.† We take the direction of the flow as the x-axis, and the plane of the surface as the xz-plane, so that y is the distance from the surface. The y and z components of the mean velocity are zero: $u_x = u$, $u_y = u_z = 0$. There is no pressure gradient, and all quantities depend on y only.

We denote by σ the frictional force on unit area of the surface; this force is clearly in the x-direction. The quantity σ is just the momentum transmitted by the fluid to the surface per unit time; it is the constant flux of the x-component of momentum, which is in the negative y-direction, and gives the amount of momentum transmitted from the layers of fluid remote from the surface to those nearer it.

The existence of this momentum flux is due, of course, to the presence of a gradient, in the y-direction, of the mean velocity u. If the fluid moved with the same velocity at every point, there would be no momentum flux. The converse problem can also be stated: given some definite value of σ, what must be the motion of a fluid of given density ρ to give rise to a momentum flux σ? For large Reynolds numbers, the viscosity ν is, as usual, unimportant; it becomes important only for small distances y (see below). Thus the value of the velocity gradient du/dy at each point must be determined by the constant parameters ρ, σ and, of course, the distance y itself. The dimensions of these quantities are respectively g/cm³, g/cm sec² and cm. The dimensions of the derivative du/dy are 1/sec. The only combination of ρ, σ and y that has the right dimensions is $\sqrt{(\sigma/\rho y^2)}$. Hence we must have

$$du/dy = \sqrt{(\sigma/\rho)}/by, \qquad (42.1)$$

where b is a numerical constant ; b cannot be calculated theoretically, and must be determined experimentally. It is found to be‡

$$b = 0 \cdot 417. \qquad (42.2)$$

We introduce the more convenient notation $v_* = \sqrt{(\sigma/\rho)}$, so that

$$\sigma = \rho v_*^2. \qquad (42.3)$$

The quantity v_* has the dimensions cm/sec and acts as a characteristic velocity for the turbulent flow considered; then (42.1) becomes $du/dy = v_*/by$, whence

$$u = (v_*/b)(\log y + c), \qquad (42.4)$$

where c is a constant of integration. To determine this constant we cannot use the ordinary boundary conditions at the surface, since for $y = 0$ the first term in (42.4) becomes infinite. The reason for this is that the above expression is really inapplicable at very small distances from the surface, since the effect of the viscosity then becomes important, and cannot be neglected.

---

† The results given in §§42–44 are due to T. von Kármán and L. Prandtl.
‡ The value of this constant, and of one in formula (42.8) below, are obtained from measurements of the velocity distribution near the walls of a pipe in which there is turbulent flow.

There are also no conditions at infinity, since for $y = \infty$ the expression (42.4) again becomes infinite. This is because, in the idealised conditions which we have imposed, the surface is unbounded, and its influence therefore extends to infinitely great distances.

Before determining the constant c, we may first point out the following important property of the flow considered: contrary to what usually happens, it has no characteristic constant parameters of length which might give the external scale of the turbulence. This scale is therefore determined by the distance y itself: the scale of turbulent flow at a distance y from the surface is of the order of y. The fluctuating velocity of the turbulence is of the order of v_*. This also follows at once from dimensional arguments, since v_* is the only quantity having the dimensions of velocity which can be formed from the quantities σ, ρ, y at our disposal. It should be emphasised that, whereas the mean velocity decreases with y, the fluctuating velocity remains of the same order of magnitude at all distances from the surface. This result is in accordance with the general rule that the order of magnitude of the fluctuating velocity is determined by the variation Δu of the mean velocity (§31). In the present case, there is no characteristic length l over which the variation of the mean velocity could be taken; Δu must now be defined, reasonably, as the change in u when the distance y changes appreciably. According to (42.4), such a change in y causes a change in the velocity u that is just of the order of v_*.

At sufficiently small distances from the surface, the viscosity of the fluid begins to be important; we denote the order of magnitude of these distances by y_0, which can be determined as follows. The scale of the turbulence at these distances is of the order of y_0, and the velocity is of the order of v_*. Hence the Reynolds number which characterises the flow at distances of the order of y_0 is $R \sim v_* y_0 / \nu$. The viscosity begins to be important when R becomes of the order of unity. Hence we find that

$$y_0 \sim \nu/v_*, \tag{42.5}$$

and this determines y_0.

At distances from the surface small compared with y_0, the flow is determined by ordinary viscous friction. The velocity distribution here can be obtained directly from the usual formula for viscous friction: $\sigma = \rho\nu\, du/dy$, whence

$$u = \sigma y/\rho\nu = v_*^2 y/\nu. \tag{42.6}$$

Thus, immediately adjoining the wall, there is a thin layer of fluid in which the mean velocity varies linearly with y; the velocity is small throughout this layer, varying from zero at the surface itself to values of the order of v_* for $y \sim y_0$. We shall call this layer the *viscous sublayer*.

It must be emphasised that the flow here is turbulent, and in this respect the customary name "laminar sublayer" is unsuitable. The resemblance to laminar flow lies only in the fact that the mean velocity is distributed according to the same law as the true velocity would be for laminar flow under the same conditions. There is, of course, no sharp boundary between the viscous sublayer and the remainder of the flow, and the concept of the viscous sublayer is therefore to some extent qualitative.

The longitudinal component v'_x of the fluctuating velocity in the viscous sublayer is of the same order of magnitude as the mean velocity, and in particular is proportional to y ($\sim v_* y/y_0$). It therefore follows from the equation of continuity that the derivative $\partial v'_y/\partial y = -\partial v'_x/\partial x$ is proportional to y, and so the transverse component v'_y of the fluctuating velocity varies as y^2 ($\sim v_* y^2/y_0^2$). Next, it follows from the linearity of the equations of motion in the viscous sublayer (the non-linear terms being there small compared with the viscosity terms) that the periods of the turbulent eddies are the same throughout the thickness of the sublayer. Multiplying these periods by the fluctuating velocity, we find that the longitudinal distances traversed by the fluid particles in their fluctuating motion are proportional to y, in order of magnitude, and the transverse distances are proportional to y^2 ($\sim y^2/y_0$).

We shall not be further interested in the flow in the viscous sublayer. Its presence has to be taken into account only in making the appropriate choice of the constant of integration in (42.4). This constant must be chosen so that the velocity becomes of the order of v_* at distances of the order of y_0. For this to be so, we must take $c = -\log y_0$, so that $u = (v_*/b) \log(y/y_0)$, or

$$u = (v_*/b) \log(y v_*/\nu). \tag{42.7}$$

This formula determines (for a certain range of y) the velocity distribution in the turbulent stream which flows along the surface. This distribution is called the *logarithmic velocity profile*.

The argument of the logarithm in formula (42.7) should include a numerical coefficient. However, in the formulae which we shall derive we shall require only "logarithmic" accuracy. This means that the argument of the logarithm is supposed large, and we neglect not only terms proportional to lower powers of the argument but also those involving the logarithm to lower powers than in the principal term. The introduction of a small numerical coefficient in the argument of the logarithm in (42.7) is equivalent to adding a term of the form constant $\times v_*$, where the constant is of the order of unity; this term does not contain the logarithm, and therefore we neglect it. However, it must be borne in mind that the argument of the logarithm in the formulae derived here is not so large that its logarithm is also very large, and so the accuracy of the formulae is not very high.

These formulae can be made more exact by introducing a numerical coefficient in the argument of the logarithm, or, what is the same thing, adding a constant to the logarithm. These constants, however, cannot be calculated theoretically, and have to be determined from experimental results. For example, a more exact formula for the velocity distribution can be written in the form

$$u = v_*[2 \cdot 40 \log(y v_*/\nu) + 5 \cdot 84]. \tag{42.8}$$

It is not difficult to determine the energy dissipation ϵ per unit mass of fluid. σ is the mean value of the component Π_{xy} of the momentum flux density tensor $\Pi_{ik} = \rho v_i v_k - \eta(\partial v_i/\partial x_k + \partial v_k/\partial x_i)$. Outside the viscous sublayer, the viscosity term may be omitted, so that $\sigma = \overline{\rho v_x v_y}$. Introducing the fluctuating velocity $\mathbf{v}'$, we can write $v_x = u + v'_x$; the velocity v_y is itself the fluctuating velocity v'_y, since its mean value is zero. The result is

$$\sigma = \overline{\rho v_x v_y} = \overline{\rho v'_x v'_y} + \overline{\rho u v'_y} = \overline{\rho v'_x v'_y}.$$

Next, the energy flux density in the y-direction is $(p + \tfrac{1}{2}\rho v^2)v_y$, the viscosity term being again omitted. Putting in the second term

$$v^2 = (u + v'_x)^2 + v'^2_y + v'^2_z$$

and averaging, we obtain

$$\overline{pv'_y} + \tfrac{1}{2}\rho(\overline{v'^2_x v'_y} + \overline{v'^3_y} + \overline{v'^2_z v'_y}) + \rho\overline{uv'_x v'_y}.$$

Here only the last term need be retained. The reason is that the fluctuating velocity is of the order of v_*, and hence, to logarithmic accuracy, it is small compared with u. The turbulent fluctuations of the pressure p are of the order of ρv_*^2 (cf. (31.4)), and so we can, to the same accuracy, neglect the corresponding term in the energy flux. Thus we have for the mean energy flux density $\rho\overline{uv'_x v'_y} = u\sigma$. As we approach the surface, this flux decreases, because the energy is dissipated. The decrease in the energy flux density on approaching the surface by a distance dy is $\sigma(du/dy)dy$. This is the amount of energy converted into heat in a fluid layer of thickness dy and of unit area. Hence we conclude that the energy dissipation per unit mass is $(\sigma/\rho)du/dy$, or

$$\epsilon = v_*^3/by = (\sigma/\rho)^{\frac{3}{2}}/by. \tag{42.9}$$

§43. Turbulent flow in pipes

Let us now apply the above results to turbulent flow in a pipe. Near the walls of the pipe (at distances small compared with its radius a), the surface may be approximately regarded as plane, and the velocity distribution must be given by formula (42.7) or (42.8). Since the function $\log y$ varies only slowly, we can use formula (42.7) to logarithmic accuracy to give the mean velocity U of the flow in the pipe if we replace y in that formula by a:

$$U = (v_*/b) \log(av_*/\nu). \tag{43.1}$$

By U we mean the volume of fluid that passes through a cross-section of the pipe per unit time, divided by the cross-sectional area: $U = Q/\rho\pi a^2$.

In order to relate the velocity U to the pressure gradient $\Delta p/l$ which maintains the flow (Δp being the pressure difference between the ends of the pipe, and l its length), we notice that the force on a cross-section of the flow is $\pi a^2 \Delta p$. This force overcomes the friction at the walls. Since the frictional force per unit area of the wall is $\sigma = \rho v_*^2$, the total frictional force is $2\pi a l \rho v_*^2$. Equating the two forces, we have

$$\Delta p/l = 2\rho v_*^2/a. \tag{43.2}$$

Equations (43.1) and (43.2) determine, through the parameter v_*, the relation between the velocity of flow in the pipe and the pressure gradient. This relation is called the *resistance law* of the pipe. Expressing v_* in terms of $\Delta p/l$ by (43.2), and substituting in (43.1), we obtain the resistance law in the form

$$U = \sqrt{(a\Delta p/2b^2\rho l)} \log[(a/\nu)\sqrt{(a\Delta p/2\rho l)}]. \tag{43.3}$$

In this formula it is customary to introduce what is called the *resistance coefficient* of the pipe, a dimensionless quantity defined as

$$\lambda = \frac{2a\Delta p/l}{\frac{1}{2}\rho U^2}. \tag{43.4}$$

The dependence of λ on the dimensionless Reynolds number $R = 2aU/\nu$ is given in implicit form by the equation

$$1/\sqrt{\lambda} = 0.85 \log(R\sqrt{\lambda}) - 0.55. \tag{43.5}$$

We have here substituted for b the value (42.2) and added to the logarithm an empirically determined constant.† The resistance coefficient determined by this formula is a slowly decreasing function of the Reynolds number. For comparison, we give the resistance law for laminar flow in a pipe. Introducing the resistance coefficient in formula (17.10), we obtain

$$\lambda = 64/R. \tag{43.6}$$

In laminar flow the resistance coefficient diminishes with increasing Reynolds number more rapidly than in turbulent flow.

Fig. 22 shows a logarithmic graph of λ as a function of R. The steep straight line corresponds to laminar flow (formula (43.6)), and the less steep curve (which is almost a straight line also) to turbulent flow. The transition from the first line to the second occurs, as the Reynolds number increases, at the point where the flow becomes turbulent; this may occur for various Reynolds numbers, depending on the actual conditions (the intensity of the perturbations; see §29). The resistance coefficient increases abruptly at the transition point.

---

† The coefficient of the logarithm in this formula is given to correspond with that in formula (42.8) for the logarithmic velocity profile. Only in this case does formula (43.5) have the theoretical significance of being a limiting formula for turbulent flow at sufficiently large values of the Reynolds number. If the values of the two constants appearing in formula (43.5) are chosen arbitrarily, it can only be a purely empirical formula for the dependence of λ on R. In that case, however, there would be no reason to prefer it to any other simpler empirical formula which adequately represents the experimental results.

So far we have assumed that the wall surface is fairly smooth. If it is rough, the formulae obtained above may be somewhat changed. As a measure of the roughness of the wall, we can take the order of magnitude of the projections, which we shall denote by d. The relative magnitudes of d and the thickness y_0 of the sublayer are of importance. If y_0 is large compared with d, the roughness is unimportant; this is what is meant by saying that the surface is fairly smooth. If y_0 and d are of the same order of magnitude, no general formulae can be obtained.

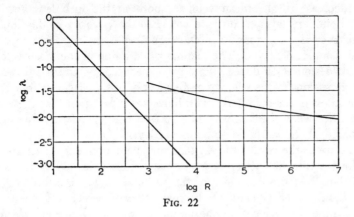

FIG. 22

In the opposite limiting case of extreme roughness $(d \gg y_0)$, some general relations can again be established. In this case we clearly cannot speak of a viscous sublayer. Turbulent flow occurs around the projections from the surface, and this flow is characterised by the quantities ρ, σ, d; the viscosity ν, as usual, cannot appear directly. The velocity of this flow is of the order of magnitude of v_*, the only quantity at our disposal having the dimensions of velocity. Thus we see that, in flow along a rough surface, the velocity becomes small $(\sim v_*)$ at distances $y \sim d$, instead of $y \sim y_0$ as for flow along a smooth surface. Hence it is clear that the velocity distribution is given by a formula which is obtained from (42.7) by substituting d for ν/v_*. Thus

$$u = (v_*/b) \log(y/d). \tag{43.7}$$

The formulae for flow in a pipe must be changed similarly. It is sufficient simply to replace ν/v_* in them by d. For the resistance law we have, instead of (43.3), the formula

$$U = \sqrt{(a\Delta p/2b^2\rho l)} \log(a/d). \tag{43.8}$$

The argument of the logarithm is now a constant, and does not involve the pressure gradient as (43.3) did. We see that the mean velocity is now simply proportional to the square root of the pressure gradient in the pipe. If we introduce the resistance coefficient, (43.8) becomes

$$\lambda = 8b^2/\log^2(a/d) = 1.4/\log^2(a/d), \tag{43.9}$$

i.e. λ is a constant and does not depend on the Reynolds number.

§44. The turbulent boundary layer

The fact that we have obtained a logarithmic velocity distribution which formally holds in all space for plane-parallel turbulent flow is due to our having considered flow along a surface of infinite area. In flow along the surface of a finite body, only the motion at short distances from the surface—in the boundary layer—has a logarithmic profile.† We may mention also that a turbulent boundary layer can exist both under a fluid moving turbulently in the main stream and under a laminar flow.

The decrease in the mean velocity, both in the turbulent and in the laminar boundary layer, is due ultimately to the viscosity of the fluid. The effect of the viscosity appears in the turbulent boundary layer in a rather unusual manner, however. The manner of variation of the mean velocity in the layer does not itself depend directly on the viscosity; the viscosity appears in the expression for the velocity gradient only in the viscous sublayer. The total thickness of the boundary layer, however, is determined by the viscosity, and vanishes when the viscosity is zero (see below). If the viscosity were exactly zero, there would be no boundary layer.

Let us apply the results of §43 to a turbulent boundary layer formed in flow along a thin flat plate, such as was discussed in §39 with respect to laminar flow. At the boundary of the turbulent layer, the fluid velocity is almost equal to the velocity of the main stream, which we denote by U. To determine this velocity at the boundary we can, however, use formula (42.7) with logarithmic accuracy, putting the thickness δ of the boundary layer instead of y. Equating the two expressions, we obtain

$$U = (v_*/b) \log(v_*\delta/\nu). \tag{44.1}$$

Here U is a constant parameter for a given flow; the thickness δ, however, varies along the plate, and v_* is therefore also a slowly varying function of x. Formula (44.1) is inadequate to determine these functions; we need some other equation, relating v_* and δ to x.

To obtain this, we use the same arguments as in deriving formula (36.3) for the width of the turbulent wake. As there, the derivative $d\delta/dx$ must be of the order of the ratio of the velocity along the y-axis to that along the x-axis at the boundary of the layer. The latter velocity is of the order of U, while the former is due to the fluctuating velocity, and is therefore of the order of v_*. Thus $d\delta/dx \sim v_*/U$, whence

$$\delta \sim v_* x/U. \tag{44.2}$$

---

† The thickness of the boundary layer increases along the surface of the body in the direction of flow, according to a law which we shall determine below. This explains why, for flow in a pipe, the logarithmic profile holds for the whole cross-section of the pipe. The thickness of the boundary layer at the wall of the pipe increases away from the point of entry of the fluid. At some finite distance from this point, the boundary layer fills almost the whole cross-section of the pipe. Hence, if we suppose the pipe sufficiently long and ignore its inlet section, the flow in the whole pipe will be of the same kind as in the turbulent boundary layer. We may recall that a similar situation occurs for laminar flow in a pipe. Such a flow obeys Poiseuille's formula for all Reynolds numbers. In Poiseuille flow the viscosity is important at all distances from the walls, and its effect is never limited to a thin layer adjoining them.

Formula (44.1) and (44.2) together determine v_* and δ as functions of the distance x.† These functions, however, cannot be written explicitly. We shall express δ in terms of an auxiliary quantity. Since v_* is a slowly varying function of x, it is seen from (44.2) that the thickness of the layer varies essentially as x. We may recall that the thickness of the laminar boundary layer increases as $\sqrt{x}$, i.e. more slowly than that of the turbulent boundary layer.

Let us determine the dependence on x of the frictional force σ acting on unit area of the plate. This dependence is given by two formulae:

$$\sigma = \rho v_*^2, \qquad U = (v_*/b) \log(v_*^2 x/U\nu).$$

The latter is obtained by substituting (44.2) in (44.1), and is valid to logarithmic accuracy. We introduce a drag coefficient c (referred to unit area of the plate), defined as the dimensionless ratio

$$c = 2\sigma/\rho U^2 = 2(v_*/U)^2. \tag{44.3}$$

Then, eliminating v_* from the two equations given, we obtain the following equation, which gives (to logarithmic accuracy) c as an implicit function of x:

$$\sqrt{(2b^2/c)} = \log(c\mathrm{R}_x), \qquad \mathrm{R}_x = Ux/\nu. \tag{44.4}$$

To increase the accuracy of this formula, we may add an empirical numerical constant to the logarithm. Such a formula is,

$$1/\sqrt{c} = 1.7 \log(c\mathrm{R}_x) + 3.0. \tag{44.5}$$

The drag coefficient c given by this formula is a slowly decreasing function of the distance x.

Finally, let us express the thickness of the boundary layer in terms of the function $c(x)$. We have $v_* = \sqrt{(\sigma/\rho)} = U\sqrt{(\tfrac{1}{2}c)}$. Substituting in (44.2), we find

$$\delta = \text{constant} \times x\sqrt{c}. \tag{44.6}$$

This formula may be written with the equality sign, of course, only in cases of a turbulent boundary layer under a laminar flow, when δ has an exact significance (the turbulent region being, as always, sharply distinct from the laminar region). The constant factor in (44.6) has to be determined from experimental results.

PROBLEMS

PROBLEM 1. Determine from formula (44.5) the total force acting on the two sides of the plate.

SOLUTION. The required force per unit length of the edge of the plate is

$$F = 2 \int_0^l \sigma \, dx,$$

---

† If there is a laminar boundary layer of considerable extent on the plate, then x must, strictly speaking, be reckoned as approximately the distance from the point where the laminar layer becomes turbulent.

where l is the length of the plate. Introducing in place of F the drag coefficient

$$C = F/\tfrac{1}{2}\rho U^2 . 2l,$$

we find

$$C = \frac{1}{l}\int\limits_0^l c\,dx.$$

If we take only terms containing the logarithm to the highest (first) power, then the above integral is simply $c(l)$, the value of c for $x = l$. In order to obtain a more exact value for C, corresponding to formula (44.5), we must effect the integration taking account of terms of the next order, which contain the logarithm to the zero power. To do so, we write

$$\int\limits_0^l c\,dx = [xc]_0^l - \int\limits_0^l x\frac{dc}{dx}dx.$$

The derivative dc/dx is calculated by means of formula (44.5), which we write in the form $c = 1/A^2 \log^2 Bxc$, obtaining to the necessary accuracy

$$C = c(l) + \frac{2}{A^2\log^3 Blc} = c(l)\left[1 + \frac{2}{\log Blc}\right],$$

and so

$$\frac{1}{\sqrt{C}} = \frac{1}{\sqrt{c}}\left(1 - \frac{1}{\log Blc}\right) = A(\log Blc - 1) = A\log(Blc/e)$$
$$\approx A\log(BlC/e).$$

Substituting the values of A and B from (44.5), we obtain the following formula, which gives the total drag coefficient C as a function of the Reynolds number $R = Ul/\nu$:

$$1/\sqrt{C} = 1\cdot 7\log(CR) + 1\cdot 3.$$

For large R, the drag coefficient given by this formula decreases as $1/\log^2 R$. For the laminar boundary layer, C decreases as $1/\sqrt{R}$ (see (39.16)), i.e. more rapidly. Thus we can say that, for large Reynolds numbers, the frictional force in a turbulent boundary layer is greater than in a laminar one.

PROBLEM 2. Determine the drag coefficient of a rough plate as a function of the Reynolds number, for a turbulent boundary layer.

SOLUTION. Substituting in place of the thickness y_0 ($\sim \nu/v_*$) of the laminar sublayer the dimension d of the projections, we obtain from (44.1) and (44.2) $U = (v_*/b)\log(xv_*/Ud)$. Introducing the drag coefficient c, we hence have $0\cdot 59/\sqrt{c} = \log(x\sqrt{c}/d)$. Similarly, the total drag coefficient for the plate is (again to logarithmic accuracy) $0\cdot 59/\sqrt{C} = \log(l\sqrt{C}/d)$. We may point out that the drag coefficient for a rough plate is independent of the Reynolds number.

§45. The drag crisis

From the results obtained in the previous sections we can draw important conclusions concerning the law of drag for large Reynolds numbers, i.e. the relation between the drag force acting on the body and the value of R when the latter is large.

The flow pattern for large R (the only case we shall discuss) has already been described, and is as follows. Throughout the main body of the fluid (i.e. everywhere except in the boundary layer, which does not here concern us) the fluid may be regarded as ideal, with potential flow everywhere except in the turbulent wake. The width of the wake depends on the position of the line of separation on the surface of the body. It is important to note that, although this position is determined by the properties of the boundary layer, it is found to be independent of the Reynolds number, as we have seen in §40. Thus we can say that the whole flow pattern for large Reynolds numbers is almost independent of the viscosity, i.e. of R (so long as the boundary layer remains laminar; see below).

Hence it follows that the drag also must be independent of the viscosity. There remain at our disposal only three quantities: the velocity U of the main stream, the fluid density ρ and the dimension l of the body. From these we can construct only one quantity having the dimensions of force, namely $\rho U^2 l^2$. Instead of the squared linear dimension of the body l^2, we introduce, as is customarily done, the proportional quantity S, the area of a cross-section transverse to the direction of flow, putting

$$F = \text{constant} \times \rho U^2 S, \tag{45.1}$$

where the constant is a number depending only on the shape of the body. Thus the drag must be (for large R) proportional to the cross-sectional area of the body and to the square of the main-stream velocity. We may recall for comparison that, for very small R ($\ll 1$), the drag is proportional to the linear dimension of the body and to the velocity itself ($F \sim \nu \rho l U$; see §20).[†]

It is customary, as we have said, to introduce, in place of the drag force F, the drag coefficient C defined by $C = F/\frac{1}{2}\rho U^2 S$. This is a dimensionless quantity, and can depend only on R. Formula (45.1) becomes

$$C = \text{constant}, \tag{45.2}$$

i.e. the drag coefficient depends only on the shape of the body.

The above behaviour of the drag force cannot continue to arbitrarily large Reynolds numbers. The reason is that, for sufficiently large R, the laminar boundary layer (on the surface of the body as far as the line of separation) becomes unstable and hence turbulent. However, the whole boundary layer does not become turbulent, but only some part of it. The surface of the body may therefore be divided into three parts: at the front there is a laminar boundary layer, then a turbulent layer, and finally the region beyond the line of separation.

The onset of turbulence in the boundary layer has an important effect on the whole pattern of flow in the main stream. It leads to a considerable displacement of the line of separation towards the rear of the body (i.e.

---

† The flow past a bubble of gas is a special case, where the drag remains proportional to U even for large R; see Problem.

downstream), so that the turbulent wake beyond the body is contracted, as shown in Fig. 23, where the wake region is shaded.† The contraction of the turbulent wake leads to a reduction of the drag force. Thus the onset of turbulence in the boundary layer at large Reynolds numbers is accompanied by a decrease in the drag coefficient, which falls off by a considerable factor over a relatively narrow range of Reynolds numbers near 10^5. We shall call this phenomenon the *drag crisis*. The decrease in the drag coefficient is so great that the drag itself, which for constant C is proportional to the square of the velocity, actually diminishes with increasing velocity in this range of Reynolds numbers.

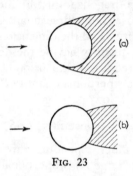

FIG. 23

It may be mentioned that the degree of turbulence in the main stream affects the drag crisis; the greater the incident turbulence, the sooner the boundary layer becomes turbulent (i.e. the smaller is R when this happens). The decrease in the drag coefficient therefore begins at a smaller Reynolds number, and extends over a wider range of R.

Figs. 24 and 25 give experimentally obtained graphs showing the drag coefficient as a function of the Reynolds number $R = Ud/\nu$ for a sphere; Fig. 24 is plotted logarithmically. For very small R ($\ll 1$), the drag coefficient decreases according to $C = 24/R$ (Stokes' formula). The decrease in C continues more slowly as far as $R \approx 5 \times 10^3$, where C reaches a minimum, beyond which it increases somewhat. In the range of Reynolds numbers 2×10^4 to 2×10^5, the law (45.2) holds, i.e. C is almost constant. The drag crisis occurs for R between 2×10^5 and 3×10^5, and the drag coefficient diminishes by a factor of 4 or 5.

For comparison, we may give an example of flow in which there is no critical Reynolds number. Let us consider flow past a flat disk in the direction perpendicular to its plane. In this case the location of the separation is obvious from purely geometrical considerations: it is clear that separation occurs at the edge of the disk and does not move from there. Hence, as R increases, the

---

† For example, in transverse flow past a long cylinder, the onset of turbulence in the boundary layer moves the point of separation from 95° to 60° (where the azimuthal angle on the cylinder is measured from the direction of flow).

drag coefficient of the disk remains constant, and there is no drag crisis.

It must be borne in mind that, for the high velocities at which the drag crisis occurs, the compressibility of the fluid may begin to have a noticeable effect. The parameter which characterises the extent of this effect is the *Mach number* $M = U/c$, where c is the velocity of sound; if $M \ll 1$, the fluid may be regarded as incompressible (§10). Since, of the two numbers M and R, only one contains the dimension of the body, these two numbers can vary independently.

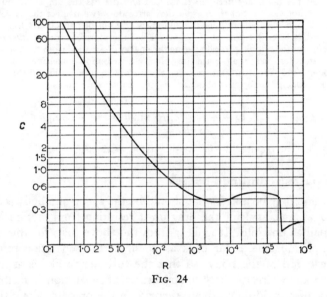

FIG. 24

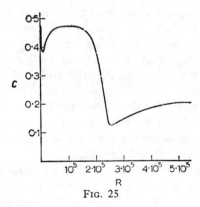

FIG. 25

The experimental data indicate that the compressibility has in general a stabilising effect on the flow in the laminar boundary layer. When M increases, the critical value of R increases. For example, when M for a sphere changes from 0·3 to 0·7, the drag crisis is postponed from $R \approx 4 \times 10^5$ to $R \approx 8 \times 10^5$.

We may also mention that, when M increases, the position of the point of separation in the laminar boundary layer moves upstream, towards the front of the body, and this must lead to some increase in the drag.

PROBLEM

Determine the drag force on a gas bubble moving in a liquid at large Reynolds numbers (V. G. LEVICH 1949).

SOLUTION. At the boundary between the liquid and the gas the tangential fluid velocity component does not vanish, but its normal derivative does (we neglect the viscosity of the gas). Hence the velocity gradient near the boundary will not be particularly high, and there will be no boundary layer in the sense of §39; there will therefore be no separation over almost the whole surface of the bubble. In calculating the energy dissipation from the volume integral (16.3) we can therefore use in all space the velocity distribution corresponding to potential flow past a sphere (§10, Problem 2), neglecting the surface layer of liquid and the very narrow turbulent wake. Using the formula obtained in §16, Problem, we find

$$ \dot{E}_{\text{kin}} = -\eta \int \left(\frac{\partial v^2}{\partial r} \right)_{r=R} 2\pi R^2 \sin \theta \, d\theta = -12\pi\eta \, RU^2. $$

Hence we see that the required dissipative drag is† $F = 12\pi\eta RU$.

§46. Flow past streamlined bodies

The question may be asked what should be the shape of a body (of a given cross-sectional area, say) for the drag on it resulting from motion in a fluid to be as small as possible. It is clear from the above that, for this to be so, the separation must be as far back as possible: the separation must occur near the rear end of the body, so that the turbulent wake is as narrow as possible. We know already that the appearance of separation is facilitated by the presence of a rapid downstream increase in the pressure along the body. Hence the body must have a shape such that the variation in pressure along it, where the pressure is increasing, takes place as slowly and smoothly as possible. This can be achieved by giving the body a shape elongated in the direction of flow, tapering smoothly to a point downstream, so that the flows along the two sides of the body meet smoothly without having to go round any corners or turn through a considerable angle from the direction of the main stream. At the front end the body must be rounded; if there were an angle here, the fluid velocity at its vertex would become infinite (see §10, Problem 6), and consequently the pressure would increase rapidly downstream, with separation inevitably resulting.

All these requirements are closely satisfied by shapes of the kind shown in Fig. 26. The profile shown in Fig. 26b may be, for example, the cross-section of an elongated solid of revolution, or the cross-section of a body with

---

† The range of applicability of this formula is actually not large, since, when the velocity increases sufficiently, the bubble ceases to be spherical.

a large "span" (we conventionally call such a body a *wing*). The cross-sectional profile of a wing may be unsymmetrical, as in Fig. 26a. In flow past a body of this shape, separation occurs only in the immediate neighbourhood of the pointed end, and consequently the drag coefficient is relatively small. Such bodies are said to be *streamlined*.

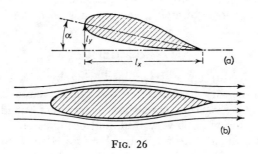

FIG. 26

The direct friction of the fluid on the surface in the boundary layer is important in the drag on streamlined bodies. This effect for non-streamlined bodies (which were considered in the previous section) is relatively small and therefore, in practice, of no significance. In the opposite limiting case of flow parallel to a flat disk, the effect becomes the only source of drag (§39).

In flow past a streamlined wing inclined to the main stream at a small angle α, called the *angle of attack* (Fig. 26), a large lift force F_y is developed, while the drag F_x remains small, and the ratio F_y/F_x may therefore reach large values ($\sim 10-100$). This continues, however, only while the angle of attack is small ($\leqslant 10°$). For larger angles the drag rises very rapidly, and the lift decreases. This is explained by the fact that, at large angles of attack, the body ceases to be streamlined: the point of separation moves a considerable way towards the front of the body, and the wake consequently becomes wider. It must be borne in mind that the limiting case of a very thin body, i.e. a flat plate, is streamlined only for a very small angle of attack; separation occurs at the leading edge of the plate when it is inclined at even a small angle to the main stream.

The angle of attack α is, by definition, measured from the position of the wing for which the lift force is zero. For small angles of attack, we can expand the lift as a series of powers of α. Taking only the first term, we can suppose that the force F_y is proportional to α. Next, by the same dimensional arguments as for the drag force, the lift must be proportional to ρU^2. Introducing also the span l_z of the wing, we can write

$$F_y = \text{constant} \times \rho U^2 \alpha l_x l_z, \tag{46.1}$$

where the numerical constant depends only on the shape of the wing and not, in particular, on the angle of attack. For very long wings, the lift may be supposed proportional to the span, in which case the constant depends only on the shape of the cross-section of the wing.

Instead of the lift on the wing, the *lift coefficient* is often used; it is defined as

$$C_y = F_y/\tfrac{1}{2}\rho U^2 l_x l_z. \tag{46.2}$$

For very long wings, according to what was said above, the lift coefficient is proportional to the angle of attack, and depends on neither the velocity nor the span:

$$C_y = \text{constant} \times \alpha. \tag{46.3}$$

To calculate the lift on a streamlined wing by means of Zhukovskiĭ's formula, it is necessary to determine the velocity circulation Γ. This is done as follows. We have potential flow everywhere outside the wake. In the present case, the wake is very thin, and occupies on the surface of the wing only a very small area near its pointed trailing edge. Hence, to determine the velocity distribution (and therefore the circulation Γ), we can solve the problem of potential flow of an ideal fluid round a wing. The existence of the wake is taken into account by the presence of a tangential discontinuity, extending into the fluid from the sharp trailing edge of the wing, where the potential has a discontinuity $\phi_2 - \phi_1 = \Gamma$. As has been shown in §37, the derivative $\partial\phi/\partial z$ also has a discontinuity on this surface, while the derivatives $\partial\phi/\partial x$ and $\partial\phi/\partial y$ are continuous. For a wing of finite span, the problem in this form has a unique solution. The finding of the exact solution is very complicated, however. The problem has been solved by N. E. KOCHIN[†] for a wing in the form of a circular disk inclined at a small angle of attack.

If the wing is very long (and has a uniform cross-section), then, regarding it as infinite in the z-direction, we may regard the flow as two-dimensional (in the xy-plane). It is evident from symmetry that the velocity $v_z = \partial\phi/\partial z$ along the wing must be zero. In this case, therefore, we must seek a solution in which only the potential has a discontinuity, its derivatives being continuous; in other words, there is no surface of tangential discontinuity, and we have simply a many-valued function $\phi(x, y)$, which receives a finite increment Γ when we go round a closed contour enclosing the profile of the wing. In this form, however, the problem of two-dimensional flow has no unique solution, since it admits solutions for any given discontinuity of the potential. To obtain a unique result, we must require the fulfilment of another condition, first formulated by S. A. CHAPLYGIN in 1909.

This condition, called the *Zhukovskiĭ–Chaplygin condition*, consists in requiring that the fluid velocity does not become infinite at the sharp trailing edge of the wing; in this connection we may recall that, when an ideal fluid flows round an angle, the fluid velocity in general becomes infinite, according to a power law, at the vertex of the angle (§10, Problem 6). We can say that

---

[†] *Prikladnaya matematika i mekhanika* **4**, 3, 1940; **9**, 13, 1945.

the condition stated implies that the jets coming from the two sides of the wing must meet smoothly without turning through an angle. When this condition is fulfilled, of course, the solution of the problem of potential flow gives a pattern very like the true one, where the velocity is everywhere finite and separation occurs only at the trailing edge. The solution now becomes unique and, in particular, the circulation Γ needed to calculate the lift force has a definite value.

§47. Induced drag

An important part of the drag on a streamlined wing (of finite span) is formed by the drag due to the dissipation of energy in the thin turbulent wake. This is called the *induced drag*.

It has been shown in §21 how we may calculate the drag force due to the wake by considering the flow far from the body. Formula (21.1), however, is not applicable in the present case. According to that formula, the drag is given by the integral of v_x over the cross-section of the wake, i.e. the discharge through the wake. On account of the thinness of the wake beyond a stream lined wing, however, the discharge is small in the present case, and may be neglected in the approximation used below.

As in §21, we write the force F_x as the difference between the total fluxes of the x-component of momentum through the planes $x = x_1$ and $x = x_2$ passing respectively far behind and far in front of the body. Writing the three velocity components as $U+v_x$, v_y, v_z, we have for the component Π_{xx} of the momentum flux density the expression $\Pi_{xx} = p+\rho(U+v_x)^2$, so that the drag force is

$$F_x = \left(\iint_{x=x_2} - \iint_{x=x_1} \right)[p+\rho(U+v_x)^2]dy\,dz. \qquad (47.1)$$

On account of the thinness of the wake, we can neglect, in the integral over the plane $x = x_1$, the integral over the cross-section of the wake, and so integrate only over the region outside the wake. In that region, however, we have potential flow, and Bernoulli's equation $p+\frac{1}{2}\rho(U+v)^2 = p_0+\frac{1}{2}\rho U^2$ holds, whence

$$p = p_0-\rho Uv_x-\tfrac{1}{2}\rho(v_x^2+v_y^2+v_z^2). \qquad (47.2)$$

Here we cannot neglect the quadratic terms as we did in §21, since it is these terms which determine the required drag force in the case under considera-tion. Substituting (47.2) in (47.1), we obtain

$$F_x = \left(\iint_{x=x_2} - \iint_{x=x_1} \right)[p_0+\rho U^2+\rho Uv_x+\tfrac{1}{2}\rho(v_x^2-v_y^2-v_z^2)]dy\,dz.$$

The difference of the integrals of the constant $p_0+\rho U^2$ is zero; the difference of the integrals of $\rho U v_x$ is likewise zero, since the mass fluxes

$$\iint \rho v_x \, dy \, dz$$

through the front and back planes must be the same (we neglect the discharge through the wake in the approximation here considered). Next, if we take the plane $x = x_2$ sufficiently far in front of the body, the velocity v on this plane is very small, so that the integral of $\frac{1}{2}\rho(v_x{}^2 - v_y{}^2 - v_z{}^2)$ over this plane may be neglected. Finally, in flow past a streamlined wing, the velocity v_x outside the wake is small compared with v_y and v_z. Hence we can neglect $v_x{}^2$ compared with $v_y{}^2 + v_z{}^2$ in the integral over the plane $x = x_1$. Thus we obtain

$$F_x = \frac{1}{2}\rho \iint (v_y{}^2 + v_z{}^2) dy \, dz, \tag{47.3}$$

where the integration is over a plane $x = $ constant lying at a great distance behind the body, the cross-section of the wake being excluded from the region of integration.†

The drag on a streamlined wing calculated in this way can be expressed in terms of the velocity circulation Γ which determines the lift also. To do this, we first of all notice that, at sufficiently great distances from the body, the velocity depends only slightly on the co-ordinate x, and so we can regard $v_y(y, z)$ and $v_z(y, z)$ as the velocity of a two-dimensional flow, supposed independent of x. It is convenient to use as an auxiliary quantity the stream function (§10), so that $v_z = \partial\psi/\partial y$, $v_y = -\partial\psi/\partial z$. Then

$$F_x = \frac{1}{2}\rho \iint \left[\left(\frac{\partial\psi}{\partial y} \right)^2 + \left(\frac{\partial\psi}{\partial z} \right)^2 \right] dy \, dz,$$

where the integration over the vertical co-ordinate y is from $+\infty$ to y_1 and from y_2 to $-\infty$, where y_1 and y_2 are the co-ordinates of the upper and lower boundaries of the wake (see Fig. 18, §37). Since we have potential flow (**curl v** $= 0$) outside the wake, $\partial^2\psi/\partial y^2 + \partial^2\psi/\partial z^2 = 0$. Using the two-dimensional Green's formula, we thus find

$$F_x = -\frac{1}{2}\rho \oint \psi(\partial\psi/\partial n) dl,$$

where the integral is taken along a contour bounding the region of integration in the original integral, and $\partial/\partial n$ denotes differentiation in the direction of the outward normal to the contour. At infinity $\psi = 0$, and so the integral is taken

---

† To avoid misunderstanding we should point out the following. Formula (47.3) may give the impression that the velocities v_y, v_z do not decrease in order of magnitude as x increases. This is true so long as the thickness of the wake is small compared with its width, as we have assumed in deriving formula (47.3). At very large distances behind the wing, the wake finally becomes so thick that it becomes approximately circular in cross-section. At this point, formula (47.3) is invalid, and v_y, v_z diminish rapidly with increasing x.

round the cross-section of the wake by the yz-plane, giving

$$F_x = \tfrac{1}{2}\rho \int \psi \left[\left(\frac{\partial \psi}{\partial y} \right)_2 - \left(\frac{\partial \psi}{\partial y} \right)_1 \right] dz.$$

Here the integration is over the width of the wake, and the difference in the brackets is the discontinuity of the derivative $\partial \psi / \partial y$ across the wake. Since $\partial \psi / \partial y = v_z = \partial \phi / \partial z$, we have

$$\left(\frac{\partial \psi}{\partial y} \right)_2 - \left(\frac{\partial \psi}{\partial y} \right)_1 = \left(\frac{\partial \phi}{\partial z} \right)_2 - \left(\frac{\partial \phi}{\partial z} \right)_1 = \frac{d\Gamma}{dz},$$

so that

$$F_x = \tfrac{1}{2}\rho \int \psi (d\Gamma/dz) dz.$$

Finally, we use a formula from potential theory,

$$\psi = -\frac{1}{2\pi} \int \left[\left(\frac{\partial \psi}{\partial n} \right)_2 - \left(\frac{\partial \psi}{\partial n} \right)_1 \right] \log r \, dl,$$

where the integration is along a plane contour, r is the distance from dl to the point where ψ is to be found, and the expression in brackets is the given discontinuity of the derivative of ψ in the direction normal to the contour.† In our case the contour of integration is a segment of the z-axis, so that we can write the value of the function $\psi(y, z)$ on the z-axis as

$$\psi(0, z) = \frac{1}{2\pi} \left[\left(\frac{\partial \psi}{\partial y} \right)_1 - \left(\frac{\partial \psi}{\partial y} \right)_2 \right] \log |z - z'| dz'$$

$$= -\frac{1}{2\pi} \int \frac{d\Gamma(z')}{dz'} \log |z - z'| dz'.$$

Finally, substituting this in F_x, we obtain the following formula for the induced drag:

$$F_x = -\frac{\rho}{4\pi} \int_0^l \int_0^l \frac{d\Gamma(z)}{dz} \frac{d\Gamma(z')}{dz'} \log |z - z'| dz \, dz' \qquad (47.4)$$

(L. PRANDTL, 1918). The span of the wing is here denoted by $l_z = l$, and the origin of z is at one end of the wing.

If all the dimensions in the z-direction are increased by some factor (Γ

---

† This formula gives, in two-dimensional potential theory, the potential due to a charged plane contour with a charge density

$$[(\partial \psi / \partial n)_2 - (\partial \psi / \partial n)_1]/2\pi.$$

remaining constant), the integral (47.4) remains constant.† This shows that the total induced drag on the wing remains of the same order of magnitude when its span is increased. In other words, the induced drag per unit length of the wing decreases with increasing length.‡ Unlike the drag, the total lift force

$$F_y = -\rho U \int \Gamma \, dz \tag{47.5}$$

increases almost linearly with the span of the wing, and the lift per unit length is constant.

The following method is convenient for the actual calculation of the integrals (47.4) and (47.5). Instead of the co-ordinate z, we introduce a new variable θ, defined by

$$z = \tfrac{1}{2}l(1 - \cos\theta) \qquad (0 \leqslant \theta \leqslant \pi). \tag{47.6}$$

The distribution of the velocity circulation is written as a Fourier series:

$$\Gamma = -2Ul \sum_{n=1}^{\infty} A_n \sin n\theta. \tag{47.7}$$

The condition that $\Gamma = 0$ at the ends of the wing ($z = 0$ and l, or $\theta = 0$ and π) is then fulfilled.

Substituting the expression (47.7) in (47.5) and effecting the integration (using the orthogonality of the functions $\sin\theta$ and $\sin n\theta$ for $n \neq 1$), we obtain $F_y = \tfrac{1}{2}\rho U^2 \pi l^2 A_1$. Thus the lift force depends only on the first coefficient in the expansion (47.7). For the lift coefficient (46.2) we have

$$C_y = \pi\lambda A_1, \tag{47.8}$$

where we have introduced the ratio $\lambda = l/l_x$ of span to width of the wing.

To calculate the drag, we rewrite formula (47.4), integrating once by parts:

$$F_x = \frac{\rho}{4\pi} \int_0^l \int_0^l \Gamma(z) \frac{d\Gamma(z')}{dz'} \frac{dz' \, dz}{z - z'}. \tag{47.9}$$

---

† To avoid misunderstanding, we should mention that it does not matter that the logarithm in the integrand is increased by a constant when the unit of length is changed. For the integral which differs from that in (47.4) by having a constant instead of $\log|z - z'|$ is zero, since

$$\int (d\Gamma/dz)\,dz = \Gamma,$$

and the definite integral is zero because Γ vanishes at the edges of the wake.

‡ In the limit of infinite span, the induced drag per unit length is zero. In reality, a small amount of drag remains, determined by the discharge through the wake (i.e. the integral $\iint v_x \, dy \, dz$), which we have neglected in deriving formula (47.3). This drag includes both the frictional drag and the remaining part due to dissipation in the wake.

It is easily seen that the integral over z' must be taken as a principal value. An elementary calculation, with the substitution (47.7),† leads to the following formula for the induced drag coefficient:

$$C_x = \pi\lambda \sum_{n=1}^{\infty} nA_n{}^2. \tag{47.10}$$

The drag coefficient for a wing is defined as

$$C_x = F_x/\tfrac{1}{2}\rho U^2 l_x l_z, \tag{47.11}$$

being referred, like the lift coefficient, to unit area in the xz-plane.

PROBLEM

Determine the least value of the induced drag for a given lift and a given span $l_z = l$.

SOLUTION. It is clear from formulae (47.8) and (47.10) that the least value of C_x for given C_y (i.e. for given A_1) is obtained if all A_n for $n \neq 1$ are zero. Then

$$C_{x,\min} = C_y{}^2/\pi\lambda. \tag{1}$$

The distribution of velocity circulation over the span is given by the formula

$$\Gamma = -\frac{4}{\pi l} U l_x C_y \sqrt{[z(l-z)]}. \tag{2}$$

If the span is sufficiently large, then the flow round any cross-section of the wing is approximately two-dimensional flow round a wing of infinite length and the same cross-section. In this case we can say that the circulation distribution (2) is obtained for a wing whose shape in the xz-plane is an ellipse with semi-axes $\tfrac{4}{3}l_x$ and $\tfrac{1}{2}l$.

§48. The lift of a thin wing

The problem of calculating the lift force on a wing amounts, by Zhukovskii's theorem, to that of finding the velocity circulation Γ. A general solution of the latter problem can be given for a thin streamlined wing of infinite span, the cross-section being the same at every point.‡ The elegant method of

---

† In integrating over z' we need the integral

$$P\int_0^{\pi} \frac{\cos n\theta'}{\cos\theta' - \cos\theta}\,d\theta' = \frac{\pi\sin n\theta}{\sin\theta}.$$

In integrating over z we use the fact that

$$\int_0^{\pi} \sin n\theta \sin m\theta\,d\theta = \tfrac{1}{2}\pi \quad (m = n),$$

$$= 0 \quad (m \neq n).$$

‡ A more detailed account of the theory of two-dimensional incompressible flow past a wing is given by N. E. KOCHIN, I. A. KIBEL' and N. V. ROZE, *Theoretical Hydromechanics* (*Teoreticheskaya gidromekhanika*), Part 1, 4th ed., Moscow 1948; L. I. SEDOV, *Two-dimensional Problems of Hydrodynamics and Aerodynamics* (*Ploskie zadachi gidrodinamiki i aërodinamiki*), Moscow 1950.

solution given below is due to M. V. KELDYSH and L. I. SEDOV (1939).

Let $y = \zeta_1(x)$ and $y = \zeta_2(x)$ be the equations of the lower and upper parts of the cross-sectional profile (Fig. 27). We suppose this profile to be thin, only slightly curved, and inclined at a small angle of attack to the main stream (the x-axis); that is, both ζ_1, ζ_2 themselves and their derivatives ζ_1', ζ_2' are small, i.e. the normal to the profile contour is everywhere almost parallel

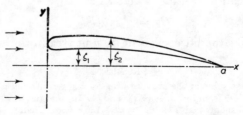

FIG. 27

to the y-axis. Under these conditions, we may suppose the perturbation **v** in the fluid velocity, caused by the presence of the wing, to be everywhere[†] small compared with the main-stream velocity U. The boundary condition at the surface of the wing is $v_y/U = \zeta'$ for $y = \zeta$. By virtue of the assumptions made, we can suppose this condition to hold for $y = 0$, and not for $y = \zeta$. Then we must have on the axis of abscissae between $x = 0$ and $x = l_x \equiv a$

$$v_y = U\zeta_2'(x) \text{ for } y \to 0+, \qquad v_y = U\zeta_1'(x) \text{ for } y \to 0-. \qquad (48.1)$$

In order to apply the methods of the theory of functions of a complex variable, we introduce the complex velocity $dw/dz = v_x - iv_y$ (cf. §10), which is an analytic function of the variable $z = x + iy$. In the present case this function must satisfy the conditions

$$\text{im}(dw/dz) = -U\zeta_2'(x) \quad \text{for} \quad y \to 0+,$$
$$\text{im}(dw/dz) = -U\zeta_1'(x) \quad \text{for} \quad y \to 0-, \qquad (48.2)$$

on the segment $(0, a)$ of the axis of abscissae.

To solve the above problem, we first represent the required velocity distribution $\mathbf{v}(x, y)$ as a sum $\mathbf{v} = \mathbf{v}^+ + \mathbf{v}^-$ of two distributions having the following symmetry properties:

$$v^-_x(x, -y) = v^-_x(x, y), \qquad v^-_y(x, -y) = -v^-_y(x, y),$$
$$v^+_x(x, -y) = -v^+_x(x, y), \qquad v^+_y(x, -y) = v^+_y(x, y). \qquad (48.3)$$

These properties of the separate distributions $\mathbf{v}^-$ and $\mathbf{v}^+$ do not violate the equation of continuity or that of potential flow, and, since the problem is linear, the two distributions may be sought separately.

---

† Except in a small region near the rounded leading edge of the wing.

The complex velocity is correspondingly represented as a sum

$$w' = w'_+ + w'_-,$$

and the boundary conditions on the segment $(0, a)$ for the two terms of the sum are

$$[\operatorname{im} w'_+]_{y\to 0+} = [\operatorname{im} w'_+]_{y\to 0-} = -\tfrac{1}{2}U(\zeta_1' + \zeta_2'),$$
$$[\operatorname{im} w'_-]_{y\to 0+} = -[\operatorname{im} w'_-]_{y\to 0-} = \tfrac{1}{2}U(\zeta_1' - \zeta_2'). \tag{48.4}$$

The function w'_- can be determined at once by Cauchy's formula:

$$w'_-(z) = \frac{1}{2\pi i} \oint_L \frac{w'_-(\xi)}{\xi - z} d\xi,$$

where the integration in the plane of the complex variable ξ is along a circle L of small radius centred at the point $\xi = z$ (Fig. 28). The contour L can

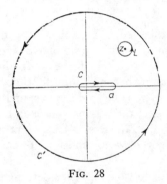

Fig. 28

be replaced by a circle C' of infinite radius and a contour C traversed clockwise; the latter can be deformed into the segment $(0, a)$ twice over. The integral along C' is zero, since $w'(z)$ vanishes at infinity. The integral along C gives

$$w'_- = -\frac{U}{2\pi} \int_0^a \frac{\zeta_2'(\xi) - \zeta_1'(\xi)}{\xi - z} d\xi. \tag{48.5}$$

Here we have used the boundary values (48.4) of the imaginary part of w'_- on the segment $(0, a)$, and the fact that, by the symmetry conditions (48.3), the real part of w'_- is continuous across this segment.

To find the function w'_+, we have to apply Cauchy's formula, not to this function itself, but to the product $w'_+(z)g(z)$, where $g(z) = \sqrt{[z/(z-a)]}$, and the square root is taken with the plus sign for $z = x > a$. On the segment $(0, a)$ of the real axis, the function $g(z)$ is purely imaginary and discontinuous: $g(x+i0) = -g(x-i0) = -i\sqrt{[x/(a-x)]}$. It is clear from these properties of the function $g(z)$ that the imaginary part of the product gw'_+

is discontinuous across the segment $(0, a)$, while the real part is continuous, as with the function w'_-. Hence we have, exactly as in the derivation of formula (48.5),

$$w'_+(z)g(z) = -\frac{U}{2\pi}\int_0^a \frac{\zeta_1'(\xi)+\zeta_2'(\xi)}{\xi-z}g(\xi+i0)d\xi.$$

Collecting the above expressions, we have the following formula for the velocity distribution in flow past a thin wing:

$$\frac{dw}{dz} = -\frac{U}{2\pi i}\sqrt{\frac{z-a}{z}}\int_0^a \frac{\zeta_1'(\xi)+\zeta_2'(\xi)}{\xi-z}\sqrt{\frac{\xi}{a-\xi}}d\xi -$$

$$-\frac{U}{2\pi}i\int_0^a \frac{\zeta_2'(\xi)-\zeta_1'(\xi)}{\xi-z}d\xi. \tag{48.6}$$

Near the rounded leading edge (i.e. for $z \to 0$), this expression in general becomes infinite, the approximation used above being invalid in this region. Near the pointed trailing edge (i.e. for $z \to a$), the first term in (48.6) is finite, but the second term becomes infinite, though only logarithmically.[†] This logarithmic singularity is due to the approximation used, and is removed by a more exact treatment; there is no power-law divergence at the trailing edge, in accordance with the Zhukovskiĭ–Chaplygin condition. The fulfilment of this condition is achieved by an appropriate choice of the function $g(z)$ used above.

Formula (48.6) immediately enables us to determine the velocity circulation Γ round the wing profile. According to the general rule (see §10), Γ is given by the residue of the function $w'(z)$ at its simple pole $z = 0$. The required residue is easily found as the coefficient of $1/z$ in an expansion of $w'(z)$ in powers of $1/z$ about the point at infinity: $dw/dz = \Gamma/2\pi iz+\ldots$, and Γ is given by the simple formula

$$\Gamma = U\int_0^a (\zeta_1'+\zeta_2')\sqrt{\frac{\xi}{a-\xi}}d\xi. \tag{48.7}$$

We may point out that only the sum of the functions ζ_1 and ζ_2 appears here. The lift force is unchanged if the thin wing is replaced by a bent plate whose shape is given by the function $\frac{1}{2}(\zeta_1+\zeta_2)$.

For example, for a wing in the form of a thin plate of infinite length, inclined at a small angle of attack α, we have $\zeta_1 = \zeta_2 = \alpha(a-x)$, and formula (48.7) gives $\Gamma = -\pi\alpha aU$. The lift coefficient for such a wing is $C_y = -\rho U\Gamma/\frac{1}{2}\rho U^2 a = 2\pi\alpha$.

---

† This divergence disappears if ζ_1 and ζ_2 vanish as $(a-x)^k$, $k > 1$, near the trailing edge, i.e. if the point at the trailing edge is a cusp.

CHAPTER V

THERMAL CONDUCTION IN FLUIDS

§49. The general equation of heat transfer

IT has been mentioned at the end of §2 that a complete system of equations of fluid dynamics must contain five equations. For a fluid in which processes of thermal conduction and internal friction occur, one of these equations is, as before, the equation of continuity, and Euler's equations are replaced by the Navier–Stokes equations. The fifth equation for an ideal fluid is the equation of conservation of entropy (2.6). In a viscous fluid this equation does not hold, of course, since irreversible processes of energy dissipation occur in it.

In an ideal fluid the law of conservation of energy is expressed by equation (6.1):

$$\frac{\partial}{\partial t}(\tfrac{1}{2}\rho v^2 + \rho\epsilon) = -\operatorname{div}[\rho\mathbf{v}(\tfrac{1}{2}v^2 + w)].$$

The expression on the left is the rate of change of the energy in unit volume of the fluid, while that on the right is the divergence of the energy flux density. In a viscous fluid the law of conservation of energy still holds, of course: the change per unit time in the total energy of the fluid in any volume must still be equal to the total flux of energy through the surface bounding that volume. The energy flux density, however, now has a different form. Besides the flux $\rho\mathbf{v}(\tfrac{1}{2}v^2 + w)$ due to the simple transfer of mass by the motion of the fluid, there is also a flux due to processes of internal friction. This latter flux is given by the vector $\mathbf{v}\cdot\boldsymbol{\sigma}'$, with components $v_i\sigma'_{ik}$ (see §16). There is, moreover, another term that must be included in the energy flux. If the temperature of the fluid is not constant throughout its volume, there will be, besides the two means of energy transfer indicated above, a transfer of heat by what is called *thermal conduction*. This signifies the direct molecular transfer of energy from points where the temperature is high to those where it is low. It does not involve macroscopic motion, and occurs even in a fluid at rest.

We denote by $\mathbf{q}$ the heat flux density due to thermal conduction. The flux $\mathbf{q}$ is related to the variation of temperature through the fluid. This relation can be written down at once in cases where the temperature gradient in the fluid is not large; in phenomena of thermal conduction we are almost always concerned with such cases. We can then expand $\mathbf{q}$ as a series of powers of the temperature gradient, taking only the first terms of the expansion. The constant term is evidently zero, since $\mathbf{q}$ must vanish when **grad** T does so. Thus we have

$$\mathbf{q} = -\kappa\,\mathbf{grad}\,T. \tag{49.1}$$

The constant κ is called the *thermal conductivity*. It is always positive, as we see at once from the fact that the energy flux must be from points at a high temperature to those at a low temperature, i.e. $\mathbf{q}$ and $\mathbf{grad}\, T$ must be in opposite directions. The coefficient κ is in general a function of temperature and pressure.

Thus the total energy flux in a fluid when there is viscosity and thermal conduction is $\rho\mathbf{v}(\tfrac{1}{2}v^2+w)-\mathbf{v}\cdot\boldsymbol{\sigma}'-\kappa\,\mathbf{grad}\, T$. Accordingly, the general law of conservation of energy is given by the equation

$$\frac{\partial}{\partial t}(\tfrac{1}{2}\rho v^2+\rho\epsilon) = -\operatorname{div}[\rho\mathbf{v}(\tfrac{1}{2}v^2+w)-\mathbf{v}\cdot\boldsymbol{\sigma}'-\kappa\,\mathbf{grad}\, T]. \tag{49.2}$$

This equation could be taken to complete the system of fluid-mechanical equations of a viscous fluid. It is convenient, however, to put it in another form by transforming it with the aid of the equations of motion. To do so, we calculate the time derivative of the energy in unit volume of fluid, starting from the equations of motion. We have

$$\frac{\partial}{\partial t}(\tfrac{1}{2}\rho v^2+\rho\epsilon) = \tfrac{1}{2}v^2\frac{\partial\rho}{\partial t} + \rho\mathbf{v}\cdot\frac{\partial\mathbf{v}}{\partial t} + \rho\frac{\partial\epsilon}{\partial t} + \epsilon\frac{\partial\rho}{\partial t}.$$

Substituting for $\partial\rho/\partial t$ from the equation of continuity and for $\partial\mathbf{v}/\partial t$ from the Navier–Stokes equation, we have

$$\frac{\partial}{\partial t}(\tfrac{1}{2}\rho v^2+\rho\epsilon) = -\tfrac{1}{2}v^2\operatorname{div}(\rho\mathbf{v})-\rho\mathbf{v}\cdot\mathbf{grad}\,\tfrac{1}{2}v^2-\mathbf{v}\cdot\mathbf{grad}\,p+$$

$$+v_i\frac{\partial\sigma'_{ik}}{\partial x_k} + \rho\frac{\partial\epsilon}{\partial t} - \epsilon\operatorname{div}(\rho\mathbf{v}).$$

Using now the thermodynamic relation $d\epsilon = T\,ds-p\,dV = T\,ds+(p/\rho^2)d\rho$, we find

$$\frac{\partial\epsilon}{\partial t} = T\frac{\partial s}{\partial t}+\frac{p}{\rho^2}\frac{\partial\rho}{\partial t} = T\frac{\partial s}{\partial t}-\frac{p}{\rho^2}\operatorname{div}(\rho\mathbf{v}).$$

Substituting this and introducing the heat function $w = \epsilon+p/\rho$, we obtain

$$\frac{\partial}{\partial t}(\tfrac{1}{2}\rho v^2+\rho\epsilon) = -(\tfrac{1}{2}v^2+w)\operatorname{div}(\rho\mathbf{v})-\rho\mathbf{v}\cdot\mathbf{grad}\,\tfrac{1}{2}v^2-\mathbf{v}\cdot\mathbf{grad}\,p+$$

$$+\rho T\frac{\partial s}{\partial t} + v_i\frac{\partial\sigma'_{ik}}{\partial x_k}.$$

Next, from the thermodynamic relation $dw = T\,ds+dp/\rho$ we have $\mathbf{grad}\,p = \rho\,\mathbf{grad}\,w-\rho T\,\mathbf{grad}\,s$. The last term on the right of the above equation can be written

$$v_i\frac{\partial\sigma'_{ik}}{\partial x_k} = \frac{\partial}{\partial x_k}(v_i\sigma'_{ik})-\sigma'_{ik}\frac{\partial v_i}{\partial x_k} \equiv \operatorname{div}(\mathbf{v}\cdot\boldsymbol{\sigma}')-\sigma'_{ik}\frac{\partial v_i}{\partial x_k}.$$

Substituting these expressions, and adding and subtracting $\mathrm{div}(\kappa\ \mathbf{grad}\ T)$, we obtain

$$\frac{\partial}{\partial t}(\tfrac{1}{2}\rho v^2 + \rho\epsilon) = -\mathrm{div}[\rho\mathbf{v}(\tfrac{1}{2}v^2 + w) - \mathbf{v}\cdot\boldsymbol{\sigma}' - \kappa\ \mathbf{grad}\ T] +$$

$$+ \rho T\left(\frac{\partial s}{\partial t} + \mathbf{v}\cdot\mathbf{grad}\,s\right) - \sigma'_{ik}\frac{\partial v_i}{\partial x_k} - \mathrm{div}(\kappa\ \mathbf{grad}\ T). \qquad (49.3)$$

Comparing this expression for the time derivative of the energy in unit volume with (49.2), we have

$$\rho T\left(\frac{\partial s}{\partial t} + \mathbf{v}\cdot\mathbf{grad}\,s\right) = \sigma'_{ik}\frac{\partial v_i}{\partial x_k} + \mathrm{div}(\kappa\ \mathbf{grad}\ T). \qquad (49.4)$$

This equation is called the *general equation of heat transfer.* If there is no viscosity or thermal conduction, the right-hand side is zero, and the equation of conservation of entropy (2.6) for an ideal fluid is obtained.

The following interpretation of equation (49.4) should be noticed. The expression on the left is just the total time derivative ds/dt of the entropy, multiplied by ρT. The quantity ds/dt gives the rate of change of the entropy of a unit mass of fluid as it moves about in space, and $T\,ds/dt$ is therefore the quantity of heat gained by this unit mass in unit time, so that $\rho T\,ds/dt$ is the quantity of heat gained per unit volume. We see from (49.4) that the amount of heat gained by unit volume of the fluid is therefore

$$\sigma'_{ik}\,\partial v_i/\partial x_k + \mathrm{div}(\kappa\ \mathbf{grad}\ T).$$

The first term here is the energy dissipated into heat by viscosity, and the second is the heat conducted into the volume concerned.

We expand the term $\sigma'_{ik}\partial v_i/\partial x_k$ in (49.4) by substituting the expression (15.3) for σ'_{ik}. We have

$$\sigma'_{ik}\frac{\partial v_i}{\partial x_k} = \eta\frac{\partial v_i}{\partial x_k}\left(\frac{\partial v_i}{\partial x_k} + \frac{\partial v_k}{\partial x_i} - \tfrac{2}{3}\delta_{ik}\frac{\partial v_l}{\partial x_l}\right) + \zeta\frac{\partial v_i}{\partial x_k}\delta_{ik}\frac{\partial v_l}{\partial x_l}.$$

It is easy to verify that the first term may be written as

$$\tfrac{1}{2}\eta\left(\frac{\partial v_i}{\partial x_k} + \frac{\partial v_k}{\partial x_i} - \tfrac{2}{3}\delta_{ik}\frac{\partial v_l}{\partial x_l}\right)^2,$$

and the second is

$$\zeta\frac{\partial v_i}{\partial x_k}\delta_{ik}\frac{\partial v_l}{\partial x_l} = \zeta\frac{\partial v_i}{\partial x_i}\frac{\partial v_l}{\partial x_l} \equiv \zeta(\mathrm{div}\,\mathbf{v})^2.$$

Thus equation (49.4) becomes

$$\rho T\left(\frac{\partial s}{\partial t} + \mathbf{v}\cdot\mathbf{grad}\,s\right) = \mathrm{div}(\kappa\ \mathbf{grad}\ T) + \tfrac{1}{2}\eta\left(\frac{\partial v_i}{\partial x_k} + \frac{\partial v_k}{\partial x_i} - \tfrac{2}{3}\delta_{ik}\frac{\partial v_l}{\partial x_l}\right)^2 +$$

$$+ \zeta(\mathrm{div}\,\mathbf{v})^2. \qquad (49.5)$$

The entropy of the fluid increases as a result of the irreversible processes of thermal conduction and internal friction. Here, of course, we mean not the entropy of each volume element of fluid separately, but the total entropy of the whole fluid, equal to the integral

$$\int \rho s \, dV.$$

The change in entropy per unit time is given by the derivative

$$d[\int \rho s \, dV]/dt = \int [\partial(\rho s)/\partial t] dV.$$

Using the equation of continuity and equation (49.5) we have

$$\frac{\partial(\rho s)}{\partial t} = \rho \frac{\partial s}{\partial t} + s \frac{\partial \rho}{\partial t} = -s \, \text{div}(\rho\mathbf{v}) - \rho\mathbf{v} \cdot \mathbf{grad} \, s + \frac{1}{T} \text{div}(\kappa \, \mathbf{grad} \, T) +$$

$$+ \frac{\eta}{2T} \left(\frac{\partial v_i}{\partial x_k} + \frac{\partial v_k}{\partial x_i} - \tfrac{2}{3}\delta_{ik} \frac{\partial v_l}{\partial x_l} \right)^2 + \frac{\zeta}{T}(\text{div}\,\mathbf{v})^2.$$

The first two terms on the right together give $-\text{div}(\rho s \mathbf{v})$. The volume integral of this is transformed into the integral of the entropy flux $\rho s \mathbf{v}$ over the surface. If we consider an unbounded volume of fluid at rest at infinity, the bounding surface can be removed to infinity; the integrand in the surface integral is then zero, and so is the integral itself. The integral of the third term on the right is transformed as follows:

$$\int \frac{1}{T} \text{div}(\kappa \, \mathbf{grad} \, T) dV = \int \text{div} \left(\frac{\kappa \, \mathbf{grad} \, T}{T} \right) dV + \int \frac{\kappa(\mathbf{grad} \, T)^2}{T^2} dV.$$

Assuming that the fluid temperature tends sufficiently rapidly to a constant value at infinity, we can transform the first integral into one over an infinitely remote surface, on which $\mathbf{grad} \, T = 0$ and the integral therefore vanishes. The result is

$$\frac{d}{dt} \int \rho s \, dV = \int \frac{\kappa(\mathbf{grad} \, T)^2}{T^2} dV + \int \frac{\eta}{2T} \left(\frac{\partial v_i}{\partial x_k} + \frac{\partial v_k}{\partial x_i} - \tfrac{2}{3}\delta_{ik} \frac{\partial v_l}{\partial x_l} \right)^2 dV +$$

$$+ \int \frac{\zeta}{T}(\text{div}\,\mathbf{v})^2 \, dV. \tag{49.6}$$

The first term on the right is the rate of increase of entropy owing to thermal conduction, and the other two terms give the rate of increase due to internal friction. The entropy can only increase, i.e. the sum on the right of (49.6) must be positive. In each term, the integrand may be non-zero even if the other two integrals vanish. Hence it follows that the second viscosity coefficient ζ is positive, as well as κ and η, which we already know are positive.

It has been tacitly assumed in the derivation of formula (49.1) that the heat flux depends only on the temperature gradient, and not on the pressure gradient. This assumption, which is not evident *a priori*, can now be justified

as follows. If **q** contained a term proportional to **grad** p, the expression
(49.6) for the rate of change of entropy would include another term having
the product **grad** $p \cdot$ **grad** T in the integrand. Since the latter might be either
positive or negative, the time derivative of the entropy would not necessarily
be positive, which is impossible.

Finally, the above arguments must also be refined in the following respect.
Strictly speaking, in a system which is not in thermodynamic equilibrium,
such as a fluid with velocity and temperature gradients, the usual definitions
of thermodynamic quantities are no longer meaningful, and must be modified.
The necessary definitions are, firstly, that ρ, ϵ and **v** are defined as before:
ρ and $\rho\epsilon$ are the mass and internal energy per unit volume, and **v** is the
momentum of unit mass of fluid. The remaining thermodynamic quantities
are then defined as being the same functions of ρ and ϵ as they are in
thermal equilibrium. The entropy $s = s(\rho, \epsilon)$, however, is no longer the true
thermodynamic entropy: the integral

$$\int \rho s \, dV$$

will not, strictly speaking, be a quantity that must increase with time. Nevertheless, it is easy to see that, for small velocity and temperature gradients, s
is the same as the true entropy in the approximation here used. For, if there
are gradients present, they in general lead to additional terms (besides
$s(\rho, \epsilon)$) in the entropy. The results given above, however, can be altered only
by terms linear in the gradients (for instance, a term proportional to the scalar
div **v**). Such terms would necessarily take both positive and negative values.
But they ought to be negative definite, since the equilibrium value $s = s(\rho, \epsilon)$
is the maximum possible value. Hence the expansion of the entropy in powers
of the small gradients can contain (apart from the zero-order term) only
terms of the second and higher orders.

Similar remarks should have been made in §15 (cf. the first footnote to that
section), since the presence of even a velocity gradient implies the absence of
thermodynamic equilibrium. The pressure p which appears in the expression
for the momentum flux density tensor in a viscous fluid must be taken to be
the same function $p = p(\rho, \epsilon)$ as in thermal equilibrium. In this case p
will not, strictly speaking, be the pressure in the usual sense, viz. the normal
force on a surface element. Unlike what happens for the entropy (see
above), there is here a resulting difference of the first order with respect to
the small gradient; we have seen that the normal component of the force
includes, besides p, a term proportional to div **v** (in an incompressible fluid,
this term is zero, and the difference is then of higher order).

Thus the three coefficients η, ζ, κ which appear in the equations of
motion of a viscous conducting fluid completely determine the fluid-mechanical properties of the fluid in the approximation considered (i.e. when the
higher-order space derivatives of velocity, temperature, etc. are neglected).
The introduction of any further terms (for example, the inclusion in the mass
flux density of terms proportional to the gradient of density or temperature)

has no physical meaning, and would mean at least a change in the definition of the basic quantities; in particular, the velocity would no longer be the momentum of unit mass of fluid.†

§50. Thermal conduction in an incompressible fluid

The general equation of thermal conduction (49.4) or (49.5) can be considerably simplified in certain cases. If the fluid velocity is small compared with the velocity of sound, the pressure variations occurring as a result of the motion are so small that the variation in the density (and in the other thermodynamic quantities) caused by them may be neglected. However, a non-uniformly heated fluid is still not completely incompressible in the sense used previously. The reason is that the density varies with the temperature; this variation cannot in general be neglected, and therefore, even at small velocities, the density of a non-uniformly heated fluid cannot be supposed constant. In determining the derivatives of thermodynamic quantities in this case, it is therefore necessary to suppose the pressure constant, and not the density. Thus we have

$$\frac{\partial s}{\partial t} = \left(\frac{\partial s}{\partial T}\right)_p \frac{\partial T}{\partial t}, \qquad \mathbf{grad}\, s = \left(\frac{\partial s}{\partial T}\right)_p \mathbf{grad}\, T,$$

and, since $T(\partial s/\partial T)_p$ is the specific heat at constant pressure c_p, we obtain $T\partial s/\partial t = c_p \partial T/\partial t$, $T\, \mathbf{grad}\, s = c_p\, \mathbf{grad}\, T$. Equation (49.4) becomes

$$\rho c_p \left(\frac{\partial T}{\partial t} + \mathbf{v}\cdot\mathbf{grad}\, T\right) = \mathrm{div}(\kappa\, \mathbf{grad}\, T) + \sigma'_{ik}\frac{\partial v_i}{\partial x_k}. \tag{50.1}$$

If the density is to be supposed constant in the equations of motion for a non-uniformly heated fluid, it is necessary that the fluid velocity should be small compared with that of sound, and also that the temperature differences in the fluid should be small. We emphasise that we mean the actual values of the temperature differences, not the temperature gradient. The fluid may then be supposed incompressible in the usual sense; in particular, the equation of continuity is simply $\mathrm{div}\, \mathbf{v} = 0$. Supposing the temperature differences small, we neglect also the temperature variation of η, κ and c_p, supposing them constant. Writing the term $\sigma'_{ik}\, \partial v_i/\partial x_k$ as in (49.5), we obtain the

---

† Worse still, the inclusion of such terms may violate the necessary conservation laws. It must be borne in mind that, whatever the definitions used, the mass flux density **j** must always be the momentum of unit volume of fluid. For **j** is defined by the equation of continuity,

$$\partial \rho/\partial t + \mathrm{div}\, \mathbf{j} = 0;$$

multiplying this by **r** and integrating over the fluid volume, we have

$$\mathrm{d}\left(\int \rho \mathbf{r}\, \mathrm{d}V\right)/\mathrm{d}t = \int \mathbf{j}\, \mathrm{d}V,$$

and since the integral $\int \rho \mathbf{r}\, \mathrm{d}V$ determines the position of the centre of mass, it is clear that the integral $\int \mathbf{j}\, \mathrm{d}V$ is the momentum.

equation of heat transfer in an incompressible fluid in the following comparatively simple form:

$$\frac{\partial T}{\partial t} + \mathbf{v} \cdot \mathbf{grad}\, T = \chi \triangle T + \frac{\nu}{2c_p} \left(\frac{\partial v_i}{\partial x_k} + \frac{\partial v_k}{\partial x_i} \right)^2, \tag{50.2}$$

where $\nu = \eta/\rho$ is the kinematic viscosity, and we have written κ in terms of the *thermometric conductivity*, defined as

$$\chi = \kappa/\rho c_p. \tag{50.3}$$

The equation of heat transfer is particularly simple for an incompressible fluid at rest, in which the transfer of energy takes place entirely by thermal conduction. Omitting the terms in (50.2) which involve the velocity, we have simply

$$\partial T/\partial t = \chi \triangle T. \tag{50.4}$$

This equation is called in mathematical physics the *equation of thermal conduction* or *Fourier's equation*. It can, of course, be obtained much more simply without using the general equation of heat transfer in a moving fluid. According to the law of conservation of energy, the amount of heat absorbed in some volume in unit time must equal the total heat flux into this volume through the surface surrounding it. As we know, such a law of conservation can be expressed as an "equation of continuity" for the amount of heat. This equation is obtained by equating the amount of heat absorbed in unit volume in unit time to minus the divergence of the heat flux density. The former is $\rho c_p\, \partial T/\partial t$; we must take the specific heat c_p, since the pressure is of course constant throughout a fluid at rest. Equating this to $-\operatorname{div} \mathbf{q} = \kappa \triangle T$, we have equation (50.4).

It must be mentioned that the applicability of the thermal conduction equation (50.4) to fluids is actually very limited. The reason is that, in fluids in a gravitational field, even a small temperature gradient usually results in considerable motion (*convection*; see §56). Hence we can actually have a fluid at rest with a non-uniform temperature distribution only if the direction of the temperature gradient is opposite to that of the gravitational force, or if the fluid is very viscous. Nevertheless, a study of the equation of thermal conduction in the form (50.4) is very important, since processes of thermal conduction in solids are described by an equation of the same form. We shall therefore consider it in more detail in §§51 and 52.

If the temperature distribution in a non-uniformly heated medium at rest is maintained constant in time (by means of some external source of heat), the equation of thermal conduction becomes

$$\triangle T = 0. \tag{50.5}$$

Thus a steady temperature distribution in a medium at rest satisfies Laplace's equation. In the more general case where κ cannot be regarded a constant, we have in place of (50.5) the equation

$$\operatorname{div}(\kappa\, \mathbf{grad}\, T) = 0. \tag{50.6}$$

If the fluid contains external sources of heat (for example, heating by an electric current), the equation of thermal conduction must correspondingly contain another term. Let Q be the quantity of heat generated by these sources in unit volume of the fluid per unit time; Q is, in general, a function of the co-ordinates and of the time. Then the heat balance equation, i.e. the equation of thermal conduction, is

$$\rho c_p \, \partial T/\partial t = \kappa \triangle T + Q. \tag{50.7}$$

Let us write down the boundary conditions on the equation of thermal conduction which hold at the boundary between two media. First of all, the temperatures of the two media must be equal at the boundary:

$$T_1 = T_2. \tag{50.8}$$

Furthermore, the heat flux out of one medium must equal the heat flux into the other medium. Taking a co-ordinate system in which the part of the boundary considered is at rest, we can write this condition as $\kappa_1 \, \mathbf{grad} \, T_1 \cdot \mathbf{df} = \kappa_2 \, \mathbf{grad} \, T_2 \cdot \mathbf{df}$ for each surface element $\mathbf{df}$. Putting $\mathbf{grad} \, T \cdot \mathbf{df} = (\partial T/\partial n)df$, where $\partial T/\partial n$ is the derivative of T along the normal to the surface, we obtain the boundary condition in the form

$$\kappa_1 \, \partial T_1/\partial n = \kappa_2 \, \partial T_2/\partial n. \tag{50.9}$$

If there are on the surface of separation external sources of heat which generate an amount of heat $Q^{(s)}$ on unit area in unit time, then (50.9) must be replaced by

$$\kappa_1 \, \partial T_1/\partial n - \kappa_2 \, \partial T_2/\partial n = Q^{(s)}. \tag{50.10}$$

In physical problems concerning the distribution of temperature in the presence of heat sources, the strength of the latter is usually given as a function of temperature. If the function $Q(T)$ increases sufficiently rapidly with T, it may be impossible to establish a steady temperature distribution in a body whose boundaries are maintained in fixed conditions (e.g. at a given temperature). The loss of heat through the outer surface of the body is proportional to some mean value of the temperature difference $T - T_0$ between the body and the external medium, regardless of the law of heat generation within the body; it is clear that, if the generation of heat increases sufficiently rapidly with temperature, the loss of heat may be inadequate to achieve an equilibrium state.

The impossibility of establishing a steady thermal state forms the basis of the *thermal theory of explosions* developed by N. N. SEMENOV (1928): if the rate of an exothermic combustion reaction increases sufficiently rapidly with temperature, the impossibility of a steady distribution leads to a rapid non-steady ignition of the substance and an acceleration of the reaction into a thermal explosion. A quantitative theory, for the case where the heat

generation is an exponential function of temperature, has been given by
D. A. FRANK–KAMENETSKIĬ (see Problem 1).†

PROBLEMS

PROBLEM 1. Heat sources of strength $Q = Q_0 e^{\alpha(T-T_0)}$ per unit volume are distributed in a layer of material bounded by two parallel infinite planes, which are kept at a constant temperature T_0. Find the condition for a steady temperature distribution to be possible.

SOLUTION. The equation for steady heat conduction is here

$$\kappa \, d^2 T/dx^2 = -Q_0 e^{\alpha(T-T_0)},$$

with the boundary conditions $T = T_0$ for $x = 0$ and $x = 2l$ ($2l$ being the thickness of the layer). We introduce the dimensionless variables $\tau = \alpha(T-T_0)$ and $\xi = x/l$. Then

$$\tau'' + \lambda e^\tau = 0, \qquad \lambda = Q_0 \alpha l^2/\kappa.$$

Integrating this equation once (after multiplying by $2\tau'$), we find

$$\tau'^2 = 2\lambda(e^{\tau_0} - e^\tau),$$

where τ_0 is a constant, which is evidently the maximum value of τ; by symmetry, this value must be attained half-way through the layer, i.e. for $\xi = 1$. Hence a second integration, with the condition $\tau = 0$ for $\xi = 0$, gives

$$\frac{1}{\sqrt{(2\lambda)}} \int_0^{\tau_0} \frac{d\tau}{\sqrt{(e^{\tau_0} - e^\tau)}} = \int_0^1 d\xi = 1.$$

Effecting the integration, we have

$$e^{-\frac{1}{2}\tau_0} \cosh^{-1} e^{\frac{1}{2}\tau_0} = \sqrt{(\tfrac{1}{2}\lambda)}. \tag{1}$$

The function $\lambda(\tau_0)$ determined by this equation has a maximum $\lambda = \lambda_{cr}$ for a definite value $\tau_0 = \tau_{0,cr}$; if $\lambda > \lambda_{cr}$, there is no solution satisfying the boundary conditions.‡ The numerical values are $\lambda_{cr} = 0\cdot88$, $\tau_{0,cr} = 1\cdot2$.††

PROBLEM 2. A sphere is immersed in a fluid at rest, in which a constant temperature gradient is maintained. Determine the resulting steady temperature distribution in the fluid and the sphere.

SOLUTION. The temperature distribution satisfies the equation $\triangle T = 0$ in all space, with the boundary conditions

$$T_1 = T_2, \qquad \kappa_1 \, \partial T_1/\partial r = \kappa_2 \, \partial T_2/\partial r$$

for $r = R$ (where R is the radius of the sphere; quantities with the suffixes 1 and 2 refer to the sphere and the fluid respectively), and $\mathbf{grad}\, T = \mathbf{A}$ at infinity, where $\mathbf{A}$ is the given

---

† The rate of explosive combustion reactions, and therefore the rate of heat generation, depend on temperature roughly as $e^{-U/RT}$, the constant U being large. FRANK-KAMENETSKIĬ has shown that, to investigate the conditions for a thermal explosion to occur, we must consider the course of the reaction when the ignition of the substance is comparatively slow, and therefore replace $e^{-U/RT}$ by $e^{-U/RT_0} e^{U(T-T_0)/RT_0^2}$, where T_0 is the external temperature. A more detailed discussion is given in the book by D. A. FRANK-KAMENETSKIĬ, *Diffusion and Heat Exchange in Chemical Kinetics*, Princeton 1955.

‡ Only the smaller of the two roots of equation (1) for $\lambda < \lambda_{cr}$ corresponds to a stable temperature distribution.

†† The corresponding values for a spherical region (of radius l) are $\lambda_{cr} = 3\cdot32$, $\tau_{0,cr} = 1\cdot47$, and for an infinite cylinder $\lambda_{cr} = 2\cdot00$, $\tau_{0,cr} = 1\cdot36$.

temperature gradient. By the symmetry of the problem, $\mathbf{A}$ is the only vector which can determine the required solution. Such solutions of Laplace's equation are constant $\times \mathbf{A} \cdot \mathbf{r}$ and constant $\times \mathbf{A} \cdot \mathbf{grad}(1/r)$. Noticing also that the solution must remain finite at the centre of the sphere, we seek the temperatures T_1 and T_2 in the forms

$$T_1 = c_1 \mathbf{A} \cdot \mathbf{r}, \qquad T_2 = c_2 \mathbf{A} \cdot \mathbf{r} / r^3 + \mathbf{A} \cdot \mathbf{r}.$$

The constants c_1 and c_2 are determined from the conditions for $r = R$, the result being

$$T_1 = \frac{3\kappa_2}{\kappa_1 + 2\kappa_2} \mathbf{A} \cdot \mathbf{r}, \qquad T_2 = \left[1 + \frac{\kappa_2 - \kappa_1}{\kappa_1 + 2\kappa_2} \left(\frac{R}{r} \right)^3 \right] \mathbf{A} \cdot \mathbf{r}.$$

§51. Thermal conduction in an infinite medium

Let us consider thermal conduction in an infinite medium at rest. The most general problem of this kind is as follows. The temperature distribution is given in all space at the initial instant $t = 0$:

$$T = T_0(x, y, z) \text{ for } t = 0,$$

where T_0 is a given function of the co-ordinates. It is required to determine the temperature distribution at all subsequent instants.

We expand the required function T as a Fourier integral with respect to the co-ordinates:

$$T = \int T_{\mathbf{k}}(t) \exp(i\mathbf{k} \cdot \mathbf{r}) d^3\mathbf{k}, \qquad d^3\mathbf{k} = dk_x dk_y dk_z, \tag{51.1}$$

where the expansion coefficients are given by

$$T_{\mathbf{k}}(t) = (2\pi)^{-3} \int T(x', y', z', t) \exp(-\mathbf{k}i \cdot \mathbf{r}') dV', \qquad dV' = dx' dy' dz'.$$

Substituting the expression (51.1) in equation (50.4), we obtain

$$\int \left(\frac{dT_{\mathbf{k}}}{dt} + k^2 \chi T_{\mathbf{k}} \right) \exp(i\mathbf{k} \cdot \mathbf{r}) d^3\mathbf{k} = 0,$$

whence

$$dT_{\mathbf{k}} / dt + k^2 \chi T_{\mathbf{k}} = 0.$$

This equation gives $T_{\mathbf{k}}$ as a function of time:

$$T_{\mathbf{k}} = \exp(-k^2 \chi t) T_{0\mathbf{k}}.$$

Substituting this in (51.1), we find

$$T = \int T_{0\mathbf{k}} \exp(-k^2 \chi t) \exp(i\mathbf{k} \cdot \mathbf{r}) d^3\mathbf{k}. \tag{51.2}$$

Since we must have $T = T_0(x, y, z)$ for $t = 0$, it is clear that the $T_{0\mathbf{k}}$ are the expansion coefficients of the function $T_0(x, y, z)$ as a Fourier integral:

$$T_{0\mathbf{k}} = (2\pi)^{-3} \int T_0(x', y', z') \exp(-i\mathbf{k} \cdot \mathbf{r}') dV'.$$

Finally, substituting this in (51.2), we obtain

$$T = (2\pi)^{-3} \iint T_0(x', y', z') \exp(-k^2\chi t) \exp[i\mathbf{k}\cdot(\mathbf{r}-\mathbf{r}')]\,dV'\,d^3\mathbf{k}.$$

The integral over $\mathbf{k}$ is the product of three simple integrals, each of the form

$$\int_{-\infty}^{\infty} \exp(-k_x{}^2\chi t)\exp[ik_x(x-x')]dk_x = \int_{-\infty}^{\infty} \exp(-k_x{}^2\chi t)\cos k_x(x-x')\,dk_x;$$

the similar integral with sin in place of cos is zero, since the sine function is odd. Using the formula

$$\int_{-\infty}^{\infty} \exp(-\alpha x^2)\cos \beta x\, dx = \sqrt{(\pi/\alpha)}\exp(-\beta^2/4\alpha) \qquad (\alpha > 0),$$

we have finally

$$T(x,y,z,t) = \frac{1}{8(\pi\chi t)^{\frac{3}{2}}} \int T_0(x',\,y',\,z') \times$$
$$\exp\{-[(x-x')^2+(y-y')^2+(z-z')^2]/4\chi t\}\,dV'. \tag{51.3}$$

This formula gives the complete solution of the problem; it determines the temperature distribution at any instant in terms of the given initial distribution.

If the initial temperature distribution is a function of only one co-ordinate, x, then we can integrate over y' and z' in (51.3) and obtain

$$T(x, t) = \frac{1}{2\sqrt{(\pi\chi t)}} \int_{-\infty}^{\infty} T_0(x')\exp[-(x-x')^2/4\chi t]\,dx'. \tag{51.4}$$

At time $t = 0$, let the temperature be zero in all space except for an infinitely thin layer at the plane $x = 0$, where it is infinite in such a way that the total quantity of heat (proportional to $\int T_0(x)dx$) is finite. Such a distribution can be represented by a delta function: $T_0(x) = $ constant $\times \delta(x)$. The integration in formula (51.4) then amounts to replacing x' by zero, the result of which is

$$T(x, t) = \text{constant} \times \frac{1}{2\sqrt{(\pi\chi t)}}\exp(-x^2/4\chi t). \tag{51.5}$$

Similarly, if at the initial instant a finite quantity of heat is concentrated at a point (the origin), the temperature distribution at subsequent instants is given by the formula

$$T(r, t) = \text{constant} \times \frac{1}{8(\pi\chi t)^{\frac{3}{2}}}\exp(-r^2/4\chi t), \tag{51.6}$$

where r is the distance from the origin. In the course of time, the temperature at the point $r = 0$ decreases as $t^{-\frac{3}{2}}$. The temperature in the surrounding space rises correspondingly, and the region where the temperature is appreciably different from zero expands (Fig. 29). The manner of this expansion is determined principally by the exponential factor in (51.6). We see that the order of magnitude l of the dimension of this region is given by $l^2/\chi t \sim 1$, whence

$$l \sim \sqrt{(\chi t)}, \tag{51.7}$$

i.e. l increases as the square root of the time.

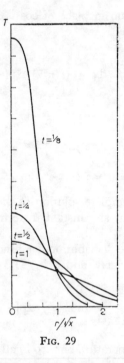

Fig. 29

Formula (51.7) can also be interpreted in a somewhat different way. Let l be the order of magnitude of the dimension of a body. Then we can say that, if the body is heated non-uniformly, the order of magnitude τ of the time required for the temperature to become more or less the same throughout the body is

$$\tau \sim l^2/\chi. \tag{51.8}$$

The time τ, which may be called the *relaxation time* for thermal conduction, is proportional to the square of the dimension of the body, and inversely proportional to the thermometric conductivity.

The thermal conduction process described by the formulae obtained above has the property that the effect of any perturbation is propagated instantaneously through all space. It is seen from formula (51.5) that the heat from a point source is propagated in such a manner that, even at the next instant, the temperature of the medium is zero only at infinity. This property holds also for a medium in which the thermometric conductivity χ depends on the temperature, provided that χ does not vanish anywhere. If, however, χ is a function of temperature which vanishes when $T = 0$, the propagation of heat is retarded, and at each instant the effect of a given perturbation extends only to a finite region of space (we suppose that the temperature outside this region can be taken as zero). This result, as well as the solution of the following Problems, is due to YA. B. ZEL'DOVICH and A. S. KOMPANEETS (1950).

PROBLEMS

PROBLEM 1. The specific heat and thermal conductivity of a medium vary as powers of the temperature, while its density is constant. Determine the manner in which the temperature tends to zero near the boundary of the region which, at a given instant, has received heat propagated from an arbitrary source (the temperature outside that region being zero).

SOLUTION. If κ and c_p vary as powers of the temperature, the same is true of the thermometric conductivity χ and of the heat function

$$w = \int c_p \, dT$$

(we omit a constant in w). Hence we can put $\chi = aW^n$, where we denote by $W = \rho w$ the heat function per unit volume. Then the thermal conduction equation

$$\rho c_p \, \partial T/\partial t = \operatorname{div}(\kappa \, \mathbf{grad} \, T)$$

becomes

$$\partial W/\partial t = a \operatorname{div}(W^n \, \mathbf{grad} \, W). \tag{1}$$

During a short interval of time, a small portion of the boundary of the region may be regarded as plane, and its rate of displacement in space, v, may be supposed constant. Accordingly, we seek a solution of equation (1) in the form $W = W(x - vt)$, where x is the co-ordinate in the direction perpendicular to the boundary. We have

$$-v \, dW/dx = a d(W^n \, dW/dx)/dx, \tag{2}$$

whence we find, after two integrations, that W vanishes as

$$W \sim |x|^{1/n}, \tag{3}$$

where $|x|$ is the distance from the boundary of the heated region. This also confirms our conclusion that, if $n > 0$, the heated region has a boundary outside which W and T are zero. If $n \leqslant 0$, then equation (2) has no solution vanishing at a finite distance, i.e. the heat is distributed through all space at every instant.

PROBLEM 2. A medium like that described in Problem 1 has, at the initial instant, an amount of heat Q per unit area concentrated on the plane $x = 0$, while $T = 0$ everywhere else. Determine the temperature distribution at subsequent instants.

SOLUTION. In the one-dimensional case, equation (1) of Problem 1 is

$$\frac{\partial W}{\partial t} = a \frac{\partial}{\partial x} \left(W^n \frac{\partial W}{\partial x} \right). \tag{1}$$

From the parameters Q and a and variables x and t at our disposal, we can form only one dimensionless combination,

$$\xi = x/(Q^n a t)^{1/(2+n)}; \tag{2}$$

Q and a have the dimensions erg/cm^2 and (cm^2/sec)(cm^3/erg)^n. Hence the required function $W(x, t)$ must be of the form

$$W = (Q^2/at)^{1/(2+n)} f(\xi),\tag{3}$$

where the dimensionless function $f(\xi)$ is multiplied by a quantity having the dimensions erg/cm^3. With this substitution, equation (1) gives

$$(2+n)\frac{d}{d\xi}\left(f^n\frac{df}{d\xi}\right)+\xi\frac{df}{d\xi}+f = 0.$$

This ordinary differential equation has a simple solution which satisfies the conditions of the problem, namely

$$f(\xi) = [\tfrac{1}{2}n(\xi_0{}^2-\xi^2)/(2+n)]^{1/n},\tag{4}$$

where ξ_0 is a constant of integration.

For $n > 0$, this formula gives the temperature distribution in the region between the planes $x = \pm x_0$ corresponding to the equation $\xi = \pm\xi_0$; outside this region, $W = 0$. Hence it follows that the heated region expands with time in a manner given by $x_0 = \text{constant} \times t^{1/(2+n)}$. The constant ξ_0 is determined by the condition that the total amount of heat is constant:

$$Q = \int\limits_{-x_0}^{x_0} W\,dx = Q\int\limits_{-\xi_0}^{\xi_0} f(\xi)\,d\xi,\tag{5}$$

whence we have

$$\xi_0{}^{2+n} = \frac{(2+n)^{1+n}\,2^{1-n}}{n\pi^{n/2}}\,\frac{\Gamma^n\left(\tfrac{1}{2}+1/n\right)}{\Gamma^n\left(1/n\right)}.\tag{6}$$

For $n = -\nu < 0$, we write the solution in the form

$$f(\xi) = \left[\frac{\nu}{2(2-\nu)}(\xi_0{}^2+\xi^2)\right]^{-1/\nu}.\tag{7}$$

Here the heat is distributed through all space, and at large distances W decreases as $x^{-2/\nu}$. This solution is valid only for $\nu < 2$; for $\nu \geqslant 2$, the normalisation integral (5) (which now extends to $\pm\infty$) diverges, which means physically that the heat is conducted instantaneously to infinity. For $\nu < 2$, the constant ξ_0 in (7) is given by

$$\xi_0{}^{2-\nu} = \frac{2(2-\nu)\pi^{\nu/2}}{\nu}\,\frac{\Gamma^\nu\left(1/\nu-\tfrac{1}{2}\right)}{\Gamma^\nu\left(1/\nu\right)}.\tag{8}$$

Finally, for $n \to 0$ we have $\xi_0 \to 2/\sqrt{n}$, and the solution given by formula (3) of Problem 1, (1), and (4) is

$$W = \lim_{n\to 0}\left\{\frac{Q}{2\sqrt{(\pi at)}}\left(1-n\frac{x^2}{4at}\right)^{1/n}\right\} = \frac{Q}{2\sqrt{(\pi at)}}\exp(-x^2/4at),$$

in agreement with formula (51.5).

§52. Thermal conduction in a finite medium

In problems of thermal conduction in a finite medium, the initial temperature distribution does not suffice to determine a unique solution, and the boundary conditions at the surface of the medium must also be given.

Let us consider thermal conduction in a half-space ($x > 0$), beginning with the case where a given constant temperature is maintained on the bounding plane $x = 0$. We may arbitrarily take this temperature as zero. At the initial instant, the temperature distribution throughout the medium is given, as before. The boundary and initial conditions are therefore

$$T = 0 \text{ for } x = 0; \quad T = T_0(x, y, z) \text{ for } t = 0 \text{ and } x > 0. \quad (52.1)$$

The solution of the thermal conduction equation with these conditions can, by means of the following device, be reduced to the solution for a medium infinite in all directions. We imagine the medium to extend on both sides of the plane $x = 0$, the temperature distribution for $t = 0$ and $x < 0$ being given by $-T_0$. That is, the temperature distribution at the initial instant is given in all space by an odd function of x:

$$T_0(-x, y, z) = -T_0(x, y, z). \quad (52.2)$$

It follows from equation (52.2) that $T_0(0, y, z) = -T_0(0, y, z) = 0$, i.e. the necessary boundary condition (52.1) is automatically satisfied for $t = 0$, and it is evident from symmetry that it will continue to be satisfied for all t.

Thus the problem is reduced to the solution of equation (50.4) in an infinite medium with an initial function $T_0(x, y, z)$ which satisfies (52.2), and without boundary conditions. Hence we can use the general formula (51.3). We divide the range of integration over x' in (51.3) into two parts, from $-\infty$ to 0 and from 0 to ∞. Using the relation (52.2), we then have

$$T(x, y, z, t) = \frac{1}{8(\pi\chi t)^{\frac{3}{2}}} \int\limits_{-\infty}^{\infty} \int\limits_{-\infty}^{\infty} \int\limits_{0}^{\infty} T_0(x', y', z') \times$$

$$\{\exp[-(x-x')^2/4\chi t] - \exp[-(x+x')^2/4\chi t]\} \times$$

$$\exp\{-[(y-y')^2 + (z-z')^2]/4\chi t\} \, dx' \, dy' \, dz'. \quad (52.3)$$

This formula gives the solution of the problem, since it determines the temperature throughout the medium, i.e. for all $x > 0$.

If the initial temperature distribution is a function of x only, formula (52.3) becomes

$$T(x, t) = \frac{1}{2\sqrt{(\pi\chi t)}} \int\limits_{0}^{\infty} T_0(x')\{\exp[-(x-x')^2/4\chi t] - \exp[-(x+x')^2/4\chi t]\} \, dx'. \quad (52.4)$$

As an example, let us consider the case where the initial temperature is a given constant everywhere except at $x = 0$. Without loss of generality, this constant may be taken as -1. The temperature on the plane $x = 0$ is always zero. The appropriate solution is obtained at once by substituting

$T_0(x) = -1$ in (52.4). The integral in (52.4) is the sum of two integrals, in each of which we change the variables as in $\xi = (x'-x)/2\sqrt{(\chi t)}$. We then obtain for $T(x, t)$ the expression

$$T(x, t) = \tfrac{1}{2}\{\text{erf}\,[-x/2\sqrt{(\chi t)}] - \text{erf}\,[x/2\sqrt{(\chi t)}]\},$$

where the function erf x is defined as

$$\text{erf}\,x = \frac{2}{\sqrt{\pi}} \int_0^x e^{-\xi^2} d\xi, \tag{52.5}$$

and is called the *error function* (we notice that erf $\infty = 1$). Since erf $(-x)$ $= -\text{erf}\,x$, we have finally

$$T(x, t) = -\text{erf}\,[x/2\sqrt{(\chi t)}]. \tag{52.6}$$

Fig. 30 shows a graph of the function erf x. The temperature distribution becomes more uniform in space in the course of time. This occurs in such a

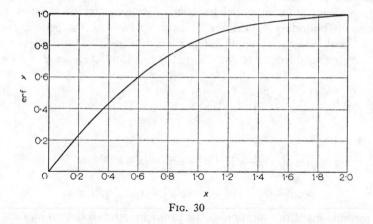

Fig. 30

way that any given value of the temperature "moves" proportionally to $\sqrt{t}$. This last result is obviously true. For the problem under consideration is characterised by only one parameter, the initial temperature difference T_0 between the boundary plane and the remaining space; in the above discussion, this difference was arbitrarily taken as unity. From the parameters T_0 and χ and variables x and t at our disposal we can form only one dimensionless combination, $x/\sqrt{(\chi t)}$; hence it is clear that the required temperature distribution must be given by a function of the form $T = T_0 f(x/\sqrt{(\chi t)})$.

Let us now consider a case where the surface bounding the medium is a thermal insulator. That is, there is no heat flux at the plane $x = 0$, so that we must have $\partial T/\partial x = 0$. We thus have the following boundary and initial conditions:

$$\partial T/\partial x = 0 \text{ for } x = 0; \quad T = T_0(x, y, z) \text{ for } t = 0, \ x > 0. \quad (52.7)$$

To find the solution we proceed as in the previous problem. That is, we again imagine the medium to extend on both sides of the plane $x = 0$, the initial temperature distribution being this time symmetrical about the plane. In other words, we now suppose that $T_0(x, y, z)$ is an even function of x:

$$T_0(-x, y, z) = T_0(x, y, z). \quad (52.8)$$

Then $\partial T_0(x, y, z)/\partial x = -\partial T_0(-x, y, z)/\partial x$, and $\partial T_0/\partial x = 0$ for $x = 0$. It is evident from symmetry that this condition will continue to be satisfied for all t.

Repeating the calculations given above, but using (52.8) in place of (52.2), we have the general solution of the problem in the form

$$T(x, y\ z, t) = \frac{1}{8(\pi\chi t)^{\frac{3}{2}}} \int\limits_{-\infty}^{\infty} \int\limits_{-\infty}^{\infty} \int\limits_{0}^{\infty} T_0(x', y', z') \times$$

$$\{\exp[-(x'-x)^2/4\chi t] + \exp[-(x'+x)^2/4\chi t]\} \times$$
$$\exp\{-[(y'-y)^2 + (z'-z)^2]/4\chi t\}\, dx'\, dy'dz'. \quad (52.9)$$

If T_0 is a function of x only, then

$$T(x, t) = \frac{1}{2\sqrt{(\pi\chi t)}} \int\limits_{0}^{\infty} T_0(x')\{\exp[-(x'-x)^2/4\chi t] + \exp[-(x'+x)^2/4\chi t]\}\, dx'.$$
$$(52.10)$$

Let us now consider problems with boundary conditions of a different type, which also enable the equation of thermal conduction to be solved in a general form. Let a heat flux (a given function of time) enter a medium through its bounding plane $x = 0$. The boundary and initial conditions are

$$-\kappa\, \partial T/\partial x = q(t) \text{ for } x = 0; \ T = 0 \text{ for } t = -\infty, \ x > 0, \quad (52.11)$$

where $q(t)$ is a given function.

We first solve an auxiliary problem, in which $q(t) = \delta(t)$. It is easy to see that this problem is physically equivalent to that of the propagation of heat in an infinite medium from a point source which generates a given amount of heat. For the boundary condition $-\kappa\partial T/\partial x = \delta(t)$ for $x = 0$ signifies physically that a unit of heat enters through each unit area of the plane $x = 0$ at the instant $t = 0$. In the problem where the condition is $T = 2\delta(x)/\rho c_p$ for $t = 0$, an amount of heat

$$\int \rho c_p\, T\, dx = 2$$

is concentrated on this area at time $t = 0$; half of this is then propagated in the positive x-direction, and the other half in the negative x-direction. Hence it is clear that the solutions of the two problems are identical, and we find from (51.5) $\kappa T(x, t) = \sqrt{(\chi/\pi t)} \exp(-x^2/4\chi t)$.

Since the equations are linear, the effects of the heat entering at different moments are simply additive, and therefore the required general solution of the equation of thermal conduction with the conditions (52.11) is

$$\kappa T(x, t) = \int_{-\infty}^{t} \sqrt{\frac{\chi}{\pi(t-\tau)}} q(\tau) \exp[-x^2/4\chi(t-\tau)] \, d\tau. \qquad (52.12)$$

In particular, the temperature on the plane $x = 0$ varies according to

$$\kappa T(0, t) = \int_{-\infty}^{t} \sqrt{\frac{\chi}{\pi(t-\tau)}} q(\tau) \, d\tau. \qquad (52.13)$$

Using these results, we can obtain at once the solution of another problem, in which the temperature T on the plane $x = 0$ is a given function of time:

$$T = T_0(t) \text{ for } x = 0; \qquad T = 0 \text{ for } t = -\infty, \qquad x > 0. \qquad (52.14)$$

To do so, we notice that, if some function $T(x, t)$ satisfies the equation of thermal conduction, then so does its derivative $\partial T/\partial x$. Differentiating (52.12) with respect to x, we obtain

$$-\kappa \frac{\partial T(x, t)}{\partial x} = \int_{-\infty}^{t} \frac{xq(\tau)}{2\sqrt{[\pi\chi(t-\tau)^3]}} \exp[-x^2/4\chi(t-\tau)] \, d\tau.$$

This function satisfies the equation of thermal conduction and (by (52.11)) its value for $x = 0$ is $q(t)$; it therefore gives the required solution of the problem whose conditions are (52.14). Writing $T(x, t)$ instead of $-\kappa \partial T/\partial x$, and $T_0(t)$ instead of $q(t)$, we thus have

$$T(x, t) = \frac{x}{2\sqrt{(\pi\chi)}} \int_{-\infty}^{t} \frac{T_0(\tau)}{(t-\tau)^{\frac{3}{2}}} \exp[-x^2/4\chi(t-\tau)] \, d\tau. \qquad (52.15)$$

The heat flux $q = -\kappa \partial T/\partial x$ through the bounding plane $x = 0$ is found by a simple calculation to be

$$q(t) = \frac{\kappa}{\sqrt{(\pi\chi)}} \int_{-\infty}^{t} \frac{dT_0(\tau)}{d\tau} \frac{d\tau}{\sqrt{(t-\tau)}}. \qquad (52.16)$$

This formula is the inverse of (52.13).

The solution is easily obtained for the important problem where the temperature on the bounding plane $x = 0$ is a given periodic function of time: $T = T_0 e^{-i\omega t}$ for $x = 0$. It is clear that the temperature distribution in all space will also depend on the time through a factor $e^{-i\omega t}$. Since the one-dimensional equation of thermal conduction is formally identical with the equation (24.3) which determines the motion of a viscous fluid above an oscillating plane, we can immediately write down the required temperature distribution by analogy with (24.4):

$$T = T_0 \exp[-x\sqrt{(\omega/2\chi)}] \exp\{i[x\sqrt{(\omega/2\chi)} - \omega t]\}. \tag{52.17}$$

We see that the oscillations of the temperature on the bounding surface are propagated from it as *thermal waves* which are rapidly damped in the interior of the medium.

Another kind of thermal-conduction problem comprises those concerning the rate at which the temperature is equalised in a non-uniformly heated finite body whose surface is maintained in given conditions. To solve these problems by general methods, we seek a solution of the equation of thermal conduction in the form $T = T_n(x, y, z)e^{-\lambda_n t}$, with λ_n a constant. For the function T_n we have the equation

$$\chi \triangle T_n = -\lambda_n T_n. \tag{52.18}$$

This equation, with given boundary conditions, has non-zero solutions only for certain λ_n, its *eigenvalues*. All the eigenvalues are real and positive, and the corresponding functions $T_n(x, y, z)$ form a complete set of orthogonal functions. Let the temperature distribution at the initial instant be given by the function $T_0(x, y, z)$. Expanding this as a series of functions T_n,

$$T_0(x, y, z) = \sum c_n T_n(x, y, z),$$

we obtain the required solution in the form

$$T(x, y, z, t) = \sum c_n T_n(x, y, z) \exp(-\lambda_n t). \tag{52.19}$$

The rate of equalisation of the temperature is evidently determined by the term corresponding to the smallest λ_n, which we call λ_1. The "equalisation time" may be defined as $\tau = 1/\lambda_1$.

PROBLEMS

PROBLEM 1. Determine the temperature distribution around a spherical surface (of radius R) whose temperature is a given function $T_0(t)$ of time.

SOLUTION. The thermal-conduction equation for a centrally symmetrical temperature distribution is, in spherical co-ordinates, $\partial T/\partial t = (\chi/r)\partial^2(rT)/\partial r^2$. The substitution $rT(r, t) = F(r, t)$ reduces this to $\partial F/\partial t = \chi \partial^2 F/\partial r^2$, which is the ordinary one-dimensional thermal-conduction equation. Hence the required solution can be found at once from (52.15), and is

$$T(r, t) = \frac{R(r-R)}{2r\sqrt{(\pi\chi)}} \int_{-\infty}^{t} \frac{T_0(\tau)}{(t-\tau)^{\frac{3}{2}}} \exp[-(r-R)^2/4\chi(t-\tau)]\, d\tau.$$

PROBLEM 2. The same as Problem 1, but for the case where the temperature of the spherical surface is $T_0 e^{-i\omega t}$.

SOLUTION. Similarly to (52.17), we obtain

$$T = T_0 \exp(-i\omega t)(R/r)\exp[-(1-i)(r-R)\sqrt{(\omega/2\chi)}].$$

PROBLEM 3. Determine the temperature equalisation time for a cube of side a whose surface is (a) maintained at a temperature $T = 0$, (b) an insulator.

SOLUTION. In case (a) the smallest value of λ is given by the following solution of equation (52.18):

$$T_1 = \sin(\pi x/a)\sin(\pi y/a)\sin(\pi z/a)$$

(the origin being at one corner of the cube), when $\tau = 1/\lambda_1 = a^2/3\pi^2\chi$. In case (b) we have $T_1 = \cos(\pi x/a)$ (or the same function of y or z), when $\tau = a^2/\pi^2\chi$.

PROBLEM 4. The same as Problem 3, but for a sphere of radius R.

SOLUTION. The smallest value of λ is given by the centrally symmetrical solution of (52.18) $T_1 = (1/r)\sin kr$; in case (a), $k = \pi/R$, and $\tau = 1/\chi k^2 = R^2/\chi\pi^2$. In case (b) k is the smallest non-zero root of the equation $kR = \tan kR$, whence we find $kR = 4\cdot493$ and $\tau = 0\cdot050\ R^2/\chi$.

§53. The similarity law for heat transfer

The processes of heat transfer in a fluid are more complex than those in solids, because the fluid may be in motion. A heated body immersed in a moving fluid cools considerably more rapidly than one in a fluid at rest, where the heat transfer is accomplished only by conduction. The motion of a non-uniformly heated fluid is called *convection*.

We shall suppose that the temperature differences in the fluid are so small that its physical properties may be supposed independent of temperature, but are at the same time so large that we can neglect in comparison with them the temperature changes caused by the heat from the energy dissipation by internal friction (see §55). Then the viscosity term in equation (50.2) may be omitted, leaving

$$\partial T/\partial t + \mathbf{v}\cdot\mathbf{grad}\,T = \chi\triangle T, \tag{53.1}$$

where $\chi = \kappa/\rho c_p$ is the thermometric conductivity. This equation, together with the Navier–Stokes equation and the equation of continuity, completely determines the convection in the conditions considered.

In what follows we shall be interested only in steady convective flow.† Then all the time derivatives are zero, and we have the following fundamental equations:

$$\mathbf{v}\cdot\mathbf{grad}\,T = \chi\triangle T, \tag{53.2}$$

$$(\mathbf{v}\cdot\mathbf{grad})\mathbf{v} = -\mathbf{grad}(p/\rho)+\nu\triangle\mathbf{v}, \quad \mathrm{div}\,\mathbf{v} = 0. \tag{53.3}$$

---

† In order that the convection should be steady, it is, strictly speaking, necessary that the solid bodies adjoining the fluid should contain sources of heat which maintain these bodies at a constant temperature.

This system of equations, in which the unknown functions are $\mathbf{v}$, T and p/ρ, contains only two constant parameters, ν and χ. Furthermore, the solution of these equations depends also, through the boundary conditions, on some characteristic length l, velocity U, and temperature difference $T_1 - T_0$. The first two of these are given by the dimension of the solid bodies which appear in the problem and the velocity of the main stream, while the third is given by the temperature difference between the fluid and these bodies.

In forming dimensionless quantities from the parameters at our disposal, the question arises of the dimensions to be ascribed to the temperature. To resolve this, we notice that the temperature is determined by equation (53.2), which is linear and homogeneous in T. Hence the temperature can be multiplied by any constant and still satisfy the equations. In other words, the unit of measurement of temperature can be chosen arbitrarily. The possibility of this transformation of the temperature can be formally allowed for by giving it a dimension of its own, unrelated to those of the other quantities. This can be measured in degrees, the usual unit of temperature.

Thus convection in the above-mentioned conditions is characterised by five parameters, whose dimensions are $\nu = \chi = \text{cm}^2/\text{sec}$, $U = \text{cm/sec}$, $l = \text{cm}$, $T_1 - T_0 = \text{deg}$. From these we can form two independent dimensionless combinations. These may be the Reynolds number $R = Ul/\nu$ and the *Prandtl number*, defined as

$$P = \nu/\chi. \tag{53.4}$$

Any other dimensionless combination can be expressed in terms of R and P.†

The Prandtl number is just a constant of the material, and does not depend on the properties of the flow. For gases it is always of the order of unity. The value of P for liquids varies more widely. For very viscous liquids, it may be very large. The following are values of P at 20°C for various substances:

Air	0·733
Water	6·75
Alcohol	16·6
Glycerine	7250
Mercury	0·044

As in §19, we can now conclude that, in steady convection (of the type described), the temperature and velocity distributions are of the form

$$\frac{T - T_0}{T_1 - T_0} = f\left(\frac{\mathbf{r}}{l}, R, P\right), \qquad \frac{\mathbf{v}}{U} = \mathbf{f}\left(\frac{\mathbf{r}}{l}, R\right). \tag{53.5}$$

The dimensionless function which gives the temperature distribution depends on both R and P as parameters, but the velocity distribution depends only on R, since it is determined by equations (53.3), which do not involve the conductivity. Two convective flows are similar if their Reynolds and Prandtl numbers are the same.

---

† The *Péclet number* is sometimes used; it is defined as $Ul/\chi = RP$.

The heat transfer between solid bodies and the fluid is usually characterised by the *heat transfer coefficient* α, defined by

$$\alpha = q/(T_1 - T_0), \tag{53.6}$$

where q is the heat flux density through the surface and $T_1 - T_0$ is a characteristic temperature difference between the solid body and the fluid. If the temperature distribution in the fluid is known, the heat transfer coefficient is easily found by calculating the heat flux density $q = -\kappa \partial T/\partial n$ at the boundary of the fluid (the derivative being taken along the normal to the surface).

The heat transfer coefficient is not a dimensionless quantity. A dimensionless quantity which characterises the heat transfer is what is called the *Nusselt number* :†

$$N = \alpha l/\kappa. \tag{53.7}$$

It follows from similarity arguments that, for any given type of convective flow, the Nusselt number is a definite function of the Reynolds and Prandtl numbers only:

$$N = f(R, P). \tag{53.8}$$

This function is very simple for convection at sufficiently small Reynolds numbers. These correspond to small velocities. Hence, in the first approximation, we can neglect the velocity term in equation (53.2), so that the temperature distribution is determined by the equation $\triangle T = 0$, i.e. the ordinary equation of steady thermal conduction in a medium at rest. The heat transfer coefficient can then depend on neither the velocity nor the viscosity and so we must have simply

$$N = \text{constant}, \tag{53.9}$$

and in calculating the constant the fluid may be supposed at rest.

PROBLEM

Determine the temperature distribution in a fluid moving in Poiseuille flow along a pipe of circular cross-section, when the temperature of the walls varies linearly along the pipe.

SOLUTION. The conditions of the flow are the same at every cross-section of the pipe, and we can look for the temperature distribution in the form $T = Az + f(r)$, where Az is the wall temperature; we use cylindrical co-ordinates, with the z-axis along the axis of the pipe. For the velocity we have, by (17.9), $v_z = v = 2v_m(1 - r^2/R^2)$, where v_m is the mean velocity. Substituting in (53.2) we find

$$\frac{1}{r}\frac{d}{dr}\left(r\frac{df}{dr}\right) = \frac{2v_m A}{\chi}\left[1 - \left(\frac{r}{R}\right)^2\right].$$

The solution of this equation which is finite for $r = 0$ and zero for $r = R$ is

$$f(r) = -\frac{v_m A r^2}{2\chi}\left[\frac{3}{4} - \left(\frac{r}{R}\right)^2 + \frac{1}{4}\left(\frac{r}{R}\right)^4\right].$$

---

† The dimensionless "heat transfer number", defined as $K_h = \alpha/\rho c_p U = N/RP$, is also used.

The heat flux density is

$$q = \kappa[\partial T/\partial r]_R = \tfrac{1}{2}\rho c_p v_m R A.$$

It is independent of the thermal conductivity.

§54. Heat transfer in a boundary layer

The temperature distribution in a fluid at very high Reynolds numbers exhibits properties similar to those of the velocity distribution. Very large values of R are equivalent to a very small viscosity. But since the number $P = \nu/\chi$ is not small, the thermometric conductivity χ must be supposed small, as well as ν. This corresponds to the fact that, for sufficiently high velocities, the fluid may be approximately regarded as an ideal fluid, and in an ideal fluid both internal friction and thermal conduction are absent.

This viewpoint, however, must again be abandoned in a boundary layer, since neither the boundary condition of no slip nor that of equal temperatures would be satisfied. In the boundary layer, therefore, there occurs both a rapid decrease of the velocity and a rapid change of the fluid temperature to a value equal to the temperature of the solid surface. The boundary layer is characterised by the presence of large gradients of both velocity and temperature.

It is easy to see that, in flow past a heated body (with R large), the heating of the fluid occurs almost exclusively in the wake, while outside the wake the fluid temperature does not change. For, when R is large, the processes of thermal conduction in the main stream are unimportant. Hence the temperature varies only in the region reached by fluid that has been heated in the boundary layer. We know (see §34) that the streamlines from the boundary layer enter the main stream only beyond the line of separation, where they go into the region of the turbulent wake. From the wake, however, the streamlines do not emerge at all. Thus the fluid which flows past the surface of the heated body in the boundary layer goes entirely into the wake and remains there. We see that the heat becomes distributed through the regions where the vorticity is non-zero.

In the turbulent region itself, a very considerable exchange of heat occurs, which is due to the intensive mixing of the fluid characteristic of any turbulent flow. This mechanism of heat transfer may be called *turbulent conduction* and characterised by a coefficient κ_{turb}, in the same way as we introduced the turbulent viscosity ν_{turb} in §31. The turbulent thermometric conductivity is defined, in order of magnitude, by the same formula as ν_{turb} (31.2): $\chi_{\text{turb}} \sim l\Delta u$.

Thus the processes of heat transfer in laminar and in turbulent flow are fundamentally different. In the limiting case of very small viscosity and thermal conductivity, in laminar flow, the processes of heat transfer are absent, and the fluid temperature is constant at every point in space. In turbulent flow, however, even in the same limiting case, heat transfer occurs and rapidly equalises the temperatures in various parts of the stream.

It should be mentioned that, when we speak of the temperature of a fluid in turbulent motion, we mean the time average of the fluid temperature. The actual temperature at any point in space undergoes very irregular variations with time, similar to those of the velocity.

Let us begin by considering heat transfer in a laminar boundary layer. The equations of motion (39.10) are unaltered. A similar simplification must now be performed for equation (53.2). Written explicitly, this equation is (since all quantities are independent of the co-ordinate z)

$$v_x \frac{\partial T}{\partial x} + v_y \frac{\partial T}{\partial y} = \chi \left(\frac{\partial^2 T}{\partial x^2} + \frac{\partial^2 T}{\partial y^2} \right).$$

On the right-hand side we may neglect the derivative $\partial^2 T/\partial x^2$ in comparison with $\partial^2 T/\partial y^2$, leaving

$$v_x \frac{\partial T}{\partial x} + v_y \frac{\partial T}{\partial y} = \chi \frac{\partial^2 T}{\partial y^2}. \tag{54.1}$$

By comparing this equation with the first of (39.10) we see that, if the Prandtl number is of the order of unity, then the order of magnitude δ of the thickness of the layer in which the velocity v_x decreases and the temperature T varies will again be given by the formulae obtained in §39, i.e. it will be inversely proportional to $\sqrt{R}$. The heat flux $q = -\kappa \partial T/\partial n$ is equal, in order of magnitude, to $\kappa (T_1 - T_0)/\delta$. Hence we conclude that q, and therefore the Nusselt number, are proportional to $\sqrt{R}$. The dependence of N on P is not determined. Thus we have

$$N = \sqrt{R} f(P). \tag{54.2}$$

From this it follows, in particular, that the heat transfer coefficient α is inversely proportional to the square root of the dimension l of the body.

Let us now consider heat transfer in a turbulent boundary layer. Here it is convenient, as in §42, to take an infinite plane-parallel turbulent stream flowing along an infinite plane surface. The transverse temperature gradient dT/dy in such a flow can be determined from the same kind of dimensional argument as we used to find the velocity gradient du/dy. We denote by q the heat flux density along the y-axis caused by the temperature gradient. This flux is a constant (independent of y), like the momentum flux σ, and can likewise be regarded as a given parameter which determines the properties of the flow. Furthermore, we have as parameters also the density ρ and the specific heat c_p per unit mass. Instead of σ we use as parameter v_*; q and c_p have the dimensions erg/cm^2 sec = g/sec^3 and erg/g deg = cm^2/sec^2 deg. The viscosity and thermal conductivity cannot appear explicitly in dT/dy when R is sufficiently large.

Because of the homogeneity of the equations as regards the temperature, already mentioned in §53, the temperature can be changed by any factor without violating the equations. When the temperature is changed in this way, however, the heat flux must change by the same factor. Hence q and T

must be proportional. From q, v_*, ρ, c_p and y we can form only one quantity proportional to q and having the dimensions deg/cm, namely $q/\rho c_p v_* y$. Thus we must have $dT/dy = \beta q/b\rho c_p v_* y$, where β is a numerical constant which must be determined by experiment.† Hence

$$T = (\beta q/b\rho c_p v_*)(\log y + c). \tag{54.3}$$

Thus the temperature, like the velocity, varies logarithmically. The constant of integration c which appears here must be determined from the conditions in the viscous sublayer, as in the derivation of (42.7). The temperature difference between the fluid at a given point and the wall (which we arbitrarily take to be at zero temperature) is composed of the temperature change across the turbulent layer and that across the viscous sublayer. The logarithmic law (54.3) is determined by only the first of these. Hence, if we write (54.3) in the form $T = (\beta q/b\rho c_p v_*)[\log(yv_*/\nu) + \text{constant}]$, including in the argument of the logarithm a factor equal to the thickness y_0, then the constant (multiplied by the coefficient in parentheses) must be the change in temperature across the viscous sublayer. This change, of course, depends on the coefficients ν and χ also. Since the constant is dimensionless, it must be some function of P, which is the only dimensionless combination of the quantities ν, χ, ρ, v_* and c_p (q cannot appear, since T must be proportional to q, which already occurs in the coefficient). Thus we find the temperature distribution to be

$$T = (\beta q/b\rho c_p v_*)[\log(\nu v_*/\nu) + f(\text{P})]. \tag{54.4}$$

Using this formula, we can calculate the heat transfer for turbulent flow in a pipe, along a flat plate, etc. We shall not pause to do this here.

PROBLEMS

PROBLEM 1. Determine the limiting form of the dependence of the Nusselt number on the Prandtl number in a laminar boundary layer when P and R are large.

SOLUTION. For large P, the distance δ' over which the temperature changes is small compared with the thickness δ of the layer in which the velocity v_x diminishes. δ' may be called the thickness of the temperature boundary layer. The order of magnitude of δ' may be obtained from an estimate of the terms in equation (54.1). Over the distance from $y = 0$ to $y \sim \delta'$, the temperature varies by an amount of the order of the total temperature difference $T_1 - T_0$ between the fluid and the solid body, while the velocity v_x varies over this distance by an amount of the order of $U\delta'/\delta$ (since the total change, of the order of U, occurs over a distance δ). Hence, for $y \sim \delta'$, the terms in equation (54.1) are, in order of magnitude,

$$\chi \partial^2 T/\partial y^2 \sim \chi(T_1 - T_0)/\delta'^2 \text{ and } v_x \partial T/\partial x \sim U\delta'(T_1 - T_0)/l\delta.$$

If the two expressions are comparable, we have $\delta'^3 \sim \chi l\delta/U$. Substituting $\delta \sim l/\sqrt{R}$, we obtain $\delta' \sim l/R^{\frac{1}{2}}P^{\frac{1}{3}} \sim \delta/P^{\frac{1}{3}}$. Thus, for large P, the thickness of the temperature boundary layer decreases, relative to that of the velocity boundary layer, inversely as the cube root of P.

---

† Here b is the constant appearing in the logarithmic velocity profile (42.4). With this definition, β is the ratio $\nu_{\text{turb}}/\chi_{\text{turb}}$, where ν_{turb} and χ_{turb} are the coefficients in $q = \rho c_p \chi_{\text{turb}} dT/dy$, $\sigma = \rho \nu_{\text{turb}} du/dy$. From simultaneous determinations of the velocity and temperature profiles in pipes and in flow along flat plates, β is found to be about 0·7. We should mention that similar measurements in the turbulent wake behind a heated body give a value of about 0·5 for the ratio $\nu_{\text{turb}}/\chi_{\text{turb}}$ in a free turbulent flow.

The heat flux $q = -\kappa \partial T/\partial y \sim \kappa(T_1 - T_0)/\delta'$, and the required limiting law of heat transfer is found to be†

$$N = \text{constant} \times R^{\frac{1}{4}}P^{\frac{1}{4}}.$$

PROBLEM 2. Determine the limiting form of the function $f(P)$, in the logarithmic temperature distribution (54.4), for large values of P.

SOLUTION. According to what was said in §42, the transverse velocity in the viscous sublayer is of the order of $v_*(y/y_0)^2$, while the scale of the turbulence is of the order of y^2/y_0. The turbulent thermometric conductivity χ_{turb} is therefore of the order of

$$v_* y_0 (y/y_0)^4 \sim \nu(y/y_0)^4$$

(where we have used the relation (42.5)); χ_{turb} is comparable in magnitude with the ordinary coefficient χ at distances of the order of $y_1 \sim y_0 P^{-\frac{1}{4}}$. Since χ_{turb} increases very rapidly with y, it is clear that most of the temperature change in the viscous sublayer occurs over distances from the wall of the order of y_1, and may be supposed proportional to y_1, being in order of magnitude $qy_1/\kappa \sim qy_0/\kappa P^{\frac{1}{4}} \sim qP^{\frac{3}{4}}/\rho c_p v_*$. Comparing with formula (54.4), we see that the function $f(P)$ is a numerical constant times $P^{\frac{3}{4}}.\ddagger$

PROBLEM 3. Determine the temperature differences T_λ in a non-uniformly heated turbulent fluid over distances λ which are small compared with the external scale of the turbulence (A. M. OBUKHOV 1949).

SOLUTION. The equalisation of temperature in a non-uniformly heated turbulent fluid occurs similarly to the dissipation of mechanical energy. Turbulent eddies of size $\lambda \gg \lambda_0$ (where λ_0 is the internal scale of the turbulence) lead to an equalisation of temperature by purely mechanical mixing of fluid particles which are at different temperatures. Considerable true temperature gradients in regions of size $\lambda \sim \lambda_0$, on the other hand, are equalised by dissipative thermal conduction.

The dissipation by thermal conduction (increase of entropy) is determined by the quantity $\chi(\mathbf{grad}\, T)^2/T^2$ (see (49.6)); supposing the turbulent fluctuations of temperature to be relatively small, we can replace T^2 in the denominator by a constant, the square of the mean temperature. According to the method described in §32 (see the first footnote to that section), we write $\chi_{\text{turb}} (T_\lambda/\lambda)^2 = \text{constant}$. Substituting $\chi_{\text{turb},\lambda} \sim \nu_{\text{turb},\lambda} \sim \lambda v_\lambda$, $v_\lambda \sim (\epsilon\lambda)^{\frac{1}{3}}$ (see (32.1)), we find the required relation to be $T_\lambda \sim \lambda^{\frac{1}{3}}$. Thus for $\lambda \gg \lambda_0$ the temperature fluctuations, like the velocity fluctuations, are proportional to the cube root of the distance. At distances $\lambda \ll \lambda_0$, however, by the same arguments as for the velocity, the differences T_λ are simply proportional to λ.

PROBLEM 4. Derive a relation between the local correlation functions

$$B_{TT} = \overline{(T_2 - T_1)^2}, \qquad B_{iTT} = \overline{(v_{2i} - v_{1i})(T_2 - T_1)^2}$$

in a non-uniformly heated turbulent flow (A. M. YAGLOM 1949).

SOLUTION. The calculations are similar to those used in deriving formula (33.18). From

---

† For the values of the thermal conductivity actually found, the Prandtl number does not reach the values for which this limiting law holds. Such laws can, however, be applied to convective diffusion; this obeys the same equations as convective heat transfer, but with the temperature replaced by the concentration of the solute, and the heat flux by the flux of solute, the "diffusion Prandtl number" being defined as $P_D = \nu/D$, where D is the diffusion coefficient. For example, for solutions in water and similar liquids, P_D reaches values of the order of 10^3, while for very viscous solvents it is 10^6 or more.

‡ The calculation of the constant in this formula for various particular cases is facilitated by the fact that, by virtue of the inequality $\delta' \ll \delta$, we need take only the first terms of an expansion, in powers of y, of the fluid velocity components in integrating equation (54.1) across the temperature boundary layer. Calculations for convective diffusion in various particular cases are given by V. G. LEVICH, *Physico-chemical Hydrodynamics (Fiziko-khimicheskaya gidrodinamika)*, Moscow 1952.

the equations

$$\frac{\partial T}{\partial t} + v_i \frac{\partial T}{\partial x_i} = \chi \triangle T, \qquad \frac{\partial v_i}{\partial x_i} = 0$$

we find

$$\frac{\partial}{\partial t} \overline{(T_1 T_2)} = -2 \frac{\partial}{\partial x_{1i}} \overline{(v_{1i} T_1 T_2)} + 2\chi \triangle_1 \overline{(T_1 T_2)}.$$

On the left-hand side we put $\mathbf{r} = \mathbf{r}_2 - \mathbf{r}_1$, and on the right we express the mean values in terms of the correlation functions, using the homogeneity and isotropy of the flow:

$$\frac{\partial}{\partial t} \overline{T^2} = -\frac{1}{2} \frac{\partial}{\partial x_{1i}} B_{iTT} - \chi \triangle_1 B_{TT}.$$

Writing $B_{iTT} = n_i B_{rTT}$ and changing to derivatives with respect to r, we obtain an equation which, on integration over r, gives the required relation

$$B_{rTT} - 2\chi \mathrm{d}B_{TT}/\mathrm{d}r = -\tfrac{4}{3}r\phi,$$

where

$$\phi \equiv -\partial(\overline{T^2})/\partial t = -\partial \overline{(T-\overline{T})^2}/\partial t.$$

Using the results of Problem 3, we then find that, for $r \gg \lambda_0$, $B_{rTT} \cong -\tfrac{4}{3}r\phi$, while for $r \ll \lambda_0$ we have $B_{TT} \cong r^2\phi/9\chi$.

§55. Heating of a body in a moving fluid

A thermometer immersed in a fluid at rest indicates a temperature equal to that of the fluid. If the fluid is in motion, however, the thermometer indicates a somewhat higher temperature. This is because the fluid brought to rest at the surface of the thermometer is heated by internal friction.

The general problem may be formulated as follows. A body of arbitrary shape is immersed in a moving fluid; thermal equilibrium is established after a sufficient length of time, and it is required to determine the temperature difference $T_1 - T_0$ then existing between the body and the fluid.

The solution of this problem is given by equation (50.2), in which, however, we cannot now neglect the term containing the viscosity as we did in (53.1); it is this term which is responsible for the effect under consideration. Thus we have for a steady state

$$\mathbf{v} \cdot \mathbf{grad}\, T = \chi \triangle T + \frac{\nu}{2c_p} \left(\frac{\partial v_i}{\partial x_k} + \frac{\partial v_k}{\partial x_i} \right)^2. \tag{55.1}$$

This must be supplemented by the equations of motion (53.3) of the fluid itself and also, strictly speaking, by the equation of thermal conduction in the body. In the limiting case where the body has a sufficiently small thermal conductivity, we can neglect the latter and suppose the temperature at any point on the surface of the body to be simply equal to the fluid temperature at that point, obtained by solving equation (55.1) with the boundary condition

$\partial T/\partial n = 0$, i.e. the condition that there is no heat flux through the surface of the body. In the opposite limiting case where the body has a sufficiently large thermal conductivity, we can use the approximate condition that the temperature should be the same at every point of its surface; the derivative $\partial T/\partial n$ will not then in general vanish over the whole surface, and we must require only that the total heat flux through the surface of the body (i.e. the integral of $\partial T/\partial n$ over the surface) should be zero. In both these limiting cases the thermal conductivity of the body does not appear explicitly in the solution of the problem, and we shall suppose in what follows that one of these cases holds.[†]

Equations (55.1) and (53.3) contain the constant parameters χ, ν and c_p, and their solutions involve also the dimension l of the body and the velocity U of the main stream. (The temperature difference $T_1 - T_0$ is not now an arbitrary parameter, but must itself be determined by solving the equations.) From these parameters we can construct two independent dimensionless quantities, which we take to be R and P. Then we can say that the required temperature difference $T_1 - T_0$ is equal to some quantity having the dimensions of temperature (which we take to be U^2/c_p), multiplied by a function of R and P:

$$T_1 - T_0 = (U^2/c_p)f(R, P). \tag{55.2}$$

It is easy to determine the form of this function for very small Reynolds numbers, i.e. for sufficiently small velocities U. In this case the term $\mathbf{v} \cdot \mathbf{grad}\, T$ in (55.1) is small compared with $\chi \triangle T$, so that this equation becomes

$$\chi \triangle T = -\frac{\nu}{2c_p}\left(\frac{\partial v_i}{\partial x_k} + \frac{\partial v_k}{\partial x_i}\right)^2. \tag{55.3}$$

The temperature and velocity vary considerably over distances of the order of l. Hence an estimate of the two sides of equation (55.3) gives $\chi(T_1 - T_0)/l^2 \sim \nu U^2/c_p l^2$, or $T_1 - T_0 \sim \nu U^2/\chi c_p$. Thus we conclude that, for small R,

$$T_1 - T_0 = \text{constant} \times PU^2/c_p, \tag{55.4}$$

where the numerical constant depends on the shape of the body. It should be noticed that the temperature difference is proportional to the square of the velocity U.

Some general conclusions concerning the form of the function $f(P, R)$ in (55.2) can be drawn in the opposite limiting case of large R, when the velocity and the temperature vary only in a narrow boundary layer. Let δ and δ' be the distances over which the velocity and temperature respectively vary; δ and δ' differ by a factor depending on P. The amount of heat evolved in unit area of the boundary layer in unit time owing to the viscosity of the fluid

---

[†] I. A. KIBEL' has obtained an exact solution for the rotation of a heated disk in a viscous fluid, similar to the solution given in §23 for a constant temperature; see *Prikladnaya matematika i mekhanika* 11, 611, 1947.

is the integral of $\frac{1}{2}\nu\rho(\partial v_i/\partial x_k + \partial v_k/\partial x_i)^2$ over the thickness of the layer (see (16.3)). This integral is of the order of $\nu\rho(U^2/\delta^2)\delta = \nu\rho U^2/\delta$. The same amount of heat must be lost to the body, and it is therefore equal to the heat flux $q = -\kappa \partial T/\partial n \sim \chi c_p \rho (T_1 - T_0)/\delta'$. Comparing the two expressions, we find

$$T_1 - T_0 = (U^2/c_p)f(P). \tag{55.5}$$

Thus, in this case, the function f is independent of R, but its dependence on P remains undetermined.

PROBLEMS

PROBLEM 1. Determine the temperature distribution in a fluid moving in Poiseuille flow in a pipe of circular cross-section whose walls are maintained at a constant temperature T_0.

SOLUTION. In cylindrical co-ordinates, with the z-axis along the axis of the pipe, we have $v_z = v = 2v_m[1 - (r/R)^2]$, where v_m is the mean velocity of the flow. Substitution in (55.3) gives the equation

$$\frac{1}{r}\frac{d}{dr}\left(r\frac{dT}{dr}\right) = -\frac{16v_m^2}{R^4}\frac{\nu}{\chi c_p}r^2.$$

The solution finite at $r = 0$ and equal to T_0 for $r = R$ is

$$T - T_0 = v_m^2\frac{P}{c_p}\left[1 - \left(\frac{r}{R}\right)\right]^4.$$

PROBLEM 2. Determine the temperature difference between a solid sphere and a fluid moving past it at small Reynolds numbers. The thermal conductivity of the sphere is supposed large.

SOLUTION. We take spherical co-ordinates r, θ, ϕ, with the origin at the centre of the sphere and the polar axis in the direction of the velocity of the main stream. Calculating the components of the tensor $\partial v_i/\partial x_k + \partial v_k/\partial x_i$ by means of formulae (15.17) and (20.9), we obtain equation (55.3) in the form

$$\frac{1}{r^2}\frac{\partial}{\partial r}\left(r^2\frac{\partial T}{\partial r}\right) + \frac{1}{r^2\sin\theta}\frac{\partial}{\partial\theta}\left(\sin\theta\frac{\partial T}{\partial\theta}\right)$$
$$= -A(R/r)^4[\cos^2\theta\{3 - 6(R/r)^2 + 2(R/r)^4\} + (R/r)^4],$$

where $A = 9u^2P/4c_p$. We look for $T(r, \theta)$ in the form $T = f(r)\cos^2\theta + g(r)$, and, separating the part which depends on θ, find two equations for f and g:

$$r^2 f'' + 2rf' - 6f = -A[3(R/r)^2 - 6(R/r)^4 + 2(R/r)^6],$$
$$r^2 g'' + 2rg' + 2f = -A(R/r)^6.$$

From the first we obtain

$$f = A[\tfrac{3}{4}(R/r)^2 + (R/r)^4 - \tfrac{1}{12}(R/r)^6] + c_1(R/r)^3;$$

the term of the form constant $\times r^2$ is omitted, since it does not vanish at infinity. The second equation then gives

$$g = -\tfrac{1}{2}A[\tfrac{3}{2}(R/r)^2 + \tfrac{1}{3}(R/r)^4 + \tfrac{1}{18}(R/r)^6] - \tfrac{1}{3}c_1(R/r)^3 + c_2 R/r + c_3.$$

The constants c_1, c_2, c_3 are determined from the conditions

$$T = \text{constant} \quad \text{and} \quad \int (\partial T/\partial r)r^2\sin\theta\,d\theta = 0$$

for $r = R$, which are equivalent to $f(R) = 0$ and $g'(R)+\frac{1}{3}f'(R) = 0$; also $T = T_0$ at infinity. Thus $c_1 = -5A/3$, $c_2 = 2A/3$, $c_3 = T_0$. The temperature difference between $T_1 = T(R)$ and T_0 is found to be $T_1-T_0 = 5u^2P/8c_p$. It may be noted that the temperature distribution obtained actually satisfies the condition $\partial T/\partial r = 0$ for $r = R$, i.e. $f'(R) = g'(R) = 0$. Hence it is also the solution of the same problem for a sphere of small thermal conductivity.

§56. Free convection

We have seen in §3 that, if there is mechanical equilibrium in a fluid in a gravitational field, the temperature distribution can depend only on the altitude z: $T = T(z)$. If the temperature distribution does not satisfy this condition, but is a function of the other co-ordinates also, then mechanical equilibrium in the fluid is not possible. Furthermore, even if $T = T(z)$, mechanical equilibrium may still be impossible if the vertical temperature gradient is directed downwards and its magnitude exceeds a certain value (§4).

The absence of mechanical equilibrium results in the appearance of internal currents in the fluid, which tend to mix the fluid and bring it to a constant temperature. Such motion in a gravitational field is called *free convection*.

Let us derive the equations describing this convection. We shall suppose the fluid incompressible. This means that the pressure is supposed to vary only slightly through the fluid, so that the density change due to changes in pressure may be neglected. For example, in the atmosphere, where the pressure varies with height, this assumption means that we shall not consider columns of air of great height, in which the density varies considerably over the height of the column. The density change due to the non-uniform heating of the fluid, of course, can not be neglected; it results in the forces which bring about the convection.

We write the variable temperature $T(x, y, z, t)$ in the form $T = T_0 + T'$, where T_0 is some constant mean temperature from which the variation T' is reckoned. We shall suppose that T' is small compared with T_0.

We write the fluid density also in the form $\rho = \rho_0 + \rho'$, with ρ_0 a constant. Since the temperature variation T' is small, the resulting density change ρ' is also small, and we can write

$$\rho' = (\partial \rho_0/\partial T)_p T' = -\rho_0\beta T'. \qquad (56.1)$$

Here $\beta = -(1/\rho)\partial\rho/\partial T$ is the thermal-expansion coefficient of the fluid.

In the pressure $p = p_0 + p'$, p_0 is not constant. It is the pressure corresponding to mechanical equilibrium, when the temperature and density are constant and equal to T_0 and ρ_0 respectively. It varies with height according to the hydrostatic equation

$$p_0 = \rho_0\,\mathbf{g\cdot r}+\text{constant.} \qquad (56.2)$$

We start by transforming the Navier–Stokes equation, which has, in the presence of a gravitational field, the form

$$\partial\mathbf{v}/\partial t+(\mathbf{v\cdot grad})\mathbf{v} = -(1/\rho)\mathbf{grad}\,p+\nu\triangle\mathbf{v}+\mathbf{g};$$

this is obtained by adding the force $\mathbf{g}$ per unit mass to the right-hand side of equation (15.7). We now substitute $p = p_0+p'$, $\rho = \rho_0+\rho'$; to the first order of small quantities, we have

$$\frac{\operatorname{grad} p}{\rho} = \frac{\operatorname{grad} p_0}{\rho_0} + \frac{\operatorname{grad} p'}{\rho_0} - \frac{\operatorname{grad} p_0}{\rho_0{}^2}\rho',$$

or, substituting (56.1) and (56.2),

$$\frac{\operatorname{grad} p}{\rho} = \mathbf{g} + \frac{\operatorname{grad} p'}{\rho_0} + \mathbf{g}T'\beta.$$

With this expression, the Navier–Stokes equation gives

$$\partial\mathbf{v}/\partial t+(\mathbf{v}\cdot\operatorname{grad})\mathbf{v} = -(1/\rho)\operatorname{grad} p'+\nu\triangle\mathbf{v}-\beta T'\mathbf{g}, \tag{56.3}$$

where the suffix has been dropped from ρ_0. In the thermal conduction equation (50.2), the viscosity term can be shown to be small in free convection compared with the other terms, and may therefore be omitted. We thus obtain

$$\partial T'/\partial t+\mathbf{v}\cdot\operatorname{grad} T' = \chi\triangle T'. \tag{56.4}$$

Equations (56.3) and (56.4), together with the equation of continuity $\operatorname{div}\mathbf{v} = 0$, form a complete system of equations governing free convection.

For steady flow, the equations of convection become

$$(\mathbf{v}\cdot\operatorname{grad})\mathbf{v} = -(1/\rho)\operatorname{grad} p'-\beta T'\mathbf{g}+\nu\triangle\mathbf{v}, \tag{56.5}$$

$$\mathbf{v}\cdot\operatorname{grad} T' = \chi\triangle T', \tag{56.6}$$

$$\operatorname{div}\mathbf{v} = 0. \tag{56.7}$$

This system of five equations for the unknown functions $\mathbf{v}$, p'/ρ and T' contains three parameters, ν, χ and βg. Moreover, the solution will involve a characteristic length l and the temperature difference $T_1 - T_0$ between the solid body and the fluid at a great distance. There is here no characteristic velocity, since there is no flow due to external forces, and the whole motion of the fluid is due to its non-uniform heating.

Thus steady free convection in a gravitational field is characterised by five parameters, which have the following dimensions: $\chi = \nu = \text{cm}^2/\text{sec}$, $T_1-T_0 = \deg$, $l = \text{cm}$, $\beta g = \text{cm}/\text{sec}^2\,\deg$. From these we can form two independent dimensionless quantities, which we take to be the Prandtl number $P = \nu/\chi$ and the *Grashof number*

$$G = \beta g l^3 (T_1-T_0)/\nu^2. \tag{56.8}$$

The similarity law for free convection is therefore

$$\mathbf{v} = (\nu/l)\mathbf{f}(\mathbf{r}/l,\,G), \qquad T = (T_1-T_0)f(\mathbf{r}/l,\,P,\,G). \tag{56.9}$$

Two flows are similar if their Prandtl and Grashof numbers are the same. Convective heat transfer caused by gravity is again characterised by the

Nusselt number, which is now a function of P and G only:

$$N = f(P, G). \tag{56.10}$$

The value of the Grashof number is an important characteristic of convective flow. When G is sufficiently small, the free convection is unimportant in the heat transfer in the fluid, which is then due mainly to ordinary conduction.

Convective flow may be either laminar or turbulent. There is no Reynolds number for free convection (since there is no characteristic velocity parameter), and the onset of turbulence is determined by the Grashof number: the convection becomes turbulent when G is very large.

A very curious case of convection is the flow which occurs in a fluid between two infinite horizontal planes at different temperatures, that of the lower plane (T_2) being greater than that of the upper plane (T_1). If the temperature difference $T_2 - T_1$ is small, the fluid remains at rest and there is pure thermal conduction, the fluid temperature and density being functions only of the vertical co-ordinate z; the density increases upward. If the difference $T_2 - T_1$ exceeds a certain critical value, however, which depends on the distance l between the planes, such a state becomes unstable and steady convection occurs. The onset of instability can be determined theoretically (see Problem 5). The critical value of the difference $T_2 - T_1$ appears as a factor in the product

$$GP = \beta g l^3 (T_2 - T_1)/\nu\chi. \tag{56.11}$$

In a layer of fluid between two solid planes at constant temperatures, convection must occur if GP > 1710. If the upper surface is free, but still at a constant temperature, then convection occurs for GP > 1100.†

The convective flow which occurs is somewhat unusual. Since the fluid is unbounded in the horizontal plane, it is evident that the flow must be periodic in that plane. In other words, the space between the bounding planes must be divided into similar right prisms in each of which the fluid moves in a similar way. The horizontal cross-sections of these prisms form a network in the horizontal plane. The theoretical determination of the nature of this network is very difficult, but experimental results seem to indicate that there is a hexagonal pattern with cells in the form of hexagonal prisms, the fluid moving up in the middle and down at the edges, or else *vice versa*.

For very large values of G, the steady convection in turn becomes unstable; turbulence sets in for G ~ 50,000.

Another similar case of instability is that of convection in a vertical cylindrical pipe along which a constant temperature gradient is maintained.

---

† These conditions (for a given difference $T_2 - T_1$) are always fulfilled if l is sufficiently large. To avoid misunderstanding, we should mention that we are speaking here of values of l for which the variation in the fluid density under the action of gravity is unimportant. Hence the above criteria cannot be applied to gas columns of great height. In this case we have to use the criterion derived in §4, from which we see that convection need not occur for a column of any height if the temperature gradient is small enough.

Here again there is a critical value of the product GP beyond which the fluid at rest is unstable; see Problem 6.

PROBLEMS

PROBLEM 1. Determine the Nusselt number for free convection on a flat vertical plate. It is assumed that the velocity and the temperature difference $T' = T - T_0$ (where T_0 is the fluid temperature at infinity) are appreciably different from zero only in a thin boundary layer adjoining the surface of the plate (K. POHLHAUSEN).

SOLUTION. We take the origin on the lower edge of the plate, the x-axis vertical, and the y-axis perpendicular to the plate. The pressure in the boundary layer does not vary along the y-axis (cf. §39), and therefore is everywhere equal to the hydrostatic pressure $p_0(x)$, i.e. $p' = 0$. With the usual accuracy of boundary-layer theory, equations (56.5)–(56.7) become

$$v_x \frac{\partial v_x}{\partial x} + v_y \frac{\partial v_x}{\partial y} = \nu \frac{\partial^2 v_x}{\partial y^2} + \beta g(T - T_0), \qquad (1)$$

$$v_x \frac{\partial T}{\partial x} + v_y \frac{\partial T}{\partial y} = \chi \frac{\partial^2 T}{\partial y^2}, \qquad (2)$$

$$\frac{\partial v_x}{\partial x} + \frac{\partial v_y}{\partial y} = 0, \qquad (3)$$

with the boundary conditions $v_x = v_y = 0$ and $T = T_1$ for $y = 0$ (T_1 being the temperature of the plate), $v_x = 0$ and $T = T_0$ for $y = \infty$. These equations can be converted into ordinary differential equations by introducing as the independent variable

$$\xi = Cy/x^{\frac{1}{4}}, \qquad C = [\beta g(T_1 - T_0)/4\nu^2]^{\frac{1}{4}}. \qquad (4)$$

We put

$$v_x = 4\nu C^2 \sqrt{x} \phi'(\xi), \qquad T - T_0 = (T_1 - T_0)\theta(\xi). \qquad (5)$$

Then (3) gives $v_y = \nu C x^{-\frac{1}{4}}(\xi\phi' - 3\phi)$, and (1) and (2) give equations for ϕ and θ:

$$\phi''' + 3\phi\phi'' - 2\phi'^2 + \theta = 0, \qquad \theta'' + 3P\phi\theta' = 0, \qquad (6)$$

with the boundary conditions $\phi(0) = \phi'(0) = 0$, $\theta(0) = 1$, $\phi'(\infty) = 0$, $\theta(\infty) = 0$. It follows from (4) and (5) that the thickness of the boundary layer is of the order $\delta \sim x^{\frac{1}{4}}/C$. The condition for the solution to be valid is therefore $\delta \ll l$ (where l is the height of the plate), or $G^{\frac{1}{4}} \gg 1$. The total heat flux per unit area of the plate is

$$q = -\frac{1}{l} \int_0^l \kappa \left(\frac{\partial T}{\partial y}\right)_{y=0} dx = -\tfrac{4}{3}\kappa\theta'(0, P)C(T_1 - T_0)l^{-\frac{1}{4}}.$$

The Nusselt number is $N = f(P)G^{\frac{1}{4}}$, where the function $f(P)$ is determined by solving the equations (6).

PROBLEM 2. A hot turbulent submerged jet of gas is bent round by a gravitational field; find its shape (G. N. ABRAMOVICH 1938).

SOLUTION. Let T' be some mean value (over the cross-section of the jet) of the temperature difference between the jet and the surrounding gas, u some mean velocity of the gas in the jet, and l the distance along the jet from its point of entry; l is supposed large compared with the dimensions of the aperture by which the jet enters. The condition of constant heat flux Q along the jet is $Q \sim \rho c_p T' u R^2 = \text{constant}$ and, since the radius of a turbulent jet is proportional to l (cf. §35), we have

$$T'ul^2 = \text{constant} \sim Q/\rho c_p; \qquad (1)$$

we notice that, in the absence of the gravitational field, $u \sim 1/l$ (see (35.3)) and it then follows from (1) that $T' \sim 1/l$.

The momentum flux vector through the cross-section of the jet is proportional to $\rho u^2 R^2 \mathbf{n}$ $\sim g u^2 l^2 \mathbf{n}$, where $\mathbf{n}$ is a unit vector along the jet. Its horizontal component is constant along the jet:

$$u^2 l^2 \cos \theta = \text{constant}, \tag{2}$$

where θ is the angle between $\mathbf{n}$ and the horizontal, while the change in the vertical component is due to the "lift force" on the jet. This force is proportional to

$$\rho \beta g T' R^2 \sim \rho \beta g T' l^2 \sim \beta g Q / c_p u.$$

Hence we have

$$d(l^2 u^2 \sin \theta)/dl \sim \beta g Q / \rho c_p u. \tag{3}$$

It then follows from (2) that $d(\tan \theta)/dl = \text{constant} \times l \cos^{\frac{1}{2}} \theta$, whence we obtain finally

$$\int_{\theta_0}^{\theta} \frac{d\theta}{\cos^{5/2}\theta} = \text{constant} \times l^2, \tag{4}$$

where θ_0 gives the direction of the emergent jet.

In particular, if θ does not vary appreciably along the jet, (4) gives $\theta - \theta_0 = \text{constant} \times l^2$. This means that the jet is a cubical parabola, in which the deviation d from a straight line is $d = \text{constant} \times l^3$.

PROBLEM 3. A turbulent jet of heated gas (i.e. one with a large Grashof number) rises from a fixed hot body. Determine the variation of the velocity and temperature in the jet with height (YA. B. ZEL'DOVICH 1937).

SOLUTION. As in the preceding case, the radius of the jet is proportional to the distance from its source, and we have, analogously to (1) of Problem 2, $T' u z^2 = \text{constant}$, and instead of (3) $d(z^2 u^2)/dz = \text{constant}/u$, where z is the height above the body, supposed large compared with the dimension of the body. Integrating, we find $u \sim z^{-\frac{1}{3}}$, and for the temperature $T' \sim z^{-5/3}$.

PROBLEM 4. The same as Problem 3, but for a laminar convective jet rising freely (YA. B. ZEL'DOVICH 1937).

SOLUTION. Together with the relation $T' u R^2 = \text{constant}$, which expresses the constancy of the heat flux, we have $u^2/z \sim \nu u/R^2 \sim \beta g T'$, which follows from equation (56.5). From these relations we find the following variation of the radius, velocity and temperature with height: $R \sim \sqrt{z}$, $u = \text{constant}$, $T' \sim 1/z$. It may be noticed that the number $G \sim T' R^3$ $\sim \sqrt{z}$, i.e. increases with height, and the jet must therefore become turbulent at a certain altitude.

PROBLEM 5. Derive the equations governing the onset of steady convection between two horizontal planes maintained at given temperatures (RAYLEIGH 1916).

SOLUTION. A perturbation proportional to $e^{-i\omega t}$ is applied to a fluid at rest with a constant vertical temperature gradient $\partial T/\partial z = -A < 0$. The state of rest is unstable if there is any possible value of ω whose imaginary part is positive. Hence the onset of instability is determined by the appearance of a solution for which the imaginary part of ω is zero. In this case we are concerned with the appearance of steady convection as a result of instability; hence we must seek solutions for which the real part of ω is also zero, that is, solutions independent of time.

In equations (56.5)–(56.7), the velocity $\mathbf{v}$ of the perturbing motion and the resulting pressure variation p' are small quantities. We write the temperature as $T' = -Az + \tau$, where the perturbation τ is small; we suppose the pressure variation resulting from the constant temperature gradient to be included in p_0. Then we find, omitting second-order terms,

$$\nu \triangle \mathbf{v} = \mathbf{grad}(p'/\rho) + \beta \tau \mathbf{g},$$

$$\chi \triangle \tau = -Av_z, \qquad \mathrm{div}\,\mathbf{v} = 0. \tag{1}$$

Eliminating $\mathbf{v}$ and p'/ρ, we obtain an equation for τ:

$$\triangle^3 \tau = \frac{\gamma}{l^4}\left(\frac{\partial^2 \tau}{\partial x^2} + \frac{\partial^2 \tau}{\partial y^2}\right), \tag{2}$$

where $\gamma = l^4 \beta g A / \nu \chi = \mathrm{GP}$, and l is the distance between the planes.

The boundary conditions on equations (1) at a solid surface are $\tau = 0$, $v_z = 0$, $\partial v_z/\partial z = 0$. The last of these follows from the equation of continuity, since we must have $v_x = v_y = 0$ for all x and y. By the second equation (1), the conditions on v_z can be replaced by conditions on higher derivatives of τ, v_z being replaced by $\partial^2 \tau/\partial z^2$.

We look for τ in the form $e^{i\mathbf{k}\cdot\mathbf{r}}f(z)$, where $\mathbf{k}$ is a vector in the xy-plane, and obtain for $f(z)$ the equation

$$\left(\frac{d^2}{dz^2} - k^2\right)^3 f + \frac{\gamma k^2}{l^4}f = 0.$$

The general solution of this equation is a linear combination of the functions $\cosh(\mu z/l)$ and $\sinh(\mu z/l)$, where $\mu^2 = k^2 l^2 - \gamma^{\frac{1}{3}}(kl)^{\frac{2}{3}}\sqrt[3]{1}$ with the three different values of $\sqrt[3]{1}$. The coefficients are determined by the boundary conditions, which lead to a system of algebraic equations; the compatibility condition then determines the function $kl(\gamma)$. The inverse function $\gamma = \gamma(kl)$ has a minimum for some value of kl; the corresponding $\gamma = \mathrm{GP}$ determines the required criterion for the appearance of instability, and the value of k determines the periodicity in the xy-plane, but not the symmetry, of the resulting motion.†

PROBLEM 6. Determine the onset of steady convection in a fluid at rest in a vertical cylindrical pipe along which a constant temperature gradient is maintained (G. A. OSTROUMOV 1946).

SOLUTION. We seek a solution of the equations (1) of Problem 5 in which the convective velocity $\mathbf{v}$ is everywhere parallel to the axis of the pipe (the z-axis), and the flow pattern does not vary along this axis, i.e. $v_z = v$, τ and $\partial p'/\partial z$ depend only on the co-ordinates x and y. Then the equations become $\partial p'/\partial x = 0$, $\partial p'/\partial y = 0$, $\nu \triangle_2 v = -\beta g \tau + (1/\rho)\partial p'/\partial z$, $\chi \triangle_2 \tau = -Av$, where $\triangle_2 \equiv \partial^2/\partial x^2 + \partial^2/\partial y^2$. The first two equations show that $\partial p'/\partial z = \mathrm{constant}$, and, eliminating τ from the other equations, we have

$$\triangle_2^2 v = \gamma v/R^4, \tag{1}$$

where we have again put $\gamma = AR^4 \beta g/\chi \nu = \mathrm{GP}$, and R is the radius of the pipe. At the surface of the pipe we must have $v = 0$ and the heat flux continuous. Moreover, the total mass flux through a cross-section of the pipe must be zero.

Equation (1) has solutions of the form $J_n(kr)\cos n\phi$ and $I_n(kr)\cos n\phi$, where J_n and I_n are Bessel functions of real and imaginary argument respectively, r and ϕ are polar co-ordinates in the cross-section, and $kR = \gamma^{\frac{1}{4}}$. The onset of convection corresponds to the solution for which γ is least. It is found that this is the solution with $n = 1$:

$$v = v_0 \cos\phi[J_1(kr)I_1(kR) - I_1(kr)J_1(kR)],$$

$$\tau = v_0(\nu k^2/\beta g)\cos\phi[J_1(kr)I_1(kR) + I_1(kr)J_1(kR)].$$

---

† A detailed account of the calculations is given by A. PELLEW and R. V. SOUTHWELL, *Proceedings of the Royal Society* A**176**, 312, 1940.

The pressure gradient $\partial p'/\partial z$ does not appear. The condition $v = 0$ for $r = R$ is satisfied identically, and the total mass flux through the cross-section of the pipe is zero. In the limiting case of thermally insulating walls, we must have also $\partial \tau/\partial r = 0$ for $r = R$, or

$$\frac{J_0(kR)}{J_1(kR)} + \frac{I_0(kR)}{I_1(kR)} = \frac{2}{kR}.$$

The smallest root of this equation gives the required critical value of $\gamma = (kR)^4 = 67\cdot4$. In the opposite limiting case of walls of infinite thermal conductivity, we must have $\tau = 0$ for $r = R$; then $J_1(kR) = 0$, whence the critical value is $\gamma = 215\cdot8$.†

---

† For a more detailed discussion see G. A. OSTROUMOV, *Free Convection in a Confined Medium* (*Svobodnaya konvektsiya v usloviyakh vnutrennei zadachi*), Moscow 1952.

CHAPTER VI

DIFFUSION

§57. The equations of fluid dynamics for a mixture of fluids

THROUGHOUT the above discussion it has been assumed that the fluid is completely homogeneous. If we are concerned with a mixture of fluids whose composition is different at different points, then the equations of fluid dynamics are considerably modified.

We shall discuss here only mixtures with two components. The composition of the mixture is described by the concentration c, defined as the ratio of the mass of one component to the total mass of the fluid in a given volume element.

In the course of time, the distribution of the concentration through the fluid will in general change. This change occurs in two ways. Firstly, when there is macroscopic motion of the fluid, any given small portion of it moves as a whole, its composition remaining unchanged. This results in a purely mechanical mixing of the fluid; although the composition of each moving portion of it is unchanged, the concentration of the fluid at any point in space varies with time. If we ignore any processes of thermal conduction and internal friction which may also be taking place, this change in concentration is a thermodynamically reversible process, and does not result in the dissipation of energy.

Secondly, a change in composition can occur by the molecular transfer of the components from one part of the fluid to another. The equalisation of the concentration by this direct change of composition of every small portion of fluid is called *diffusion*. Diffusion is an irreversible process, and is, like thermal conduction and viscosity, one of the sources of energy dissipation in a mixture of fluids.

We denote by ρ the total density of the fluid. The equation of continuity for the total mass of the fluid is, as before,

$$\partial\rho/\partial t + \operatorname{div}(\rho\mathbf{v}) = 0. \tag{57.1}$$

It signifies that the total mass of fluid in any volume can vary only by the movement of fluid into or out of that volume. It must be emphasised that, strictly speaking, the concept of velocity itself must be redefined for a mixture of fluids. By writing the equation of continuity in the form (57.1), we have defined the velocity, as before, as the total momentum of unit mass of fluid.

The Navier–Stokes equation (15.5) is also unchanged. We shall now derive the remaining equations of fluid dynamics for a mixture of fluids.

In the absence of diffusion, the composition of any given fluid element would remain unchanged as it moved about. This means that the total derivative dc/dt would be zero, i.e. the equation $dc/dt = \partial c/\partial t + \mathbf{v} \cdot \mathbf{grad}\, c = 0$ would hold. This equation can be written, using (57.1), as

$$\partial(\rho c)/\partial t + \operatorname{div}(\rho c \mathbf{v}) = 0,$$

i.e. as an equation of continuity for one of the components of the mixture (ρc being the mass of that component in unit volume). In the integral form

$$\frac{\partial}{\partial t} \int \rho c\, dV = - \oint \rho c \mathbf{v} \cdot \mathbf{df}$$

it shows that the rate of change of the amount of this component in any volume is equal to the amount of the component transported through the surface of that volume by the motion of the fluid.

When diffusion occurs, besides the flux $\rho c \mathbf{v}$ of the component in question as it moves with the fluid, there is another flux which results in the transfer of the components even when the fluid as a whole is at rest. Let $\mathbf{i}$ be the density of this diffusion flux, i.e. the amount of the component transported by diffusion through unit area in unit time.† Then we have for the rate of change of the amount of the component in any volume

$$\frac{\partial}{\partial t} \int \rho c\, dV = - \oint \rho c \mathbf{v} \cdot \mathbf{df} - \oint \mathbf{i} \cdot \mathbf{df},$$

or, in differential form,

$$\partial(\rho c)/\partial t = - \operatorname{div}(\rho c \mathbf{v}) - \operatorname{div} \mathbf{i}. \tag{57.2}$$

Using (57.1), we can rewrite this "equation of continuity" for one component in the form

$$\rho(\partial c/\partial t + \mathbf{v} \cdot \mathbf{grad}\, c) = - \operatorname{div} \mathbf{i}. \tag{57.3}$$

To derive another equation, we repeat the arguments given in §49, bearing in mind that the thermodynamic quantities for the fluid are now functions of the concentration also. In calculating the derivative $\partial(\tfrac{1}{2}\rho v^2 + \rho \epsilon)/\partial t$ (in §49) by means of the equations of motion, we had to transform the terms $\rho \partial \epsilon / \partial t$ and $- \mathbf{v} \cdot \mathbf{grad}\, p$. This transformation must now be modified, because the thermodynamic identities for the energy and the heat function now contain an additional term involving the differential of the concentration:

$$d\epsilon = T\, ds + (p/\rho^2)d\rho + \mu\, dc,$$

$$dw = T\, ds + (1/\rho)dp + \mu\, dc,$$

---

† The sum of the flux densities for the two components must be $\rho \mathbf{v}$. If the flux density for one component is $\rho c \mathbf{v} + \mathbf{i}$, that for the other component is therefore $\rho(1-c)\mathbf{v} - \mathbf{i}$.

where μ is an appropriately defined chemical potential of the mixture.†
Accordingly, an additional term $\rho\mu\partial c/\partial t$ appears in the derivative $\rho\partial\epsilon/\partial t$.
Writing the second thermodynamic relation in the form

$$dp = \rho\,dw - \rho T\,ds - \rho\mu\,dc,$$

we see that the term $-\mathbf{v}\cdot\mathbf{grad}\,p$ will contain an additional term $\rho\mu\mathbf{v}\cdot\mathbf{grad}\,c$.

Thus we must add $\rho\mu(\partial c/\partial t + \mathbf{v}\cdot\mathbf{grad}\,c)$ to the expression (49.3). By
equation (57.3), this can be written $-\mu\,\mathrm{div}\,\mathbf{i}$. The result is

$$\frac{\partial}{\partial t}(\tfrac{1}{2}\rho v^2 + \rho\epsilon) = -\mathrm{div}[\rho\mathbf{v}(\tfrac{1}{2}v^2 + w) - \mathbf{v}\cdot\mathbf{\sigma}' + \mathbf{q}] +$$

$$+ \rho T\left(\frac{\partial s}{\partial t} + \mathbf{v}\cdot\mathbf{grad}\,s\right) - \sigma'_{ik}\frac{\partial v_i}{\partial x_k} + \mathrm{div}\,\mathbf{q} - \mu\,\mathrm{div}\,\mathbf{i}. \quad (57.4)$$

We have replaced $-\kappa\,\mathbf{grad}\,T$ by a heat flux $\mathbf{q}$, which may depend not only
on the temperature gradient but also on the concentration gradient (see the
next section). The sum of the last two terms on the right can be written

$$\mathrm{div}\,\mathbf{q} - \mu\,\mathrm{div}\,\mathbf{i} = \mathrm{div}(\mathbf{q} - \mu\mathbf{i}) + \mathbf{i}\cdot\mathbf{grad}\,\mu.$$

The expression $\rho\mathbf{v}(\tfrac{1}{2}v^2 + w) - \mathbf{v}\cdot\mathbf{\sigma}' + \mathbf{q}$ which is the operand of the divergence operator in (57.4) is, by the definition of $\mathbf{q}$, the total energy flux in
the fluid. The first term is the reversible energy flux, due simply to the
movement of the fluid as a whole, while the sum $-\mathbf{v}\cdot\mathbf{\sigma}' + \mathbf{q}$ is the irreversible
flux. When there is no macroscopic motion, the viscosity flux $\mathbf{v}\cdot\mathbf{\sigma}'$ is zero,
and the thermal flux is simply $\mathbf{q}$.

The equation of conservation of energy is

$$\frac{\partial}{\partial t}(\tfrac{1}{2}\rho v^2 + \rho\epsilon) = -\mathrm{div}[\rho\mathbf{v}(\tfrac{1}{2}v^2 + w) - \mathbf{v}\cdot\mathbf{\sigma}' + \mathbf{q}]. \quad (57.5)$$

Subtracting from (57.4), we obtain the required equation

$$\rho T\left(\frac{\partial s}{\partial t} + \mathbf{v}\cdot\mathbf{grad}\,s\right) = \sigma'_{ik}\frac{\partial v_i}{\partial x_k} - \mathrm{div}(\mathbf{q} - \mu\mathbf{i}) - \mathbf{i}\cdot\mathbf{grad}\,\mu, \quad (57.6)$$

which is a generalisation of (49.4).

---

† It is known from thermodynamics that, for a mixture of two substances, the thermodynamic
identity is

$$d\epsilon = T\,ds - p\,dV + \mu_1\,dn_1 + \mu_2\,dn_2,$$

where n_1, n_2 are the numbers of particles of the two substances in 1 g of the mixture, and μ_1, μ_2 are
the chemical potentials of the substances. The numbers n_1, n_2 satisfy the relation $n_1 m_1 + n_2 m_2 = 1$,
where m_1 and m_2 are the masses of the two kinds of particle. If we introduce as a variable the
concentration $c = n_1 m_1$, we have

$$d\epsilon = T\,ds - p\,dV + \left(\frac{\mu_1}{m_1} - \frac{\mu_2}{m_2}\right)dc.$$

Comparing this with the relation given in the text, we see that the chemical potential μ is related
to μ_1 and μ_2 by

$$\mu = \frac{\mu_1}{m_1} - \frac{\mu_2}{m_2}.$$

We have thus obtained a complete system of equations of fluid mechanics for a mixture of fluids. The number of equations in this system is one more than for a single fluid, since there is one more unknown function, namely the concentration. The equations are the equation of continuity (57.1), the Navier–Stokes equations, the "equation of continuity" (57.2) for one component, and equation (57.6), which determines the change in entropy. It must be noticed that equations (57.2) and (57.6) as they stand determine only the form of the corresponding equations of fluid dynamics, since they involve the undetermined fluxes **i** and **q**. These equations become determinate only when **i** and **q** are replaced by expressions in terms of the gradients of concentration and temperature. The corresponding expressions will be obtained in §58.

For the rate of change of the total entropy of the fluid, a calculation entirely similar to that of §49, but using (57.6) in place of (49.4), gives the result

$$\frac{\partial}{\partial t} \int \rho s \, dV = - \int \frac{(\mathbf{q} - \mu \mathbf{i}) \cdot \mathbf{grad}\, T}{T^2} dV - \int \frac{\mathbf{i} \cdot \mathbf{grad}\, \mu}{T} dV + \dots, \quad (57.7)$$

where we have omitted, for brevity, the viscosity terms.

§58. Coefficients of mass transfer and thermal diffusion

The diffusion flux **i** and the heat flux **q** are due to the presence of concentration and temperature gradients in the fluid. It should not be thought, however, that **i** depends only on the concentration gradient and **q** only on the temperature gradient. On the contrary, each of these fluxes depends, in general, on both gradients.

If the concentration and temperature gradients are small, we can suppose that **i** and **q** are linear functions of **grad** μ and **grad** T.[†] Accordingly, we write **i** and **q** as

$$\mathbf{i} = -\alpha\, \mathbf{grad}\, \mu - \beta\, \mathbf{grad}\, T, \quad \mathbf{q} = -\delta\, \mathbf{grad}\, \mu - \gamma\, \mathbf{grad}\, T + \mu \mathbf{i}.$$

There is a simple relation between the coefficients β and δ, which is a consequence of a *symmetry principle for the kinetic coefficients*. This symmetry principle is as follows.[‡]

Let us consider some closed system, and let x_1, x_2, ... be some quantities characterising the state of the system. Their equilibrium values are determined by the fact that, in statistical equilibrium, the entropy S of the whole system must be a maximum, i.e. we must have $X_a = 0$ for all a, where X_a denotes the derivative

$$X_a = -\partial S/\partial x_a. \quad (58.1)$$

We assume that the system is in a state near to equilibrium. This means that

---

[†] The fluxes **q** and **i** are independent of the pressure gradient (for given **grad** μ and **grad** T), for the same reason as that given with regard to **q** in §49.

[‡] See *Statistical Physics*, §119, Pergamon Press, London 1958.

all the x_a are very little different from their equilibrium values, and the X_a are small. Processes will occur in the system which tend to bring it into equilibrium. The quantities x_a are functions of time, and their rate of change is given by the time derivatives $\dot{x}_a$; we express the latter as functions of X_a, and expand these functions in series. As far as terms of the first order we have

$$\dot{x}_a = -\sum_b \gamma_{ab} X_b. \tag{58.2}$$

The symmetry principle for the kinetic coefficients states that the γ_{ab} (called the *kinetic coefficients*) are symmetrical with respect to the suffixes a and b:

$$\gamma_{ab} = \gamma_{ba}. \tag{58.3}$$

The rate of change of the entropy S is

$$\dot{S} = -\sum X_a \dot{x}_a. \tag{58.4}$$

Now let the x_a themselves be different at different points of the system, i.e. each volume element have its own values of the x_a. That is, we suppose the x_a to be functions of the co-ordinates. Then, in the expression for $\dot{S}$, besides summing over a we must integrate over the volume of the system:

$$\dot{S} = -\int \sum_a X_a \dot{x}_a \, dV. \tag{58.4a}$$

It is usually true that the values of the $\dot{x}_a$ at any given point depend only on the values of the X_a at that point. In this case we can write down the relation between $\dot{x}_a$ and X_a for each point in the system, and obtain the same formulae as previously.†

In the problem under consideration we take as the $\dot{x}_a$ the components of the vectors $\mathbf{i}$ and $\mathbf{q} - \mu \mathbf{i}$. Then we see from a comparison of (57.7) and (58.4a) that the X_a are respectively the components of the vectors $(1/T)\,\mathbf{grad}\,\mu$ and $(1/T^2)\,\mathbf{grad}\,T$. The kinetic coefficients γ_{ab} are the coefficients of these vectors in the equations

$$\mathbf{i} = -\alpha T\left(\frac{\mathbf{grad}\,\mu}{T}\right) - \beta T^2\left(\frac{\mathbf{grad}\,T}{T^2}\right),$$

$$\mathbf{q} - \mu \mathbf{i} = -\delta T\left(\frac{\mathbf{grad}\,\mu}{T}\right) - \gamma T^2\left(\frac{\mathbf{grad}\,T}{T^2}\right).$$

By the symmetry of the kinetic coefficients, we must have $\beta T^2 = \delta T$, or $\delta = \beta T$. This is the required relation.

---

† Strictly speaking, in order to apply the relations obtained for a discrete set of quantities to a continuous distribution, we should write the integral (58.4a) as a sum over small but finite regions ΔV of the body (cf. §132); then the definition of the coefficients γ_{ab} also involves ΔV. In the present case, however, this procedure is unnecessary, since we use only the symmetry of the kinetic coefficients, and not their actual values.

We can therefore write the fluxes **i** and **q** as

$$\mathbf{i} = -\alpha \, \mathbf{grad}\,\mu - \beta \, \mathbf{grad}\,T,$$
$$\mathbf{q} = -\beta T \, \mathbf{grad}\,\mu - \gamma \, \mathbf{grad}\,T + \mu \mathbf{i}, \tag{58.5}$$

with only three independent coefficients α, β, γ. It is convenient to eliminate **grad** μ from the expression for the heat flux, replacing it by **i** and **grad** T. Then we have

$$\mathbf{i} = -\alpha \, \mathbf{grad}\,\mu - \beta \, \mathbf{grad}\,T, \tag{58.6}$$
$$\mathbf{q} = (\mu + \beta T/\alpha)\mathbf{i} - \kappa \, \mathbf{grad}\,T, \tag{58.7}$$

where

$$\kappa = \gamma - \beta^2 \, T/\alpha. \tag{58.8}$$

If the diffusion flux **i** is zero, we have *pure thermal conduction*. For this to be so, T and μ must satisfy the equation $\alpha \, \mathbf{grad}\,\mu + \beta \, \mathbf{grad}\,T = 0$, or $\alpha \, d\mu + \beta \, dT = 0$. The integration of this equation gives a relation of the form $f(c, T) = 0$ which does not contain the co-ordinates explicitly. (The chemical potential is a function of the pressure, as well as of c and T, but in equilibrium the pressure is constant.) This relation determines the dependence of the concentration on the temperature which must hold if there is no diffusion flux. Moreover, for $\mathbf{i} = 0$ we have from (58.7)

$$\mathbf{q} = -\kappa \, \mathbf{grad}\,T,$$

so that κ is just the thermal conductivity.

Let us now change to the usual variables p, T and c. We have

$$\mathbf{grad}\,\mu = (\partial\mu/\partial c)_{p,T} \, \mathbf{grad}\,c + (\partial\mu/\partial T)_{c,p} \, \mathbf{grad}\,T + (\partial\mu/\partial p)_{c,T} \, \mathbf{grad}\,p.$$

In the last term we can replace the derivative $(\partial\mu/\partial p)_{c,T}$ by $(\partial V/\partial c)_{p,T}$, where V is the specific volume.† Substituting in (58.6) and (58.7), and putting

$$D = \frac{\alpha}{\rho}\left(\frac{\partial\mu}{\partial c}\right)_{T,p},$$
$$\rho k_T D/T = \alpha(\partial\mu/\partial T)_{c,p} + \beta, \tag{58.9}$$
$$k_p = p(\partial V/\partial c)_{p,T}/(\partial\mu/\partial c)_{p,T}, \tag{58.10}$$

we obtain

$$\mathbf{i} = -\rho D[\mathbf{grad}\,c + (k_T/T) \, \mathbf{grad}\,T + (k_p/p) \, \mathbf{grad}\,p], \tag{58.11}$$
$$\mathbf{q} = [k_T(\partial\mu/\partial c)_{p,T} - T(\partial\mu/\partial T)_{p,c} + \mu]\mathbf{i} - \kappa \, \mathbf{grad}\,T. \tag{58.12}$$

The coefficient D is called the *diffusion coefficient* or *mass transfer coefficient*;

---

† The equality of these two derivatives follows from the thermodynamic identity
$$d\phi = -s\,dT + V\,dp + \mu\,dc,$$
where ϕ is the thermodynamic potential per unit mass;
$$(\partial\mu/\partial p)_{c,T} = \partial^2\phi/\partial p\,\partial c = (\partial V/\partial c)_{p,T}.$$

it gives the diffusion flux when only a concentration gradient is present. The diffusion flux due to the temperature gradient is given by the *thermal diffusion coefficient* $k_T D$; the dimensionless quantity k_T is called the *thermal diffusion ratio*.

The last term in (58.11) need be taken into account only when there is a considerable pressure gradient in the fluid (caused by an external field, say). The coefficient $k_p D$ may be called the *barodiffusion coefficient*. It should be noticed that, by formula (58.10), the dimensionless quantity k_p is entirely determined by thermodynamic properties alone.

In a single fluid there is, of course, no diffusion flux. Hence it is clear that k_T and k_p must vanish in each of the two limiting cases $c = 0$ and $c = 1$.

The condition that the entropy must increase places certain restrictions on the coefficients in formulae (58.6) and (58.7). Substituting these formulae in the expression (57.7) for the rate of change of the entropy, we find

$$\frac{\partial}{\partial t} \int \rho s \, dV = \int \frac{\kappa (\mathbf{grad}\, T)^2}{T^2} dV + \int \frac{\mathbf{i}^2}{\alpha T} dV + \dots . \qquad (58.13)$$

Hence it is clear that, besides the condition $\kappa > 0$ which we already know, we must have also $\alpha > 0$. Bearing in mind that the derivative $(\partial \mu / \partial c)_{p,T}$ is always positive,† we therefore find that the diffusion coefficient must be positive: $D > 0$. The quantities k_T and k_p, however, may be either positive or negative.

We shall not pause to write out the lengthy general equations obtained by substituting the above expressions for $\mathbf{i}$ and $\mathbf{q}$ in (57.3) and (57.6). We shall take only the case where there is no significant pressure gradient, while the concentration and temperature of the fluid vary so little that the coefficients in the expressions (58.11) and (58.12) may be supposed constant, although they are in general functions of c and T. Furthermore, we shall suppose that there is no macroscopic motion in the fluid except that which may be caused by the temperature and concentration gradients. The velocity of this motion is proportional to the gradients, and the terms in equations (57.3) and (57.6) which involve the velocity are therefore quantities of the second order, and may be neglected. The term $-\mathbf{i} \cdot \mathbf{grad}\, \mu$ in (57.6) is also of the second order. Thus we have $\rho \partial c / \partial t + \text{div}\,\mathbf{i} = 0$, $\rho T \partial s / \partial t + \text{div}(\mathbf{q} - \mu \mathbf{i}) = 0$.

Substituting for $\mathbf{i}$ and $\mathbf{q}$ the expressions (58.11) and (58.12) (without the term in $\mathbf{grad}\, p$), and transforming the derivative $\partial s / \partial t$ as follows:‡

$$\frac{\partial s}{\partial t} = \left(\frac{\partial s}{\partial T} \right)_{c,p} \frac{\partial T}{\partial t} + \left(\frac{\partial s}{\partial c} \right)_{T,p} \frac{\partial c}{\partial t} = \frac{c_p}{T} \frac{\partial T}{\partial t} - \left(\frac{\partial \mu}{\partial T} \right)_{p,c} \frac{\partial c}{\partial t},$$

---

† See *Statistical Physics*, §95.
‡ For

$$(\partial s / \partial c)_{p,T} = -\partial^2 \phi / \partial c\, \partial T = -(\partial \mu / \partial T)_{p,c}.$$

we obtain after a simple calculation

$$\partial c/\partial t = D[\triangle c + (k_T/T)\triangle T], \tag{58.14}$$

$$\partial T/\partial t - (k_T/c_p)(\partial\mu/\partial c)_{p,T}\, \partial c/\partial t = \chi\triangle T. \tag{58.15}$$

This system of linear equations determines the temperature and concentration distributions in the fluid.

There is a particularly important case where the concentration is small. When the concentration tends to zero, the diffusion coefficient tends to a finite constant, but the thermal diffusion coefficient tends to zero. Hence k_T is small for small concentrations, and we can neglect the term $k_T\triangle T$ in (58.14), which then becomes the diffusion equation

$$\partial c/\partial t = D\triangle c. \tag{58.16}$$

The boundary conditions on the solution of (58.16) are different in different cases. At the surface of a body insoluble in the fluid the normal component of the diffusion flux $\mathbf{i} = -\rho D\,\mathbf{grad}\,c$ must vanish, i.e. we must have $\partial c/\partial n = 0$. If, however, there is diffusion from a body which dissolves in the fluid, equilibrium is rapidly established near its surface, and the concentration in the fluid adjoining the body is the saturation concentration c_0; the diffusion out of this layer takes place more slowly than the process of solution. The boundary condition at such a surface is therefore $c = c_0$. Finally, if a solid surface absorbs the diffusing substance incident on it, the boundary condition is $c = 0$; an example of such a case is found in the study of chemical reactions at the surface of a solid.

Since the equations of pure diffusion (58.16) and of thermal conduction (50.4) are of exactly the same form, we can immediately apply all the formulae derived in §§51 and 52 to the case of diffusion, simply replacing T by c and χ by D. The boundary condition for a thermally insulating surface corresponds to that for an insoluble surface, while a surface maintained at a constant temperature corresponds to a soluble surface from which diffusion takes place.

In particular, we can write down, by analogy with (51.6), the following solution of the diffusion equation:

$$c(r) = \frac{M}{8\rho(\pi Dt)^{\frac{3}{2}}}\exp(-r^2/4Dt). \tag{58.17}$$

This gives the distribution of the solute at any time, if at time $t = 0$ it is all concentrated at the origin (M being the total amount of the solute).

PROBLEM

Determine the barodiffusion coefficient for a mixture of two perfect gases.

SOLUTION. We have for the specific volume $V = kT(n_1+n_2)/p$ (the notation is that used in the second footnote to §57), and the chemical potentials are[†]

$$\mu_1 = f_1(p, T) + kT\log[n_1/(n_1+n_2)],$$
$$\mu_2 = f_2(p, T) + kT\log[n_2/(n_1+n_2)].$$

---

† See *Statistical Physics*, §92.

The numbers n_1 and n_2 are expressed in terms of the concentration of the first component by $n_1 m_1 = c$, $n_2 m_2 = 1 - c$. A calculation using formula (58.10) gives

$$k_p = (m_2 - m_1)c(1-c)\left[\frac{1-c}{m_2} + \frac{c}{m_1}\right].$$

§59. Diffusion of particles suspended in a fluid

Under the influence of the molecular motion in a fluid, particles suspended in the fluid move in an irregular manner (called the *Brownian motion*). Let one such particle be at the origin at the initial instant. Its subsequent motion may be regarded as a diffusion, in which the concentration is represented by the probability of finding the particle in any particular volume element. To determine this probability, therefore, we can use the solution (58.17) of the diffusion equation. The possibility of this procedure is due to the fact that, for diffusion in weak solutions (i.e. when $c \ll 1$, which is when the diffusion equation can be used in the form (58.16)), the particles of the solute hardly affect one another, and so the motion of each particle can be considered independently.

Let $w(r, t)\mathrm{d}r$ be the probability of finding the particle at a distance between r and $r + \mathrm{d}r$ from the origin at time t. Putting in (58.17) $M/\rho = 1$ and multiplying by the volume $4\pi r^2 \mathrm{d}r$ of the spherical shell, we find

$$w(r, t)\mathrm{d}r = \frac{1}{2\sqrt{(\pi D^3 t^3)}} \exp(-r^2/4Dt) r^2 \,\mathrm{d}r. \tag{59.1}$$

Let us determine the mean square distance from the origin at time t. We have

$$\overline{r^2} = \int_0^\infty r^2 w(r, t)\mathrm{d}r. \tag{59.2}$$

The result, using (59.1), is

$$\overline{r^2} = 6Dt. \tag{59.3}$$

Thus the mean distance travelled by the particle during any time is proportional to the square root of the time.

The diffusion coefficient for particles suspended in a fluid can be calculated from what is called their *mobility*. Let us suppose that some constant external force $\mathbf{f}$ (the force of gravity, for example) acts on the particles. In a steady state, the force acting on each particle must be balanced by the drag force exerted by the fluid on a moving particle. When the velocity is small, the drag force is proportional to it and is $\mathbf{v}/b$, say, where b is a constant. Equating this to the external force $\mathbf{f}$, we have

$$\mathbf{v} = b\mathbf{f}, \tag{59.4}$$

i.e. the velocity acquired by the particle under the action of the external force is proportional to that force. The constant b is called the *mobility*, and can in principle be calculated from the equations of fluid dynamics. For example, for spherical particles of radius R, the drag force is $6\pi\eta Rv$ (see (20.14)), and therefore the mobility is

$$b = 1/6\pi\eta R. \tag{59.5}$$

For non-spherical particles, the drag depends on the direction of motion; it can be written in the form $a_{ik}v_k$, where a_{ik} is a symmetrical tensor (see (20.15)). To calculate the mobility we have to average over all orientations of the particle; if a_1, a_2, a_3 are the principal values of the symmetrical tensor a_{ik}, then we have

$$b = \frac{1}{3}\left(\frac{1}{a_1} + \frac{1}{a_2} + \frac{1}{a_3}\right). \tag{59.6}$$

The mobility b is simply related to the diffusion coefficient D. To derive this relation, we write down the diffusion flux $\mathbf{i}$, which contains the usual term $-\rho D \, \mathbf{grad} \, c$ due to the concentration gradient (we suppose the temperature constant), and also a term involving the velocity acquired by the particle owing to the external forces. This latter term is evidently $\rho c\mathbf{v}$. Thus

$$\mathbf{i} = -\rho D \, \mathbf{grad} \, c + \rho cb\mathbf{f}, \tag{59.7}$$

where we have used the expression (59.4). In a state of thermodynamic equilibrium, there is no diffusion, and the flux $\mathbf{i}$ must be zero. The equilibrium distribution of the concentration of particles suspended in a fluid, in an external field, is determined by Boltzmann's formula, according to which $c = \text{constant} \times e^{-U/kT}$, U being the potential energy of the particle in the external field. Since $\mathbf{f} = -\mathbf{grad} \, U$, we find the equilibrium concentration gradient to be $\mathbf{grad} \, c = c\mathbf{f}/kT$. Substituting this in (59.7) and equating $\mathbf{i}$ to zero, we have

$$D = kTb. \tag{59.8}$$

This is *Einstein's relation* between the diffusion coefficient and the mobility.

Substituting (59.5) in (59.8), we find the following expression for the diffusion coefficient for spherical particles:

$$D = kT/6\pi\eta R. \tag{59.9}$$

Besides the translatory Brownian motion and diffusion of suspended particles, we may consider also their rotary Brownian motion and diffusion. Just as the translatory diffusion coefficient is calculated in terms of the drag force, so the rotary diffusion coefficient can be expressed in terms of the forces on a particle executing a rotary movement in the fluid.†

---

† If (non-spherical) particles are suspended in a plane-parallel stream with a transverse velocity gradient, a definite distribution of the particles as regards their orientation in space is established as a result of the simultaneous action of the orienting forces of fluid dynamics and the disorienting Brownian motion. For the solution of this problem for ellipsoidal particles, see A. PETERLIN and H. A. STUART, *Zeitschrift für Physik* 112, 1, 1939.

PROBLEMS

PROBLEM 1. Particles execute Brownian motion in a fluid bounded on one side by a plane wall; particles incident on the wall "adhere" to it. Determine the probability that a particle which is at a distance x_0 from the wall at time $t = 0$ will have "adhered" to it after a time t.

SOLUTION. The probability distribution $w(x, t)$ (where x is the distance from the wall) is determined by the diffusion equation, with the boundary condition $w = 0$ for $x = 0$ and the initial condition $w = \delta(x - x_0)$ for $t = 0$. Such a solution is given by formula (52.4) when T is replaced by w, χ by D, and $T_0(x')$ in the integrand by $\delta(x' - x_0)$. We then obtain

$$w(x, t) = \frac{1}{2\sqrt{(\pi Dt)}}\{\exp[-(x - x_0)^2/4Dt] - \exp[-(x + x_0)^2/4Dt]\}.$$

The probability of "adhering" to the wall per unit time is given by the diffusion flux $D \partial w/\partial x$ for $x = 0$, and the required probability $W(t)$ over the time t is

$$W(t) = D \int_0^t [\partial w/\partial x]_{x=0} \, dt.$$

Substituting for w, we find

$$W(t) = 1 - \mathrm{erf}[x_0/2\sqrt{(Dt)}].$$

PROBLEM 2. Determine the order of magnitude of the time τ during which a particle suspended in a fluid turns through a large angle about its axis.

SOLUTION. The required time τ is that during which a particle in Brownian motion moves over a distance of the order of its linear dimension a. According to (59.3) we have $\tau \sim a^2/D$, and by (59.9) $D \sim kT/\eta a$. Thus $\tau \sim \eta a^3/kT$.

CHAPTER VII

SURFACE PHENOMENA

§60. Laplace's formula

IN this chapter we shall study the phenomena which occur near the surface separating two continuous media (in reality, of course, the media are separated by a narrow transitional layer, but this is so thin that it may be regarded as a surface). If the surface of separation is curved, the pressures near it in the two media are different. To determine the pressure difference (called the *surface pressure*), we write down the condition that the two media are in thermodynamic equilibrium together, taking into account the properties of the surface of separation.

Let the surface of separation undergo an infinitesimal displacement. At each point of the undisplaced surface we draw the normal. The length of the segment of the normal lying between the points where it intersects the displaced and undisplaced surfaces is denoted by $\delta\zeta$. Then a volume element between the two surfaces is $\delta\zeta\,df$, where df is a surface element. Let p_1 and p_2 be the pressures in the two media, and let $\delta\zeta$ be reckoned positive if the displacement of the surface is towards medium 2 (say). Then the work necessary to bring about the above change in volume is

$$\int (-p_1+p_2)\delta\zeta\,df.$$

The total work δR done in displacing the surface is obtained by adding to this the work connected with the change in area of the surface. This part of the work is proportional to the change δf in the area of the surface, and is $\alpha\delta f$, where α is called the *surface-tension coefficient*.† Thus the total work is

$$\delta R = - \int (p_1-p_2)\delta\zeta\,df + \alpha\delta f. \tag{60.1}$$

The condition of thermodynamic equilibrium is, of course, that δR is zero.

Next, let R_1 and R_2 be the principal radii of curvature at a given point of the surface; we reckon R_1 and R_2 as positive if they are drawn into medium 1. Then the elements of length dl_1 and dl_2 on the surface in its principal sections receive increments $(\delta\zeta/R_1)dl_1$ and $(\delta\zeta/R_2)dl_2$ respectively when the surface undergoes an infinitesimal displacement; here dl_1 and dl_2 are regarded as

---

† For an air–water interface $\alpha = 72 \cdot 5$ erg/cm² at 20° C; for air and paraffin $\alpha = 24$ at 20° C. The surface tension of liquid metals is very large; for instance, at an air–mercury interface $\alpha = 547$ at 175° C; for air and liquid platinum $\alpha = 1820$ at 2000° C. The surface tension between liquid helium and its vapour is very small, $\alpha = 0 \cdot 24$ at −270° C.

elements of the circumference of circles with radii R_1 and R_2. Hence the surface element $df = dl_1 dl_2$ becomes, after the displacement,

$$dl_1(1 + \delta\zeta/R_1)dl_2(1 + \delta\zeta/R_2) \approx dl_1 dl_2(1 + \delta\zeta/R_1 + \delta\zeta/R_2),$$

i.e. it changes by $\delta\zeta df(1/R_1 + 1/R_2)$. Hence we see that the total change in area of the surface of separation is

$$\delta f = \int \delta\zeta\left(\frac{1}{R_1} + \frac{1}{R_2}\right)df. \tag{60.2}$$

Substituting these expressions in (60.1) and equating to zero, we obtain the equilibrium condition in the form

$$\int \delta\zeta\left\{(p_1 - p_2) - \alpha\left(\frac{1}{R_1} + \frac{1}{R_2}\right)\right\}df = 0.$$

This condition must hold for every infinitesimal displacement of the surface, i.e. for all $\delta\zeta$. Hence the expression in braces must be identically equal to zero:

$$p_1 - p_2 = \alpha\left(\frac{1}{R_1} + \frac{1}{R_2}\right). \tag{60.3}$$

This is *Laplace's formula*, which gives the surface pressure. We see that, if R_1 and R_2 are positive, $p_1 - p_2 > 0$. This means that the pressure is greater in the medium whose surface is convex. If $R_1 = R_2 = \infty$, i.e. the surface of separation is plane, the pressure is the same in either medium, as we should expect.

Let us apply formula (60.3) to investigate the mechanical equilibrium of two adjoining media. We assume that no external forces act, either on the surface of separation or on the media themselves. Using formula (60.3), we can then write the equation of equilibrium as

$$\frac{1}{R_1} + \frac{1}{R_2} = \text{constant}. \tag{60.4}$$

Thus the sum of the curvatures must be a constant over any free surface of separation. If the whole surface is free, the condition (60.4) means that it must be spherical (for instance, the surface of a small drop, for which the effect of gravity may be neglected). If, however, the surface is supported along some curve (for instance, a film of liquid on a solid frame), its shape is less simple.

When the condition (60.4) is applied to the equilibrium of thin films supported on a solid frame, the constant on the right must be zero. For the sum $1/R_1 + 1/R_2$ must be the same everywhere on the free surface of the film, while on opposite sides of the film it must have opposite signs, since, if one side is convex, the other side is concave, and the radii of curvature are the same with opposite signs. Hence it follows that the equilibrium condition for a thin film is

$$\frac{1}{R_1} + \frac{1}{R_2} = 0. \tag{60.5}$$

Let us now consider the equilibrium condition on the surface of a medium in a gravitational field. We assume for simplicity that medium 2 is simply the atmosphere, whose pressure may be regarded as constant over the surface, and that medium 1 is an incompressible fluid. Then we have $p_2 = $ constant, while p_1, the fluid pressure, is by (3.2) $p_1 = $ constant $- \rho g z$, the co-ordinate z being measured vertically upwards. Thus the equilibrium condition becomes

$$\frac{1}{R_1} + \frac{1}{R_2} + \frac{g\rho z}{\alpha} = \text{constant}. \tag{60.6}$$

It should be mentioned that, to determine the equilibrium form of the surface of the fluid in particular cases, it is usually convenient to use the condition of equilibrium, not in the form (60.6), but by directly solving the variational problem of minimising the total free energy. The internal free energy of an incompressible fluid depends only on the volume of the fluid, and not on the shape of its surface. The latter affects, firstly, the surface free energy $\int \alpha \, df$ and, secondly, the energy in the external field (gravity), which is $g\rho \int z \, dV$. Thus the equilibrium condition can be written

$$\alpha \int df + g\rho \int z \, dV = \text{minimum}. \tag{60.7}$$

The minimum is to be determined subject to the condition

$$\int dV = \text{constant}, \tag{60.8}$$

which expresses the fact that the volume of the fluid is constant.

The constants α, ρ and g appear in the equilibrium conditions (60.6) and (60.7) only in the form $\alpha/g\rho$. This ratio has the dimensions cm². The length

$$a = \sqrt{(2\alpha/g\rho)} \tag{60.9}$$

is called the *capillary constant* for the substance concerned.[†] The shape of the fluid surface is determined by this quantity alone. If the capillary constant is large compared with the dimension of the medium, we may neglect gravity in determining the shape of the surface.

In order to find the shape of the surface from the condition (60.4) or (60.6), we need formulae which determine the radii of curvature, given the shape of the surface. These formulae are obtained in differential geometry,

---

† For water (e.g.), $a = 0{\cdot}122$ cm at 20° C.

but in the general case they are somewhat complicated. They are considerably simplified when the surface deviates only slightly from a plane. We shall derive the appropriate formula directly, without using the general results of differential geometry.

Let $z = \zeta(x, y)$ be the equation of the surface; we suppose that ζ is everywhere small, i.e. that the surface deviates only slightly from the plane $z = 0$. As is well known, the area f of the surface is given by the integral

$$f = \int \sqrt{\left[1 + \left(\frac{\partial \zeta}{\partial x}\right)^2 + \left(\frac{\partial \zeta}{\partial y}\right)^2\right]}\, dx\, dy,$$

or, for small ζ, approximately by

$$f = \int \left[1 + \frac{1}{2}\left(\frac{\partial \zeta}{\partial x}\right)^2 + \frac{1}{2}\left(\frac{\partial \zeta}{\partial y}\right)^2\right]\, dx\, dy. \tag{60.10}$$

The variation δf is

$$\delta f = \int \left\{\frac{\partial \zeta}{\partial x} \frac{\partial \delta \zeta}{\partial x} + \frac{\partial \zeta}{\partial y} \frac{\partial \delta \zeta}{\partial y}\right\}\, dx\, dy.$$

Integrating by parts, we find

$$\delta f = -\int \left(\frac{\partial^2 \zeta}{\partial x^2} + \frac{\partial^2 \zeta}{\partial y^2}\right) \delta \zeta\, dx\, dy.$$

Comparing this with (60.2), we obtain

$$\frac{1}{R_1} + \frac{1}{R_2} = -\left(\frac{\partial^2 \zeta}{\partial x^2} + \frac{\partial^2 \zeta}{\partial y^2}\right). \tag{60.11}$$

This is the required formula; it determines the sum of the curvatures of a slightly curved surface.

When three adjoining media are in equilibrium, the surfaces of separation are such that the resultant of the surface-tension forces is zero on the common line of intersection. This condition implies that the surfaces of separation must intersect at angles (called *angles of contact*) determined by the values of the surface-tension coefficients.[†]

Finally, let us consider the question of the boundary conditions that must be satisfied at the boundary between two fluids in motion, when the surface-tension forces are taken into account. If the latter forces are neglected, we have at the boundary between the fluids $n_k(\sigma_{2,ik} - \sigma_{1,ik}) = 0$, which expresses the equality of the forces of viscous friction on the surface of each fluid. When the surface tension is included, we have to add on the right-hand side a force determined in magnitude by Laplace's formula and directed along the normal:

$$n_k \sigma_{2,ik} - n_k \sigma_{1,ik} = \alpha \left(\frac{1}{R_1} + \frac{1}{R_2}\right) n_i. \tag{60.12}$$

---

† See, for instance, *Statistical Physics*, §145, Pergamon Press, London 1958.

This equation can also be written

$$(p_1 - p_2)n_i = (\sigma'_{1,ik} - \sigma'_{2,ik})n_k + \alpha\left(\frac{1}{R_1} + \frac{1}{R_2}\right)n_i. \qquad (60.13)$$

If the two fluids are both ideal, the viscous stresses σ'_{ik} are zero, and we return to the simple equation (60.3).

The condition (60.13), however, is still not completely general. The reason is that the surface-tension coefficient α may not be constant over the surface (for example, on account of a variation in temperature). Then, besides the normal force (which is zero for a plane surface), there is another force tangential to the surface. Just as there is a volume force $-\mathbf{grad}\, p$ per unit volume (see §2) in cases where the pressure is not uniform, so we have here a tangential force $\mathbf{f}_t = \mathbf{grad}\, \alpha$ per unit area of the surface of separation. In this case we take the positive gradient, because the surface-tension forces tend to reduce the area of the surface, whereas the pressure forces tend to increase the volume. Adding this force to the right-hand side of equation (60.13), we obtain the boundary condition

$$\left[p_1 - p_2 - \alpha\left(\frac{1}{R_1} + \frac{1}{R_2}\right)\right]n_i = (\sigma'_{1,ik} - \sigma'_{2,ik})n_k + \frac{\partial\alpha}{\partial x_i}; \qquad (60.14)$$

the unit normal vector $\mathbf{n}$ is directed into medium 1. We notice that this condition can be satisfied only for a viscous fluid: in an ideal fluid, $\sigma'_{ik} = 0$ and the left-hand side of equation (60.14) is a vector along the normal, while the right-hand side is in this case a tangential vector. This equality cannot hold, except of course in the trivial case where both sides are zero.

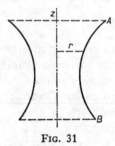

FIG. 31

PROBLEMS

PROBLEM 1. Determine the shape of a film of liquid supported on two circular frames with their centres on a line perpendicular to their planes, which are parallel; Fig. 31 shows a cross-section of the film.

SOLUTION. The problem amounts to that of finding the surface having the smallest area that can be formed by the revolution about the line $r = 0$ of a curve $r = r(z)$ which passes between two given points A and B. The area of a surface of rotation is

$$f = 2\pi \int_{z_1}^{z_2} r\sqrt{\left[1 + \left(\frac{dr}{dz}\right)^2\right]}\, dz.$$

It is well known that the minimum of an integral of the form

$$\int_{t_1}^{t_2} L(x, \dot{x})\, dt$$

is given by the equation $L - \dot{x}\, \partial L/\partial \dot{x} = \text{constant}$. In the present case this leads to

$$r = c_1 \sqrt{[1 + (dr/dz)^2]},$$

whence we have by integration $r = c_1 \cosh[(z - c_2)/c_1]$. Thus the required surface (called a *catenoid*) is that formed by the revolution of a catenary. The constants c_1 and c_2 must be chosen so that the curve $r(z)$ passes through the given points A and B. The value of c_2 depends only on the choice of the origin of z. For the constant c_1, however, two values are obtained, of which the larger must be chosen (the smaller does not give a minimum of the integral).

When the distance h between the frames increases, it reaches a value for which the equation for the constant c_1 no longer has a real root. For greater values of h, only the shape consisting of one film on each frame is stable. For example, for two frames of equal radius R the catenoid form is impossible for a distance h between the frames greater than $1 \cdot 33 R$.

PROBLEM 2. Determine the shape of the surface of a fluid in a gravitational field and bounded on one side by a vertical plane wall. The angle of contact between the fluid and the wall is θ (Fig. 32).

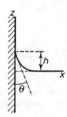

FIG. 32

SOLUTION. We take the co-ordinate axes as shown in Fig. 32. The plane $x = 0$ is the plane of the wall, and $z = 0$ is the plane of the fluid surface far from the wall. The radii of curvature of the surface $z = z(x)$ are $R_1 = \infty$, $R_2 = -(1 + z'^2)^{\frac{3}{2}}/z''$, so that equation (60.6) becomes

$$\frac{2z}{a^2} - \frac{z''}{(1 + z'^2)^{\frac{3}{2}}} = \text{constant}, \tag{1}$$

where a is the capillary constant. For $x = \infty$ we must have $z = 0$, $1/R_2 = 0$, and the constant is therefore zero. A first integral of the resulting equation is

$$\frac{1}{\sqrt{(1 + z'^2)}} = A - \frac{z^2}{a^2}. \tag{2}$$

From the condition at infinity ($z = z' = 0$ for $x = \infty$) we have $A = 1$. A second integration gives

$$x = -\frac{a}{\sqrt{2}} \cosh^{-1} \frac{\sqrt{2}a}{z} + a\sqrt{\left(2 - \frac{z^2}{a^2}\right)} + x_0.$$

The constant x_0 must be chosen so that, at the surface of the wall ($x = 0$), we have $z' = -\cot\theta$ or, by (2), $z = h$, where $h = a\sqrt{(1 - \sin\theta)}$ is the height to which the fluid rises at the wall itself.

PROBLEM 3. Determine the shape of the surface of a fluid rising between two parallel vertical flat plates (Fig. 33).

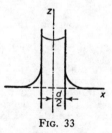

FIG. 33

SOLUTION. We take the yz-plane half-way between the two plates, and the xy-plane to coincide with the fluid surface far from the plates. In equation (1) of Problem 2, which gives the condition of equilibrium and is therefore valid everywhere on the surface of the fluid (both between the plates and elsewhere), the conditions at $x = \infty$ again give the constant as zero. In the integral (2), the constant A is now different according as $|x| > \frac{1}{2}d$ or $|x| < \frac{1}{2}d$ (the function $z(x)$ having a discontinuity for $|x| = \frac{1}{2}d$). For the space between the plates, the conditions are $z' = 0$ for $x = 0$ and $z' = \cot\theta$ for $x = \frac{1}{2}d$, where θ is the angle of contact. According to (2) we have for the heights $z_0 = z(0)$ and $z_1 = z(\frac{1}{2}d)$: $z_0 = a\sqrt{(A-1)}$, $z_1 = a\sqrt{(A-\sin\theta)}$. Integrating (2), we obtain

$$x = \int_{z_0}^{z} \frac{(A - z^2/a^2)\mathrm{d}z}{\sqrt{[1-(A-z^2/a^2)^2]}} = \frac{1}{2}a \int_{0}^{a\sqrt{(A-\cos\xi)=z}} \frac{\cos\xi\,\mathrm{d}\xi}{\sqrt{(A-\cos\xi)}},$$

where ξ is a new variable related to z by $z = a\sqrt{(A-\cos\xi)}$. This is an elliptic integral, and cannot be expressed in terms of elementary functions. The constant A is found from the condition that $z = z_1$ for $x = \frac{1}{2}d$, or

$$d = a \int_{0}^{\frac{1}{2}\pi-\theta} \frac{\cos\xi\,\mathrm{d}\xi}{\sqrt{(A-\cos\xi)}}.$$

The formulae obtained above give the shape of the fluid surface in the space between the plates. As $d \to 0$, A tends to infinity. Hence we have for $d \ll a$

$$d \approx \frac{a}{\sqrt{A}} \int_{0}^{\frac{1}{2}\pi-\theta} \cos\xi\,\mathrm{d}\xi = \frac{a}{\sqrt{A}}\cos\theta,$$

or $A = (a^2/d^2)\cos^2\theta$. The height to which the fluid rises is $z_0 \approx z_1 \approx (a^2/d)\cos\theta$; this formula can also be obtained directly, of course.

PROBLEM 4. A thin non-uniformly heated layer of fluid rests on a horizontal plane solid surface; its temperature is a given function of the co-ordinate x in the plane, and (because the layer is thin) may be supposed independent of the co-ordinate z across the layer. The non-uniform heating results in the occurrence of a steady flow, and its thickness ζ consequently varies in the x-direction. Determine the function $\zeta(x)$.

SOLUTION. The fluid density ρ and the surface tension α are, together with the temperature, known functions of x. The fluid pressure $p = p_0+\rho g(\zeta-z)$, where p_0 is the atmospheric pressure (the pressure on the free surface); the variation of pressure due to the curvature of the surface may be neglected. The fluid velocity in the layer may be supposed everywhere parallel to the x-axis. The equation of motion is

$$\eta \partial^2 v/\partial z^2 = \partial p/\partial x = g[d(\rho\zeta)/dx - z\, dp/dx].\tag{1}$$

On the solid surface ($z = 0$) we have $v = 0$, while on the free surface ($z = \zeta$) the boundary condition (60.14) must be fulfilled; in this case it is $\eta[dv/dz]_{z=\zeta} = d\alpha/dx$. Integrating equation (1) with these conditions, we obtain

$$\eta v = gz(\zeta - \tfrac{1}{2}z)d(\rho\zeta)/dx - \tfrac{1}{6}gz(3\zeta^2 - z^2)dp/dx - z\, d\alpha/dx.\tag{2}$$

Since the flow is steady, the total mass flux through a cross-section of the layer must be zero:

$$\int_0^\zeta v\, dz = 0.$$

Substituting (2), we find

$$\tfrac{1}{3}\rho\frac{d\zeta^2}{dx} + \tfrac{1}{4}\zeta^2\frac{d\rho}{dx} = \frac{1}{g}\frac{d\alpha}{dx},$$

which determines the function $\zeta(x)$. Integrating, we obtain

$$g\zeta^2 = 3\rho^{-\frac{3}{4}}[\int \rho^{-\frac{1}{4}}d\alpha + \text{constant}].\tag{3}$$

If the temperature (and therefore ρ and α) varies only slightly, then (3) can be written

$$\zeta^2 = \zeta_0^2(\rho_0/\rho)^{\frac{3}{4}} + 3(\alpha - \alpha_0)/\rho g,$$

where ζ_0 is the value of ζ at a point where $\rho = \rho_0$ and $\alpha = \alpha_0$.

§61. Capillary waves

Fluid surfaces tend to assume an equilibrium shape, both under the action of the force of gravity and under that of surface-tension forces. In studying waves on the surface of a fluid in §§12 and 13, we did not take the latter forces into account. We shall see below that capillarity has an important effect on gravity waves of small wavelength.

As in §12, we suppose the amplitude of the oscillations small compared with the wavelength. For the velocity potential we have as before the equation $\triangle\phi = 0$. The condition at the surface of the fluid is now different, however: the pressure difference between the two sides of the surface is not zero, as we supposed in §12, but is given by Laplace's formula (60.3).

We denote by ζ the z co-ordinate of a point on the surface. Since ζ is small, we can use the expression (60.11), and write Laplace's formula as

$$p - p_0 = -\alpha\left(\frac{\partial^2\zeta}{\partial x^2} + \frac{\partial^2\zeta}{\partial y^2}\right).$$

Here p is the pressure in the fluid near the surface, and p_0 is the constant external pressure. For p we substitute, according to (12.2),

$$p = -\rho g\zeta - \rho\partial\phi/\partial t,$$

obtaining

$$\rho g \zeta + \rho \frac{\partial \phi}{\partial t} - \alpha \left(\frac{\partial^2 \zeta}{\partial x^2} + \frac{\partial^2 \zeta}{\partial y^2} \right) = 0;$$

for the same reasons as in §12, we can omit the constant p_0 if we redefine ϕ. Differentiating this relation with respect to t, and replacing $\partial \zeta / \partial t$ by $\partial \phi / \partial z$, we obtain the boundary condition on the potential ϕ:

$$\rho g \frac{\partial \phi}{\partial z} + \rho \frac{\partial^2 \phi}{\partial t^2} - \alpha \frac{\partial}{\partial z} \left(\frac{\partial^2 \phi}{\partial x^2} + \frac{\partial^2 \phi}{\partial y^2} \right) = 0 \quad \text{for} \quad z = 0. \qquad (61.1)$$

Let us consider a plane wave propagated in the direction of the x-axis. As in §12, we obtain a solution in the form $\phi = A e^{kz} \cos(kx - \omega t)$. The relation between k and ω is now obtained from the boundary condition (61.1), and is

$$\omega^2 = gk + \alpha k^3 / \rho. \qquad (61.2)$$

We see that, for long wavelengths such that $k \ll \sqrt{(g\rho/\alpha)}$, or $k \ll 1/a$ (where a is the capillary constant), the effect of capillarity may be neglected, and we have a pure gravity wave. In the opposite case of short wavelengths, the effect of gravity may be neglected. Then

$$\omega^2 = \alpha k^3 / \rho. \qquad (61.3)$$

Such waves are called *capillary waves* or *ripples*. Intermediate cases are referred to as *capillary gravity waves*.

Let us also determine the nature of the oscillations of a spherical drop of incompressible fluid under the action of capillary forces. The oscillations cause the surface of the drop to deviate from the spherical form. As usual, we shall suppose the amplitude of the oscillations to be small.

We begin by determining the value of the sum $1/R_1 + 1/R_2$ for a surface slightly different from that of a sphere. Here we proceed as in the derivation of formula (60.11). The area of a surface given in spherical co-ordinates[†] r, θ, ϕ by a function $r = r(\theta, \phi)$ is

$$f = \int_0^{2\pi} \int_0^{\pi} \sqrt{\left[r^2 + \left(\frac{\partial r}{\partial \theta} \right)^2 + \frac{1}{\sin^2 \theta} \left(\frac{\partial r}{\partial \phi} \right)^2 \right]} \, r \sin \theta \, d\theta \, d\phi. \qquad (61.4)$$

A spherical surface is given by $r = \text{constant} \equiv R$ (where R is the radius of the sphere), and a neighbouring surface by $r = R + \zeta$, where ζ is small. Substituting in (61.4), we obtain

$$f = \int_0^{2\pi} \int_0^{\pi} \left\{ (R + \zeta)^2 + \frac{1}{2} \left[\left(\frac{\partial \zeta}{\partial \theta} \right)^2 + \frac{1}{\sin^2 \theta} \left(\frac{\partial \zeta}{\partial \phi} \right)^2 \right] \right\} \sin \theta \, d\theta \, d\phi.$$

---

† In the remainder of this section ϕ denotes the azimuthal angle, and we denote the velocity potential by ψ.

Let us find the variation δf in the area when ζ changes. We have

$$\delta f = \int_0^{2\pi} \int_0^\pi \left\{ 2(R+\zeta)\delta\zeta + \frac{\partial\zeta}{\partial\theta}\frac{\partial\delta\zeta}{\partial\theta} + \frac{1}{\sin^2\theta}\frac{\partial\zeta}{\partial\phi}\frac{\partial\delta\zeta}{\partial\phi} \right\} \sin\theta\, d\theta\, d\phi.$$

Integrating the second term by parts with respect to θ, and the third by parts with respect to ϕ, we obtain

$$\delta f = \int_0^{2\pi} \int_0^\pi \left\{ 2(R+\zeta) - \frac{1}{\sin\theta}\frac{\partial}{\partial\theta}\left(\sin\theta\frac{\partial\zeta}{\partial\theta}\right) - \frac{1}{\sin^2\theta}\frac{\partial^2\zeta}{\partial\phi^2} \right\} \delta\zeta \sin\theta\, d\theta\, d\phi$$

If we divide the expression in braces by $R(R+2\zeta)$, the resulting coefficient of $\delta\zeta\delta f \approx \delta\zeta R(R+2\zeta)\sin\theta\, d\theta\, d\phi$ in the integrand is, by formula (60.2), just the required sum of the curvatures, correct to terms of the first order in ζ. Thus we find

$$\frac{1}{R_1} + \frac{1}{R_2} = \frac{2}{R} - \frac{2\zeta}{R^2} - \frac{1}{R^2}\left\{ \frac{1}{\sin^2\theta}\frac{\partial^2\zeta}{\partial\phi^2} + \frac{1}{\sin\theta}\frac{\partial}{\partial\theta}\left(\sin\theta\frac{\partial\zeta}{\partial\theta}\right) \right\}. \quad (61.5)$$

The first term corresponds to a spherical surface, for which $R_1 = R_2 = R$.

The velocity potential ψ satisfies Laplace's equation $\triangle\psi = 0$, with a boundary condition at $r = R$ like that for a plane surface:

$$\rho\frac{\partial\psi}{\partial t} + \alpha\left\{ \frac{2}{R} - \frac{2\zeta}{R^2} - \frac{1}{R^2}\left[\frac{1}{\sin\theta}\frac{\partial}{\partial\theta}\left(\sin\theta\frac{\partial\zeta}{\partial\theta}\right) + \frac{1}{\sin^2\theta}\frac{\partial^2\zeta}{\partial\phi^2} \right] \right\} + p_0 = 0.$$

The constant $p_0 + 2\alpha/R$ can again be omitted; differentiating with respect to time and putting $\partial\zeta/\partial t = v_r = \partial\psi/\partial r$, we have finally the boundary condition on ψ:

$$\rho\frac{\partial^2\psi}{\partial t^2} - \frac{\alpha}{R^2}\left\{ 2\frac{\partial\psi}{\partial r} + \frac{\partial}{\partial r}\left[\frac{1}{\sin\theta}\frac{\partial}{\partial\theta}\left(\sin\theta\frac{\partial\psi}{\partial\theta}\right) + \frac{1}{\sin^2\theta}\frac{\partial^2\psi}{\partial\phi^2} \right] \right\} = 0$$

$$\text{for} \quad r = R. \quad (61.6)$$

We shall seek a solution in the form of a stationary wave: $\psi = e^{-i\omega t} f(r, \theta, \phi)$, where the function f satisfies Laplace's equation, $\triangle f = 0$. As is well known, any solution of Laplace's equation can be represented as a linear combination of what are called *volume spherical harmonic functions* $r^l Y_{lm}(\theta, \phi)$, where $Y_{lm}(\theta, \phi)$ are Laplace's spherical harmonics: $Y_{lm}(\theta, \phi) = P_l^m(\cos\theta)e^{im\phi}$. Here $P_l^m(\cos\theta) = \sin^m\theta\, d^m P_l(\cos\theta)/d(\cos\theta)^m$ is what is called an *associated Legendre function*, $P_l(\cos\theta)$ being the Legendre polynomial of order l. As is well known, l takes all integral values from zero upwards, while m takes the values $0, \pm 1, \pm 2, ..., \pm l$.

Accordingly, we seek a particular solution of the problem in the form

$$\psi = Ae^{-i\omega t} r^l P_l^m(\cos\theta)e^{im\phi}. \quad (61.7)$$

The frequency ω must be such as to satisfy the boundary condition (61.6). Substituting the expression (61.7) and using the fact that the spherical harmonics Y_{lm} satisfy

$$\frac{1}{\sin\theta}\frac{\partial}{\partial\theta}\left(\sin\theta\frac{\partial Y_{lm}}{\partial\theta}\right) + \frac{1}{\sin^2\theta}\frac{\partial^2 Y_{lm}}{\partial\phi^2} + l(l+1)Y_{lm} = 0,$$

we find (cancelling ψ)

$$\rho\omega + l\alpha[2 - l(l+1)]/R^3 = 0,$$

or

$$\omega^2 = \alpha l(l-1)(l+2)/\rho R^3. \tag{61.8}$$

This formula gives the eigenfrequencies of capillary oscillations of a spherical drop. We see that it depends only on l, and not on m. To a given l, however, there correspond $2l+1$ different functions (61.7). Thus each of the frequencies (61.8) corresponds to $2l+1$ different oscillations. Independent oscillations having the same frequency are said to be *degenerate*; in this case we have $(2l+1)$-fold degeneracy.

The expression (61.8) vanishes for $l = 0$ and $l = 1$. The value $l = 0$ would correspond to radial oscillations, i.e. to spherically symmetrical pulsations of the drop; in an incompressible fluid such oscillations are clearly impossible. For $l = 1$ the motion is simply a translatory motion of the drop as a whole. The smallest possible frequency of oscillations of the drop corresponds to $l = 2$, and is

$$\omega_{\min} = \sqrt{(8\alpha/\rho R^3)}. \tag{61.9}$$

A peculiar wave motion due to surface tension is observed when a thin layer of viscous fluid flows down a vertical wall. P. L. KAPITZA has shown that these waves must be due to an instability of the original flow that sets in at comparatively small Reynolds numbers.†

PROBLEMS

PROBLEM 1. Determine the frequency as a function of the wave number for capillary gravity waves on the surface of a fluid of depth h.

SOLUTION. Substituting in the condition (61.1) $\phi = A\cos(kx - \omega t)\cosh k(z+h)$ (cf. §12, Problem 1), we obtain $\omega^2 = (gk + \alpha k^3/\rho)\tanh kh$. For $kh \gg 1$ we return to formula (61.2), while for long waves ($kh \ll 1$) we have $\omega^2 = ghk^2 + \alpha hk^4/\rho$.

PROBLEM 2. Determine the damping coefficient for capillary waves.

SOLUTION. Substituting (61.3) in (25.5), we find $\gamma = 2\eta k^2/\rho = 2\eta\omega^{4/3}/\rho^{1/3}\alpha^{2/3}$.

PROBLEM 3. Find the condition for the stability of a horizontal tangential discontinuity in a gravitational field, taking account of surface tension (the fluids on the two sides of the surface of discontinuity being supposed different).

SOLUTION. Let U be the velocity of the upper fluid relative to the lower. On the original flow we superpose a perturbation periodic in the horizontal direction, and seek the velocity

---

† See P. L. KAPITZA, *Zhurnal èksperimental'noĭ i teoreticheskoĭ fiziki* **18**, 3, 1948.

potential in the form

$$\phi = Ae^{kz} \cos(kx - \omega t) \text{ in the lower fluid,}$$

$$\phi' = A'e^{-kz} \cos(kx - \omega t) + Ux \text{ in the upper fluid.}$$

For the lower fluid we have on the surface of discontinuity $v_z = \partial\phi/\partial z = \partial\zeta/\partial t$, where ζ is a vertical co-ordinate in the surface of discontinuity, and for the upper fluid

$$v'_z = \partial\phi'/\partial z = U\partial\zeta/\partial x + \partial\zeta/\partial t.$$

The condition of equal pressures in the two fluids at the surface of discontinuity is

$$\rho\,\partial\phi/\partial t + \rho g\zeta - \alpha\partial^2\zeta/\partial x^2 = \rho'\,\partial\phi'/\partial t + \rho'g\zeta + \tfrac{1}{2}\rho'(v'^2 - U^2);$$

only terms of the first order in A' need be retained in expanding the expression $v'^2 - U^2$. We seek the displacement ζ in the form $\zeta = a\sin(kx - \omega t)$. Substituting ϕ, ϕ' and ζ in the above three conditions for $z = 0$, we obtain three equations from which a, A and A' can be eliminated, leaving

$$\omega = \frac{k\rho'U}{\rho + \rho'} \pm \sqrt{\left[\frac{kg(\rho - \rho')}{\rho + \rho'} - \frac{k^2\rho\rho'U^2}{(\rho + \rho')^2} + \frac{\alpha k^3}{\rho + \rho'}\right]}.$$

In order that this expression should be real for all k, it is necessary that

$$U^4 \leqslant 4\alpha g(\rho - \rho')(\rho + \rho')^2/\rho^2\rho'^2.$$

If this condition does not hold, there are complex ω with a positive imaginary part, and the motion is unstable.

§62. The effect of adsorbed films on the motion of a liquid

The presence on the surface of a liquid of a film of adsorbed material may have a considerable effect on the hydrodynamical properties of the surface. The reason is that, when the shape of the surface changes with the motion of the liquid, the film is stretched or compressed, i.e. the surface concentration of the adsorbed substance is changed. These changes result in the appearance of additional forces which have to be taken into account in the boundary conditions at the free surface.

Here we shall consider only adsorbed films of substances which may be regarded as insoluble in the liquid. This means that the substance is entirely on the surface, and does not penetrate into the liquid. If the adsorbed substance is appreciably soluble, it is necessary to take into account the it diffusion of between the surface film and the volume of the liquid when the concentration of the film varies.

When the adsorbed material is present, the surface-tension coefficient α is a function of the surface concentration of the material (the amount of it per unit surface area), which we denote by γ. If γ varies over the surface, then the coefficient α is also a function of the co-ordinates in the surface. The boundary condition at the surface of the liquid therefore includes a tangential force, which we have already discussed at the end of §60 (equation (60.14)). In the present case, the gradient of α can be expressed in terms of the surface concentration gradient, so that the tangential force on the surface is

$$\mathbf{f}_i = (\partial\alpha/\partial\gamma)\mathbf{grad}\,\gamma. \tag{62.1}$$

It has been mentioned in §60 that the boundary condition (60.14), in which this force is taken into account, can be satisfied only for a viscous fluid. Hence it follows that, in cases where the viscosity of the liquid is small, and unimportant as regards the phenomenon under consideration, the presence of the film can be ignored.

To determine the motion of a liquid covered by a film we must add to the equations of motion, with the boundary condition (60.14), a further equation, since we now have another unknown quantity, the surface concentration γ. This further equation is an "equation of continuity", expressing the fact that the total amount of adsorbed material in the film is unchanged. The actual form of the equation depends on the shape of the surface. If the latter is plane, then the equation is evidently

$$\partial\gamma/\partial t + \partial(\gamma v_x)/\partial x + \partial(\gamma v_y)/\partial y = 0, \tag{62.2}$$

where all quantities have their values at the surface (taken as the xy-plane).

The solution of problems of the motion of a liquid covered by an adsorbed film is considerably simplified in cases where the film may be supposed incompressible, i.e. we may assume that the area of any surface element of the film remains constant during the motion.

An example of the important hydrodynamic effects of an adsorbed film is given by the motion of a gas bubble in a viscous liquid. If there is no film on the surface of the bubble, the gas inside it moves also, and the drag force exerted on the bubble by the liquid is not the same as the drag on a solid sphere of the same radius (see §20, Problem 2). If, however, the bubble is covered by a film of adsorbed material, it is clear from symmetry that the film remains at rest when the bubble moves. For a motion in the film could occur only along meridian lines on the bubble surface, and the result would be that material would continually accumulate at one of the poles (since the adsorbed material does not penetrate into the liquid or the gas); this is impossible. Besides the velocity of the film, the gas velocity at the surface of the bubble must also be zero, and with this boundary condition the gas in the bubble must be entirely at rest. Thus a bubble covered by a film moves like a solid sphere and, in particular, the drag on it (for small Reynolds numbers) is given by Stokes' formula. This result is due to V. G. Levich, who also gave the solutions to the following Problems.†

PROBLEMS

PROBLEM 1. Two vessels are joined by a long deep channel of width a and length l with plane parallel walls. The surface of the liquid in the system is covered by an adsorbed film, and the surface concentrations γ_1 and γ_2 of the film in the two vessels are different. There results a motion near the surface of the liquid in the channel. Determine the amount of film material transported by this motion.

SOLUTION. We take the plane of one wall of the channel as the xz-plane, and the surface of the liquid as the xy-plane, so that the x-axis is along the channel; the liquid is in the region

---

† For a more detailed account see V. G. LEVICH, *Physico-chemical Hydrodynamics (Fiziko-khimicheskaya gidrodinamika)*, Moscow 1952.

$z < 0$. There is no pressure gradient, so that the equation of steady flow is (cf. §17)

$$\frac{\partial^2 v}{\partial y^2} + \frac{\partial^2 v}{\partial z^2} = 0, \tag{1}$$

where v is the liquid velocity, which is evidently in the x-direction. There is a concentration gradient $d\gamma/dx$ along the channel. At the surface of the liquid in the channel we have the boundary condition

$$\eta\, \partial v/\partial z = d\alpha/dx \quad \text{for} \quad z = 0. \tag{2}$$

At the channel walls the liquid must be at rest, i.e.

$$v = 0 \quad \text{for} \quad y = 0 \quad \text{and} \quad y = a. \tag{3}$$

The channel depth is supposed infinite, and so

$$v = 0 \quad \text{for} \quad z \to -\infty. \tag{4}$$

Particular solutions of equation (1) which satisfy the conditions (3) and (4) are

$$v_n = \text{constant} \times \exp[(2n+1)\pi z/a]\, \sin(2n+1)\pi y/a,$$

with n integral. The condition (2) is satisfied by the sum

$$v = \frac{4a}{\eta\pi^2} \frac{d\alpha}{dx} \sum_{n=0}^{\infty} \frac{\exp[(2n+1)\pi z/a]\, \sin(2n+1)\pi y/a}{(2n+1)^2}.$$

The amount of film material transferred per unit time is

$$Q = \int_0^a \gamma[v]_{z=0}\, dy = \frac{8a^2}{\eta\pi^3}\left(\sum_{n=0}^{\infty} \frac{1}{(2n+1)^3}\right)\gamma\frac{d\alpha}{dx},$$

the motion being in the direction of α increasing. The value of Q must obviously be constant along the channel. Hence we can write

$$\gamma\frac{d\alpha}{dx} = \text{constant} \equiv \frac{1}{l}\int_0^l \gamma\frac{d\alpha}{dx}dx = \frac{1}{l}\int_{\alpha_2}^{\alpha_1} \gamma\, d\alpha,$$

where $\alpha_1 = \alpha(\gamma_1)$, $\alpha_2 = \alpha(\gamma_2)$, and we assume that $\alpha_1 > \alpha_2$. Thus we have finally

$$Q = \frac{8a^2}{\eta l\pi^3}\left(\sum_{n=0}^{\infty} \frac{1}{(2n+1)^3}\right)\int_{\alpha_2}^{\alpha_1} \gamma\, d\alpha = 0{\cdot}27\frac{a^2}{\eta l}\int_{\alpha_2}^{\alpha_1} \gamma\, d\alpha.$$

PROBLEM 2. Determine the damping coefficient for capillary waves on the surface of a liquid covered by an adsorbed film.

SOLUTION. If the viscosity of the liquid is not too great, the stretching (tangential) forces exerted on the film by the liquid are small, and the film may therefore be regarded as incompressible. Accordingly, we can calculate the energy dissipation as if it took place at a solid wall, i.e. from formula (24.14). Writing the velocity potential in the form

$$\phi = \phi_0\, e^{ikx-i\omega t}\, e^{-kz},$$

we obtain for the dissipation per unit area of the surface

$$\bar{\dot{E}}_{\text{kin}} = -\sqrt{(\tfrac{1}{8}\rho\eta\omega)}|k\phi_0|^2.$$

The total energy (also per unit area) is

$$\bar{E} = \rho \int \overline{v^2}\,\mathrm{d}z = \tfrac{1}{2}\rho|k\phi_0|^2/k.$$

The damping coefficient is (using (61.3))

$$\gamma = \frac{\omega^{7/6}\eta^{1/2}}{2\sqrt{2}\,\alpha^{1/3}\rho^{1/6}} = \frac{k^{7/4}\eta^{1/2}\alpha^{1/4}}{2\sqrt{2}\,\rho^{3/4}}.$$

The ratio of this quantity to the damping coefficient for capillary waves on a clean surface (§61, Problem 2) is $(\alpha\rho/k\eta^2)^{1/4}/4\sqrt{2}$, and is large compared with unity unless the wavelength is extremely small. Thus the presence of an adsorbed film on the surface of a liquid leads to a marked increase in the damping coefficient.

CHAPTER VIII

SOUND

§63. Sound waves

WE proceed now to the study of the flow of compressible fluids, and begin by investigating small oscillations; an oscillatory motion of small amplitude in a compressible fluid is called a *sound wave*. At each point of the fluid, a sound wave causes alternate compression and rarefaction.

Since the oscillations are small, the velocity $\mathbf{v}$ is small also, so that the term $(\mathbf{v} \cdot \mathbf{grad})\mathbf{v}$ in Euler's equation may be neglected. For the same reason, the relative changes in the fluid density and pressure are small. We can write the variables p and ρ in the form

$$p = p_0 + p', \quad \rho = \rho_0 + \rho', \tag{63.1}$$

where ρ_0 and p_0 are the constant equilibrium density and pressure, and ρ' and p' are their variations in the sound wave ($\rho' \ll \rho_0, p' \ll p_0$). The equation of continuity $\partial\rho/\partial t + \operatorname{div}(\rho\mathbf{v}) = 0$, on substituting (63.1) and neglecting small quantities of the second order (ρ', p' and $\mathbf{v}$ being of the first order), becomes

$$\partial\rho'/\partial t + \rho_0 \operatorname{div}\mathbf{v} = 0. \tag{63.2}$$

Euler's equation

$$\partial\mathbf{v}/\partial t + (\mathbf{v} \cdot \mathbf{grad})\mathbf{v} = -(1/\rho)\mathbf{grad}\,p$$

reduces, in the same approximation, to

$$\partial\mathbf{v}/\partial t + (1/\rho_0)\mathbf{grad}\,p' = 0. \tag{63.3}$$

The condition that the linearised equations of motion (63.2) and (63.3) should be applicable to the propagation of sound waves is that the velocity of the fluid particles in the wave should be small compared with the velocity of sound: $v \ll c$. This condition can be obtained, for example, from the requirement that $\rho' \ll \rho_0$ (see formula (63.12) below).

Equations (63.2) and (63.3) contain the unknown functions $\mathbf{v}$, p' and ρ'. To eliminate one of these, we notice that a sound wave in an ideal fluid is, like any other motion in an ideal fluid, adiabatic. Hence the small change p' in the pressure is related to the small change ρ' in the density by

$$p' = (\partial p/\partial\rho_0)_s \rho'. \tag{63.4}$$

Replacing ρ' according to this equation in (63.2), we find

$$\partial p'/\partial t + \rho_0(\partial p/\partial\rho_0)_s \operatorname{div}\mathbf{v} = 0. \tag{63.5}$$

The two equations (63.3) and (63.5), with the unknowns $\mathbf{v}$ and p', give a complete description of the sound wave.

In order to express all the unknowns in terms of one of them, it is convenient to introduce the velocity potential by putting $\mathbf{v} = \mathbf{grad}\ \phi$. We have from equation (63.3)

$$p' = -\rho\ \partial\phi/\partial t, \tag{63.6}$$

which relates p' and ϕ (here, and henceforward, we omit for brevity the suffix in p_0 and ρ_0). We then obtain from (63.5) the equation

$$\partial^2\phi/\partial t^2 - c^2\triangle\phi = 0, \tag{63.7}$$

which the potential ϕ must satisfy; here we have introduced the notation

$$c = \sqrt{(\partial p/\partial\rho)_s}. \tag{63.8}$$

An equation of the form (63.7) is called a *wave equation*. Applying the gradient operator to (63.7), we find that each of the three components of the velocity $\mathbf{v}$ satisfies an equation of the same form, and on differentiating (63.7) with respect to time we see that the pressure p' (and therefore ρ') also satisfies the wave equation.

Let us consider a sound wave in which all quantities depend on only one co-ordinate (x, say). That is, the flow is completely homogeneous in the yz-plane. Such a wave is called a *plane wave*. The wave equation (63.7) becomes

$$\partial^2\phi/\partial x^2 - (1/c^2)\partial^2\phi/\partial t^2 = 0. \tag{63.9}$$

To solve this equation, we replace x and t by the new variables $\xi = x - ct$, $\eta = x + ct$. It is easy to see that in these variables (63.9) becomes $\partial^2\phi/\partial\eta\partial\xi = 0$. Integrating this equation with respect to ξ, we find $\partial\phi/\partial\eta = F(\eta)$, where $F(\eta)$ is an arbitrary function of η. Integrating again, we obtain $\phi = f_1(\xi) + f_2(\eta)$, where f_1 and f_2 are arbitrary functions of their arguments. Thus

$$\phi = f_1(x - ct) + f_2(x + ct). \tag{63.10}$$

The distribution of the other quantities (p', ρ', $\mathbf{v}$) in a plane wave is given by functions of the same form.

For definiteness, we shall discuss the density, $\rho' = f_1(x - ct) + f_2(x + ct)$. For example, let $f_2 = 0$, so that $\rho' = f_1(x - ct)$. The meaning of this solution is evident: in any plane $x = $ constant the density varies with time, and at any given time it is different for different x, but it is the same for co-ordinates x and times t such that $x - ct = $ constant, or $x = $ constant $+ ct$. This means that, if at some instant $t = 0$ and at some point the fluid density has a certain value, then after a time t the same value of the density is found at a distance ct along the x-axis from the original point. The same is true of all the other quantities in the wave. Thus the pattern of motion is propagated through the medium in the x-direction with a velocity c; c is called the *velocity of sound*.

Thus $f_1(x - ct)$ represents what is called a *travelling plane wave* propagated in the positive direction of the x-axis. It is evident that $f_2(x + ct)$ represents a wave propagated in the opposite direction.

Of the three components of the velocity $\mathbf{v} = \mathbf{grad}\,\phi$ in a plane wave, only $v_x = \partial\phi/\partial x$ is not zero. Thus the fluid velocity in a sound wave is in the direction of propagation. For this reason sound waves in a fluid are said to be *longitudinal*.

In a travelling plane wave, the velocity $v_x = v$ is related to the pressure p' and the density ρ' in a simple manner. Putting $\phi = f(x - ct)$, we find $v = \partial\phi/\partial x = f'(x - ct)$ and $p' = -\rho\partial\phi/\partial t = \rho c f'(x - ct)$. Comparing the two expressions, we find

$$v = p'/\rho c. \tag{63.11}$$

Substituting here from (63.4) $p' = c^2\rho'$, we find the relation between the velocity and the density variation:

$$v = c\rho'/\rho. \tag{63.12}$$

We may mention also the relation between the velocity and the temperature oscillations in a sound wave. We have $T' = (\partial T/\partial p)_s p'$ and, using the well-known thermodynamic formula $(\partial T/\partial p)_s = (T/c_p)(\partial V/\partial T)_p$ and formula (63.11), we obtain

$$T' = c\beta Tv/c_p, \tag{63.13}$$

where $\beta = (1/V)(\partial V/\partial T)_p$ is the coefficient of thermal expansion.

Formula (63.8) gives the velocity of sound in terms of the adiabatic compressibility of the fluid. This is related to the isothermal compressibility by the thermodynamic formula

$$(\partial p/\partial \rho)_s = (c_p/c_v)(\partial p/\partial \rho)_T. \tag{63.14}$$

Let us calculate the velocity of sound in a perfect gas. The equation of state is $pV = p/\rho = RT/\mu$, where R is the gas constant and μ the molecular weight. We obtain for the velocity of sound the expression

$$c = \sqrt{(\gamma RT/\mu)}, \tag{63.15}$$

where γ denotes the ratio c_p/c_v.† Since γ usually depends only slightly on the temperature, the velocity of sound in the gas may be supposed proportional to the square root of the temperature. For a given temperature it does not depend on the pressure.

What are called *monochromatic waves* are a very important case. Here all quantities are just periodic (harmonic) functions of the time. It is usually convenient to write such functions as the real part of a complex quantity (see the beginning of §24). For example, we put for the velocity potential

$$\phi = \mathrm{re}[\phi_0(x, y, z)e^{-i\omega t}], \tag{63.16}$$

where ω is the frequency of the wave. The function ϕ_0 satisfies the equation

$$\triangle\phi_0 + (\omega^2/c^2)\phi_0 = 0, \tag{63.17}$$

which is obtained by substituting (63.16) in (63.7).

---

† It is useful to note that the velocity of sound in a gas is of the same order of magnitude as the mean thermal velocity of the molecules.

Let us consider a monochromatic travelling plane wave, propagated in the positive direction of the x-axis. In such a wave, all quantities are functions of $x - ct$ only, and so the potential is of the form

$$\phi = \text{re}\{A \exp[-i\omega(t - x/c)]\}, \tag{63.18}$$

where A is a constant called the *complex amplitude*. Writing this as $A = ae^{i\alpha}$ with real constants a and α, we have

$$\phi = a \cos(\omega x/c - \omega t + \alpha). \tag{63.19}$$

The constant a is called the *amplitude* of the wave, and the argument of the cosine is called the *phase*. We denote by $\mathbf{n}$ a unit vector in the direction of propagation. The vector

$$\mathbf{k} = (\omega/c)\mathbf{n} = (2\pi/\lambda)\mathbf{n} \tag{63.20}$$

is called the *wave vector*. In terms of this vector (63.18) can be written

$$\phi = \text{re}\{A \exp[i(\mathbf{k}\cdot\mathbf{r} - \omega t)]\}. \tag{63.21}$$

Monochromatic waves are very important, because any wave whatsoever can be represented as a sum of superposed monochromatic plane waves with various wave vectors and frequencies. This decomposition of a wave into monochromatic waves is simply an expansion as a Fourier series or integral (called also *spectral resolution*). The terms of this expansion are called the *monochromatic components* or *Fourier components* of the wave.

PROBLEMS

PROBLEM 1. Determine the velocity of sound in a nearly homogeneous two-phase system consisting of a vapour with small liquid droplets suspended in it (a "wet vapour"), or a liquid with small vapour bubbles in it. The wavelength of the sound is supposed large compared with the size of the inhomogeneities in the system.

SOLUTION. In a two-phase system, p and T are not independent variables, but are related by the equation of equilibrium of the phases. A compression or rarefaction of the system is accompanied by a change from one phase to the other. Let x be the fraction (by mass) of phase 2 in the system. We have

$$
\begin{aligned}
s &= (1-x)s_1 + xs_2, \\
V &= (1-x)V_1 + xV_2,
\end{aligned} \tag{1}
$$

where the suffixes 1 and 2 distinguish quantities pertaining to the pure phases 1 and 2. To calculate the derivative $(\partial V/\partial p)_s$, we transform it from the variables p, s to p, x, obtaining $(\partial V/\partial p)_s = (\partial V/\partial p)_x - (\partial V/\partial x)_p(\partial s/\partial p)_x/(\partial s/\partial x)_p$. The substitution (1) then gives

$$\left(\frac{\partial V}{\partial p}\right)_s = x\left[\frac{dV_2}{dp} - \frac{V_2 - V_1}{s_2 - s_1}\frac{ds_2}{dp}\right] + (1-x)\left[\frac{dV_1}{dp} - \frac{V_2 - V_1}{s_2 - s_1}\frac{ds_1}{dp}\right]. \tag{2}$$

The velocity of sound is obtained from (1) and (2), using formula (63.8).

Expanding the total derivatives with respect to the pressure, introducing the latent heat of the transition from phase 1 to phase 2 ($q = T(s_2 - s_1)$), and using the Clapeyron–Clausius equation for the derivative dp/dT along the curve of equilibrium ($dp/dT = p/T(V_2 - V_1)$), we obtain the expression in the first brackets in (2) in the form

$$\left(\frac{\partial V_2}{\partial p}\right)_T + \frac{2T}{q}\left(\frac{\partial V_2}{\partial T}\right)_p (V_2 - V_1) - \frac{Tc_p^2}{q^2}(V_2 - V_1)^2.$$

The second bracket is transformed similarly.

Let phase 1 be the liquid and phase 2 the vapour; we suppose the latter to be a perfect gas, and neglect the specific volume V_1 in comparison with V_2. If $x \ll 1$ (a liquid containing some bubbles of vapour), the velocity of sound is found to be

$$c = q\mu p V_1 / RT\sqrt{(c_{p_1}T)}, \tag{3}$$

where R is the gas constant and μ the molecular weight. This velocity is in general very small; thus, when vapour bubbles form in a liquid (*cavitation*), the velocity of sound undergoes a sudden sharp decrease.

If $1 - x \ll 1$ (a vapour containing some droplets of liquid), we obtain

$$\frac{1}{c^2} = \frac{\mu}{RT} - \frac{2}{q} + \frac{c_{p_1}T}{q^2}. \tag{4}$$

Comparing this with the velocity of sound in the pure gas (63.15), we find that here also the addition of a second phase reduces the value of c, though by no means so markedly.

As x increases from 0 to 1, the velocity of sound increases monotonically from the value (3) to the value (4). For $x = 0$ and $x = 1$ it changes discontinuously as we go from a one-phase system to a two-phase system. This has the result that, for values of x very close to zero or unity, the usual linear theory of sound is no longer applicable, even when the amplitude of the sound wave is small; the compressions and rarefactions produced by the wave are in this case accompanied by a change between a one-phase and a two-phase system, and the essential assumption of a constant velocity of sound no longer holds good.

PROBLEM 2. Determine the velocity of sound in a gas heated to such a high temperature that the pressure of equilibrium black-body radiation becomes comparable with the gas pressure.

SOLUTION. The pressure is $p = nkT + \frac{1}{3}akT^4$, and the entropy is

$$s = (k/m)\log(T^{\frac{3}{2}}/n) + akT^3/n.$$

In these expressions the first terms relate to the particles, and the second terms to the radiation; n is the number density of particles, m their mass, k Boltzmann's constant, and $a = 4\pi^2k^3/45\hbar^3c^3$.[†] The density of matter is not affected by the black-body radiation, so that $\rho = mn$. The velocity of sound, denoted here by u to distinguish it from that of light, is

$$u^2 = \frac{\partial(p, s)}{\partial(\rho, s)} = \frac{\partial(p, s)}{\partial(n, T)} \bigg/ \frac{\partial(\rho, s)}{\partial(n, T)},$$

where the derivatives have been written in Jacobian form. Evaluating the Jacobians, we have

$$u^2 = \frac{5kT}{3m}\left[1 + \frac{2a^2T^6}{5n(n+2aT^3)}\right].$$

§64. The energy and momentum of sound waves

Let us derive an expression for the energy of a sound wave. According to the general formula, the energy in unit volume of the fluid is $\rho\epsilon + \frac{1}{2}\rho v^2$. We now substitute $\rho = \rho_0 + \rho'$, $\epsilon = \epsilon_0 + \epsilon'$, where the primed letters denote the deviations of the respective quantities from their values when the fluid is at rest. The term $\frac{1}{2}\rho'v^2$ is a quantity of the third order. Hence, if we take only terms up to the second order, we have

$$\rho_0\epsilon_0 + \rho'\frac{\partial(\rho\epsilon)}{\partial\rho_0} + \frac{1}{2}\rho'^2\frac{\partial^2(\rho\epsilon)}{\partial\rho_0^2} + \frac{1}{2}\rho_0v^2.$$

---

† See, for instance, *Statistical Physics*, §60, Pergamon Press, London 1958.

The derivatives are taken at constant entropy, since the sound wave is adiabatic. From the thermodynamic relation $d\epsilon = Tds - pdV = Tds + (p/\rho^2)d\rho$ we have $[\partial(\rho\epsilon)/\partial\rho]_s = \epsilon + p/\rho = w$, and the second derivative is

$$[\partial^2(\rho\epsilon)/\partial\rho]_s = (\partial w/\partial\rho)_s = (\partial w/\partial p)_s(\partial p/\partial\rho)_s = c^2/\rho.$$

Thus the energy in unit volume of the fluid is

$$\rho_0\epsilon_0 + w_0\rho' + \tfrac{1}{2}c^2\rho'^2/\rho + \tfrac{1}{2}\rho_0 v^2.$$

The first term ($\rho_0\epsilon_0$) in this expression is the energy in unit volume when the fluid is at rest, and does not relate to the sound wave. The second term ($w_0\rho'$) is the change in energy due to the change in the mass of fluid in unit volume. This term disappears in the total energy, which is obtained by integrating the energy over the whole volume of the fluid: since the total mass of fluid is unchanged, we have

$$\int \rho\,dV = \int \rho_0\,dV, \quad \text{or} \quad \int \rho'\,dV = 0.$$

Thus the total change in the energy of the fluid caused by the sound wave is given by the integral

$$\int (\tfrac{1}{2}\rho_0 v^2 + \tfrac{1}{2}c^2\rho'^2/\rho_0)\,dV.$$

The integrand may be regarded as the density E of sound energy:

$$E = \tfrac{1}{2}\rho_0 v^2 + \tfrac{1}{2}c^2\rho'^2/\rho_0. \tag{64.1}$$

This expression takes a simpler form for a travelling plane wave. In such a wave $\rho' = \rho_0 v/c$ (see (63.12)), and the two terms in (64.1) are equal, so that

$$E = \rho_0 v^2. \tag{64.2}$$

In general this relation does not hold. A similar formula can be obtained only for the (time) average of the total sound energy. It follows immediately from a well-known general theorem of mechanics, that the mean total potential energy of a system executing small oscillations is equal to the mean total kinetic energy. Since the latter is, in the case considered,

$$\tfrac{1}{2}\int \rho_0\overline{v^2}\,dV,$$

we find that the mean total sound energy is

$$\int \overline{E}\,dV = \int \rho_0\overline{v^2}\,dV. \tag{64.3}$$

If a non-monochromatic wave is represented as a series of monochromatic waves, the mean energy is equal to the sum of the mean energies of the monochromatic components. For, if **v** is represented as a sum of terms of various frequencies, $\overline{v^2}$ will contain both the square of each term and the products of terms of different frequencies. These products contain factors of the form $e^{i(\omega-\omega')t}$, which are periodic functions of time. But the mean value of a periodic function is zero, and these terms therefore vanish. Thus the mean energy contains only terms in the mean squares of the monochromatic components.

Next, let us consider some volume of a fluid in which sound is propagated, and determine the mean flux of energy through the closed surface bounding this volume. The energy flux density in the fluid is, by (6.3), $\rho\mathbf{v}(\frac{1}{2}v^2+w)$. In the present case we can neglect the term in v^2, which is of the third order. Hence the mean energy flux density in the sound wave is $\overline{\rho w\mathbf{v}}$. Substituting $w = w_0+w'$, we have $\overline{\rho w\mathbf{v}} = w_0\overline{\rho\mathbf{v}}+\overline{\rho w'\mathbf{v}}$. For a small change w' in the heat function we have $w' = (\partial w/\partial p)_s\, p'$. Since $(\partial w/\partial p)_s = 1/\rho$, it follows that $w' = p'/\rho$ and $\overline{\rho w\mathbf{v}} = w_0\overline{\rho\mathbf{v}}+\overline{p'\mathbf{v}}$. The total energy flux through the surface in question is

$$\oint (w_0\,\overline{\rho\mathbf{v}}+\overline{'p\mathbf{v}})\cdot d\mathbf{f}.$$

However, since the total quantity of fluid in the volume considered is unchanged on the average, the time average of the mass flux through the closed surface must be zero. Hence the energy flux is simply

$$\oint \overline{p'\mathbf{v}}\cdot d\mathbf{f}.$$

We see that the mean sound energy flux is represented by the vector

$$\overline{\mathbf{q}} = \overline{p'\mathbf{v}}. \tag{64.4}$$

It is easy to verify that the relation

$$\partial E/\partial t + \mathrm{div}(p'\mathbf{v}) = 0 \tag{64.5}$$

holds. In this form the equation gives the law of conservation of the sound energy, with the vector $\mathbf{q} = p'\mathbf{v}$ taking the part of the sound energy flux. Thus the expression is valid not only for the mean flux but also for the flux at any instant.

In a travelling plane wave the pressure variation is related to the velocity by $p' = c\rho_0 v$. Introducing the unit vector $\mathbf{n}$ in the direction of propagation of the wave (which is the same as the direction of the velocity $\mathbf{v}$), we obtain $\mathbf{q} = c\rho_0 v^2\mathbf{n}$, or

$$\mathbf{q} = cE\mathbf{n}. \tag{64.6}$$

Thus the energy flux density in a plane sound wave equals the energy density multiplied by the velocity of sound, a result which was to be expected.

Let us now consider a sound wave which, at any given instant, occupies a finite region of space† (a *wave packet*), and determine the total momentum of

---

† Nowhere bounded by solid walls.

the fluid in the wave. The momentum of unit volume of fluid is equal to the mass flux density $\mathbf{j} = \rho\mathbf{v}$. Substituting $\rho = \rho_0 + \rho'$, we have $\mathbf{j} = \rho_0\mathbf{v} + \rho'\mathbf{v}$. The density change is related to the pressure change by $\rho' = p'/c^2$. Using (64.4), we therefore obtain

$$\mathbf{j} = \rho_0\mathbf{v} + \mathbf{q}/c^2. \tag{64.7}$$

Since we have potential flow in a sound wave, we can write $\mathbf{v} = \mathbf{grad}\,\phi$; it should be emphasised that this result is not a consequence of the approximations made in deriving the linear equations of motion in §63, since a solution such that $\mathbf{curl}\,\mathbf{v} = 0$ is an exact solution of Euler's equations. We therefore have $\mathbf{j} = \rho_0\,\mathbf{grad}\,\phi + \mathbf{q}/c^2$. The total momentum in the wave equals the integral $\int \mathbf{j}\,dV$ over the volume occupied by the wave. The integral of $\mathbf{grad}\,\phi$ can be transformed into a surface integral,

$$\int \mathbf{grad}\,\phi\,dV = \oint \phi\,d\mathbf{f},$$

and is zero, since ϕ is zero outside the volume occupied by the wave. Thus the total momentum of the wave is

$$\int \mathbf{j}\,dV = (1/c^2) \int \mathbf{q}\,dV. \tag{64.8}$$

This quantity is not, in general, zero. The existence of a non-zero total momentum means that there is a transfer of matter. We therefore conclude that the propagation of a sound-wave packet is accompanied by the transfer of fluid. This is a second-order effect (since $\mathbf{q}$ is a second-order quantity).

Finally, let us calculate the mean value of the pressure change p' in a sound wave. In the first approximation, corresponding to the usual linearised equations of motion, p' is a function which periodically changes sign, and the mean value of p' is zero. This result, however, ceases to hold if we go to higher approximations. If we take only second-order quantities, p' can be expressed in terms of quantities calculated from the linear sound equations, so that it is not necessary to solve directly the non-linear equations of motion obtained when terms of higher order are taken into account.

We start from Bernoulli's equation: $w + \tfrac{1}{2}v^2 + \partial\phi/\partial t = $ constant, and average it with respect to time. The mean value of the time derivative $\partial\phi/\partial t$ is zero.† Putting also $w = w_0 + w'$ and including w_0 in the constant, we obtain $\overline{w'} + \tfrac{1}{2}\overline{v^2} = $ constant. We suppose that the wave is propagated in an infinite volume of fluid but is damped at infinity, i.e. $\mathbf{v}$, w', etc. are zero at infinity.

---

† By the general definition of the mean value, we have for the mean derivative of any function $f(t)$

$$\overline{df/dt} = \lim_{T\to\infty} \frac{1}{2T} \int_{-T}^{T} \frac{df}{dt}\,dt = \lim_{T\to\infty} \frac{f(T) - f(-T)}{2T}.$$

If the function $f(t)$ remains finite for all t, the limit is zero, so that $\overline{df/dt} = 0$.

Since the constant is the same in all space, it must evidently be zero, so that

$$\overline{w' + \tfrac{1}{2}v^2} = 0. \tag{64.9}$$

We next expand w' in powers of p', and take only the terms up to the second order:

$$w' = (\partial w/\partial p)_s \, p' + \tfrac{1}{2}(\partial^2 w/\partial p^2)_s \, p'^2;$$

since $(\partial w/\partial p)_s = 1/\rho$, we have

$$w' = \frac{p'}{\rho_0} - \frac{p'^2}{2\rho_0^2}\left(\frac{\partial \rho}{\partial p}\right)_s = \frac{p'}{\rho_0} - \frac{p'^2}{2c^2\rho_0^2}.$$

Substituting this in (64.9) gives

$$\overline{p'} = -\tfrac{1}{2}\rho_0\overline{v^2} + \overline{p'^2}/2\rho_0 c^2 = -\tfrac{1}{2}\rho_0\overline{v^2} + \overline{p'^2}c^2/2\rho_0, \tag{64.10}$$

which determines the required mean value. The expression on the right is a second-order quantity, and is calculated by using the ρ' and v obtained from the solution of the linearised equations of motion. The mean density is

$$\overline{\rho'} = (\partial\rho/\partial p_0)_s \, \overline{p'} + \tfrac{1}{2}(\partial^2\rho/\partial p_0^2)_s \, \overline{p'^2}. \tag{64.11}$$

If the wave may be regarded as a travelling plane wave in the volume concerned, then $v = cp'/\rho_0$, so that $\overline{v^2} = c^2\overline{p'^2}/\rho_0^2$, and the expression (64.10) is zero, i.e. the mean pressure variation in a plane wave is an effect of higher order than the second. The density variation $\overline{\rho'} = \tfrac{1}{2}(\partial^2\rho/\partial p_0^2)_s \, \overline{p'^2}$ is not zero, however. (We may mention that the derivative $(\partial^2\rho/\partial p_0^2)_s$ is in fact always negative, and therefore $\overline{\rho'} < 0$ in a travelling wave.) In the same approximation, we have for the mean value of the momentum flux density tensor in a travelling plane wave $\overline{p}\delta_{ik} + \overline{\rho v_i v_k} = p_0\delta_{ik} + \rho_0\overline{v_i v_k}$. The first term is the equilibrium pressure and does not relate to the sound wave. In the second term, we introduce the unit vector $\mathbf{n}$ in the direction of $\mathbf{v}$ (the same as the direction of propagation of the wave), and, using (64.2), obtain for the momentum flux density in a sound wave

$$\overline{\Pi}_{ik} = \overline{E}n_i n_k. \tag{64.12}$$

If the wave is propagated in the x-direction, only the component $\overline{\Pi}_{xx} = \overline{E}$ is not zero. Thus, in this approximation, there is in the plane sound wave only an x-component of the mean momentum flux, and this is transmitted in the x-direction.

§65. Reflection and refraction of sound waves

When a sound wave is incident on the boundary between two different fluid media, it undergoes reflection and refraction. This means that, in addition to the incident wave, two more appear; one (the *reflected wave*) is propagated back into the first medium from the surface of separation, and the other (the *refracted* wave) is propagated into the second medium. Consequently, the motion in the first medium is a combination of two waves (the incident and the reflected), whereas in the second medium there is only one, the refracted wave.

The relation between these three waves is determined by the boundary conditions at the surface of separation, which require the pressures and normal velocity components to be equal.

Let us consider the reflection and refraction of a monochromatic longitudinal wave at a plane surface separating two media, which we take as the yz-plane. It is easy to see that all three waves have the same frequency ω and the same components k_y, k_z of the wave vector, but not the same component k_x perpendicular to the plane of separation. For, in an infinite homogeneous medium, a monochromatic wave with constant $\mathbf{k}$ and ω satisfies the equations of motion. The presence of a boundary introduces only some boundary conditions, which in the case considered apply at $x = 0$, i.e. do not depend on the time or on the co-ordinates y, z. Hence the dependence of the solution on t, y and z remains the same in all space and time, i.e. ω, k_y, and k_z are the same as in the incident wave.

From this result we can immediately derive the relations which give the directions of propagation of the reflected and refracted waves. Let the plane of the incident wave be the xy-plane. Then $k_z = 0$ in the incident wave, and the same must be true of the reflected and refracted waves. Thus the directions of propagation of the three waves are coplanar.

Let θ be the angle between the direction of propagation of the wave and the x-axis. Then, from the equality of $k_y = (\omega/c) \sin \theta$ for the incident and reflected waves, it follows that

$$\theta_1 = \theta_1', \tag{65.1}$$

i.e. the angle of incidence θ_1 is equal to the angle of reflection θ_1'. From a similar equation for the incident and refracted waves it follows that

$$\sin \theta_1/\sin \theta_2 = c_1/c_2, \tag{65.2}$$

which relates the angle of incidence θ_1 to the angle of refraction θ_2 (c_1 and c_2 being the velocities of sound in the two media).

In order to obtain a quantitative relation between the intensities of the three waves, we write the respective velocity potentials as

$$\phi_1 = A_1 \exp[i\omega\{(x/c_1) \cos \theta_1 + (y/c_1) \sin \theta_1 - t\}],$$

$$\phi_1' = A_1' \exp[i\omega\{(-x/c_1) \cos \theta_1 + (y/c_1) \sin \theta_1 - t\}],$$

$$\phi_2 = A_2 \exp[i\omega\{(x/c_2) \cos \theta_2 + (y/c_2) \sin \theta_2 - t\}].$$

On the surface of separation ($x = 0$) the pressure ($p = -\rho \partial\phi/\partial t$) and the normal velocities ($v_x = \partial\phi/\partial x$) in the two media must be equal; these conditions lead to the equations

$$\rho_1(A_1 + A_1') = \rho_2 A_2, \quad \frac{\cos\theta_1}{c_1}(A_1 - A_1') = \frac{\cos\theta_2}{c_2}A_2.$$

The *reflection coefficient* R is defined as the ratio of the (time) average energy flux densities in the reflected and incident waves. Since the energy flux density in a plane wave is $c\rho v^2$, we have $R = c_1\rho_1\overline{v_1'^2}/c_1\rho_1\overline{v_1^2} = |A_1'|^2/|A_1|^2$. A simple calculation gives

$$R = \left(\frac{\rho_2 \tan\theta_2 - \rho_1 \tan\theta_1}{\rho_2 \tan\theta_2 + \rho_1 \tan\theta_1}\right)^2. \tag{65.3}$$

The angles θ_1 and θ_2 are related by (65.2); expressing θ_2 in terms of θ_1, we can put the reflection coefficient in the form

$$R = \left[\frac{\rho_2 c_2 \cos\theta_1 - \rho_1\sqrt{(c_1^2 - c_2^2 \sin^2\theta_1)}}{\rho_2 c_2 \cos\theta_1 + \rho_1\sqrt{(c_1^2 - c_2^2 \sin^2\theta_1)}}\right]^2. \tag{65.4}$$

For normal incidence ($\theta_1 = 0$), this formula gives simply

$$R = \left(\frac{\rho_2 c_2 - \rho_1 c_1}{\rho_2 c_2 + \rho_1 c_1}\right)^2. \tag{65.5}$$

For an angle of incidence such that

$$\tan^2\theta_1 = \frac{\rho_2^2 c_2^2 - \rho_1^2 c_1^2}{\rho_1^2(c_1^2 - c_2^2)} \tag{65.6}$$

the reflection coefficient is zero, i.e. the wave is totally refracted. This can happen if $c_1 > c_2$ but $\rho_2 c_2 > \rho_1 c_1$, or if both inequalities are reversed.

PROBLEM

Determine the pressure exerted by a sound wave on the boundary separating two fluids.

SOLUTION. The sum of the total energy fluxes in the reflected and refracted waves must equal the incident energy flux. Taking the energy flux per unit area of the surface of separation, we can write this condition in the form $c_1 E_1 \cos\theta_1 = c_1 E_1' \cos\theta_1 + c_2 E_2 \cos\theta_2$, where E_1, E_1' and E_2 are the energy densities in the three waves. Introducing the reflection coefficient $R = \overline{E_1'}/\overline{E_1}$, we therefore have

$$\overline{E_2} = \frac{c_1 \cos\theta_1}{c_2 \cos\theta_2}(1 - R)\overline{E_1}.$$

The required pressure p is determined as the x-component of the momentum lost per unit time by the sound wave (per unit area of the boundary). Using the expression (64.12) for the momentum flux density tensor in a sound wave, we find

$$p = \overline{E_1}\cos^2\theta_1 + \overline{E_1'}\cos^2\theta_1 - \overline{E_2}\cos^2\theta_2.$$

Substituting for $\overline{E_2}$, introducing R and using (65.2), we obtain

$$p = \overline{E_1}\sin\theta_1\cos\theta_1[(1 + R)\cot\theta_1 - (1 - R)\cot\theta_2].$$

For normal incidence ($\theta_1 = 0$), we find, using (65.5),

$$p = 2\overline{E_1}\left[\frac{\rho_1^2 c_1^2 + \rho_2^2 c_2^2 - 2\rho_1\rho_2 c_1^2}{(\rho_1 c_1 + \rho_2 c_2)^2}\right].$$

§66. Geometrical acoustics

A plane wave has the distinctive property that its direction of propagation and its amplitude are the same in all space. An arbitrary sound wave, of course, does not possess this property. However, cases can occur where a sound wave that is not plane may still be regarded as plane in any small region of space. For this to be so it is evidently necessary that the amplitude and the direction of propagation should vary only slightly over distances of the order of the wavelength.

If this condition holds, we can introduce the idea of *rays*, these being lines such that the tangent to them at any point is in the same direction as the direction of propagation; and we can say that the sound is propagated along the rays, and ignore its wave nature. The study of the laws of propagation of sound in such cases is the task of *geometrical acoustics*. We may say that geometrical acoustics corresponds to the limit of small wavelengths, $\lambda \rightarrow 0$.

Let us derive the basic equation of geometrical acoustics, which determines the direction of the rays. We write the wave velocity potential as

$$\phi = ae^{i\psi}. \tag{66.1}$$

In the case where the wave is not plane but geometrical acoustics can be applied, the amplitude a is a slowly varying function of the co-ordinates and the time, while the wave phase ψ is "almost linear" (we recall that in a plane wave $\psi = \mathbf{k}\cdot\mathbf{r} - \omega t + \alpha$, with constant $\mathbf{k}$ and ω). Over small regions of space and short intervals of time, the phase ψ may be expanded in series; up to terms of the first order we have

$$\psi = \psi_0 + \mathbf{r}\cdot\mathbf{grad}\,\psi + t\,\partial\psi/\partial t.$$

In accordance with the fact that, in any small region of space (and during short intervals of time), the wave may be regarded as plane, we define the wave vector and the frequency at each point as,

$$\mathbf{k} = \partial\psi/\partial\mathbf{r} \equiv \mathbf{grad}\,\psi, \qquad \omega = -\partial\psi/\partial t. \tag{66.2}$$

The quantity ψ is called the *eikonal*.

In a sound wave we have $\omega^2/c^2 = k^2 = k_x^2 + k_y^2 + k_z^2$. Substituting (66.2), we obtain the basic equation of geometrical acoustics:

$$\left(\frac{\partial\psi}{\partial x}\right)^2 + \left(\frac{\partial\psi}{\partial y}\right)^2 + \left(\frac{\partial\psi}{\partial z}\right)^2 - \frac{1}{c^2}\left(\frac{\partial\psi}{\partial t}\right)^2 = 0. \tag{66.3}$$

If the fluid is not homogeneous, the coefficient $1/c^2$ is a function of the co-ordinates.

As we know from mechanics, the motion of material particles can be determined by means of the Hamilton–Jacobi equation, which, like (66.3), is a first-order partial differential equation. The quantity analogous to ψ is the action S of the particle, and the derivatives of the action determine the momentum $\mathbf{p} = \partial S/\partial \mathbf{r}$ and the Hamilton's function (the energy) of the particle $H = -\partial S/\partial t$; these formulae are similar to (66.2). We know, also, that the Hamilton–Jacobi equation is equivalent to Hamilton's equations

$$\dot{\mathbf{p}} = -\partial H/\partial \mathbf{r}, \quad \mathbf{v} \equiv \dot{\mathbf{r}} = \partial H/\partial \mathbf{p}.$$

From the above analogy between the mechanics of a material particle and geometrical acoustics, we can write down similar equations for rays:

$$\dot{\mathbf{k}} = -\partial \omega/\partial \mathbf{r}, \quad \dot{\mathbf{r}} = \partial \omega/\partial \mathbf{k}. \tag{66.4}$$

In a homogeneous isotropic medium $\omega = ck$ with c constant, so that $\dot{\mathbf{k}} = 0$, $\dot{\mathbf{r}} = c\mathbf{n}$ ($\mathbf{n}$ being a unit vector in the direction of $\mathbf{k}$), i.e. the rays are propagated in straight lines with a constant frequency ω, as we should expect.

The frequency, of course, remains constant along a ray in all cases where the propagation of sound occurs under steady conditions, i.e. the properties of the medium at each point in space do not vary with time. For the total time derivative of the frequency, which gives its rate of variation along a ray, is $d\omega/dt = \partial \omega/\partial t + \dot{\mathbf{r}} \cdot \partial \omega/\partial \mathbf{r} + \dot{\mathbf{k}} \cdot \partial \omega/\partial \mathbf{k}$. On substituting (66.4), the last two terms cancel, and in a steady state $\partial \omega/\partial t = 0$, so that $d\omega/dt = 0$.

In steady propagation of sound in an inhomogeneous medium at rest $\omega = ck$, where c is a given function of the co-ordinates. The equations (66.4) give

$$\dot{\mathbf{r}} = c\mathbf{n}, \quad \dot{\mathbf{k}} = -k\,\mathbf{grad}\,c. \tag{66.5}$$

The magnitude of the vector $\mathbf{k}$ varies along a ray simply according to $k = \omega/c$ (with ω constant). To determine the change in direction of $\mathbf{n}$ we put $\mathbf{k} = \omega\mathbf{n}/c$ in the second of (66.5): $\omega\dot{\mathbf{n}}/c - (\omega\mathbf{n}/c^2)(\dot{\mathbf{r}} \cdot \mathbf{grad}\,c) = -k\,\mathbf{grad}\,c$, whence $d\mathbf{n}/dt = -\mathbf{grad}\,c + \mathbf{n}(\mathbf{n} \cdot \mathbf{grad}\,c)$. Introducing the element of length along the ray $dl = c\,dt$, we can rewrite this equation

$$d\mathbf{n}/dl = -(1/c)\,\mathbf{grad}\,c + \mathbf{n}(\mathbf{n} \cdot \mathbf{grad}\,c)/c. \tag{66.6}$$

This equation determines the form of the rays; $\mathbf{n}$ is a unit vector tangential to a ray.†

If equation (66.3) is solved, and the eikonal ψ is a known function of co-ordinates and time, we can then find also the distribution of sound intensity in space. In steady conditions, it is given by the equation $\operatorname{div} \mathbf{q} = 0$ ($\mathbf{q}$ being the sound energy flux density), which must hold in all space except

---

† As we know from differential geometry, the derivative $d\mathbf{n}/dl$ along the ray is equal to $\mathbf{N}/R$, where $\mathbf{N}$ is a unit vector along the principal normal and R is the radius of curvature of the ray. The expression on the right-hand side of (66.6) is, apart from a factor $1/c$, the derivative of the velocity of sound along the principal normal; hence we can write the equation as $1/R = -(1/c)\mathbf{N} \cdot \mathbf{grad}\,c$. The rays bend towards the region where c is smaller.

at sources of sound. Putting $\mathbf{q} = cE\mathbf{n}$, where E is the sound energy density (see (64.6)), and remembering that $\mathbf{n}$ is a unit vector in the direction of $\mathbf{k} = \mathbf{grad}\,\psi$, we obtain the equation

$$\mathrm{div}(cE\,\mathbf{grad}\,\psi/|\mathbf{grad}\,\psi|) = 0, \tag{66.7}$$

which determines the distribution of E in space.

The second formula (66.4) gives the velocity of propagation of the waves from the known dependence of the frequency on the components of the wave vector. This is a very important formula, which holds not only for sound waves, but for all waves (for example, we have already applied it to gravity waves in §12). We shall give here another derivation of this formula, which puts in evidence the meaning of the velocity which it defines. Let us consider a *wave packet*, which occupies some finite region of space. We assume that its spectral composition includes monochromatic components whose frequencies lie in only a small range; the same is true of the components of their wave vectors. Let ω be some mean frequency of the wave packet, and $\mathbf{k}$ a mean wave vector. Then, at some initial instant, the wave packet is described by a function of the form

$$\phi = \exp(i\mathbf{k}\cdot\mathbf{r})f(\mathbf{r}). \tag{66.8}$$

The function $f(\mathbf{r})$ is appreciably different from zero only in a region which is small (though it is large compared with the wavelength $1/k$). Its expansion as a Fourier integral contains, by the above assumptions, components of the form $\exp(i\mathbf{r}\cdot\Delta\mathbf{k})$, where $\Delta\mathbf{k}$ is small.

Thus each monochromatic component is, at the initial instant,

$$\phi_{\mathbf{k}} = \text{constant}\times\exp[i(\mathbf{k}+\Delta\mathbf{k})\cdot\mathbf{r}]. \tag{66.9}$$

The corresponding frequency is $\omega(\mathbf{k}+\Delta\mathbf{k})$ (we recall that the frequency is a function of the wave vector). Hence the same component at time t has the form

$$\phi_{\mathbf{k}} = \text{constant}\times\exp[i(\mathbf{k}+\Delta\mathbf{k})\cdot\mathbf{r} - i\omega(\mathbf{k}+\Delta\mathbf{k})t].$$

We use the fact that $\Delta\mathbf{k}$ is small, and expand $\omega(\mathbf{k}+\Delta\mathbf{k})$ in series, taking only the first two terms: $\omega(\mathbf{k}+\Delta\mathbf{k}) = \omega + (\partial\omega/\partial\mathbf{k})\cdot\Delta\mathbf{k}$, where $\omega = \omega(\mathbf{k})$ is the frequency corresponding to the mean wave vector. Then $\phi_{\mathbf{k}}$ becomes

$$\phi_{\mathbf{k}} = \text{constant}\times\exp[i(\mathbf{k}\cdot\mathbf{r} - \omega t)]\exp[i\Delta\mathbf{k}\cdot(\mathbf{r} - t\partial\omega/\partial\mathbf{k})]. \tag{66.10}$$

If we now sum all the monochromatic components, with all the $\Delta\mathbf{k}$ that occur in the wave packet, we see from (66.9) and (66.10) that the result is

$$\phi = \exp[i(\mathbf{k}\cdot\mathbf{r} - \omega t)]f(\mathbf{r} - t\partial\omega/\partial\mathbf{k}), \tag{66.11}$$

where f is the same function as in (66.8). A comparison with (66.8) shows that, after a time t, the amplitude distribution has moved as a whole through a distance $t\partial\omega/\partial\mathbf{k}$; the exponential coefficient of f in (66.11) affects only the phase. Consequently, the velocity of the wave packet is

$$\mathbf{U} = \partial\omega/\partial\mathbf{k}. \tag{66.12}$$

This formula gives the velocity of propagation for any dependence of ω on $\mathbf{k}$.† When $\omega = ck$, with c constant, it of course gives the usual result $U = \omega/k = c$. In general, when $\omega(\mathbf{k})$ is an arbitrary function, the velocity of propagation is a function of the frequency, and the direction of propagation may not be the same as that of the wave vector.

PROBLEM

Determine the altitude variation in the amplitude of sound propagated in an isothermal atmosphere under gravity.

SOLUTION. In an isothermal atmosphere (regarded as a perfect gas) the velocity of sound is constant. The energy flux density evidently decreases along a ray in inverse proportion to the square of the distance r from the source: $c\rho\overline{v^2} \sim 1/r^2$. Hence it follows that the amplitude of the velocity fluctuations in the sound wave varies along a ray inversely as $r\sqrt{\rho}$; according to the barometric formula, $\rho \sim \exp(-\mu gz/RT)$, where z is the altitude, μ the molecular weight of the gas and R the gas constant.

§67. Propagation of sound in a moving medium

The relation $\omega = ck$ between the frequency and the wave number is valid only for a monochromatic sound wave propagated in a medium at rest. It is not difficult to obtain a similar relation for a wave propagated in a moving medium (and observed in a fixed system of co-ordinates).

Let us consider a homogeneous flow of velocity $\mathbf{u}$. We take a fixed system K of co-ordinates x, y, z, and also a system K' of co-ordinates x', y', z' moving with velocity $\mathbf{u}$ relative to K. In the system K' the fluid is at rest, and a monochromatic wave has the usual form $\phi = \text{constant} \times \exp[i(\mathbf{k} \cdot \mathbf{r}' - kct)]$. The radius vector $\mathbf{r}'$ in the system K' is related to the radius vector $\mathbf{r}$ in the system K by $\mathbf{r}' = \mathbf{r} - \mathbf{u}t$. Hence, in the fixed system of co-ordinates, the wave has the form $\phi = \text{constant} \times \exp\{i[\mathbf{k} \cdot \mathbf{r} - (kc + \mathbf{k} \cdot \mathbf{u})t]\}$. The coefficient of t in the exponent is the frequency ω of the wave. Thus the frequency in a moving medium is related to the wave vector $\mathbf{k}$ by

$$\omega = ck + \mathbf{u} \cdot \mathbf{k}. \tag{67.1}$$

The velocity of propagation is

$$\partial\omega/\partial\mathbf{k} = c\mathbf{k}/k + \mathbf{u}; \tag{67.2}$$

this is the vector sum of the velocity c in the direction of $\mathbf{k}$ and the velocity $\mathbf{u}$ with which the sound is "carried along" by the moving fluid.

Using formula (67.1), we can investigate what is called the *Doppler effect*:

---

† The velocity defined by (66.12) is called the *group velocity* of the wave, and the ratio ω/k the *phase velocity*. However, it must be borne in mind that the phase velocity does not correspond to any actual physical propagation.

Regarding the derivation given here it should be emphasised that the motion of the wave packet without change of form (i.e. without change in the spatial distribution of the amplitude), expressed by (66.11), is approximate, and results from the assumption that the range $\Delta\mathbf{k}$ is small. In general, when U depends on ω, a wave packet is "smoothed out" during its propagation, and the region of space which it occupies increases in size. It can be shown that the amount of this smoothing out is proportional to the squared magnitude of the range $\Delta\mathbf{k}$ of the wave vectors which occur in the composition of the wave packet.

the frequency of sound, as received by an observer moving relative to the source, is not the same as the frequency of oscillation of the source.

Let sound emitted by a source at rest (relative to the medium) be received by an observer moving with velocity $\mathbf{u}$. In a system K' at rest relative to the medium we have $k = \omega_0/c$, where ω_0 is the frequency of oscillation of the source. In a system K moving with the observer, the medium moves with velocity $-\mathbf{u}$, and the frequency of the sound is, by (67.1), $\omega = ck - \mathbf{u} \cdot \mathbf{k}$. Introducing the angle θ between the direction of the velocity $\mathbf{u}$ and that of the wave vector $\mathbf{k}$, and putting $k = \omega_0/c$, we find that the frequency of the sound received by the moving observer is

$$\omega = \omega_0[1 - (u/c) \cos \theta]. \tag{67.3}$$

The opposite case, to a certain extent, is the propagation in a medium at rest of a sound wave emitted from a moving source. Let $\mathbf{u}$ be now the velocity of the source. We change from the fixed system of co-ordinates to a system K' moving with the source; in the system K', the fluid moves with velocity $-\mathbf{u}$. In K', where the source is at rest, the frequency of the emitted sound wave must equal the frequency ω_0 of the oscillations of the source. Changing the sign of $\mathbf{u}$ in (67.1) and introducing the angle θ between the directions of $\mathbf{u}$ and $\mathbf{k}$, we have $\omega_0 = ck[1 - (u/c) \cos \theta]$. In the original fixed system K, however, the frequency and the wave vector are related by $\omega = ck$. Thus we find

$$\omega = \omega_0/[1 - (u/c) \cos \theta]. \tag{67.4}$$

This formula gives the relation between the frequency ω_0 of the oscillations of a moving source and the frequency ω of the sound heard by an observer at rest.

If the source is moving away from the observer, the angle θ between its velocity and the direction to the observer lies in the range $\frac{1}{2}\pi < \theta \leqslant \pi$, so that $\cos \theta < 0$. It then follows from (67.4) that, if the source is moving away from the observer, the frequency of the sound heard is less than ω_0.

If, on the other hand, the source is approaching the observer, then $0 \leqslant \theta < \frac{1}{2}\pi$, so that $\cos \theta > 0$, and the frequency $\omega > \omega_0$ increases with u. For $u \cos \theta > c$, according to formula (67.4) ω becomes negative, which means that the sound heard by the observer actually reaches him in the reverse order, i.e. sound emitted by the source at any given instant arrives earlier than sound emitted at previous instants.

As has been mentioned at the beginning of §66, the approximation of geometrical acoustics corresponds to the case of small wavelengths, i.e. large magnitudes of the wave vector. For this to be so the frequency of the sound must in general be large. In the acoustics of moving media, however, the latter condition need not be fulfilled if the velocity of the medium exceeds that of sound. For in this case k can be large even when the frequency is zero; from (67.1) we have for $\omega = 0$ the equation

$$ck = -\mathbf{u} \cdot \mathbf{k}, \tag{67.5}$$

and this has solutions if $u > c$. Thus, in a medium moving with supersonic velocities, there can be steady small perturbations described (if k is sufficiently large) by geometrical acoustics. This means that such perturbations are propagated along rays.

Let us consider, for example, a homogeneous supersonic stream moving with constant velocity $\mathbf{u}$, whose direction we take as the x-axis. The vector $\mathbf{k}$ is taken to lie in the xy-plane, and its components are related by

$$(u^2 - c^2)k_x{}^2 = c^2 k_y{}^2, \tag{67.6}$$

which is obtained by squaring both sides of equation (67.5). To determine the form of the rays, we use the equations of geometrical acoustics (66.4), according to which $\dot{x} = \partial\omega/\partial k_x$, $\dot{y} = \partial\omega/\partial k_y$. Dividing one of these equations by the other, we have $dy/dx = (\partial\omega/\partial k_y)/(\partial\omega/\partial k_x)$. This relation, however, is, by the rule of differentiation for implicit functions, just the derivative $-\partial k_x/\partial k_y$ taken at a constant frequency (in this case zero). Thus the equation which gives the form of the rays from the known relation between k_x and k_y is

$$dy/dx = -\partial k_x/\partial k_y. \tag{67.7}$$

Substituting (67.6), we obtain

$$dy/dx = \pm c/\sqrt{(u^2 - c^2)}.$$

For constant u this equation represents two straight lines intersecting the x-axis at angles $\pm\alpha$, where $\sin\alpha = c/u$.

We shall return to a detailed study of these rays in gas dynamics, where they are very important; see in particular §§79, 96 and 109.

PROBLEMS

PROBLEM 1. Derive an equation giving the form of sound rays propagated in a steadily moving homogeneous medium with a velocity distribution $\mathbf{u}(x, y, z)$, when $u \ll c$ everywhere.†

SOLUTION. Substituting (67.1) in (66.4), we obtain the equations of propagation of the rays in the form

$$\dot{\mathbf{k}} = -(\mathbf{k}\cdot\mathbf{grad})\mathbf{u} - \mathbf{k}\times\mathbf{curl}\,\mathbf{u}, \quad \dot{\mathbf{r}} \equiv \mathbf{v} = c\mathbf{k}/k + \mathbf{u}.$$

Using these equations, and also

$$d\mathbf{u}/dt \equiv \partial\mathbf{u}/\partial t + (\mathbf{v}\cdot\mathbf{grad})\mathbf{u} = (\mathbf{v}\cdot\mathbf{grad})\mathbf{u} \approx (c/k)(\mathbf{k}\cdot\mathbf{grad})\mathbf{u},$$

we calculate the derivative $d(k\mathbf{v})/dt$, retaining only terms as far as the first order in $\mathbf{u}$. The result is $d(k\mathbf{v})/dt = -kv\,\mathbf{n}\times\mathbf{curl}\,\mathbf{u}$, where $\mathbf{n}$ is a unit vector in the direction of $\mathbf{v}$. But $d(k\mathbf{v})/dt = \mathbf{n}d(kv)/dt + kv\,d\mathbf{n}/dt$. Since $\mathbf{n}$ and $d\mathbf{n}/dt$ are perpendicular (because $\mathbf{n}^2 = 1$, and therefore $\mathbf{n}\cdot\dot{\mathbf{n}} = 0$), it follows from the above equations that $\dot{\mathbf{n}} = -\mathbf{n}\times\mathbf{curl}\,\mathbf{u}$. Introducing the element of length along the ray $dl = c\,dt$, we can write finally

$$d\mathbf{n}/dl = -\mathbf{n}\times\mathbf{curl}\,\mathbf{u}/c. \tag{1}$$

This equation determines the form of the rays; $\mathbf{n}$ is a unit tangential vector (and is no longer in the same direction as $\mathbf{k}$).

---

† It is assumed that the velocity $\mathbf{u}$ varies only over distances large compared with the wavelength of the sound.

PROBLEM 2. Determine the form of sound rays in a moving medium with a velocity distribution $u_x = u(z)$, $u_y = u_z = 0$.

SOLUTION. Expanding equation (1), Problem 1, we find $dn_x/dl = (n_z/c)du/dz$, $dn_y/dl = 0$; the equation for n_z need not be written down, since $\mathbf{n}^2 = 1$. The second equation gives $n_y = $ constant $\equiv n_{y,0}$. In the first equation we write $n_z = dz/dl$, and then we have by integration $n_x = n_{x,0} + u(z)/c$. These formulae give the required solution.

Let us assume that the velocity u is zero for $z = 0$ and increases upwards ($du/dz > 0$). If the sound is propagated "against the wind" ($n_x < 0$), its path is curved upwards; if it is propagated "with the wind" ($n_x > 0$), its path is curved downwards. In the latter case a ray leaving the point $z = 0$ at a small angle to the x-axis (i.e. with $n_{x,0}$ close to unity) rises only to a finite altitude $z = z_{max}$, which can be calculated as follows. At the altitude z_{max} the ray is horizontal, i.e. $n_z = 0$. Hence we have

$$n_x^2 + n_y^2 \approx n_{x,0}^2 + n_{y,0}^2 + 2n_{x,0}u/c = 1,$$

so that $2n_{x,0}u(z_{max})/c = n_{z,0}^2$, whence we can determine z_{max} from the given function $u(z)$ and the initial direction $\mathbf{n_0}$ of the ray.

PROBLEM 3. Obtain the expression of Fermat's principle for sound rays in a steadily moving medium.

SOLUTION. Fermat's principle is that the integral

$$\oint \mathbf{k} \cdot d\mathbf{l},$$

taken along a ray between two given points, is a minimum; $\mathbf{k}$ is supposed expressed as a function of the frequency ω and the direction $\mathbf{n}$ of the ray.† This function can be found by eliminating v and k from the relations $\omega = ck + \mathbf{u} \cdot \mathbf{k}$ and $v\mathbf{n} = ck/k + \mathbf{u}$. Fermat's principle then takes the form

$$\delta \oint \{\sqrt{[(c^2 - u^2)dl^2 + (\mathbf{u} \cdot d\mathbf{l})^2]} - \mathbf{u} \cdot d\mathbf{l}\}/(c^2 - u^2) = 0.$$

In a medium at rest, this integral reduces to the usual one, $\oint dl/c$.

§68. Characteristic vibrations

Hitherto we have discussed only oscillatory motion in infinite media, and we have seen, in particular, that in such media waves of any frequency can be propagated.

The situation is very different when we consider a fluid in a vessel of finite dimensions. The equations of motion themselves (the wave equations) are of course unchanged, but they must now be supplemented by boundary conditions to be satisfied at the solid walls or at the free surface of the fluid. We shall consider here only what are called *free vibrations*, i.e. those which occur in the absence of variable external forces. Vibrations occurring as a result of external forces are called *forced vibrations*.

The equations of motion for a finite fluid do not have solutions satisfying the appropriate boundary conditions for every frequency. Such solutions exist only for a series of definite frequencies ω. In other words, in a medium of finite volume, free vibrations can occur only with certain frequencies. These are called the *characteristic frequencies* of the fluid in the vessel concerned.

The actual values of the characteristic frequencies depend on the size and

---

† See *The Classical Theory of Fields*, §7–1, Addison-Wesley Press, Cambridge (Mass.) 1951.

shape of the vessel. In any given case there is an infinite number of characteristic frequencies. To find them, it is necessary to examine the equations
of motion with the appropriate boundary conditions.

The order of magnitude of the first (i.e. smallest) characteristic frequency
can be seen at once from dimensional considerations. The only parameter
having the dimensions of length which appears in the problem is the linear
dimension l of the body. Hence it is clear that the wavelength λ_1 corresponding to the first characteristic frequency must be of the order of l, and the order
of magnitude of the frequency ω_1 itself is obtained by dividing the velocity
of sound by the wavelength. Thus

$$\lambda_1 \sim l, \quad \omega_1 \sim c/l. \tag{68.1}$$

Let us ascertain the nature of the motion in characteristic vibrations.
If we seek a solution of the wave equation for the velocity potential (say)
which is periodic in time, of the form $\phi = \phi_0(x, y, z)e^{-i\omega t}$, then we have for
ϕ_0 the equation

$$\triangle\phi_0 + (\omega^2/c^2)\phi_0 = 0. \tag{68.2}$$

In an infinite medium, where no boundary conditions need be applied, this
equation has both real and complex solutions. In particular, it has a solution
proportional to $e^{i\mathbf{k}\cdot\mathbf{r}}$, which gives a velocity potential of the form

$$\phi = \text{constant} \times \exp[i(\mathbf{k}\cdot\mathbf{r} - \omega t)].$$

Such a solution represents a wave propagated with a definite velocity—a
travelling wave.

For a medium of finite volume, however, complex solutions cannot in
general exist. This can be seen as follows. The equation satisfied by ϕ_0
is real, and the boundary conditions are real also. Hence, if $\phi_0(x, y, z)$ is a
solution of the equations of motion, the complex conjugate function ϕ_0^*
is also a solution. Since, however, the solution of the equations for given
boundary conditions is in general unique† apart from a constant factor, we
must have $\phi_0^* = \text{constant} \times \phi_0$, where the constant is complex and its
modulus is clearly unity. Thus ϕ_0 must be of the form $\phi_0 = f(x, y, z)e^{-i\alpha}$,
the function f and the constant α being real. The potential ϕ is thus of the
form (taking the real part of $\phi_0 e^{-i\omega t}$)

$$\phi = f(x, y, z)\cos(\omega t + \alpha), \tag{68.3}$$

i.e. it is the product of some function of the co-ordinates and a simple periodic
function of the time.

This solution has properties entirely different from those of a travelling
wave. In the latter, where $\phi = \text{constant} \times \cos(\mathbf{k}\cdot\mathbf{r} - \omega t + \alpha)$, the phase
$\mathbf{k}\cdot\mathbf{r} - \omega t + \alpha$ of the oscillations at different points in space is different at any
given instant, except only at points for which $\mathbf{k}\cdot\mathbf{r}$ differs by an integral

---

† This may not be true when the vessel is highly symmetrical in form (e.g. a sphere).

multiple of the wavelength. In the wave represented by (68.3), all points are oscillating in the same phase $\omega t + \alpha$ at any given instant. Such a wave is obviously not "propagated"; it is called a *stationary wave*. Thus the characteristic vibrations are stationary waves.

Let us consider a stationary plane sound wave, in which all quantities are functions of one co-ordinate only (x, say) and of time. Writing the general solution of $\partial^2 \phi_0 / \partial x^2 + {}^2 \omega \phi_0 / c^2 = 0$ in the form $\phi_0 = a \cos(\omega x/c + \beta)$, we have $\phi = a \cos(\omega t + \alpha) \cos(\omega x/c + \beta)$. By an appropriate choice of the origins of x and t, we can make α and β zero, so that

$$\phi = a \cos \omega t \cos \omega x/c. \tag{68.4}$$

For the velocity and pressure in the wave we have

$$v = \partial \phi / \partial x = -(a\omega/c) \cos \omega t \sin \omega x/c;$$
$$p' = -\rho \, \partial \phi / \partial t = \rho \omega \sin \omega t \cos \omega x/c.$$

At the points $x = 0$, $\pi c/\omega$, $2\pi c/\omega$, ..., which are at a distance $\pi c/\omega = \frac{1}{2}\lambda$ apart, the velocity v is always zero; these points are called *nodes* of the velocity. The points midway between them ($x = \pi c/2\omega$, $3\pi c/2\omega$, ...) are those at which the amplitude of the time variations of the velocity is greatest. These are called *antinodes*. The pressure p' evidently has nodes and antinodes in the reverse positions. Thus, in a stationary plane wave, the nodes of the pressure are the antinodes of the velocity, and *vice versa*.

An interesting case of characteristic vibrations is that of the vibrations of a gas in a vessel having a small aperture (a *resonator*). In a closed vessel the smallest characteristic frequency is, as we know, of the order of c/l, where l is the linear dimension of the vessel. When there is a small aperture, however, new characteristic vibrations of considerably smaller frequency appear. These are due to the fact that, if there is a pressure difference between the gas in the vessel and that outside, this difference can be equalised by the motion of gas into or out of the vessel. Thus oscillations appear which involve an exchange of gas between the resonator and the outside medium. Since the aperture is small, this exchange takes place only slowly, and hence the period of the oscillations is large, and the frequency correspondingly small (see Problem 2). The frequencies of the ordinary vibrations occurring in a closed vessel are practically unchanged by the presence of a small aperture.

PROBLEMS

PROBLEM 1. Determine the characteristic frequencies of sound waves in a fluid contained in a cuboidal vessel.

SOLUTION. We seek a solution of the equation (68.2) in the form

$$\phi_0 = \text{constant} \times \cos qx \cos ry \cos sz,$$

where $q^2 + r^2 + s^2 = \omega^2/c^2$. At the walls of the vessel we have the conditions $v_x = \partial \phi / \partial x = 0$ for $x = 0$ and a, $\partial \phi / \partial y = 0$ for $y = 0$ and b, $\partial \phi / \partial z = 0$ for $z = 0$ and c, where a, b, c are the sides of the cuboid. Hence we find $q = m\pi/a$, $r = n\pi/b$, $s = p\pi/c$, where m, n, p are any integers. Thus the characteristic frequencies are

$$\omega^2 = c^2\pi^2(m^2/a^2 + n^2/b^2 + p^2/c^2).$$

PROBLEM 2. A narrow tube of cross-sectional area S and length l is fixed to the aperture of a resonator. Determine the characteristic frequency.

SOLUTION. Since the tube is narrow, in considering oscillations accompanied by the movement of gas into and out of the resonator we can suppose that only the gas in the tube has an appreciable velocity, while the gas in the vessel is almost at rest. The mass of gas in the tube is $S\rho l$, and the force on it is $S(p_0-p)$, where p and p_0 are the gas pressures inside and outside the resonator respectively. Hence we must have $S\rho l\dot{v} = S(p-p_0)$, where v is the gas velocity in the tube. The time derivative of the pressure is given by $\dot{p} = c^2\dot{\rho}$, and the decrease per unit time in the gas density in the resonator ($-\dot{\rho}$) can be supposed equal to the mass of gas leaving the resonator per unit time ($S\rho v$) divided by the volume V of the resonator. Thus we have $\dot{p} = -c^2 S\rho v/V$, whence

$$\ddot{p} = -c^2 S\rho\dot{v}/V = -c^2 S(p-p_0)/lV.$$

This equation gives $p-p_0 = \text{constant}\times\cos\omega_0 t$, where the characteristic frequency $\omega_0 = c\sqrt{(S/lV)}$. This is small compared with c/L (where L is the linear dimension of the vessel), and the wavelength is therefore large compared with L.

In solving this problem we have supposed that the linear amplitude of the oscillations of gas in the tube is small compared with its length l. If this were not so, the oscillations would be accompanied by the outflow of a considerable fraction of the gas in the tube, and the linear equation of motion used above would be inapplicable.

§69. Spherical waves

Let us consider a sound wave in which the distribution of density, velocity, etc., depends only on the distance from some point, i.e. is spherically symmetrical. Such a wave is called a *spherical wave*.

Let us determine the general solution of the wave equation which represents a spherical wave. We take the wave equation for the velocity potential: $\triangle\phi-(1/c^2)\partial^2\phi/\partial t^2 = 0$. Since ϕ is a function only of the distance r from the centre and of the time t, we have, using the expression for the Laplacian in spherical co-ordinates,

$$\frac{\partial^2\phi}{\partial t^2} = c^2\frac{1}{r^2}\frac{\partial}{\partial r}\left(r^2\frac{\partial\phi}{\partial r}\right). \tag{69.1}$$

We seek a solution in the form $\phi = f(r, t)/r$. Substituting, we have after a simple calculation the following equation for f: $\partial^2 f/\partial t^2 = c^2\partial^2 f/\partial r^2$. This is just the ordinary one-dimensional wave equation, with the radius r as the co-ordinate. The solution of this equation is, as we know, of the form $f = f_1(ct-r)+f_2(ct+r)$, where f_1 and f_2 are arbitrary functions. Thus the general solution of equation (69.1) is of the form

$$\phi = \frac{f_1(ct-r)}{r} + \frac{f_2(ct+r)}{r}. \tag{69.2}$$

The first term is an outgoing wave, propagated in all directions from the origin. The second term is a wave coming in to the centre. Unlike a plane wave, whose amplitude remains constant, a spherical wave has an amplitude which decreases inversely as the distance from the centre. The intensity in the wave is given by the square of the amplitude, and falls off inversely as the square of the distance, as it should, since the total energy flux in the wave is distributed over a surface whose area increases as r^2.

The variable parts of the pressure and density are related to the potential by $p' = -\rho\partial\phi/\partial t$, $\rho' = -(\rho/c^2)\partial\phi/\partial t$, and their distribution is determined by formulae of the same form as (69.2). The (radial) velocity distribution, however, being given by the gradient of the potential, is of the form

$$v = \frac{\partial}{\partial r}\left\{\frac{f_1(ct-r)+f_2(ct+r)}{r}\right\}. \tag{69.3}$$

If there is no source of sound at the origin, the potential (69.2) must remain finite for $r = 0$. For this to be so we must have $f_1(ct) = -f_2(ct)$, i.e. ϕ is of the form

$$\phi = \frac{f(ct-r)-f(ct+r)}{r} \tag{69.4}$$

(a stationary spherical wave). If there is a source at the origin, on the other hand, the potential of the outgoing wave from it is $\phi = f(ct-r)/r$; it need not remain finite at $r = 0$, since the solution holds only for the region outside sources.

A monochromatic stationary spherical wave is of the form

$$\phi = Ae^{-i\omega t}\frac{\sin kr}{r}, \tag{69.5}$$

where $k = \omega/c$. An outgoing monochromatic spherical wave is given by

$$\phi = Ae^{i(kr-\omega t)}/r. \tag{69.6}$$

It is useful to note that this expression satisfies the differential equation

$$\triangle\phi+k^2\phi = -4\pi Ae^{-i\omega t}\delta(\mathbf{r}), \tag{69.7}$$

where on the right-hand side we have the delta function $\delta(\mathbf{r}) = \delta(x)\delta(y)\delta(z)$. For $\delta(\mathbf{r}) = 0$ everywhere except at the origin, and we return to the homogeneous equation (69.1); and, integrating (69.7) over the volume of a small sphere including the origin (where the expression (69.6) reduces to $Ae^{-i\omega t}/r$) we obtain $-4\pi Ae^{-i\omega t}$ on each side.

Let us consider an outgoing spherical wave, occupying a spherical shell outside which the medium is either at rest or very nearly so; such a wave can originate from a source which emits during a finite interval of time only, or from some region where there is a sound disturbance (cf. the end of §71, and §73, Problem 4). Before the wave arrives at any given point, the potential is $\phi \equiv 0$. After the wave has passed, the motion must die away; this means that ϕ must become constant. In an outgoing spherical wave, however, the potential is a function of the form $\phi = f(ct-r)/r$; such a function can tend to a constant only if the function f is zero identically. Thus the potential must be zero both before and after the passage of the wave.† From this we

---

† Unlike what happens for a plane wave, after whose passage we can have $\phi = $ constant $\neq 0$

can draw an important conclusion concerning the distribution of condensations and rarefactions in a spherical wave.

The variation of pressure in the wave is related to the potential by
$p' = -\rho\partial\phi/\partial t$. From what has been said above, it is clear that, if we integrate
p' over all time for a given r, the result is zero:

$$\int_{-\infty}^{\infty} p' \, dt = 0. \qquad (69.8)$$

This means that, as the spherical wave passes through a given point, both
condensations ($p' > 0$) and rarefactions ($p' < 0$) will be observed at that
point. In this respect a spherical wave is markedly different from a plane
wave, which may consist of condensations or rarefactions only.

A similar pattern will be observed if we consider the manner of variation of
p' with distance at a given instant; instead of the integral (69.8) we now
consider another which also vanishes, namely

$$\int_{0}^{\infty} rp' \, dr = 0. \qquad (69.9)$$

PROBLEMS

PROBLEM 1. At the initial instant, the gas inside a sphere of radius a is compressed so that
$\rho' = \text{constant} \equiv \Delta$; outside this sphere, $\rho' = 0$. The initial velocity is zero in all space.
Determine the subsequent motion.

SOLUTION. The initial conditions on the potential are $\phi = 0$ for $t = 0$, and $r < a$ or
$r > a$; $\dot{\phi} = F(r)$ for $t = 0$, where $F(r) = 0$ for $r > a$ and $F(r) = -c^2\Delta/\rho$ for $r < a$. We
seek ϕ in the form

$$\phi(r, t) = \frac{f(ct-r) - f(ct+r)}{r}.$$

From the initial conditions we obtain $f(-r) - f(r) = 0$, $f'(-r) - f'(r) = rF(r)/c$. From the
first equation we have $f'(-r) + f'(r) = 0$, which together with the second equation gives
$f'(r) = -f'(-r) = -rF(r)/2c$. Finally, substituting the value of $F(r)$, we find the following
expressions for the derivative $f'(\xi)$ and the function $f(\xi)$ itself:

$$\text{for } |\xi| > a, \quad f'(\xi) = 0, \quad f(\xi) = 0;$$

$$\text{for } |\xi| < a, \quad f'(\xi) = c\xi\Delta/2\rho, \quad f(\xi) = c(\xi^2 - a^2)\Delta/4\rho,$$

which give the solution of the problem. If we consider a point with $r > a$, i.e. outside the
region of the initial compression, we have for the density

$$\text{for } t < (r-a)/c, \quad \rho' = 0;$$
$$\text{for } (r-a)/c < t < (r+a)/c, \quad \rho' = \tfrac{1}{2}(r-ct)\Delta/r;$$
$$\text{for } t > (r+a)/c, \quad \rho' = 0.$$

The wave passes the point considered during a time interval $2a/c$; in other words, the wave
has the form of a spherical shell of thickness $2a$, which at time t lies between the spheres of
radii $ct-a$ and $ct+a$. Within this shell the density varies linearly; in the outer part ($r > ct$),
the gas is compressed ($\rho' > 0$), while in the inner part ($r < ct$) it is rarefied ($\rho' < 0$).

PROBLEM 2. Determine the characteristic frequencies of centrally symmetrical sound oscillations in a spherical vessel of radius a.

SOLUTION. From the boundary condition $\partial\phi/\partial r = 0$ for $r = a$ (where ϕ is given by (69.5)) we find $\tan ka = ka$, which determines the characteristic frequencies. The first (lowest) frequency is $\omega_1 = 4\cdot49 \, c/a$.

§70. Cylindrical waves

Let us now consider a wave in which the distribution of all quantities is homogeneous in some direction (which we take as the z-axis) and has complete axial symmetry about that direction. This is called a *cylindrical wave*, and in it we have $\phi = \phi(R, t)$, where R denotes the distance from the z-axis. Let us determine the general form of such an axisymmetric solution of the wave equation. This can be done by starting from the general spherically symmetrical solution (69.2). R is related to r by $r^2 = R^2 + z^2$, so that ϕ as given by formula (69.2) depends on z when R and t are given. A function which depends on R and t only and still satisfies the wave equation can be obtained by integrating (69.2) over all z from $-\infty$ to ∞, or equally well from 0 to ∞. We can convert the integration over z to one over r. Since $z = \sqrt{(r^2 - R^2)}$, $dz = r \, dr/\sqrt{(r^2 - R^2)}$. When z varies from 0 to ∞, r varies from R to ∞. Hence we find the general axisymmetric solution to be

$$\phi = \int_R^\infty \frac{f_1(ct-r)}{\sqrt{(r^2 - R^2)}} dr + \int_R^\infty \frac{f_2(ct+r)}{\sqrt{(r^2 - R^2)}} dr, \tag{70.1}$$

where f_1 and f_2 are arbitrary functions. The first term is an outgoing cylindrical wave, and the second an ingoing one.

Substituting in these integrals $ct \pm r = \xi$, we can rewrite formula (70.1) as

$$\phi = \int_{-\infty}^{ct-R} \frac{f_1(\xi)d\xi}{\sqrt{[(ct-\xi)^2 - R^2]}} + \int_{ct+R}^\infty \frac{f_2(\xi)d\xi}{\sqrt{[(\xi-ct)^2 - R^2]}}. \tag{70.2}$$

We see that the value of the potential at time t in the outgoing cylindrical wave is determined by the values of f_1 at times from $-\infty$ to $t - R/c$; similarly, the values of f_2 which affect the ingoing wave are those at times from $t + R/c$ to infinity.

As in the spherical case, stationary waves are obtained when $f_1(\xi) = -f_2(\xi)$. It can be shown that a stationary cylindrical wave can also be represented in the form

$$\phi = \int_{ct-R}^{ct+R} \frac{F(\xi)d\xi}{\sqrt{[R^2 - (\xi - ct)^2]}}, \tag{70.3}$$

where $F(\xi)$ is another arbitrary function.

Let us derive an expression for the potential in a monochromatic cylindrical wave. The wave equation for the potential $\phi(R, t)$ in cylindrical co-ordinates is

$$\frac{1}{R}\frac{\partial}{\partial R}\left(R\frac{\partial \phi}{\partial R}\right) - \frac{1}{c^2}\frac{\partial^2 \phi}{\partial t^2} = 0.$$

In a monochromatic wave $\phi = e^{-i\omega t}f(R)$, and we have for the function $f(R)$ the equation $f'' + f'/R + k^2 f = 0$. This is Bessel's equation of order zero. In a stationary cylindrical wave, ϕ must remain finite for $R = 0$; the appropriate solution is $J_0(kR)$, where J_0 is a Bessel function of the first kind. Thus, in a stationary cylindrical wave,

$$\phi = Ae^{-i\omega}J_0(kR). \tag{70.4}$$

For $R = 0$ the function J_0 tends to unity, so that the amplitude tends to the finite limit A. At large distances R, J_0 may be replaced by its asymptotic expression, and ϕ then takes the form

$$\phi = A\sqrt{\frac{2}{\pi}}\,\frac{\cos(kR-\tfrac14\pi)}{\sqrt{(kR)}}e^{-i\omega t}. \tag{70.5}$$

The solution corresponding to a monochromatic outgoing travelling wave is

$$\phi = Ae^{-i\omega t}H_0^{(1)}(kR), \tag{70.6}$$

where $H_0^{(1)}$ is the Hankel function of the first kind, of order zero. For $R \to 0$ this function has a logarithmic singularity:

$$\phi \cong (2iA/\pi)\log(kR)e^{-i\omega t}. \tag{70.7}$$

At large distances we have the asymptotic formula

$$\phi = A\sqrt{\frac{2}{\pi}}\,\frac{\exp[i(kR-\omega t-\tfrac14\pi)]}{\sqrt{(kR)}}. \tag{70.8}$$

We see that the amplitude of a cylindrical wave diminishes (at large distances) inversely as the square root of the distance from the axis, and the intensity therefore decreases as $1/R$. This result is obvious, since the total energy flux is distributed over a cylindrical surface, whose area increases proportionally to R as the wave is propagated.

A cylindrical outgoing wave differs from a spherical or plane wave in the important respect that it has a forward front but no backward front: once the sound disturbance has reached a given point, it does not cease, but diminishes comparatively slowly as $t \to \infty$. Suppose that the function $f_1(\xi)$ in the first term of (70.2) is different from zero only in some finite range $\xi_1 \leqslant \xi \leqslant \xi_2$. Then, at times such that $ct > R+\xi_2$, we have

$$\phi = \int_{\xi_1}^{\xi_2} \frac{f_1(\xi)d\xi}{\sqrt{[(ct-\xi)^2 - R^2]}}.$$

As $t \to \infty$, this expression tends to zero as

$$\phi = \frac{1}{ct} \int\limits_{\xi_1}^{\xi_2} f_1(\xi) \mathrm{d}\xi,$$

i.e. inversely as the time.

Thus the potential in an outgoing cylindrical wave, due to a source which operates only for a finite time, vanishes, though slowly, as $t \to \infty$. This means that, as in the spherical case, the integral of p' over all time is zero:

$$\int\limits_{-\infty}^{\infty} p' \, \mathrm{d}t = 0. \tag{70.9}$$

Hence a cylindrical wave, like a spherical wave, must necessarily include both condensations and rarefactions.

§71. The general solution of the wave equation

We shall now derive a general formula giving the solution of the wave equation in an infinite fluid for any initial conditions, i.e. giving the velocity and pressure distribution in the fluid at any instant in terms of their initial distribution.

We first obtain some auxiliary formulae. Let $\phi(x, y, z, t)$ and $\psi(x, y, z, t)$ be any two solutions of the wave equation which vanish at infinity. We consider the integral

$$I = \int (\phi\ddot{\psi} - \psi\ddot{\phi}) \, \mathrm{d}V,$$

taken over all space, and calculate its time derivative. Since ϕ and ψ satisfy the equations $\triangle\phi - \ddot{\phi}/c^2 = 0$ and $\triangle\psi - \ddot{\psi}/c^2 = 0$, we have

$$\mathrm{d}I/\mathrm{d}t = \int (\phi\dddot{\psi} - \psi\dddot{\phi}) \mathrm{d}V = c^2 \int (\phi\triangle\psi - \psi\triangle\phi) \mathrm{d}V$$

$$= c^2 \int \mathrm{div}(\phi \, \mathbf{grad}\,\psi - \psi \, \mathbf{grad}\,\phi) \mathrm{d}V.$$

The last integral can be transformed into an integral over an infinitely distant surface, and is therefore zero. Thus we conclude that $\mathrm{d}I/\mathrm{d}t = 0$, i.e. I is independent of time:

$$\int (\phi\ddot{\psi} - \psi\ddot{\phi}) \mathrm{d}V = \text{constant}. \tag{71.1}$$

Next, let us consider the following particular solution of the wave equation:

$$\psi = \delta[r - c(t_0 - t)]/r \tag{71.2}$$

(where r is the distance from some given point O, t_0 is some definite instant, and δ denotes the delta function), and calculate the integral of ψ over all space. We have

$$\int \psi \, dV = \int_0^\infty \psi \cdot 4\pi r^2 \, dr = 4\pi \int_0^\infty r\delta[r - c(t_0 - t)] \, dr.$$

The argument of the delta function is zero for $r = c(t_0 - t)$ (we assume that $t_0 > t$). Hence, from the properties of the delta function, we find

$$\int \psi \, dV = 4\pi c(t_0 - t). \tag{71.3}$$

Differentiating this equation with respect to time, we obtain

$$\int \dot{\psi} \, dV = -4\pi c. \tag{71.4}$$

We now substitute for ψ, in the integral (71.1), the function (71.2), and take ϕ to be the required general solution of the wave equation. According to (71.1), I is a constant; using this, we write down the expressions for I at the instants $t = 0$ and $t = t_0$, and equate the two. For $t = t_0$ the two functions ψ and $\dot{\psi}$ are each different from zero only for $r = 0$. Hence, on integrating, we can put $r = 0$ in ϕ and $\dot{\phi}$ (i.e. take their values at the point O), and take ϕ and $\dot{\phi}$ outside the integral:

$$I = \phi(x, y, z, t_0) \int \psi \, dV - \dot{\phi}(x, y, z, t_0) \int \psi \, dV,$$

where x, y, z are the co-ordinates of O. According to (71.3) and (71.4), the second term is zero for $t = t_0$, and the first term gives

$$I = -4\pi c\dot{\phi}(x, y, z, t_0).$$

Let us now calculate I for $t = 0$. Putting $\dot{\psi} = \partial \psi / \partial t = -\partial \psi / \partial t_0$, and denoting by ϕ_0 the value of the function ϕ for $t = 0$, we have

$$I = -\int \left(\phi_0 \frac{\partial \psi}{\partial t_0} + \dot{\phi_0}\psi \right) dV = -\frac{\partial}{\partial t_0} \int \phi_0 \psi_{t=0} \, dV - \int \dot{\phi_0}\psi_{t=0} \, dV.$$

We write the element of volume as $dV = r^2 dr do$, where do is an element of solid angle, and then we obtain, by the properties of the delta function,

$$\int \phi_0 \psi_{t=0} \, dV = \int \phi_0 r\delta(r - ct_0) dr \, do = ct_0 \int \phi_{0,\, r=ct_0} \, do;$$

the integral of $\dot{\phi_0}\psi$ is similar. Thus

$$I = -\frac{\partial}{\partial t_0}\left(ct_0 \int \phi_{0,\, r=ct_0} \, do \right) - ct_0 \int \dot{\phi}_{0,\, r=ct_0} \, do.$$

Finally, equating the two expressions for I and omitting the suffix zero in t_0, we obtain

$$\phi(x,y,z,t) = \frac{1}{4\pi}\left\{\frac{\partial}{\partial t}\left(t\int\phi_{0,\,r=ct}\,do\right) + t\int\dot{\phi}_{0,\,r=ct}\,do\right\}. \qquad (71.5)$$

This formula, called *Poisson's formula*, gives the spatial distribution of the potential at any instant in terms of the distribution of the potential and its time derivative (or, equivalently, in terms of the velocity and pressure distribution) at some initial instant. We see that the value of the potential at time t is determined by the values of ϕ and $\dot{\phi}$ at time $t = 0$ on the surface of a sphere centred at O, of radius ct.

Let us suppose that, at the initial instant, ϕ_0 and $\dot{\phi}_0$ are different from zero only in some finite region of space, bounded by a closed surface C (Fig. 34).

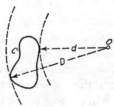

FIG. 34

We consider the values of ϕ at subsequent instants at some point O. These values are determined by the values of ϕ_0 and $\dot{\phi}_0$ at a distance ct from O. The spheres of radius ct pass through the region within the surface C only for $d/c \leqslant t \leqslant D/c$, where d and D are the least and greatest distances from the point O to the surface C. At other instants, the integrands in (71.5) are zero. Thus the motion at O begins at time $t = d/c$ and ceases at time $t = D/c$. The wave propagated from the region inside C has a forward front and a backward front. The motion begins when the forward front arrives at the point in question, while on the backward front particles previously oscillating come to rest.

PROBLEM

Derive the formula giving the potential in terms of the initial conditions for a wave depending on only two co-ordinates, x and y.

SOLUTION. An element of area of a sphere of radius ct can be written $df = c^2t^2\,do$, where do is an element of solid angle. The projection of df on the xy-plane is $dx\,dy = df\sqrt{[(ct)^2 - \rho^2]}/ct$, where ρ is the distance of the point x, y from the centre of the sphere. Comparing the two expressions, we can write $do = dx\,dy/ct\sqrt{[(ct)^2 - \rho^2]}$. Denoting by x, y the co-ordinates of the point where we seek the value of ϕ, and by ξ, η the co-ordinates of a variable point in the region of integration, we can therefore replace do in the general formula (71.5) by $d\xi\,d\eta/ct\sqrt{[(ct)^2 - (x-\xi)^2 - (y-\eta)^2]}$, doubling the resulting expression because $dx\,dy$ is the projection of two elements of area on opposite sides of the xy-plane. Thus

$$\phi(x, y, z, t) = \frac{1}{2\pi c} \frac{\partial}{\partial t} \int\int \frac{\phi_0(\xi, \eta)\mathrm{d}\xi\,\mathrm{d}\eta}{\sqrt{[(ct)^2 - (x-\xi)^2 - (y-\eta)^2]}} +$$

$$+ \frac{1}{2\pi c} \int\int \frac{\dot\phi_0(\xi, \eta)\mathrm{d}\xi\,\mathrm{d}\eta}{\sqrt{[(ct)^2 - (x-\xi)^2 - (y-\eta)^2]}},$$

where the integration is over a circle centred at O, of radius ct. If ϕ_0 and $\dot\phi_0$ are zero except in a finite region C of the xy-plane (or, more exactly, except in a cylindrical region with its generators parallel to the z-axis), the oscillations at the point O (Fig. 34) begin at time $t = d/c$, where d is the least distance from O to a point in the region. After this time, however, circles of radius $ct > d$ centred at O will always enclose part or all of the region C, and ϕ will tend only asymptotically to zero. Thus, unlike three-dimensional waves, the two-dimensional waves here considered have a forward front but no backward front (cf. §70).

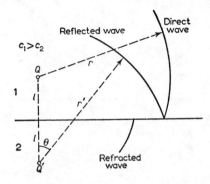

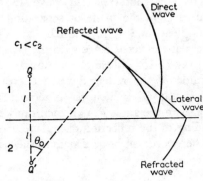

FIG. 35

§72. The lateral wave

The reflection of a spherical wave from the surface separating two media is of particular interest in that it may be accompanied by an unusual pheno-menon, the appearance of a *lateral wave*.

Let Q (Fig. 35) be the source of a spherical sound wave in medium 1, at a

distance l from the infinite plane surface separating media 1 and 2. The
distance l is arbitrary, and need not be large compared with the wavelength λ.
Let the densities of the two media be ρ_1, ρ_2, and the velocities of sound in
them c_1, c_2. We suppose first that $c_1 > c_2$; then, at distances from the source
large compared with λ, the motion in medium 1 will be a superposition of
two outgoing waves. One of these is the spherical wave emitted by the source
(the *direct wave*); its potential is

$$\phi_1{}^0 = e^{ikr}/r, \tag{72.1}$$

where r is the distance from the source, and the amplitude is arbitrarily taken
to be unity. We shall, for brevity, omit the factor $e^{-i\omega t}$ from all expressions
in the present section.

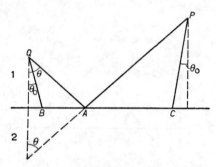

FIG. 36

The wave surfaces of the second (*reflected*) wave are spheres centred at Q',
the image of the source Q in the plane of separation; this is the locus of
points P reached at a given time by rays which leave Q simultaneously and
are reflected from the plane in accordance with the laws of geometrical acoustics (in Fig. 36, the ray QAP with angles of incidence and reflection θ is
shown). The amplitude of the reflected wave decreases inversely as the
distance r' from the point Q' (which is sometimes called an *imaginary
source*), but depends also on the angle θ, as if each ray were reflected with the
coefficient corresponding to the reflection of a plane wave at the given angle
of incidence θ. In other words, at large distances the reflected wave is given
by the formula

$$\phi_1' = \frac{e^{ikr'}}{r'}\left[\frac{\rho_2 c_2 - \rho_1\sqrt{(c_1{}^2 - c_2{}^2 \sin^2\theta)}}{\rho_2 c_2 + \rho_1\sqrt{(c_1{}^2 - c_2{}^2 \sin^2\theta)}}\right]; \tag{72.2}$$

cf. formula (65.4) for the reflection coefficient for a plane wave. This formula,
which is clearly valid for large r', can be rigorously derived by the method
shown below.

A more interesting case is that where $c_1 < c_2$. Here, besides the ordinary
reflected wave (72.2), another wave appears in the first medium. The

chief properties of this wave can be seen from the following simple considerations.

The ordinary reflected ray QAP (Fig. 36) obeys Fermat's principle in the sense that it is the quickest path from Q to P, among paths lying entirely in medium 1 and involving a single reflection. When $c_1 < c_2$, however, Fermat's principle is also satisfied by another path, where the ray is incident on the boundary at the critical angle of total internal reflection θ_0 (sin $\theta_0 = c_1/c_2$), then is propagated in medium 2 along the boundary, and finally returns to medium 1 at the angle θ_0. The path is $QBCP$ in Fig. 36, and it is evident that $\theta > \theta_0$. It is easy to see that this path also has the extremal property: the time taken to traverse it is less than for any other path from Q to P lying partly in medium 2.

The geometrical locus of points P reached at the same time by rays which simultaneously leave Q along the path QB, and then return to medium 1 at various points C, is evidently a conical surface whose generators are perpendicular to lines drawn from the imaginary source Q' at an angle θ_0.

Thus, if $c_1 < c_2$, together with the ordinary reflected wave, which has a spherical front, there is propagated in medium 1 another wave, which has a conical front extending from the plane of separation (where it meets the refracted wave front in medium 2) to the point where it touches the spherical front of the reflected wave; this occurs along the line of intersection with a cone of semi-angle θ_0 and axis QQ' (Fig. 35). This conical wave is called the *lateral wave*.

It is easy to see by a simple calculation that the time along the path $QBCP$ (Fig. 36) is less than along the path QAP to the same point P. This means that a sound signal from the source Q reaches an observer at P first as the lateral wave, and only later as the ordinary reflected wave.

It must be borne in mind that the lateral wave is an effect of wave acoustics, despite the fact that it allows the above simple interpretation in terms of the concepts of geometrical acoustics. We shall see below that the amplitude of the lateral wave tends to zero in the limit $\lambda \to 0$.

Let us now make a quantitative calculation. The propagation of a monochromatic sound wave from a point source is described by equation (69.7):

$$\triangle\phi + k^2\phi = -4\pi\delta(\mathbf{r}-\mathbf{1}), \tag{72.3}$$

where $k = \omega/c$ and $\mathbf{1}$ is the radius vector of the source. The coefficient of the delta function is chosen so that the direct wave has the form (72.1). In what follows we take a system of co-ordinates with the xy-plane as the plane of separation and the z-axis along QQ', with the first medium in $z > 0$. At the surface of separation the pressure and the z-component of the velocity, or (equivalently) $\rho\phi$ and $\partial\phi/\partial z$, must be continuous.

Using the general Fourier method, we obtain the solution in the form

$$\phi = \int\limits_{-\infty}^{\infty}\int\limits_{-\infty}^{\infty} \phi_\kappa(z)\exp[i(\kappa_x x + \kappa_y y)]\,d\kappa_x\,d\kappa_y, \tag{72.4}$$

where

$$\phi_\kappa(z) = \frac{1}{4\pi^2} \int\limits_{-\infty}^{\infty} \int\limits_{-\infty}^{\infty} \phi \exp[-i(\kappa_x x + \kappa_y y)] \, dx \, dy. \tag{72.5}$$

From the symmetry relative to the xy-plane it is evident that ϕ_κ can depend only on the quantity $|\kappa| = \surd(\kappa_x^2 + \kappa_y^2)$. Using the well-known formula

$$J_0(u) = \frac{1}{2\pi} \int\limits_{0}^{2\pi} \cos(u \sin \phi) d\phi,$$

we can therefore write (72.4) as

$$\phi = 2\pi \int\limits_{0}^{\infty} \phi_\kappa(z) J_0(\kappa R) \kappa \, d\kappa, \tag{72.6}$$

where $R = \surd(x^2 + y^2)$ is the cylindrical co-ordinate (the distance from the z-axis). It is convenient for the subsequent calculations to transform this formula into one in which the integral is taken from $-\infty$ to ∞, expressing the integrand in terms of the Hankel function $H_0^{(1)}(u)$. The latter has a logarithmic singularity at the point $u = 0$; if we agree to go from positive to negative real u by passing above the point $u = 0$ in the complex u-plane, then $H_0^{(1)}(-u) = H_0^{(1)}(ue^{i\pi}) = H_0^{(1)}(u) - 2J_0(u)$. Using this relation, we can rewrite (72.6) as

$$\phi = \pi \int\limits_{-\infty}^{\infty} \phi_\kappa(z) H_0^{(1)}(\kappa R) \kappa \, d\kappa. \tag{72.7}$$

From equation (72.3) we find for the function ϕ_κ the equation

$$\frac{d^2 \phi_\kappa}{dz^2} - \left(\kappa^2 - \frac{\omega^2}{c^2}\right) \phi_\kappa = -\frac{1}{\pi} \delta(z - l). \tag{72.8}$$

The delta function on the right-hand side of the equation can be eliminated by imposing on the function $\phi_\kappa(z)$ (satisfying the homogeneous equation) the boundary conditions at $z = l$:

$$[\phi_\kappa(z)]_{l+} - [\phi_\kappa(z)]_{l-} = 0,$$
$$[d\phi_\kappa/dz]_{l+} - [d\phi_\kappa/dz]_{l-} = -1/\pi. \tag{72.9}$$

The boundary conditions at $z = 0$ are

$$[\rho\phi_\kappa]_{0+} - [\rho\phi_\kappa]_{0-} = 0,$$
$$[d\phi_\kappa/dz]_{0+} - [d\phi_\kappa/dz]_{0-} = 0. \tag{72.10}$$

We seek a solution in the form

$$\phi_\kappa = Ae^{-\mu_1 z} \qquad \text{for } z > l,$$
$$\phi_\kappa = Be^{-\mu_1 z} + Ce^{\mu_1 z} \text{ for } l > z > 0,$$
$$\phi_\kappa = De^{\mu_1 z} \qquad \text{for } 0 > z. \qquad\qquad (72.11)$$

Here

$$\mu_1{}^2 = \kappa^2 - k_1{}^2, \quad \mu_2{}^2 = \kappa^2 - k_2{}^2 \; (k_1 = \omega/c_1, \, k_2 = \omega/c_2),$$

and we must put

$$\mu = +\sqrt{(\kappa^2 - k^2)} \text{ for } \kappa > k,$$
$$\mu = -i\sqrt{(k^2 - \kappa^2)} \text{ for } \kappa < k. \qquad\qquad (72.12)$$

The first of these is necessary so that ϕ should not increase without limit as $z \to \infty$, and the second so that ϕ should represent an outgoing wave. The conditions (72.9) and (72.10) give four equations which determine the coefficients A, B, C and D. A simple calculation gives

$$B = C\frac{\mu_1\rho_2 - \mu_2\rho_1}{\mu_1\rho_2 + \mu_2\rho_1}, \qquad C = \frac{e^{-l\mu_1}}{2\pi\mu_1},$$
$$D = C\frac{2\rho_1\mu_1}{\mu_1\rho_2 + \mu_2\rho_1}, \qquad A = B + Ce^{2l\mu_1}. \qquad (72.13)$$

For $\rho_2 = \rho_1$, $c_2 = c_1$ (i.e. when all space is occupied by one medium), B is zero and $A = Ce^{2l\mu_1}$; the corresponding term in ϕ is evidently the direct wave (72.1), and the reflected wave in which we are interested is therefore

$$\phi_1' = \pi \int_{-\infty}^{\infty} B(\kappa)e^{-z\mu_1} H_0^{(1)}(\kappa R)\kappa \, d\kappa. \qquad (72.14)$$

In this expression the path of integration has to be specified. It passes above the singular point $\kappa = 0$ (in the complex κ-plane), as we have already mentioned. The integrand also has singular points (branch points) at $\kappa = \pm k_1, \pm k_2$, where μ_1 or μ_2 vanishes. In accordance with the conditions (72.10), the contour must pass below the points $+k_1, +k_2$, and above the points $-k_1, -k_2$.

Let us investigate the resulting expression for large distances from the source. Replacing the Hankel function by its well-known asymptotic expression, we obtain

$$\phi_1' = \int_C \frac{\mu_1\rho_2 - \mu_2\rho_1}{\mu_1(\mu_1\rho_2 + \mu_2\rho_1)} \sqrt{\frac{\kappa}{2i\pi R}} \exp[-(z+l)\mu_1 + i\kappa R] \, d\kappa. \qquad (72.15)$$

Fig. 37 shows the path of integration C for the case $c_1 > c_2$. The integral can be calculated by means of the saddle-point method. The exponent $i[(z+l)\sqrt{(k_1{}^2 - \kappa^2)} + \kappa R]$ has an extremum at the point where

$$\kappa/\sqrt{(k_1{}^2 - \kappa^2)} = R/(z+l) = r'\sin\theta/r'\cos\theta = \tan\theta,$$

i.e. $\kappa = k_1 \sin \theta$, where θ is the angle of incidence (see Fig. 35). On changing to the path of integration C' which passes through this point at an angle of $\pi/4$ to the axis of abscissae, we obtain formula (72.2).

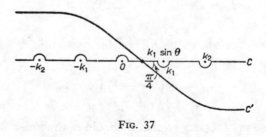

FIG. 37

In the case $c_1 < c_2$ (i.e. $k_1 > k_2$), the point $\kappa = k_1 \sin \theta$ lies between k_2 and k_1 if $\sin \theta > k_2/k_1 = c_1/c_2 = \sin \theta_0$, i.e. if $\theta > \theta_0$ (Fig. 38). In this case the contour C' must make a loop round the point k_2, and we have, besides the ordinary reflected wave (72.2), a wave ϕ_1'' given by the integral (72.15) taken around the loop, which we call C''. This is the lateral wave. The integral is easily calculated if the point $k_1 \sin \theta$ is not close to k_2, i.e. if the angle θ is not close to the internal-reflection angle θ_0.†

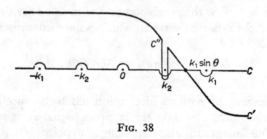

FIG. 38

Near the point $\kappa = k_2$, μ_2 is small; we expand the coefficient of the exponential in the integrand of (72.15) in powers of μ_2. The zero-order term has no singularity at $\kappa = k_2$, and its integral round C'' is zero. Hence we have

$$\phi_1'' = - \int_{C'} \frac{2\mu_2\rho_1}{\mu_1^2\rho_2} \sqrt{\frac{\kappa}{2i\pi r}} \exp[-(z+l)\mu_1 + i\kappa R]\, d\kappa. \qquad (72.16)$$

Expanding the exponent in powers of $\kappa - k_2$ and integrating round the loop

---

† For an investigation of the lateral wave for all values of θ, see L. BREKHOVSKIKH, *Zhurnal tekhnicheskoĭ fiziki* **18**, 455, 1948. This paper gives also the next term in the expansion of the ordinary reflected wave in powers of λ/R. We may mention here that, for angles θ close to θ_0 (in the case $c_1 < c_2$), the ratio of the correction term to the leading term falls off with distance as $(\lambda/R)^{\frac{1}{4}}$, and not as λ/R.

C'', we have after a simple calculation the following expression for the potential of the lateral wave:

$$\phi_1'' = \frac{2i\rho_1 k_2 \exp[ik_1 r' \cos(\theta_0 - \theta)]}{r'^2 \rho_2 k_1{}^2 \sqrt{[\cos\theta_0 \sin\theta \sin^3(\theta_0 - \theta)]}}. \tag{72.17}$$

In accordance with the previous results, the wave surfaces are the cones $r' \cos(\theta - \theta_0) = R \sin\theta_0 + (s + l)\cos\theta_0 = $ constant. In a given direction, the wave amplitude decreases inversely as the square of the distance r'. We see also that this wave disappears in the limit $\lambda \to 0$. For $\theta \to \theta_0$, the expression (72.17) ceases to be valid; in actual fact, the amplitude of the lateral wave in this range of θ decreases with distance as $r'^{-5/4}$.

§73. The emission of sound

A body oscillating in a fluid causes a periodic compression and rarefaction of the fluid near it, and thus produces sound waves. The energy carried away by these waves is supplied from the kinetic energy of the body. Thus we can speak of the emission of sound by oscillating bodies.[†]

In the general case of a body of arbitrary shape oscillating in any manner, the problem of the emission of sound waves must be solved as follows. We take the velocity potential ϕ as the fundamental quantity; it satisfies the wave equation

$$\triangle\phi - (1/c^2)\partial^2\phi/\partial t^2 = 0. \tag{73.1}$$

At the surface of the body, the normal component of the fluid velocity must be equal to the corresponding component of the velocity $\mathbf{u}$ of the body:

$$\partial\phi/\partial n = u_n. \tag{73.2}$$

At large distances from the body, the wave must become an outgoing spherical wave. The solution of equation (73.1) which satisfies these boundary conditions and the condition at infinity determines the sound wave emitted by the body.

Let us consider the two boundary conditions in more detail. We suppose first that the frequency of oscillation of the body is so large that the length of the emitted wave is very small compared with the dimension l of the body:

$$\lambda \ll l. \tag{73.3}$$

In this case we can divide the surface of the body into portions whose dimensions are so small that they may be approximately regarded as plane, but yet are large compared with the wavelength. Then we may suppose that each

---

[†] In what follows we shall always suppose that the velocity u of the oscillating body is small compared with the velocity of sound. Since $u \sim a\omega$, where a is the linear amplitude of the oscillations, this means that $a \ll \lambda$.

The amplitude of the oscillations is in general supposed small in comparison with the dimensions of the body also, since otherwise we do not have potential flow near the body (cf. §9). This condition is unnecessary only for pure pulsations, when the solution (73.7) used below is really a direct deduction from the equation of continuity.

such portion emits a plane wave, in which the fluid velocity is simply the normal component u_n of the velocity of that portion of the surface. But the mean energy flux in a plane wave is (see §64) $c\rho\overline{v^2}$, where v is the fluid velocity in the wave. Putting $v = u_n$ and integrating over the whole surface of the body, we reach the result that the mean energy emitted per unit time by the body in the form of sound waves, i.e. the total intensity of the emitted sound, is

$$I = c\rho \oint \overline{u_n{}^2}\, df. \tag{73.4}$$

It is independent of the frequency of the oscillations (for a given velocity amplitude).

Let us now consider the opposite limiting case, where the length of the emitted wave is large compared with the dimension of the body:

$$\lambda \gg l. \tag{73.5}$$

Then we can neglect the term $(1/c^2)\partial^2\phi/\partial t^2$, in the general equation (73.1), near the body (at distances small compared with the wavelength). For this term is of the order of $\omega^2\phi/c^2 \sim \phi/\lambda^2$, whereas the second derivatives with respect to the co-ordinates are, in this region, of the order of ϕ/l^2.

Thus the flow near the body satisfies Laplace's equation, $\triangle\phi = 0$. This is the equation for potential flow of an incompressible fluid. Consequently the fluid near the body moves as if it were incompressible. Sound waves proper, i.e. compression and rarefaction waves, occur only at large distances from the body.

At distances of the order of the dimension of the body and smaller, the required solution of the equation $\triangle\phi = 0$ cannot be written in a general form, but depends on the actual shape of the oscillating body. At distances large compared with l, however (though still small compared with λ, so that the equation $\triangle\phi = 0$ remains valid), we can find a general form of the solution by using the fact that ϕ must decrease with increasing distance. We have already discussed such solutions of Laplace's equation in §11. As there, we write the general form of the solution as

$$\phi = -(a/r) + \mathbf{A}\cdot\mathbf{grad}(1/r), \tag{73.6}$$

where r is the distance from an origin anywhere inside the body. Here, of course, the distances involved must be large compared with the dimension of the body, since we cannot otherwise restrict ourselves to the terms in ϕ which decrease least rapidly as r increases. We have included both terms in (73.6), although it must be borne in mind that the first term is sometimes absent (see below).

Let us ascertain in what cases this term $-a/r$ is non-zero. We found in §11 that a potential $-a/r$ results in a non-zero value $4\pi\rho a$ of the mass flux through a surface surrounding the body. In an incompressible fluid, however such a mass flux can occur only if the total volume of fluid enclosed within the surface changes. In other words, there must be a change in the volume of the body, as a result of which the fluid is either expelled from or "sucked into" the volume of space concerned. Thus the first term in (73.6) appears in cases where the emitting body undergoes pulsations during which its volume changes.

Let us suppose that this is so, and determine the total intensity of the emitted sound. The volume $4\pi a$ of the fluid which flows through the closed surface must, by the foregoing argument, be equal to the change per unit time in the volume V of the body, i.e. to the derivative dV/dt (the volume V being a given function of the time): $4\pi a = \dot{V}$. Thus, at distances r such that $l \ll r \ll \lambda$, the motion of the fluid is given by the function $\phi = -\dot{V}(t)/4\pi r$. At distances $r \gg \lambda$, however (i.e. in the "wave region"), ϕ must represent an outgoing spherical wave, i.e. must be of the form

$$\phi = -\frac{f(t-r/c)}{r}. \tag{73.7}$$

Hence we conclude at once that the emitted wave has, at all distances large compared with l, the form

$$\phi = -\frac{\dot{V}(t-r/c)}{4\pi r}, \tag{73.8}$$

which is obtained by replacing the argument t of $(t\dot{V})$ by $t-r/c$.

The velocity $\mathbf{v} = \mathbf{grad}\,\phi$ is directed at every point along the radius vector, and its magnitude is $v = \partial\phi/\partial r$. In differentiating (73.8) for distances $r \gg \lambda$, only the derivative of the numerator need be taken, since differentiation of the denominator would give a term of higher order in $1/r$, which we neglect. Since $\partial\dot{V}(t-r/c)/\partial r = -(1/c)\ddot{V}(t-r/c)$, we obtain

$$\mathbf{v} = \ddot{V}(t-r/c)\mathbf{n}/4\pi cr, \tag{73.9}$$

where $\mathbf{n}$ is a unit vector in the direction of $\mathbf{r}$.

The intensity of the sound is given by the square of the velocity, and is here independent of the direction of emission, i.e. the emission is isotropic. The mean value of the total energy emitted per unit time is

$$I = \rho c \oint \overline{v^2}\,df = (\rho/16c\pi^2) \oint (\overline{\ddot{V}^2}/r^2)df,$$

where the integration is taken over a closed surface surrounding the origin. Taking this surface to be a sphere of radius r, and noticing that the integrand depends only on the distance from the origin, we have finally

$$I = \rho\overline{\ddot{V}^2}/4\pi c. \tag{73.10}$$

This is the total intensity of the emitted sound. We see that it is given by the squared second time derivative of the volume of the body.

If the body executes harmonic pulsations of frequency ω, the second time derivative of the volume is proportional to the frequency and velocity amplitude of the oscillations, and its mean square is proportional to the square of the frequency for a given velocity amplitude of points on the surface of the body. For a given amplitude of the oscillations, however, the velocity amplitude is itself proportional to the frequency, so that the intensity of emission is proportional to ω^4.

Let us now consider the emission of sound by a body oscillating without change of volume. Only the second term then remains in (73.6); we write it $\phi = \mathrm{div}[\mathbf{A}(t)/r]$. As in the preceding case, we conclude that the general form of the solution at all distances $r \gg l$ is $\phi = \mathrm{div}[\mathbf{A}(t-r/c)/r]$. That this expression is in fact a solution of the wave equation is seen immediately, since the function $\mathbf{A}(t-r/c)/r$ is a solution, and therefore so are its derivatives with respect to the co-ordinates. Again differentiating only the numerator, we obtain (for distances $r \gg \lambda$)

$$\phi = -\dot{\mathbf{A}}(t-r/c)\cdot\mathbf{n}/cr. \tag{73.11}$$

To calculate the velocity $\mathbf{v} = \mathbf{grad}\,\phi$, we need again differentiate only $\mathbf{A}$. Hence we have, by the familiar rules of vector analysis for differentiation with respect to a scalar argument,

$$\mathbf{v} = -\frac{\ddot{\mathbf{A}}(t-r/c)\cdot\mathbf{n}}{cr}\mathbf{grad}\left(t-\frac{r}{c}\right),$$

and, substituting $\mathbf{grad}(t-r/c) = -(1/c)\mathbf{grad}\,r = -\mathbf{n}/c$, we have finally

$$\mathbf{v} = \mathbf{n}(\mathbf{n}\cdot\ddot{\mathbf{A}})/c^2r. \tag{73.12}$$

The intensity is now proportional to the squared cosine of the angle between the direction of emission (i.e. the direction of $\mathbf{n}$) and the vector $\ddot{\mathbf{A}}$; this is called *dipole emission*. The total emission is given by the integral

$$I = \frac{\rho}{c^3}\oint\frac{\overline{(\mathbf{n}\cdot\ddot{\mathbf{A}})^2}}{r}df.$$

We again take the surface of integration to be a sphere of radius r, and use spherical co-ordinates with the polar axis in the direction of the vector $\mathbf{A}$. A simple integration gives finally for the total emission per unit time

$$I = \frac{4\pi\rho}{3c^3}\overline{\ddot{\mathbf{A}}^2}. \tag{73.13}$$

The components of the vector $\mathbf{A}$ are linear functions of the components of the velocity $\mathbf{u}$ of the body (see §11). Thus the intensity is here a quadratic function of the second time derivatives of the velocity components.

If the body executes harmonic oscillations of frequency ω, we conclude (reasoning as in the previous case) that the intensity is proportional to ω^4

for a given value of the velocity amplitude. For a given linear amplitude of the oscillations of the body, the velocity amplitude is proportional to the frequency, and therefore the intensity is proportional to ω^6.

In an entirely similar manner we can solve the problem of the emission of cylindrical sound waves by a cylinder of any cross-section pulsating or oscillating perpendicularly to its axis. We shall give here the corresponding formulae, with a view to later applications.

Let us first consider small pulsations of a cylinder, and let $S = S(t)$ be its (variable) cross-sectional area. At distances r from the axis of the cylinder such that $l \ll r \ll \lambda$, where l is the transverse dimension of the cylinder, we have similarly to (73.8)

$$\phi = [\dot{S}(t)/2\pi]\log fr, \qquad (73.14)$$

where $f(t)$ is a function of time, and the coefficient of $\log fr$ is chosen so as to obtain the correct value for the mass flux through a coaxial cylindrical surface. In accordance with the formula for the potential of an outgoing cylindrical wave (the first term of formula (70.2)), we now conclude that at all distances $r \gg l$ the potential is given by

$$\phi = -\frac{c}{2\pi} \int_{-\infty}^{t-r/c} \frac{\dot{S}(t')dt'}{\sqrt{[c^2(t-t')^2 - r^2]}}. \qquad (73.15)$$

As $r \to 0$ the leading term of this expression is the same as (73.14), and the function $f(t)$ in the latter equation is automatically determined (we suppose that the derivative $\dot{S}(t)$ tends sufficiently rapidly to zero as $t \to -\infty$). For very large values of r, on the other hand (the "wave region"), the values of $t - t' \sim r/c$ are the most important in the integral (73.15). We can therefore put, in the denominator of the integrand,

$$(t-t')^2 - r^2/c^2 \approx (2r/c)(t-t'-r/c),$$

obtaining

$$\phi = -\frac{c}{2\pi\sqrt{(2r)}} \int_{-\infty}^{t-r/c} \frac{\dot{S}(t')dt'}{\sqrt{[c(t-t')-r]}}. \qquad (73.16)$$

Finally, the velocity $v = \partial\phi/\partial r$. To effect the differentiation, it is convenient to substitute $t - t' - r/c = \xi$:

$$\phi = -\frac{1}{2\pi}\sqrt{\frac{c}{2r}} \int_{0}^{\infty} \frac{\dot{S}(t-r/c-\xi)}{\sqrt{\xi}} d\xi;$$

the limits of integration are then independent of r. The factor $1/\sqrt{r}$ in front of the integral need not be differentiated, since this would give a term of higher order in $1/r$. Differentiating under the integral sign and then returning to the variable t', we obtain

$$v = \frac{1}{2\pi\sqrt{(2r)}} \int_{-\infty}^{t-r/c} \frac{\ddot{S}(t')dt'}{\sqrt{[c(t-t')-r]}}. \qquad (73.17)$$

The intensity is given by the product $2\pi r\rho c\overline{v^2}$. It should be noticed that here, unlike what happens for the spherical case, the intensity at any instant is determined by the behaviour of the function $S(t)$ at all times from $-\infty$ to $t-r/c$.

Finally, for translatory oscillations of an infinite cylinder in a direction perpendicular to its axis, the potential at distances r such that $l \ll r \ll \lambda$ has the form

$$\phi = \text{div}(\mathbf{A} \log fr), \qquad (73.18)$$

where $\mathbf{A}(t)$ is determined by solving Laplace's equation for the flow of an incompressible fluid past a cylinder. Hence we again conclude that, at all distances $r \gg l$,

$$\phi = -\text{div} \int_{-\infty}^{t-r/c} \frac{\mathbf{A}(t')dt'}{\sqrt{[(t-t')^2-r^2/c^2]}}. \qquad (73.19)$$

In conclusion, we must make the following remark. We have here entirely neglected the effect of the viscosity of the fluid, and accordingly have supposed that there is potential flow in the emitted wave. In reality, however, we do not have potential flow in a fluid layer of thickness $\sim \sqrt{(\nu/\omega)}$ round the oscillating body (see §24). Hence, if the above formulae are to be applicable, it is necessary that the thickness of this layer should be small in comparison with the dimension l of the body:

$$\sqrt{(\nu/\omega)} \ll l. \qquad (73.20)$$

This condition may not hold for small frequencies or small dimensions of the body.

PROBLEMS

PROBLEM 1. Determine the total intensity of sound emitted by a sphere executing small (harmonic) translatory oscillations of frequency ω, the wavelength being comparable in magnitude with the radius R of the sphere.

SOLUTION. We write the velocity of the sphere in the form $\mathbf{u} = \mathbf{u}_0 e^{-i\omega t}$; then ϕ depends on the time through a factor $e^{-i\omega t}$ also, and satisfies the equation $\triangle\phi+k^2\phi = 0$, where $k = \omega/c$. We seek a solution in the form $\phi = \mathbf{u} \cdot \text{grad} f(r)$, the origin being taken at the instantaneous position of the centre of the sphere. For f we obtain the equation $\mathbf{u} \cdot \text{grad}(\triangle f+k^2 f) = 0$, whence $\triangle f+k^2 f = $ constant. Apart from an unimportant additive constant, we therefore have $f = Ae^{ikr}/r$. The constant A is determined from the condition $\partial\phi/\partial r = u_r$ for $r = R$, and the result is

$$\phi = \mathbf{u} \cdot \mathbf{r} e^{ik(r-R)} \left(\frac{R}{r}\right)^3 \frac{ikr-1}{2-2ikR-k^2R^2}.$$

Thus we have dipole emission. At fairly large distances from the sphere, we can neglect unity in comparison with ikr, and ϕ takes the form (73.11), the vector $\dot{\mathbf{A}}$ being

$$\dot{\mathbf{A}} = -\mathbf{u}e^{ik(r-R)} R^3 \frac{i\omega}{2 - 2ikR - k^2R^2}.$$

Noticing that $\overline{(\text{re }\dot{\mathbf{A}})^2} = \frac{1}{2}|\dot{\mathbf{A}}|^2$, we obtain for the total emission, by (73.13),

$$I = \frac{2\pi\rho}{3c^3}|\mathbf{u}_0|^2 \frac{R^6\omega^4}{4 + (\omega R/c)^4}.$$

For $\omega R/c \ll 1$, this expression becomes $I = \pi\rho R^6|\mathbf{u}_0|^2\omega^4/6c^6$, a result which could also be obtained by directly substituting in (73.13) the expression $\mathbf{A} = \frac{1}{2}R^3\mathbf{u}$ from §11, Problem 1. For $\omega R/c \gg 1$ we have $I = 2\pi\rho cR^2|\mathbf{u}_0|^2/3$, corresponding to formula (73.4).

The drag force acting on the sphere is obtained by integrating over the surface of the sphere the component of the pressure forces ($p' = -\rho(\dot{\phi}')_{r=R}$) in the direction of $\mathbf{u}$, and is

$$\mathbf{F} = \frac{4\pi}{3}\rho\omega R^3\mathbf{u}\frac{-k^3R^3 + i(2 + k^2R^2)}{4 + k^4R^4};$$

see the end of §24 concerning the meaning of a complex drag force.

PROBLEM 2. The same as Problem 1, but for the case where the radius R of the sphere is comparable in magnitude with $\sqrt{(\nu/\omega)}$, whilst $\lambda \gg R$.

SOLUTION. If the dimension of the body is small compared with $\sqrt{(\nu/\omega)}$, then the emitted wave must be investigated not from the equation $\Delta\phi = 0$, but from the equation of motion of an incompressible viscous fluid. The appropriate solution of this equation for a sphere is given by formulae (1) and (2) in §24, Problem 5. At great distances the first term in (1), which diminishes exponentially with r, may be omitted. The second term gives the velocity $\mathbf{v} = -b(\mathbf{u} \cdot \mathbf{grad})\mathbf{grad}(1/r)$. Comparison with (73.6) shows that

$$\mathbf{A} = -b\mathbf{u} = \frac{1}{2}R^3[1 - 3/(i-1)\kappa - 3/2i\kappa^2]\mathbf{u},$$

where $\kappa = R\sqrt{(\omega/2\nu)}$, i.e. $\mathbf{A}$ differs from the corresponding expression for an ideal fluid by the factor in brackets. The result is

$$I = \frac{\pi\rho R^6}{6c^3}\omega^4\left(1 + \frac{3}{\kappa} + \frac{9}{2\kappa^2} + \frac{9}{2\kappa^3} + \frac{9}{4\kappa^4}\right)|\mathbf{u}_0|^2.$$

For $\kappa \gg 1$ this becomes the formula given in Problem 1, while for $\kappa \ll 1$ we obtain

$$I = 3\pi\rho R^2\nu^2\omega^2|\mathbf{u}_0|^2/2c^3,$$

i.e. the emission is proportional to the second, and not the fourth, power of the frequency.

PROBLEM 3. Determine the intensity of sound emitted by a sphere executing small (harmonic) pulsations of arbitrary frequency.

SOLUTION. We seek a solution of the form $\phi = (au/r)e^{ik(r-R)}$, R being the equilibrium radius of the sphere, and determine the constant a from the condition $[\partial\phi/\partial r]_{r=R} = u = u_0e^{-i\omega t}$ (where u is the radial velocity of points on the surface of the sphere):

$$a = R^2/(ikR - 1).$$

The intensity is $I = 2\pi\rho c|u_0|^2k^2R^4/(1 + k^2R^2)$. For $kR \ll 1$, $I = 2\pi\rho\omega^2R^4|u_0|^2/c$, in accordance with (73.10), while for $kR \gg 1$, $I = 2\pi\rho cR^2|u_0|^2$, in accordance with (73.4).

PROBLEM 4. Determine the nature of the wave emitted by a sphere (of radius R) executing small pulsations, when the radial velocity of points on the surface is an arbitrary function $u(t)$ of the time.

SOLUTION. We seek a solution in the form $\phi = f(t')/r$, where $t' = t-(r-R)/c$, and determine f from the boundary condition $\partial\phi/\partial r = u(t)$ for $r = R$. This gives the equation $df/dt+cf(t)/R = -Rcu(t)$. Solving this linear equation and replacing t by t' in the solution, we obtain

$$\phi(r, t') = -\frac{cR}{r}e^{-ct'/R}\int_{-\infty}^{t'} u(\tau)e^{c\tau/R}d\tau. \tag{1}$$

If the oscillations of the sphere cease at some instant, say $t = 0$ (i.e. $u(\tau) = 0$ for $\tau > 0$), then the potential at a distance r from the centre will be of the form $\phi = \text{constant} \times e^{-ct/R}$ after the instant $t = (r-R)/c$, i.e. it will diminish exponentially.

Let T be the time during which the velocity $u(t)$ changes appreciably. If $T \gg R/c$, i.e. if the wavelength of the emitted waves $\lambda \sim cT \gg R$, then we can take the slowly varying factor $u(\tau)$ outside the integral in (1), replacing it by $u(t')$. For distances $r \gg R$, we then obtain $\phi = -(R^2/r)u(t-r/c)$, in accordance with formula (73.8). If, on the other hand, $T \ll R/c$, we obtain in a similar manner

$$\phi = -\frac{cR}{r}\int_{-\infty}^{t'} u(\tau)d\tau, \quad v = \partial\phi/\partial r = (R/r)u(t'),$$

in accordance with formula (73.4).

PROBLEM 5. Determine the motion of an ideal compressible fluid when a sphere of radius R executes in it an arbitrary translatory motion, with velocity small compared with that of sound.

SOLUTION. We seek a solution in the form

$$\phi = \text{div}[\mathbf{f}(t')/r], \tag{1}$$

where r is the distance from the origin, taken at the position of the centre of the sphere at the time $t' = t-(r-R)/c$; since the velocity $\mathbf{u}$ of the sphere is small compared with the velocity of sound, the movement of the origin may be neglected. The fluid velocity is

$$\mathbf{v} = \text{grad}\,\phi = \frac{3(\mathbf{f}\cdot\mathbf{n})\mathbf{n}-\mathbf{f}}{r^3} + \frac{3(\mathbf{f}'\cdot\mathbf{n})\mathbf{n}-\mathbf{f}'}{cr^2} + \frac{(\mathbf{f}''\cdot\mathbf{n})\mathbf{n}}{c^2r}, \tag{2}$$

where $\mathbf{n}$ is a unit vector in the direction of $\mathbf{r}$, and the prime denotes differentiation with respect to the argument of $\mathbf{f}$. The boundary condition is $v_r = \mathbf{u}\cdot\mathbf{n}$ for $r = R$, whence $\mathbf{f}''(t)+(2c/R)\mathbf{f}'(t)+(2c^2/R^2)\mathbf{f}(t) = Rc^2u(t)$. Solving this equation by variation of the parameters, we obtain for the function $\mathbf{f}(t)$ the general expression

$$\mathbf{f}(t) = cR^2 e^{-ct/R}\int_{-\infty}^{t} \mathbf{u}(\tau)\sin\frac{c(t-\tau)}{R}e^{c\tau/R}\,d\tau. \tag{3}$$

In substituting in (1), we must replace t by t'. The lower limit is taken as $-\infty$ so that $\mathbf{f}$ shall be zero for $t = -\infty$.

PROBLEM 6. A sphere of radius R begins at time $t = 0$ to move with constant velocity $\mathbf{u}_0$. Determine the sound intensity emitted at the instant when the motion begins.

SOLUTION. Putting in formula (3) of Problem 5 $\mathbf{u}(\tau) = 0$ for $\tau < 0$ and $\mathbf{u}(\tau) = \mathbf{u}_0$ for $\tau > 0$, and substituting in formula (2) (retaining only the last term, which decreases least rapidly with r), we find the fluid velocity far from the sphere:

$$\mathbf{v} = -\mathbf{n}(\mathbf{n}\cdot\mathbf{u}_0)\frac{\sqrt{2}R}{r} e^{-ct'/R}\sin\left(\frac{ct'}{R} - \tfrac{1}{4}\pi\right),$$

where $t' > 0$. The total intensity diminishes with time according to

$$I = (8\pi/3)c\rho R^2 u_0^2 e^{-2ct'/R} \sin^2(ct'/R - \tfrac{1}{4}\pi).$$

The total amount of energy emitted is $\tfrac{1}{3}\pi\rho R^3 u_0^2$.

PROBLEM 7. Determine the intensity of sound emitted by an infinite cylinder, of radius R, executing harmonic pulsations of wavelength $\lambda \gg R$.

SOLUTION. According to formula (73.14), we find first of all that, at distances $r \ll \lambda$ (in Problems 7 and 8 r is the distance from the axis of the cylinder), the potential is $\phi = Ru \log kr$, where $u = u_0 e^{-i\omega t}$ is the velocity of points on the surface of the cylinder. From a comparison with formulae (70.7) and (70.8), we now find that at large distances the potential is of the form $\phi = -Ru\sqrt{(i\pi/2kr)}e^{ikr}$. The velocity is therefore

$$\mathbf{v} = Ru\sqrt{(\pi k/2ir)}\mathbf{n}e^{ikr},$$

where $\mathbf{n}$ is a unit vector perpendicular to the axis of the cylinder, and the intensity per unit length of the cylinder is $I = \tfrac{1}{2}\pi^2 \rho\omega R^2 u_0^2$.

PROBLEM 8. Determine the intensity of sound emitted by a cylinder executing harmonic translatory oscillations in a direction perpendicular to its axis.

SOLUTION. At distances $r \ll \lambda$ we have $\phi = -\mathrm{div}(R^2\mathbf{u} \log kr)$; cf. formula (73.18) and §10, Problem 3. Hence we conclude that at large distances

$$\phi = R^2\sqrt{(i\pi/2k)}\mathrm{div}(e^{ikr}\mathbf{u}/\sqrt{r}) = -R^2\mathbf{u}\cdot\mathbf{n}\sqrt{(\pi k/2ir)}e^{ikr},$$

whence the velocity is $\mathbf{v} = -kR^2\sqrt{(i\pi k/2r)}\mathbf{n}(\mathbf{u}\cdot\mathbf{n})e^{ikr}$. The intensity is proportional to the squared cosine of the angle between the directions of oscillation and emission. The total intensity is $I = (\pi^2/4c^2)\rho\omega^3 R^4|\mathbf{u}_0|^2$.

PROBLEM 9. Determine the intensity of sound emitted by a plane surface whose temperature varies periodically with frequency $\omega \ll c^2/\chi$, where χ is the thermometric conductivity of the fluid.

SOLUTION. Let the variable part of the temperature of the surface be $T'_0 e^{-i\omega t}$. These temperature oscillations cause a damped thermal wave in the fluid (52.17):

$$T' = T'_0 e^{-i\omega t}e^{-(1-i)\sqrt{(\omega/2\chi)}x},$$

and the fluid density therefore oscillates also: $\rho' = (\partial\rho/\partial T)_p T' = -\rho\beta T'$, where β is the coefficient of thermal expansion. This, in turn, results in the occurrence of a motion determined by the equation of continuity: $\rho\,\partial v/\partial x = -\partial\rho'/\partial t = -i\omega\rho\beta T'$. At the solid surface the velocity $v_x = v = 0$, and far from the surface it tends to the limit

$$v = -i\omega\beta \int_0^\infty T'\,dx = \frac{1-i}{\sqrt{2}}\beta\sqrt{(\omega\chi)}T'_0 e^{-i\omega t}.$$

This value is reached at distances $\sim\sqrt{(\chi/\omega)}$, which are small compared with c/ω, and we thus have a boundary condition on the resulting sound wave. Hence we find the intensity per unit area of the surface to be $I = \tfrac{1}{2}c\rho\beta^2\omega\chi|T'_0|^2$.

PROBLEM 10. A point source emitting a spherical wave is at a distance l from a solid wall which totally reflects sound and bounds a half-space occupied by fluid. Determine the ratio of the total intensity of sound emitted by the source to that which would be found in an infinite medium, and the dependence of the intensity on direction for large distances from the source.

SOLUTION. The sum of the direct and reflected waves is given by a solution of the wave equation such that the normal velocity component $v_n = \partial\phi/\partial n$ is zero at the wall. Such a solution is

$$\phi = \left(\frac{e^{ikr}}{r} + \frac{e^{ikr'}}{r'}\right)e^{-i\omega}$$

(we omit the constant factor, for brevity), where r is the distance from the source O (Fig. 39), and r' is the distance from a point O' which is the image of O in the wall. At large distances from the source we have $r' \approx r - 2l \cos \theta$, so that

$$\phi = \frac{e^{i(kr-\omega t)}}{r}(1 + e^{-2ikl\cos\theta}).$$

The dependence of the intensity on direction is given by a factor $\cos^2(kl \cos \theta)$.

To determine the total intensity, we integrate the energy flux $\bar{\mathbf{q}} = \overline{p'\mathbf{v}} = -\overline{\rho\dot{\phi}\,\mathbf{grad}\,\phi}$ (see (64.4)) over the surface of a sphere of arbitrarily small radius, centred at O. This gives $2\pi\rho k\omega(1 + [1/2kl] \sin 2kl)$. In an infinite medium, on the other hand, we should have simply a spherical wave $\phi = e^{i(kr-\omega t)}/r$, with a total energy flux $2\pi\rho k\omega$. Thus the required ratio of intensities is $1 + (1/2kl) \sin 2kl$.

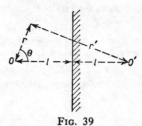

FIG. 39

PROBLEM 11. The same as Problem 10, but for a fluid bounded by a free surface.

SOLUTION. At the free surface the condition $p' = -\rho\dot{\phi} = 0$ must hold; in a monochromatic wave this is equivalent to $\phi = 0$. The corresponding solution of the wave equation is

$$\phi = \left(\frac{e^{ikr}}{r} - \frac{e^{ikr'}}{r'}\right)e^{-i\omega t}.$$

At large distances from the source, the intensity is given by a factor $\sin^2(kl \cos \theta)$. The required ratio of intensities is $1 - (1/2kl) \sin 2kl$.

§74. The reciprocity principle

In deriving the equations of a sound wave in §63, it was assumed that the wave is propagated in a homogeneous medium. In particular, the density ρ_0 of the medium and the velocity of sound in it, c, were regarded as constants. In order to obtain some general relations applicable for an arbitrary inhomogeneous medium, we shall first derive the equation for the propagation of sound in such a medium.

We write the equation of continuity in the form $d\rho/dt + \rho \, \text{div} \, \mathbf{v} = 0$. Since the propagation of sound is adiabatic, we have

$$\frac{d\rho}{dt} = \left(\frac{\partial\rho}{\partial p}\right)_s \frac{dp}{dt} = \frac{1}{c^2}\frac{dp}{dt} = \frac{1}{c^2}\left(\frac{\partial p}{\partial t} + \mathbf{v}\cdot\mathbf{grad}\,p\right),$$

and the equation of continuity becomes $\partial p/\partial t + \mathbf{v}\cdot\mathbf{grad}\,p + \rho c^2 \, \text{div} \, \mathbf{v} = 0$.

As usual, we put $\rho = \rho_0 + \rho'$, where ρ_0 is now a given function of the co-ordinates. In the equation $p = p_0 + p'$, however, we must put as before

p_0 = constant, since the pressure must be constant throughout a medium in equilibrium (in the absence of an external field, of course). Thus we have to within second-order quantities $\partial p'/\partial t + \rho_0 c^2 \operatorname{div} \mathbf{v} = 0$.

This equation is the same in form as equation (63.5), but the coefficient $\rho_0 c^2$ is a function of the co-ordinates. As in §63, we obtain Euler's equation in the form $\partial \mathbf{v}/\partial t = -(1/\rho_0) \operatorname{\mathbf{grad}} p'$. Eliminating $\mathbf{v}$, and omitting the suffix in ρ_0, we finally obtain the equation of propagation of sound in an inhomogeneous medium:

$$\operatorname{div} \frac{\operatorname{\mathbf{grad}} p'}{\rho} - \frac{1}{\rho c^2} \frac{\partial^2 p'}{\partial t^2} = 0. \tag{74.1}$$

If the wave is monochromatic, with frequency ω, we have $\ddot{p}' = -\omega^2 p'$, so that

$$\operatorname{div} \frac{\operatorname{\mathbf{grad}} p'}{\rho} + \frac{\omega^2}{\rho c^2} p' = 0. \tag{74.2}$$

Let us consider a sound wave emitted by a pulsating source of small dimension; we have seen in §73 that the emission is isotropic. We denote by A the point where the source is, and by $p_A(B)$ the pressure p' at a point B in the emitted wave.† If the same source is placed at B, it produces at A a pressure which we denote by $p_B(A)$. We shall derive the relation between $p_A(B)$ and $p_B(A)$.

To do so, we use equation (74.2), applying it first to the sound from a source at A and then to the sound from a source at B:

$$\operatorname{div} \frac{\operatorname{\mathbf{grad}} p'_A}{\rho} + \frac{\omega^2}{\rho c^2} p'_A = 0, \quad \operatorname{div} \frac{\operatorname{\mathbf{grad}} p'_B}{\rho} + \frac{\omega^2}{\rho c^2} p'_B = 0.$$

We multiply the first equation by p'_B and the second by p'_A and subtract. The result is

$$p'_B \operatorname{div} \frac{\operatorname{\mathbf{grad}} p'_A}{\rho} - p'_A \operatorname{div} \frac{\operatorname{\mathbf{grad}} p'_B}{\rho}$$

$$= \operatorname{div} \left(\frac{p'_B \operatorname{\mathbf{grad}} p'_A}{\rho} - \frac{p'_A \operatorname{\mathbf{grad}} p'_B}{\rho} \right) = 0.$$

We integrate this equation over the volume between an infinitely distant closed surface C and two small spheres C_A and C_B which enclose the points A and B respectively. The volume integral can be transformed into three surface integrals, and the integral over C is zero, since the sound field vanishes at infinity. Thus we obtain

$$\oint_{C_A + C_B} \left(p'_B \frac{\operatorname{\mathbf{grad}} p'_A}{\rho} - p'_A \frac{\operatorname{\mathbf{grad}} p'_B}{\rho} \right) \cdot \mathbf{df} = 0. \tag{74.3}$$

---

† The dimension of the source must be small compared with the distance between A and B and with the wavelength.

Inside the small sphere C_A, the pressure p'_A in the wave from a source at A falls off rapidly with the distance from A, and the gradient **grad** p'_A is therefore large. The pressure p'_B due to a source at B is a slowly varying function of the co-ordinates in the region near the point A, which is at a considerable distance from B, so that the gradient **grad** p'_B is relatively small. When the radius of the sphere C_A is sufficiently small, therefore, we can neglect the integral

$$\oint (p'_A/\rho) \, \mathbf{grad} \, p'_B \cdot d\mathbf{f}$$

over C_A in comparison with

$$\oint (p'_B/\rho) \, \mathbf{grad} \, p'_A \cdot d\mathbf{f},$$

and in the latter the almost constant quantity p'_B can be taken outside the integral and replaced by its value at the point A. Similar arguments hold for the integrals over the sphere C_B, and as a result we obtain from (74.3) the relation

$$p'_B(A) \oint_{C_A} \frac{\mathbf{grad} \, p'_A}{\rho} \cdot d\mathbf{f} = p'_A(B) \oint_{C_B} \frac{\mathbf{grad} \, p'_B}{\rho} \cdot d\mathbf{f}.$$

But $(1/\rho) \, \mathbf{grad} \, p' = -\partial \mathbf{v}/\partial t$, and this equation can therefore be rewritten

$$p'_B(A) \frac{\partial}{\partial t} \oint_{C_A} \mathbf{v}_A \cdot d\mathbf{f} = p'_A(B) \frac{\partial}{\partial t} \oint_{C_B} \mathbf{v}_B \cdot d\mathbf{f}.$$

The integral

$$\oint \mathbf{v}_A \cdot d\mathbf{f}$$

over C_A is the volume of fluid flowing per unit time through the surface of the sphere C_A, i.e. it is the rate of change of the volume of the pulsating source of sound. Since the sources at A and B are identical, it is clear that

$$\oint_{C_A} \mathbf{v}_A \cdot d\mathbf{f} = \oint_{C_B} \mathbf{v}_B \cdot d\mathbf{f},$$

and consequently

$$p'_A(B) = p'_B(A). \tag{74.4}$$

This equation constitutes the *reciprocity principle*: the pressure at B due to a source at A is equal to the pressure at A due to a similar source at B. It should be emphasised that this result holds, in particular, for the case where the medium is composed of several different regions, each of which is homogeneous. When sound is propagated in such a medium, it is reflected and refracted at the surfaces separating the various regions. Thus the reciprocity principle is valid also in cases where the wave undergoes reflection and refraction on its path from A to B.

PROBLEM

Derive the reciprocity principle for dipole emission of sound by a source which oscillates without change of volume.

SOLUTION. In this case the integral

$$\oint \mathbf{v}_A \cdot \mathbf{df}$$

over C_A is zero identically, and the next approximation must be taken in calculating the integrals in (74.3). To do so, we write, as far as the first-order terms,

$$p'_B = p'_B(A) + \mathbf{r} \cdot \mathbf{grad}\, p'_B,$$

where $\mathbf{r}$ is the radius vector from A. In the integral

$$\oint_{C_A} \left(p'_B \frac{\mathbf{grad}\, p'_A}{\rho} - p'_A \frac{\mathbf{grad}\, p'_B}{\rho} \right) \cdot \mathbf{df}, \tag{1}$$

the two terms are now of the same order of magnitude. Substituting here for p'_B the above expansion, and using the fact that the integral

$$\oint (1/\rho)\, \mathbf{grad}\, p'_A \cdot \mathbf{df}$$

over C_A is now zero, we obtain

$$\oint_{C_A} \left\{ (\mathbf{r} \cdot \mathbf{grad}\, p'_B) \frac{\mathbf{grad}\, p'_A}{\rho} - p'_A \frac{\mathbf{grad}\, p'_B}{\rho} \right\} \cdot \mathbf{df}.$$

Next, we take the almost constant quantity $\mathbf{grad}\, p'_B = -\rho \dot{\mathbf{v}}_B$ outside the integral, replacing it by its value at A:

$$\rho_A \dot{\mathbf{v}}_B(A) \cdot \oint_{C_A} \left\{ \frac{p'_A}{\rho} \mathbf{df} - \mathbf{r} \left(\frac{\mathbf{grad}\, p'_A}{\rho} \cdot \mathbf{df} \right) \right\},$$

where ρ_A is the density of the medium at the point A. To calculate this integral, we notice that near a source the fluid can be supposed incompressible (see §73), and hence we can write for the pressure inside the small sphere C_A, by (11.1), $p'_A = -\rho\phi = \rho\mathbf{A} \cdot \mathbf{r}/r^3$. In a monochromatic wave $\dot{\mathbf{v}} = -i\omega\mathbf{v}$, $\dot{\mathbf{A}} = -i\omega\mathbf{A}$; introducing also the unit vector $\mathbf{n}_A$ in the direction of the vector $\mathbf{A}$ for a source at A, we find that the integral (1) is proportional to $\rho_A \mathbf{v}_B(A) \cdot \mathbf{n}_A$. Similarly, the integral over the sphere C_B is proportional to $-\rho_B \mathbf{v}_A(B) \cdot \mathbf{n}_B$, with the same factor of proportionality. Equating the sum to zero, we find the required relation

$$\rho_A \mathbf{v}_B(A) \cdot \mathbf{n}_A = \rho_B \mathbf{v}_A(B) \cdot \mathbf{n}_B,$$

which expresses the reciprocity principle for dipole emission of sound.

§75. Propagation of sound in a tube

Let us now consider the propagation of a sound wave in a long narrow tube. By a "narrow" tube we mean one whose width is small compared with the wavelength. The cross-section of the tube may vary along its length in both shape and area. It is important, however, that this variation should occur fairly slowly: the cross-sectional area S must vary only slightly over distances of the order of the width of the tube.

Under these conditions we can suppose that all quantities (velocity, density, etc.) are constant over any transverse cross-section of the tube. The direction of propagation of the wave can be supposed to coincide with that of the axis of the tube at all points. The equation for the propagation of such a wave is most conveniently derived by a method similar to that used in §13 in deriving the equation for the propagation of gravity waves in channels.

In unit time a mass $S\rho v$ of fluid passes through a cross-section of the tube. Hence the mass of fluid in the volume between two transverse cross-sections at a distance dx apart decreases in unit time by

$$(S\rho v)_{x+dx} - (S\rho v)_x = [\partial(S\rho v)/\partial x]dx,$$

the co-ordinate x being measured along the axis of the tube. Since the volume between the two cross-sections remains constant, the decrease must be due only to the change in density of the fluid. The change in density per unit time is $\partial\rho/\partial t$, and the corresponding decrease in the mass of fluid in the volume $S\,dx$ between the two cross-sections is $-S(\partial\rho/\partial t)dx$. Equating the two expressions, we obtain

$$S\partial\rho/\partial t = -\partial(S\rho v)/\partial x, \tag{75.1}$$

which is the "equation of continuity" for flow in a pipe.

Next, we write down Euler's equation, omitting the term quadratic in the velocity:

$$\partial v/\partial t = -(1/\rho)\partial p/\partial x. \tag{75.2}$$

We differentiate (75.1) with respect to time, regarding ρ on the right-hand side as independent of time, since the differentiation of ρ gives a term which involves $v\,\partial\rho/\partial t = v\,\partial\rho'/\partial t$ and is therefore of the second order of smallness. Thus $S\,\partial^2\rho/\partial t^2 = -\partial(S\rho\partial v/\partial t)/\partial x$. Here we substitute the expression (75.2) for $\partial v/\partial t$, and express the derivative of the density on the left-hand side in terms of the derivative of the pressure by $\ddot\rho = (\partial\rho/\partial p)\ddot p = \ddot p/c^2$.

The result is the following equation for the propagation of sound in a tube:

$$\frac{1}{S}\frac{\partial}{\partial x}\left(S\frac{\partial p}{\partial x}\right) - \frac{1}{c^2}\frac{\partial^2 p}{\partial t^2} = 0. \tag{75.3}$$

In a monochromatic wave p depends on time through a factor $e^{-i\omega t}$, and (75.3) becomes

$$\frac{1}{S}\frac{\partial}{\partial x}\left(S\frac{\partial p}{\partial x}\right) + k^2 p = 0. \tag{75.4}$$

where $k = \omega/c$ is the wave number.†

---

† Here, and in the Problems, p denotes the variable part of the pressure, which we have previously denoted by p'.

Finally, let us consider the problem of the emission of sound from the open end of a tube. The pressure difference between the gas in the end of the tube and that in the space surrounding the tube is small compared with the pressure differences within the tube. Hence the boundary condition at the open end of the tube is, with sufficient accuracy, that the pressure p should vanish. The gas velocity v at the end of the tube is not zero; let its value be v_0. The product Sv_0 is the volume of gas leaving the tube per unit time.

We can now regard the open end of the tube as a "source" of gas of strength Sv_0. The problem of the emission from a tube thus becomes equivalent to that of the emission by a pulsating body, which is solved by formula (73.10). In place of the time derivative $\dot{V}$ of the volume of the body we must now put Sv_0. Thus the total intensity of the sound emitted is

$$I = \rho S^2 \overline{\dot{v}_0{}^2}/4\pi c. \tag{75.5}$$

PROBLEMS

PROBLEM 1. Determine the transmission coefficient for sound passing from a tube of cross-section S_1 into one of cross-section S_2.

SOLUTION. In the first tube we have two waves, the incident wave $p_1 = a_1 e^{i(kx-\omega t)}$ and the reflected wave $p_1' = a_1' e^{-i(kx+\omega t)}$. In the second tube we have the transmitted wave $p_2 = a_2 e^{i(kx-\omega t)}$. At the point where the tubes join ($x = 0$), the pressures must be equal, and so must the volumes Sv of gas passing from one tube to the other per unit time. These conditions give $a_1 + a_1' = a_2$, $S_1(a_1 - a_1') = S_2 a_2$, whence $a_2 = 2a_1 S_1/(S_1 + S_2)$. The ratio D of the energy flux in the transmitted wave to that in the incident wave is

$$D = S_2 \rho c \overline{|v_2|^2}/S_1 \rho c \overline{|v_1|^2} = S_2 \overline{|v_2|^2}/S_1 \overline{|v_1|^2}$$

or

$$D = \frac{4S_1 S_2}{(S_1 + S_2)^2} = 1 - \left(\frac{S_2 - S_1}{S_2 + S_1}\right)^2.$$

PROBLEM 2. Determine the amount of energy emitted from the open end of a cylindrical tube.

SOLUTION. In the boundary condition $p = 0$ at the open end of the tube, we can approximately neglect the emitted wave (we shall see that the intensity emitted from the end of the tube is small). Then we have the condition $p_1 = -p_1'$, where p_1 and p_1' are the pressures in the incident wave and in the wave reflected back into the tube; for the velocities we have correspondingly $v_1 = v_1'$, so that the total velocity at the end of the tube is $v_0 = v_1 + v_1' = 2v_1$. The energy flux in the incident wave is $cS\rho \overline{v_1{}^2} = \frac{1}{4}cS\rho \overline{v_0{}^2}$. Using (75.5), we obtain for the ratio of the emitted energy to the energy flux in the incident wave $D = S\omega^2/\pi c^2$. For a tube of circular cross-section (radius R) we have $D = R^2 \omega^2/c^2$. Since, by hypothesis, $R \ll c/\omega$, it follows that $D \ll 1$.

PROBLEM 3. One of the ends of a cylindrical pipe is covered by a membrane which executes a given oscillation and emits sound; the other end is open. Determine the way in which sound is emitted from the tube.

SOLUTION. In the general solution

$$p = (ae^{ikx} + be^{-ikx})e^{-i\omega t}$$

we determine the constants a and b from the conditions $v = u = u_0 e^{-i\omega t}$, the given velocity of the membrane, at the closed end ($x = 0$), and $p = 0$ at the open end ($x = l$). These give

$ae^{ikl}+be^{-ikl}=0$, $a-b=c\rho u_0$. Determining a and b, we find the gas velocity at the open end of the tube to be $v_0 = u/\cos kl$. If the tube were absent, the intensity of the sound emitted by the oscillating membrane would be given by the mean square $S^2|\dot{u}|^2 = S^2\omega^2|u|^2$, according to formula (73.10) with Su in place of V; S is the cross-sectional area of the membrane. The emission from the end of the tube is proportional to $S^2|\overline{v_0}|^2\omega^2$. Defining the "amplification coefficient" of the pipe as $A = S^2|\overline{v_0}|^2/S^2|\overline{u}|^2$, we obtain $A = 1/\cos^2 kl$. This becomes infinite for frequencies of oscillation of the membrane equal to the characteristic frequencies of the tube (*resonance*); in reality, of course, it remains finite because of effects which we have neglected (such as friction due to the emission of sound).

PROBLEM 4. The same as Problem 3, but for a conical tube, with the membrane covering the smaller end.

SOLUTION. The cross-section of the tube is $S = S_0 x^2$; let the values of the co-ordinate x which correspond to the smaller and larger ends be x_1, x_2, so that the length of the tube is $l = x_2 - x_1$. The general solution of equation (75.4) is $p = (1/x)(ae^{ikx}+be^{-ikx})e^{-i\omega t}$; a and b are determined from the conditions $v = u$ for $x = x_1$ and $p = 0$ for $x = x_2$. The amplification coefficient is found to be

$$A = \frac{S_0 x_2^2 |\overline{v_2}|^2}{S_0 x_1^2 |\overline{u}|^2} = \frac{k^2 x_1^2}{(\sin kl + kx_1 \cos kl)^2}.$$

PROBLEM 5. The same as Problem 3, but for a tube whose cross-section varies exponentially along its length: $S = S_0 e^{\alpha x}$.

SOLUTION. Equation (75.4) becomes $\partial^2 p/\partial x^2 + \alpha \partial p/\partial x + k^2 p = 0$, whence

$$p = e^{-\frac{1}{2}\alpha x}(ae^{imx} + be^{-imx})e^{-i\omega t},$$

with $m = \sqrt{(k^2 - \frac{1}{4}\alpha^2)}$. Determining a and b from the conditions $v = u$ for $x = 0$ and $p = 0$ for $x = l$, we find the amplification coefficient

$$A = \frac{S_0 e^{\alpha l} |\overline{v_0}|^2}{S_0 |\overline{u}|^2} = \frac{e^{\alpha l}}{[\frac{1}{2}(\alpha/m) \sin ml + \cos ml]^2}$$

for $k > \frac{1}{2}\alpha$ and

$$A = \frac{e^{\alpha}}{[\frac{1}{2}(\alpha/m') \sinh m'l + \cosh m'l]^2}, \qquad m' = \sqrt{(\frac{1}{4}\alpha^2 - k^2)},$$

for $k < \frac{1}{2}\alpha$.

§76. Scattering of sound

If there is some body in the path of propagation of a sound wave, then the sound is *scattered*: besides the incident wave there appear other (scattered) waves, which are propagated in all directions from the scattering body. The scattering of a sound wave occurs simply on account of the presence of the body in its path. In addition, the incident wave causes the body itself to move, and this in turn brings about additional emission of sound by the body, i.e. further scattering. If, however, the density of the body is large compared with that of the medium in which the sound is propagated, and its compressibility is small, then the scattering due to the motion of the body forms only a small correction to the main scattering caused by the mere presence of the body. In what follows we shall neglect this correction, and therefore suppose the scattering body immovable.

We assume that the wavelength λ of the sound is large compared with the dimension l of the body; to calculate the properties of the scattered wave, we can then use formulae (73.8) and (73.11).† In doing so, we regard the scattered wave as being emitted by the body; the only difference is that, instead of a motion of the body in the fluid, we now have a motion of the fluid relative to the body. The two problems are clearly equivalent.

For the potential of the emitted wave we have obtained the expression $\phi = -\dot{V}/4\pi r - \dot{\mathbf{A}}\cdot\mathbf{r}/cr^2$. In this formula V was the volume of the body. In the present case, however, the volume of the body itself remains unchanged, and $\dot{V}$ must be taken not as the rate of change of the volume of the body, but as the volume of fluid which would enter, per unit time, the volume V_0 occupied by the body if the body were absent. For, in the presence of the body, this volume of fluid does not penetrate into V_0, which is equivalent to the emission of the same volume of fluid from V_0. The coefficient of $1/4\pi r$ in the first term of ϕ must, as we have seen in §73, be just the volume of fluid emitted from the origin per unit time. This volume is easily found. The change per unit time in the mass of fluid in a volume equal to that of the body is $V_0\dot{\rho}$, where $\dot{\rho}$ gives the rate of change of the fluid density in the incident sound wave (since the wavelength is large compared with the dimension of the body, the density ρ may be supposed constant over distances of the order of this dimension; hence we can write the rate of change of the mass of fluid in V_0 as $V_0\dot{\rho}$ simply, where $\dot{\rho}$ is the same throughout the volume V_0). The change in volume corresponding to a mass change $V_0\dot{\rho}$ is evidently $V_0\dot{\rho}/\rho$. Thus $\dot{V}$ in the expression for ϕ must be replaced by $V_0\dot{\rho}/\rho$. In an incident plane wave, the variable part ρ' of the density is related to the velocity by $\rho' = \rho v/c$; hence $\dot{\rho} = \dot{\rho}' = \rho\dot{v}/c$, and we can replace $V_0\dot{\rho}/\rho$ by $V_0\dot{v}/c$.

When the body moves in the fluid, the vector $\mathbf{A}$ is determined by formulae (11.5), (11.6): $4\pi\rho A_i = m_{ik}u_k + \rho V_0 u_i$. We must now replace the velocity $\mathbf{u}$ of the body by the reversed velocity $\mathbf{v}$ of the fluid in the incident wave which it would have at the position of the body if the latter were absent. Thus

$$A_i = -m_{ik}v_k/4\pi\rho - V_0 v_i/4\pi. \tag{76.1}$$

We finally obtain for the potential of the scattered wave

$$\phi_{\mathrm{sc}} = -V_0\dot{v}/4\pi cr - \dot{\mathbf{A}}\cdot\mathbf{r}/cr^2, \tag{76.2}$$

the vector $\mathbf{A}$ being given by formula (76.1). Hence we have for the velocity distribution in the scattered wave

$$\mathbf{v}_{\mathrm{sc}} = V_0\ddot{v}\mathbf{n}/4\pi rc^2 + \mathbf{n}(\mathbf{n}\cdot\ddot{\mathbf{A}})/rc^2 \tag{76.3}$$

(see §73), $\mathbf{n}$ being a unit vector in the direction of scattering.

The mean amount of energy scattered per unit time into a given solid angle element do is given by the energy flux, which is $c\rho\overline{\mathbf{v}_{\mathrm{sc}}^2}r^2 do$. The total scattered intensity I_{sc} is obtained by integrating this expression over all directions.

---

† At the same time, the dimension of the body must be large in comparison with the displacement amplitude of fluid particles in the wave, since otherwise the fluid is not in general in potential flow.

The integration of twice the product of the two terms in (76.3) gives zero, since this product is proportional to the cosine of the angle between the direction of scattering and the direction of propagation of the incident wave, and there remains (cf. (73.10) and (73.13))

$$I_{sc} = \frac{V_0^2 \rho}{4\pi c^3}\overline{\ddot{v}^2} + \frac{4\pi \rho}{3c^3}\overline{\ddot{\mathbf{A}}^2}. \tag{76.4}$$

The scattering is generally characterised by what is called the *effective cross-section* $d\sigma$, which is the ratio of the (time) average energy scattered into a given solid-angle element to the mean energy flux density in the incident wave. The *total effective cross-section* σ is the integral of $d\sigma$ over all directions of scattering, i.e. it is the ratio of the total scattered intensity to the incident energy flux density, and evidently has the dimensions of area.

The mean energy flux density in the incident wave is $c\rho \overline{\mathbf{v}^2}$. Hence the differential effective scattering cross-section is $(c\rho\overline{\mathbf{v}_{sc}^2}/c\rho\overline{\mathbf{v}^2})r^2 do$, i.e.

$$d\sigma = (\overline{\mathbf{v}_{sc}^2}/\overline{\mathbf{v}^2})r^2 \, do. \tag{76.5}$$

The total effective cross-section is

$$\sigma = \frac{V_0^2}{4\pi c^4}\frac{\overline{\ddot{\mathbf{v}}^2}}{\overline{\mathbf{v}^2}} + \frac{4\pi}{3c^4}\frac{\overline{\ddot{\mathbf{A}}^2}}{\overline{\mathbf{v}^2}}. \tag{76.6}$$

For a monochromatic incident wave, the mean square second time derivative of the velocity is proportional to the fourth power of the frequency. Thus the effective cross-section for the scattering of sound by a body which is small compared with the wavelength is proportional to ω^4.

Finally, let us briefly discuss the opposite limiting case, where the wavelength of the scattered sound is small compared with the dimension of the body. In this case all the scattering, except for the scattering through very small angles, amounts to simple reflection from the surface of the body. The corresponding part of the total effective scattering cross-section is clearly equal to the area S of the cross-section of the body by a plane perpendicular to the direction of the incident wave. The scattering through small angles (of the order of λ/l), however, constitutes *diffraction* from the edges of the body. We shall not pause here to expound the theory of this phenomenon, which is entirely analogous to that of the diffraction of light.† We shall only mention that, by Babinet's principle, the total intensity of diffracted sound is equal to the total intensity of reflected sound. Hence the diffraction part of the effective scattering cross-section is also equal to S, and the total cross-section is therefore $2S$.

PROBLEMS

PROBLEM 1. Determine the effective cross-section for the scattering of a plane sound wave by a solid sphere of radius R small compared with the wavelength.

---

† See *The Classical Theory of Fields*, §§7–7 to 7–9.

SOLUTION. The velocity at a given point in a plane wave is $v = a \cos \omega t$. In the case of a sphere (see §11, Problem 1), the vector $\mathbf{A}$ is $-\frac{1}{2}R^3\mathbf{v}$. For the differential effective cross-section we obtain

$$d\sigma = \frac{\omega^4 R^6}{9c^4}(1 - \tfrac{3}{2}\cos\theta)^2\, do,$$

where θ is the angle between the direction of the incident wave and the direction of scattering. The scattered intensity is greatest in the direction $\theta = \pi$, which is opposite to the direction of incidence. The total effective cross-section is

$$\sigma = (7\pi/9)(R^3\omega^2/c^2)^2. \qquad (1)$$

Here (and also in Problems 3 and 4 below) it is assumed that the density ρ_0 of the sphere is large compared with the density ρ of the gas; if this were not so, it would be necessary to take account of the movement of the sphere by the pressure forces exerted on it by the oscillating gas.

PROBLEM 2. Determine the effective cross-section for the scattering of sound by a drop of fluid, taking into account the compressibility of the fluid and the motion of the drop caused by the incident wave.

SOLUTION. When the pressure of the gas surrounding the drop changes adiabatically by p', the volume of the drop is reduced by $(V_0/\rho_0)(\partial\rho_0/\partial p)_s p'$, where ρ_0 is the density of the drop. But $(\partial p/\partial\rho_0)_s$ is the square of the velocity of sound c_0 in the fluid, and the pressure change in a plane wave is related to the velocity by $p' = vc\rho$, where ρ is the density of the gas. Thus the decrease in the volume of the drop is $V_0\dot{v}c\rho/c_0^2\rho_0$ per unit time. In the expressions (76.2) and (76.3), we must now replace $V_0\dot{v}/c$ by the difference $V_0(\dot{v}/c - \dot{v}c\rho/c_0^2\rho_0)$. Moreover, in the expression for $\mathbf{A}$ we must replace $-\mathbf{v}$ by the difference $\mathbf{u}-\mathbf{v}$, where $\mathbf{u}$ is the velocity acquired by the drop as a result of the action of the incident wave. For a sphere we have, using the results of §11, Problem 1, $\mathbf{A} = R^3\mathbf{v}(\rho - \rho_0)/(2\rho_0 + \rho)$. Substituting these expressions, we have the effective cross-section

$$d\sigma = \frac{\omega^4 R^6}{9c^4}\left\{\left(1 - \frac{c^2\rho}{c_0^2\rho_0}\right) - 3\cos\theta\frac{\rho_0 - \rho}{2\rho_0 + \rho}\right\}^2\, do.$$

The total effective cross-section is

$$\sigma = \frac{4\pi\omega^4 R^6}{9c^4}\left\{\left(1 - \frac{c^2\rho}{c_0^2\rho_0}\right)^2 + \frac{3(\rho_0 - \rho)^2}{(2\rho_0 + \rho)^2}\right\}.$$

PROBLEM 3. Determine the effective cross-section for the scattering of sound by a solid sphere of radius R small compared with $\sqrt{(\nu/\omega)}$. The specific heat of the sphere is supposed so large that its temperature can be regarded as a constant.

SOLUTION. In this case we have to take into account the effect of the gas viscosity on the motion of the sphere, and the vector $\mathbf{A}$ must be modified as shown in §73, Problem 2. For $R\sqrt{(\omega/\nu)} \ll 1$ we have $\mathbf{A} = -3iR\nu\mathbf{v}/2\omega$.

The thermal conductivity of the gas also results in scattering of the same order. Let $T'_0 e^{-i\omega t}$ be the temperature variation at a given point in the sound wave. The temperature distribution near a sphere is (see §52, Problem 2)

$$T' = T'_0 e^{-i\omega t}[1 - (R/r)e^{-(1-i)(r-R)\sqrt{(\omega/2\chi)}}]$$

(for $r = R$ we must have $T' = 0$). The amount of heat transferred from the gas to the sphere per unit time is (for $R\sqrt{(\omega/\chi)} \ll 1$) $q = 4\pi R^2\kappa[dT'/dr]_{r=R} = 4\pi R\kappa T'_0 e^{-i\omega t}$. This transfer of heat results in a change in the volume of the gas, which can be taken to affect the scattering like a corresponding effective change in the volume of the sphere, $\dot{V} = -4\pi R\chi\beta T'_0 e^{-i\omega t} = -4\pi R\chi(\gamma - 1)v/c$, where β is the coefficient of thermal expansion of the gas and $\gamma = c_p/c_v$; we have used also formulae (63.13) and (77.2).

Taking account of both effects, we obtain the differential effective scattering cross-section

$$d\sigma = (\omega R/c^2)^2 [\chi(\gamma-1) - \tfrac{3}{2}\nu \cos\theta]^2 \, do.$$

The total effective cross-section is

$$\sigma = 4\pi(\omega R/c^2)^2 [\chi^2(\gamma-1)^2 + \tfrac{3}{4}\nu^2].$$

These formulae are valid only if the Stokes frictional force is small compared with the inertia force, i.e. $\eta R \ll M\omega$, where $M = 4\pi R^3 \rho_0/3$ is the mass of the sphere; otherwise, the movement of the sphere by viscosity forces becomes important.

PROBLEM 4. Determine the mean force on a solid sphere which scatters a plane sound wave ($\lambda \gg R$).

SOLUTION. The momentum transmitted per unit time from the incident wave to the sphere, i.e. the required force, is the difference between the momentum in the incident wave and the total momentum flux in the scattered wave. From the incident wave an energy flux $\sigma c\overline{E_0}$ is scattered, where E_0 is the energy density in the incident wave; the corresponding momentum flux is obtained by dividing by c, and is therefore $\sigma\overline{E_0}$. In the scattered wave, the momentum flux into the solid angle element do is $\overline{E_{sc}}r^2 do = \overline{E_0}d\sigma$; projecting this on the direction of propagation of the incident wave (which is obviously the direction of the required force), and integrating over all angles, we obtain

$$\overline{E_0} \int \cos\theta \, d\sigma.$$

Thus the force on the sphere is

$$F = \overline{E_0} \int (1 - \cos\theta) d\sigma.$$

Substituting for $d\sigma$ from Problem 1, we obtain $F = 11\pi\omega^4 R^6 \overline{E_0}/9c^4$.

§77. Absorption of sound

The existence of viscosity and thermal conductivity results in the dissipation of energy in sound waves, and the sound is consequently *absorbed*, i.e. its intensity progressively diminishes. To calculate the rate of energy dissipation $\dot{E}_{mech}$, we use the following general arguments. The mechanical energy is just the maximum amount of work that can be done in passing from a given non-equilibrium state to one of thermodynamic equilibrium. As we know from thermodynamics,[†] the maximum work is obtained when the transition is reversible (i.e. without change of entropy), and is then $E_{mech} = E_0 - E(S)$, where E_0 is the given initial value of the energy, and $E(S)$ is the energy in the equilibrium state with the same entropy S as the system had initially. Differentiating with respect to time, we obtain $\dot{E}_{mech} = -\dot{E}(S) = -(\partial E/\partial S)\dot{S}$. The derivative of the energy with respect to the entropy is the temperature. Hence $\partial E/\partial S$ is the temperature which the system would have if it were in thermodynamic equilibrium (with the given value of the entropy). Denoting this temperature by T_0, we therefore have $\dot{E}_{mech} = -T_0\dot{S}$.

---

† See, for instance, *Statistical Physics*, §19.

We use for $\dot{S}$ the expression (49.6), which gives the rate of change of the entropy due to both thermal conduction and viscosity. Since the temperature T varies only slightly through the fluid, and differs little from T_0, it can be taken outside the integral, and T_0 can be written as T simply:

$$\dot{E}_{\text{mech}} = -\frac{\kappa}{T} \int (\mathbf{grad}\,T)^2\,dV - \tfrac{1}{2}\eta \int \left(\frac{\partial v_i}{\partial x_k} + \frac{\partial v_k}{\partial x_i} - \tfrac{2}{3}\delta_{ik}\frac{\partial v_l}{\partial x_l}\right)^2 dV -$$

$$-\zeta \int (\text{div}\,\mathbf{v})^2\,dV. \tag{77.1}$$

This formula generalises formula (16.3) to the case of a compressible fluid which conducts heat.

Let the x-axis be in the direction of propagation of the sound wave. Then $v_x = v_0 \cos(kx - \omega t)$, $v_y = v_z = 0$. The last two terms in (77.1) give

$$-(\tfrac{4}{3}\eta + \zeta) \int \left(\frac{\partial v_x}{\partial x}\right)^2 dV = -k^2(\tfrac{4}{3}\eta + \zeta)v_0^2 \int \sin^2(kx - \omega t)\,dV.$$

We are, of course, interested only in the time average; taking this average, we have $-k^2(\tfrac{4}{3}\eta + \zeta)\cdot\tfrac{1}{2}v_0^2 V_0$, where V_0 is the volume of the fluid.

Next we calculate the first term in (77.1). The deviation T' of the temperature in the sound wave from its equilibrium value is related to the velocity by formula (63.13), so that the temperature gradient is

$$\partial T/\partial x = (\beta c T/c_p)\,\partial v/\partial x = -(\beta c T/c_p)v_0 k \sin(kx - \omega t).$$

For the time average of the first term in (77.1) we obtain $-\kappa c^2 T\beta^2 v_0^2 k^2 V_0/2c_p^2$. Using the well-known thermodynamic formulae

$$c_p - c_v = T\beta^2(\partial p/\partial \rho)_T = T\beta^2(c_v/c_p)(\partial p/\partial \rho)_s = T\beta^2 c^2 c_v/c_p, \tag{77.2}$$

we can rewrite this expression as $-\tfrac{1}{2}\kappa(1/c_v - 1/c_p)k^2 v_0^2 V_0$.

Collecting the above results, we find the mean value of the energy dissipation:

$$\dot{E}_{\text{mech}} = -\tfrac{1}{2}k^2 v_0^2 V_0[(\tfrac{4}{3}\eta + \zeta) + \kappa(1/c_v - 1/c_p)]. \tag{77.3}$$

The total energy of the sound wave is

$$\bar{E} = \tfrac{1}{2}\rho v_0^2 V_0. \tag{77.4}$$

The damping coefficient derived in §25 for gravity waves gives the manner of decrease of the intensity with time. For sound, however, the problem is usually stated somewhat differently: a sound wave is propagated through a fluid, and its intensity decreases with the distance x traversed. It is evident that this decrease will occur according to a law $e^{-2\gamma x}$, and the amplitude will decrease as $e^{-\gamma x}$, where the *absorption coefficient* γ is defined by

$$\gamma = |\dot{E}_{\text{mech}}|/2c\bar{E}. \tag{77.5}$$

Substituting here (77.3) and (77.4), we find the following expression for the sound absorption coefficient:

$$\gamma = \frac{\omega^2}{2\rho c^3}\left[\left(\tfrac{4}{3}\eta+\zeta\right)+\kappa\left(\frac{1}{c_v}-\frac{1}{c_p}\right)\right]. \tag{77.6}$$

We may point out that it is proportional to the square of the frequency of the sound.†

This formula is applicable so long as the absorption coefficient determined by it is small: the amplitude must decrease relatively little over distances of the order of a wavelength (i.e. we must have $\gamma c/\omega \ll 1$). The above derivation is essentially founded on this assumption, since we have calculated the energy dissipation by using the expression for an undamped sound wave. For gases this condition is in practice always satisfied. Let us consider, for example, the first term in (77.6). The condition $\gamma c/\omega \ll 1$ means that $v\omega/c^2 \ll 1$. It is known from the kinetic theory of gases, however, that the viscosity coefficient v for a gas is of the order of the product of the mean free path l and the mean thermal velocity of the molecules; the latter is of the same order as the velocity of sound in the gas, so that $v \sim lc$. Hence we have

$$v\omega/c^2 \sim l\omega/c \sim l/\lambda \ll 1, \tag{77.7}$$

since we know that $l \ll \lambda$. The thermal-conduction term in (77.6) gives the same result, since $\chi \sim v$.

In liquids, the condition of small absorption is always fulfilled when the problem of sound absorption, as stated here, is significant at all. The absorption over one wavelength can become large only if the viscosity forces are comparable with the pressure forces which occur when the substance is compressed. In these conditions, however, the Navier–Stokes equation itself (with the viscosity coefficients independent of frequency) becomes invalid and a considerable dispersion of sound, due to processes of internal friction, occurs.‡

For absorption of sound, the relation between the wave number and the frequency can evidently be written

$$k = \omega/c + ia\omega^2, \tag{77.8}$$

where a denotes the coefficient of ω^2 in the absorption coefficient $\gamma = a\omega^2$.

---

† M. A. ISAKOVICH has shown that there must be a special absorption when sound is propagated in a two-phase system (an *emulsion*). Because of the different thermodynamic properties of the two components, their temperature changes during the passage of the sound wave will in general be different. The resulting heat exchange between the components leads to an additional absorption of sound. On account of the relative slowness of this heat exchange, a considerable dispersion of the sound takes place comparatively quickly. For detailed calculations see M. A. ISAKOVICH, *Zhurnal éksperimental'noĭ i teoreticheskoĭ fiziki* **18**, 907, 1948.

‡ A special case where strong absorption is possible but can be discussed by the usual methods is that of a gas with a thermal conductivity which is unusually large compared with its viscosity, on account of effects such as radiative transfer at very high temperatures (see Problem 3).

It is easy to see from this how the equation for a travelling sound wave must be modified in order to take absorption into account. To do so, we notice that, in the absence of absorption, the differential equation for (say) the pressure $p' = p'(x - ct)$ can be written $\partial p'/\partial x = -(1/c)\partial p'/\partial t$. The equation whose solution is $e^{i(kx-\omega t)}$, with k given by (77.8), must clearly be

$$\frac{\partial p'}{\partial x} = -\frac{1}{c}\frac{\partial p'}{\partial t} + a\frac{\partial^2 p'}{\partial t^2}. \tag{77.9}$$

If we replace t by $\tau + x/c$, this equation becomes

$$\partial p'/\partial x = a\,\partial^2 p'/\partial \tau^2,$$

i.e. a one-dimensional equation of thermal conduction.

The general solution of this equation can be written (see §51)

$$p'(x,\tau) = \frac{1}{2\sqrt{(\pi a x)}} \int p'_0(\tau') \exp[-(\tau'-\tau)^2/4ax]\,d\tau', \tag{77.10}$$

where $p'_0(\tau) = p'(0,\tau)$. If the wave is emitted during a finite time interval, this expression becomes, at sufficiently large distances from the source,

$$p'(x,\tau) = \frac{1}{2\sqrt{(\pi a x)}}\exp(-\tau^2/4ax)\int p'_0(\tau')\,d\tau'. \tag{77.11}$$

In other words, the wave profile at large distances is Gaussian. Its "width" is of the order of $\sqrt{(ax)}$, i.e. it increases as the square root of the distance travelled by the wave, while the amplitude falls off inversely as $\sqrt{x}$. Hence we at once conclude that the total energy of the wave decreases as $1/\sqrt{x}$.

It is easy to derive analogous formulae for spherical waves; to do so, we must use the fact that for such a wave

$$\int p'\,dt = 0$$

(see §69). Instead of (77.11) we now have

$$p'(r,\tau) = \text{constant} \times \frac{1}{r}\frac{\partial}{\partial \tau}\frac{\exp(-\tau^2/4ar)}{\sqrt{r}},$$

or

$$p'(r,\tau) = \text{constant} \times \frac{\tau}{r^{\frac{3}{2}}}\exp(-\tau^2/4ar). \tag{77.12}$$

Strong absorption must occur when a sound wave is reflected from a solid wall (K. F. HERZFELD, 1938; B. P. KONSTANTINOV, 1939). The reason for this is the following. In a sound wave not only the density and the pressure, but also the temperature, undergo periodic oscillations about their mean values. Near a solid wall, therefore, there is a periodically fluctuating temperature difference between the fluid and the wall, even if the mean fluid temperature is equal to the wall temperature. At the wall itself, however, the temperatures of the wall and the adjoining fluid must be the same. As a result, a large temperature gradient is formed in a thin boundary layer of fluid, where the temperature changes rapidly from its value in the sound wave to the wall temperature. The presence of large temperature gradients, however, results in a large dissipation of energy by thermal conduction. For a similar reason, the fluid viscosity leads to strong absorption of sound when the wave is incident in an oblique direction. In this case the fluid velocity in the wave (in the direction of propagation) has a non-zero component tangential to the surface. At the surface itself, however, the fluid must completely "adhere". Hence a large tangential-velocity gradient† must occur in the boundary layer of fluid, resulting in a large viscous dissipation of energy (see Problem 1).

PROBLEMS

PROBLEM 1. Determine the fraction of energy that is absorbed when a sound wave is reflected from a solid wall. The density of the wall is supposed so large that the sound does not penetrate it, and the specific heat so large that the temperature of the wall may be supposed constant.

SOLUTION. We take the plane of the wall as the plane $x = 0$, and the plane of incidence as the xy-plane. Let the angle of incidence (which equals the angle of reflection) be θ. The change in density in the incident wave at any given point on the surface ($x = y = 0$, say) is $\rho'_1 = Ae^{-i\omega t}$. The reflected wave has the same amplitude, so that $\rho'_2 = \rho'_1$ at the wall. The actual change in the fluid density, since both waves (incident and reflected) are propagated simultaneously, is $\rho' = 2Ae^{-i\omega t}$. The fluid velocity in the wave is given by $\mathbf{v}_1 = c\rho'_1\mathbf{n}_1/\rho$, $\mathbf{v}_2 = c\rho'_2\mathbf{n}_2/\rho$. The total velocity on the wall, $\mathbf{v} = \mathbf{v}_1 + \mathbf{v}_2$, is therefore $v = v_y = 2A \sin \theta \times ce^{-i\omega t}/\rho$ (or, more precisely, this is what the velocity is found to be when the correct boundary conditions at the wall in the presence of viscosity are not applied). The actual variation of the velocity v_y along the wall is determined by formula (24.13), and the energy dissipation due to viscosity by formula (24.14), in which the above expression for v must be substituted for $v_0 e^{-i\omega t}$.

The deviation T' of the temperature from its mean value (which is the temperature of the wall), if calculated without using the correct boundary conditions at the wall, would be found to be (see (63.13)) $T' = 2Ac^2T\beta e^{-i\omega t}/c_p\rho$. In reality, however, the temperature distribution is determined by the equation of thermal conduction, with the boundary condition $T' = 0$ for $x = 0$, and is accordingly given by a formula entirely similar to (24.13).

On calculating the energy dissipation due to thermal conduction as the first term in formula (77.1), we obtain for the total energy dissipation per unit area of the wall

$$\bar{E}_{\text{mech}} = -\frac{A^2c^2\sqrt{(2\omega)}}{\rho}\left[\sqrt{\chi}\left(\frac{c_p}{c_v} - 1\right) + \sqrt{\nu}\sin^2\theta\right].$$

The mean energy flux density incident on unit area of the wall from the incident wave is $c\rho v_1^2 \cos \theta = (c^3A^2/2\rho)\cos \theta$. Hence the fraction of energy absorbed on reflection is

$$\frac{2\sqrt{(2\omega)}}{c \cos \theta}\left[\sqrt{\nu}\sin^2\theta + \sqrt{\chi}\left(\frac{c_p}{c_v} - 1\right)\right].$$

This expression is valid only if its value is small (since in deriving it we have supposed the amplitudes of the incident and reflected waves to be the same). This condition means that the angle of incidence θ must not be too near $\tfrac{1}{2}\pi$.‡

---

† The normal velocity component is zero at the boundary because of the boundary conditions, whether or not viscosity is present.

‡ A calculation of the absorption of sound on reflection at any angle is given by B. P. KONSTANTINOV, *Zhurnal tekhnicheskoĭ fiziki* 9, 226, 1939.

PROBLEM 2. Determine the coefficient of absorption of sound propagated in a cylindrical pipe.

SOLUTION. The main contribution to the absorption is due to the presence of the walls. The absorption coefficient γ is equal to the energy dissipated at the walls per unit time and per unit length of the pipe, divided by twice the total energy flux through a cross-section of the pipe. A calculation entirely similar to that given in Problem 1 leads to the result

$$\gamma = \frac{\sqrt{\omega}}{\sqrt{2Rc}} \left[\sqrt{\nu} + \sqrt{\chi} \left(\frac{c_p}{c_v} - 1 \right) \right],$$

where R is the radius of the pipe.

PROBLEM 3. Find the dispersion relation for sound propagated in a medium of very high thermal conductivity.

SOLUTION. In the presence of a large thermal conductivity the flow in a sound wave is not adiabatic. Hence, instead of the condition of constant entropy, we now have

$$\dot{s} = \kappa \triangle T' / \rho T, \tag{1}$$

which is the linearised form of equation (49.4) without the viscosity terms. As a second equation we take

$$\ddot{\rho}' = \triangle p', \tag{2}$$

which is obtained by eliminating $\mathbf{v}$ from equations (63.2) and (63.3). Taking as the fundamental variables p' and T', we write ρ' and s' in the form

$$\rho' = (\partial \rho / \partial T)_p T' + (\partial \rho / \partial p)_T p', \quad s' = (\partial s / \partial T)_p T' + (\partial s / \partial p)_T p'.$$

We substitute these expressions in (1) and (2), and then seek T' and p' in a form proportional to $e^{i(kx-\omega t)}$. The compatibility condition for the resulting two equations for p' and T' can (by using various relations between the derivatives of thermodynamic quantities) be brought to the form

$$k^4 - k^2 \left(\frac{\omega^2}{c_T{}^2} + \frac{i\omega}{\chi} \right) + \frac{i\omega^3}{\chi c_s{}^2} = 0, \tag{3}$$

which gives the required relation between k and ω. We have here used the notation

$$c_s{}^2 = (\partial p / \partial \rho)_s, \quad c_T{}^2 = (\partial p / \partial \rho)_T = c_s{}^2 / \gamma,$$

where $\gamma = c_p / c_v$ is the ratio of specific heats.

In the limiting case of small frequencies ($\omega \ll c^2/\chi$), equation (3) gives

$$k = \frac{\omega}{c_s} + i \frac{\omega^2 \chi}{2c_s} \left(\frac{1}{c_T{}^2} - \frac{1}{c_s{}^2} \right),$$

which corresponds to the propagation of sound with the ordinary "adiabatic" velocity c_s and a small absorption coefficient which is the second term in (77.6). This is as it should be, since the condition $\omega \ll c^2/\chi$ means that, during one period, heat can be transmitted only over a distance $\sim \sqrt{(\chi/\omega)}$ (cf. (51.7)) which is small compared with the wavelength c/ω.

In the opposite limiting case of large frequencies, we find from (3)

$$k = \frac{\omega}{c_T} + i \frac{c_T}{2\chi c_s{}^2} (c_s{}^2 - c_T{}^2).$$

In this case the sound is propagated with the "isothermal" velocity c_T, which is always less than c_s. The absorption coefficient is again small compared with the reciprocal of the wavelength, and is independent of the frequency and inversely proportional to the thermal conductivity.[†]

PROBLEM 4. Determine the additional absorption, due to diffusion, of sound propagated in a mixture of two substances (I. G. SHAPOSHNIKOV and Z. A. GOL'DBERG 1952).

SOLUTION. The mixture contains an additional source of absorption of sound because the temperature and pressure gradients occurring in the sound wave result in irreversible processes of thermal diffusion and barodiffusion (but there is evidently no mass-concentration gradient, and therefore no mass transfer). This absorption is given by the term

$$(1/T\rho D)(\partial\mu/\partial C)_{p,T} \int \mathbf{i}^2 \, dV$$

in the rate of change of entropy (58.13); we here denote the concentration by C to distinguish it from c, the velocity of sound. The diffusion flux is

$$\mathbf{i} = -\rho D[(k_T/T)\,\mathbf{grad}\,T + (k_p/p)\,\mathbf{grad}\,p],$$

with k_p given by (58.10). A calculation similar to that given in §77, using various relations between the derivatives of thermodynamic quantities, leads to the result that there must be added to the expression (77.6) for the absorption coefficient a term

$$\gamma_D = \frac{D\omega^2}{2c\rho^2(\partial\mu/\partial C)_{p,T}}\left\{\left(\frac{\partial\rho}{\partial C}\right)_{p,T} + \frac{k_T}{c_p}\left(\frac{\partial\rho}{\partial T}\right)_{p,C}\left(\frac{\partial\mu}{\partial C}\right)_{p,T}\right\}^2.$$

PROBLEM 5. Determine the effective cross-section for the absorption of sound by a sphere of radius small compared with $\sqrt{(\nu/\omega)}$.

SOLUTION. The total absorption is composed of the effects of the viscosity and thermal conductivity of the gas. The former is given by the work done by the Stokes frictional force when gas moving in a sound wave flows round a sphere; as in §76, Problem 3, it is assumed that the sphere is not moved by this force. The effect of conductivity is given by the amount of heat q transferred from the gas to the sphere per unit time (§76, Problem 3): the energy dissipation when an amount of heat q is transferred, the temperature difference between the gas (far from the sphere) and the sphere being T', is qT'/T. The total effective absorption cross-section is found to be

$$\sigma = \frac{2\pi R}{c}\left[3\nu + 2\chi\left(\frac{c_p}{c_v} - 1\right)\right].$$

§78. Second viscosity

The second viscosity coefficient ζ (which we shall call simply the *second viscosity*) is usually of the same order of magnitude as the viscosity coefficient η. There are, however, cases where ζ can take values considerably exceeding η. As we know, the second viscosity appears in processes which are accompanied by a change in volume (i.e. in density) of the fluid. In compression or expansion, as in any rapid change of state, the fluid ceases to be in thermodynamic equilibrium, and internal processes are set up in it which tend to

---

[†] The second root of equation (3), which is quadratic in k^2, corresponds to "thermal waves" which are rapidly damped with increasing x. In the limit $\omega\chi \ll c^2$ this root gives

$$k = \sqrt{(i\omega/\chi)} = (1+i)\sqrt{(\omega/2\chi)},$$

in agreement with (52.17). In the case $\omega\chi \gg c^2$ we have

$$k = (1+i)\sqrt{(\omega c_v/2\chi c_p)}.$$

restore this equilibrium. These processes are usually so rapid (i.e. their relaxation time is so short) that the restoration of equilibrium follows the change in volume almost immediately unless, of course, the rate of change of volume is very large.

It may happen, nevertheless, that the relaxation times of the processes of restoration of equilibrium are long, i.e. they take place comparatively slowly. For instance, if we are concerned with a liquid or gas which is a mixture of substances between which a chemical reaction occurs, there is a state of chemical equilibrium, characterised by the concentrations of the substances in the mixture, for any given density and temperature. If, for example, we compress the fluid, the state of equilibrium is destroyed, and a reaction begins, as a result of which the concentrations of the substances tend to take the equilibrium values corresponding to the new density and temperature. If this reaction is not rapid, the restoration of equilibrium occurs relatively slowly and does not immediately follow the compression. The latter process is then accompanied by internal processes which tend towards the equilibrium state. But the processes which establish equilibrium are irreversible; they increase the entropy, and therefore involve energy dissipation. Hence, if the relaxation time of these processes is long, a considerable dissipation of energy occurs when the fluid is compressed or expanded, and, since this dissipation must be determined by the second viscosity, we reach the conclusion that ζ is large.†

The intensity of the dissipative processes, and therefore the value of ζ, depend of course on the relation between the rate of compression or expansion and the relaxation time. If, for example, we have compression or expansion due to a sound wave, the second viscosity will depend on the frequency of the wave. Thus the second viscosity is not just a constant characteristic of the material concerned, but depends on the frequency of the motion in which it appears. The dependence of ζ on the frequency is called its *dispersion*.

The following general method of discussing all these phenomena is due to L. I. MANDEL'SHTAM and M. A. LEONTOVICH (1937). Let ξ be some physical quantity characterising the state of a body, and ξ_0 its value in the equilibrium state; ξ_0 is a function of density and temperature. For instance, in fluid mixtures ξ may be the concentration of one component, and then ξ_0 is the concentration in chemical equilibrium.

If the body is not in equilibrium, ξ will vary with time, tending to the value ξ_0. In states close to equilibrium the difference $\xi - \xi_0$ is small, and we can expand the rate of change $\dot{\xi}$ of ξ in a series of powers of this difference. The zero-order term is absent, since $\dot{\xi}$ must be zero in the equilibrium state, i.e. when $\xi = \xi_0$. Hence, as far as the first-order term, we have

$$\dot{\xi} = -(\xi - \xi_0)/\tau. \tag{78.1}$$

The proportionality coefficient must be negative, since otherwise ξ would not

---

† A slow process which results in a large ζ is often the transfer of energy from translatory degrees of freedom of a molecule to vibrational (intramolecular) degrees of freedom.

tend to a finite limit. The positive constant τ is of the dimensions of time, and may be regarded as the relaxation time for the process in question; the greater is τ, the more slowly the approach to equilibrium takes place.

In what follows we shall consider processes in which the fluid is subjected to a periodic adiabatic† compression and expansion, so that the variable part of the density (and of the other thermodynamic quantities) depends on the time through a factor $e^{-i\omega t}$; we are considering a sound wave in the fluid. Together with the density and other quantities, the position of equilibrium also varies, so that ξ_0 can be written as $\xi_0 = \xi_{00} + \xi_0'$, where ξ_{00} is the constant value of ξ_0 corresponding to the mean density, and ξ_0' is a periodic part, proportional to $e^{-i\omega t}$. Writing the true value ξ in the form $\xi = \xi_{00} + \xi'$, we conclude from equation (78.1) that ξ' also is a periodic function of time, related to ξ_0' by

$$\xi' = \xi_0'/(1 - i\omega\tau). \tag{78.2}$$

Let us calculate the derivative of the pressure with respect to the density for the process in question. The pressure must now be regarded as a function of the density and of the value of ξ in the state concerned, and also of the entropy, which we suppose constant and, for brevity, omit. Then

$$\partial p/\partial\rho = (\partial p/\partial\rho)_\xi + (\partial p/\partial\xi)_\rho \, \partial\xi/\partial\rho.$$

In accordance with (78.2), we substitute here

$$\frac{\partial\xi}{\partial\rho} = \frac{\partial\xi'}{\partial\rho} = \frac{1}{1 - i\omega\tau} \frac{\partial\xi_0'}{\partial\rho} = \frac{1}{1 - i\omega\tau} \frac{\partial\xi_0}{\partial\rho},$$

obtaining

$$\frac{\partial p}{\partial\rho} = \frac{1}{1 - i\omega\tau}\left\{\left(\frac{\partial p}{\partial\rho}\right)_\xi + \left(\frac{\partial p}{\partial\xi}\right)_\rho \frac{\partial\xi_0}{\partial\rho} - i\omega\tau\left(\frac{\partial p}{\partial\rho}\right)_\xi\right\}.$$

The sum $(\partial p/\partial\rho)_\xi + (\partial p/\partial\xi)_\rho \partial\xi_0/\partial\rho$ is just the derivative of p with respect to ρ for a process which is so slow that the fluid remains in equilibrium; denoting it by $(\partial p/\partial\rho)_{\text{eq}}$, we have finally

$$\frac{\partial p}{\partial\rho} = \frac{1}{1 - i\omega\tau}\left[\left(\frac{\partial p}{\partial\rho}\right)_{\text{eq}} - i\omega\tau\left(\frac{\partial p}{\partial\rho}\right)_\xi\right]. \tag{78.3}$$

Next, let p_0 be the pressure in a state of thermodynamic equilibrium; p_0 is related to the other thermodynamic quantities by the equation of state of the fluid, and is entirely determined when the density and entropy are given. The pressure p in a non-equilibrium state, however, differs from p_0, and is a function of ξ also. If the density is adiabatically increased by $\delta\rho$, the equilibrium pressure changes by $\delta p_0 = (\partial p/\partial\rho)_{\text{eq}}\delta\rho$, while the total increase in the pressure is $(\partial p/\partial\rho)\delta\rho$, with $\partial p/\partial\rho$ given by formula (78.3).

---

† The change in the entropy (in states close to equilibrium) is of the second order of smallness. Hence, to this order of accuracy, we can speak of an adiabatic process.

Hence the difference $p-p_0$ between the true pressure and the equilibrium pressure, in a state where the density is $\rho+\delta\rho$, is

$$p-p_0 = \left[\frac{\partial p}{\partial \rho} - \left(\frac{\partial p}{\partial \rho}\right)_{\text{eq}}\right]\delta\rho = \frac{i\omega\tau}{1-i\omega\tau}\left[\left(\frac{\partial p}{\partial \rho}\right)_{\text{eq}} - \left(\frac{\partial p}{\partial \rho}\right)_{\xi}\right]\delta\rho. \quad (78.3a)$$

We are here interested in the density changes due to the motion of the fluid. Then $\delta\rho$ is related to the velocity by the equation of continuity, which we write in the form $d(\delta\rho)/dt + \rho\,\text{div }\mathbf{v} = 0$, where d/dt denotes the total time derivative. In a periodic motion we have $d(\delta\rho)/dt = -i\omega\delta\rho$, and therefore $\delta\rho = (\rho/i\omega)\,\text{div }\mathbf{v}$. Substituting this expression in (78.3a), we obtain

$$p-p_0 = \frac{\tau\rho}{1-i\omega\tau}(c_0{}^2 - c_\infty{}^2)\,\text{div }\mathbf{v}, \quad (78.4)$$

where we have used the notation

$$c_0{}^2 = (\partial p/\partial\rho)_{\text{eq}}, \quad c_\infty{}^2 = (\partial p/\partial\rho)_\xi, \quad (78.5)$$

the significance of which will be explained below.

In order to relate these expressions to the viscosity of the fluid, we write down the stress tensor σ_{ik}. In this tensor the pressure appears in the term $-p\delta_{ik}$. Subtracting the pressure p_0 determined by the equation of state, we find that in a non-equilibrium state σ_{ik} contains an additional term

$$-(p-p_0)\delta_{ik} = \frac{\tau\rho}{1-i\omega\tau}(c_\infty{}^2 - c_0{}^2)\delta_{ik}\,\text{div }\mathbf{v}.$$

Comparing this with the general expression (15.2) and (15.3) for the stress tensor, in which div $\mathbf{v}$ appears in the term $\zeta\,\text{div }\mathbf{v}$, we conclude that the presence of slow processes tending to establish equilibrium is macroscopically equivalent to the presence of a second viscosity given by

$$\zeta = \tau\rho(c_\infty{}^2 - c_0{}^2)/(1-i\omega\tau). \quad (78.6)$$

These processes do not affect the ordinary viscosity η. For processes so slow that $\omega\tau \ll 1$, ζ is

$$\zeta_0 = \tau\rho(c_\infty{}^2 - c_0{}^2); \quad (78.7)$$

it increases with the relaxation time τ, in accordance with what was said above. For large frequencies, ζ depends on the frequency, i.e. it exhibits dispersion.

Let us now consider the question of how the presence of processes with large relaxation times (for definiteness, we shall speak of chemical reactions) affects the propagation of sound in a fluid. To do so, we might start from the equation of motion of a viscous fluid, with ζ given by formula (78.6). It is simpler, however, to consider a motion in which viscosity is neglected but the pressure p is given by the above formulae instead of by the equation of state. The general relations which we obtained in §63 then remain formally applicable. In particular, the wave number and the frequency are still related by $k = \omega/c$, where $c = \sqrt{(\partial p/\partial \rho)}$, and the derivative $\partial p/\partial \rho$ is now given by (78.3); the quantity c, however, no longer denotes the velocity of sound, being complex. Thus we obtain

$$k = \omega\sqrt{[(1 - i\omega\tau)/(c_0^2 - c_\infty^2 i\omega\tau)]}. \tag{78.8}$$

The "wave number" given by this formula is complex. The meaning of this fact is easily seen. In a plane wave, all quantities depend on the coordinate x (the x-axis being in the direction of propagation) through a factor e^{ikx}. Writing k in the form $k = k_1 + ik_2$ with k_1, k_2 real, we have $e^{ikx} = e^{ik_1x} e^{-k_2x}$, i.e. besides the periodic factor e^{ik_1x} we have a damping factor e^{-k_2x} (k_2 must, of course, be positive). Thus the complex nature of the wave number formally expresses the fact that the wave is damped, i.e. there is absorption of sound. The real part of the complex wave number gives the variation in phase of the wave with distance, and the imaginary part is the absorption coefficient.

It is not difficult to separate the real and imaginary parts of (78.8). In the general case of arbitrary ω the expressions for k_1 and k_2 are rather cumbersome, and we shall not write them out here. It is important that k_1 is a function of the frequency (as is k_2). Thus, if chemical reactions can occur in the fluid, the propagation of sound at sufficiently high frequencies is accompanied by dispersion.

In the limiting case of low frequencies ($\omega\tau \ll 1$), formula (78.8) gives to a first approximation $k = \omega/c_0$, corresponding to the propagation of sound with velocity c_0. This is as it should be, of course: the condition $\omega\tau \ll 1$ means that the period $1/\omega$ of the sound wave is large compared with the relaxation time, i.e. the establishment of chemical equilibrium follows the variations of density in the sound wave, and the velocity of sound is determined by the equilibrium value of the derivative $\partial p/\partial \rho$. In the second approximation we have

$$k = \frac{\omega}{c_0} + \frac{i\omega^2\tau}{2c_0^3}(c_\infty^2 - c_0^2), \tag{78.9}$$

i.e. damping occurs, with a coefficient proportional to the square of the frequency. Using (78.7), we can write the imaginary part of k in the form $k_2 = \omega^2\zeta_0/2\rho c_0^3$; this agrees with the ζ-dependent part of the absorption coefficient γ as given by (77.6), which was obtained without taking account of the dispersion.

In the opposite limiting case of high frequencies ($\omega\tau \gg 1$), we have in the first approximation $k = \omega/c_\infty$, i.e. the propagation of sound with velocity c_∞—again a natural result, since for $\omega\tau \gg 1$ we can suppose that no reaction occurs during a single period, and the velocity of sound must therefore be determined by the derivative $(\partial p/\partial \rho)_\xi$ taken at constant concentration. The second approximation gives

$$k = \frac{\omega}{c_\infty} + i\frac{c_\infty^2 - c_0^2}{2\tau c_\infty^3}. \tag{78.10}$$

The damping coefficient is independent of the frequency. As we go from $\omega \ll 1/\tau$ to $\omega \gg 1/\tau$, this coefficient increases monotonically to the constant value given by formula (78.10). It should be noted that the quantity k_2/k_1, which represents the amount of absorption over a distance of one wavelength, is small in both limiting cases ($k_2/k_1 \ll 1$); it has a maximum at some intermediate frequency, namely $\omega = \sqrt{(c_0/c_\infty)}/\tau$.

It is seen from (78.7) (e.g.) that

$$c_\infty > c_0, \tag{78.11}$$

since we must have $\zeta > 0$. The same result can be obtained by simple arguments based on Le Chatelier's principle. Let us suppose that the volume of the system is reduced, and the density increased, by some external agency. The system is thereby brought out of equilibrium, and according to Le Chatelier's principle processes must begin which tend to reduce the pressure. This means that $\partial p/\partial \rho$ will decrease, and, when the system returns to equilibrium, the value of $\partial p/\partial \rho = c^2$ will be less than in the non-equilibrium state.

In deriving all the above formulae we have assumed that there is only a single slow internal process of relaxation. Cases are also possible where several different such processes occur simultaneously. All the formulae can easily be generalised to cover such cases. Instead of a single quantity ξ, we now have several quantities $\xi_1, \xi_2, \ldots$ which characterise the state of the system, and a corresponding series of relaxation times $\tau_1, \tau_2, \ldots$. We choose the quantities ξ_n in such a way that each of the derivatives $\dot{\xi}_n$ depends only on the corresponding ξ_n, i.e. so that

$$\dot{\xi}_n = -(\xi_n - \xi_{n0})/\tau_n. \tag{78.12}$$

Calculations entirely similar to the above then give

$$c^2 = c_\infty^2 + \sum_n a_n/(1 - i\omega\tau_n), \tag{78.13}$$

where $c_\infty^2 = (\partial p/\partial \rho)_\xi$, and the constants a_n are

$$a_n = (\partial p/\partial \xi_n)(\partial \xi_n/\partial \rho)_{\text{eq}}. \tag{78.14}$$

If there is only one quantity ξ, formula (78.13) becomes (78.3), as it should.

CHAPTER IX

SHOCK WAVES

§79. Propagation of disturbances in a moving gas

WHEN the velocity of a fluid in motion becomes comparable with or exceeds that of sound, effects due to the compressibility of the fluid become of prime importance. Such motions are in practice met with in gases. The dynamics of high-speed flow is therefore usually called *gas dynamics*.

It should be mentioned first of all that, in gas dynamics, the Reynolds numbers involved are almost always very large. For the kinematic viscosity of a gas is, as we know from the kinetic theory of gases, of the order of the mean free path l of the molecules multiplied by the mean velocity of their thermal motion; the latter is of the same order as the velocity of sound, so that $\nu \sim cl$. If the characteristic velocity in a problem of gas dynamics is also of the order of c, then the Reynolds number $R \sim Lc/\nu \sim L/l$, i.e. it is determined by the ratio of the dimension L to the mean free path l, which we know is very large.† As always occurs when R is very large, the viscosity has an important effect on the motion of the gas only in a very small region, and in what follows we shall (except where the contrary is specifically stated) regard the gas as an ideal fluid.

The flow of a gas is entirely different in nature according as it is *subsonic* or *supersonic*, i.e. the velocity is less than or greater than that of sound. One of the most important distinctive features of supersonic flow is the fact that there can occur in it what are called *shock waves*, whose properties we shall examine in detail in the following sections. Here we shall consider another characteristic property of supersonic flow, relating to the manner of propagation of small disturbances in the gas.

If a gas in steady motion receives a slight perturbation at any point, the effect of the perturbation is subsequently propagated through the gas with the velocity of sound (relative to the gas itself). The rate of propagation of the disturbance relative to a fixed system of co-ordinates is composed of two parts: firstly, the perturbation is "carried along" by the gas flow with velocity $\mathbf{v}$ and, secondly, it is propagated relative to the gas with velocity c in any direction $\mathbf{n}$. Let us consider, for simplicity, a uniform flow of gas with constant velocity $\mathbf{v}$, subjected to a small perturbation at some point O (fixed in space). The velocity $\mathbf{v} + c\mathbf{n}$ with which the perturbation is propagated from O (relative to the fixed system of co-ordinates) has different values for different directions of the unit vector $\mathbf{n}$. We obtain

---

† We shall not consider the problem of the motion of bodies in very rarefied gases, where the mean free path of the molecules is comparable with the dimension of the body. This problem is in essence not one of fluid dynamics, and must be examined by means of the kinetic theory of gases.

all its possible values by placing one end of the vector **v** at the point O and drawing a sphere of radius c centred at the other end. The vectors from O to points on this sphere give the possible magnitudes and directions of the velocity of propagation of the perturbation. Let us first suppose that $v < c$. Then the vector $\mathbf{v} + c\mathbf{n}$ can have any direction in space (Fig. 40a). That is, a disturbance which starts from any point in a subsonic flow will eventually reach every point in the gas. If, on the other hand, $v > c$, the direction of the vector $\mathbf{v} + c\mathbf{n}$ can lie, as we see from Fig. 40b, only in a cone with its vertex at O, which touches the sphere with its centre at the other end of the vector **v**. If the aperture of the cone is 2α, then, as is seen from the figure,

$$\sin \alpha = c/v. \tag{79.1}$$

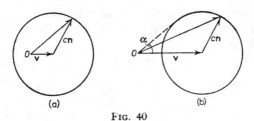

(a)   (b)

Fig. 40

Thus a disturbance starting from any point in a supersonic flow is propagated only downstream within a cone whose aperture is the smaller, the smaller the ratio c/v. A disturbance starting from O does not affect the flow outside this cone.

The angle α determined by equation (79.1) is called the *Mach angle*. The ratio v/c itself, which often occurs in gas dynamics, is the *Mach number* M:

$$\mathrm{M} = v/c. \tag{79.2}$$

The surface bounding the region reached by a disturbance starting from a given point is called the *Mach surface* or *characteristic surface*.

In the general case of an arbitrary steady flow, the Mach surface is not a cone throughout the volume. However, it can be asserted that, as before, this surface cuts the streamline through any point on it at the Mach angle. The value of the Mach angle varies from point to point with the velocities v and c. It should be emphasised here, incidentally, that, in flow with high velocities, the velocity of sound is different at different points: it varies with the thermodynamic quantities (pressure, density, etc.) of which it is a function.[†] The velocity of sound as a function of the co-ordinates is sometimes called the *local velocity of sound*.

---

† In the discussion of sound waves given in Chapter VIII, the velocity of sound could be regarded as constant.

The properties of supersonic flow described above give it a character quite different from that of subsonic flow. If a subsonic gas flow meets any obstacle (if, for instance, it flows past a body), the presence of this obstacle affects the flow in all space, both upstream and downstream; the effect of the obstacle is zero only asymptotically at an infinite distance from it. A supersonic flow, however, is incident "blindly" on an obstacle; the effect of the latter extends only downstream,† and in all the remaining part of space upstream the gas flows as if the obstacle were absent.

In the case of steady plane flow of a gas, the characteristic surfaces can be replaced by *characteristic lines* (or simply *characteristics*) in the plane of the flow. Through any point O in this plane there pass two characteristics (AA' and BB' in Fig. 41), which intersect the streamline through this point at the Mach angle. The downstream branches OA and OB of the characteristics may be said to *leave* the point O; they bound the region AOB of the flow where perturbations starting from O can take effect. The branches $B'O$ and $A'O$ may be said to *reach* the point O; the region $A'OB'$ between them is that which can affect the flow at O.

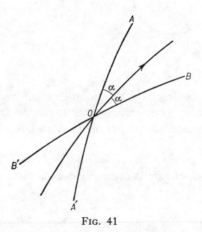

FIG. 41

The concept of characteristics (surfaces in the three-dimensional case) has also a somewhat different aspect. They are rays along which disturbances are "propagated" which satisfy the conditions of geometrical acoustics. If, for example, a steady supersonic gas flow meets a fairly small obstacle, then a steady perturbation of the gas flow will be found along the characteristics which leave this obstacle. The same result was reached in §67 from a study of the geometrical acoustics of moving media.

When we speak of a perturbation of the state of the gas, we mean a slight change in any of the quantities characterising its state: the velocity, pressure,

---

† To avoid misunderstanding, we should mention that, if a shock wave is formed in front of the obstacle, this region is somewhat enlarged (see §114).

density, etc. The following remark should be made on this point. Perturbations in the values of the entropy of the gas (for constant pressure) and of
its vorticity are not propagated with the velocity of sound. These perturbations, once having arisen, do not move relative to the gas; relative to a fixed
system of co-ordinates they move with the gas at the velocity appropriate to
each point. For the entropy, this is an immediate consequence of the law of
conservation (in an ideal fluid),

$$ds/dt \equiv \partial s/\partial t + \mathbf{v} \cdot \mathbf{grad}\, s = 0,$$

which shows that the entropy of any given volume element in the gas remains
constant as the element moves about, i.e. each value of s moves with the
point to which it belongs. The same result for the vorticity follows from the
conservation of circulation.

Thus we can say that, for perturbations of entropy and vorticity, the
characteristics are the streamlines. This, of course, does not affect the general
validity of the statements made above about regions of influence, since
they were based only on the existence of a maximum velocity of propagation
(that of sound) of disturbances relative to the gas itself.

§80. Steady flow of a gas

We can obtain immediately from Bernoulli's equation a number of general
results concerning adiabatic steady flow of a gas. The equation is, for steady
flow, $w + \tfrac{1}{2}v^2 =$ constant along each streamline; if we have potential flow,
then the constant is the same for every streamline, i.e. at every point in the
fluid. If there is a point on some streamline at which the gas velocity is zero,
then we can write Bernoulli's equation as

$$w + \tfrac{1}{2}v^2 = w_0, \tag{80.1}$$

where w_0 is the value of the heat function at the point where $v = 0$.

The equation of conservation of entropy for steady flow is $\mathbf{v} \cdot \mathbf{grad}\, s$
$= v \partial s/\partial l = 0$, i.e. s is constant along each streamline. We can write this in a
form analogous to (80.1):

$$s = s_0. \tag{80.2}$$

We see from equation (80.1) that the velocity v is greater at points where
the heat function w is smaller. The maximum value of the velocity (on
the streamline considered) is found at the point where w is least. For constant entropy, however, we have $dw = dp/\rho$; since $\rho > 0$, the differentials
dw and dp have like signs, and therefore w and p vary in the same sense.
We can therefore say that the velocity increases along a streamline when the
pressure decreases, and *vice versa*.

The smallest possible values of the pressure and the heat function (in
adiabatic flow) are obtained when the absolute temperature $T = 0$. The
corresponding pressure is $p = 0$, and the value of w for $T = 0$ can be
arbitrarily taken as the zero of energy; then $w = 0$ for $T = 0$. We can

now deduce from (80.1) that the greatest possible value of the velocity (for given values of the thermodynamic quantities at the point where $v = 0$) is

$$v_{\max} = \sqrt{(2w_0)}. \tag{80.3}$$

This velocity can be attained when a gas flows steadily out into a vacuum.†

Let us now consider how the mass flux density $j = \rho v$ varies along a streamline. From Euler's equation $(\mathbf{v \cdot grad})\mathbf{v} = -(1/\rho)\,\mathbf{grad}\,p$, we find that the relation $v\,dv = dp/\rho$ between the differentials dv and dp holds along a streamline. Putting $dp = c^2 d\rho$, we have

$$d\rho/dv = -\rho v/c^2 \tag{80.4}$$

and, substituting in $d(\rho v) = \rho\,dv + v\,d\rho$, we obtain

$$d(\rho v)/dv = \rho(1 - v^2/c^2). \tag{80.5}$$

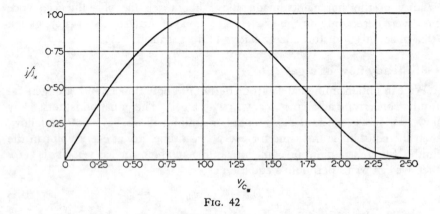

FIG. 42

From this we see that, as the velocity increases along a streamline, the mass flux density increases as long as the flow remains subsonic. In the supersonic range, however, the mass flux density diminishes with increasing velocity, and vanishes together with ρ when $v = v_{\max}$ (Fig. 42). This important difference between subsonic and supersonic steady flows can be simply interpreted as follows. In a subsonic flow, the streamlines approach in the direction of increasing velocity. In a supersonic flow, however, they diverge in that direction.

The flux j has its maximum value j_* at the point where the gas velocity is equal to the local velocity of sound:

$$j_* = \rho_* c_*, \tag{80.6}$$

where the asterisk suffix indicates values corresponding to this point. The

---

† In reality, of course, when there is a sharp fall in temperature the gas must condense and form a two-phase "fog". This, however, does not essentially affect the results given.

velocity $v_* = c_*$ is called the *critical velocity*. In the general case of an arbitrary gas, the critical values of quantities can be expressed in terms of their values at the point $v = 0$, by solving the simultaneous equations

$$s_* = s_0, \qquad w_* + \tfrac{1}{2}c_*^2 = w_0. \tag{80.7}$$

It is evident that, whenever $M = v/c < 1$, we have also $v/c_* < 1$, and if $M > 1$ then $v/c_* > 1$. Hence the ratio $M_* = v/c_*$ serves in this case as a criterion analogous to M, and is more convenient, since c_* is a constant, unlike c, which varies along the stream.

In applications of the general equations of gas dynamics, the case of a perfect gas is of particular importance. For a perfect gas we know from thermodynamics all the relations between the various thermodynamic quantities, and these relations are very simple. This makes it possible to give a complete solution of the equations of gas dynamics in many cases.

We shall give here, for reference, the relations between the various thermodynamic quantities for a perfect gas, since they will often be needed in what follows. We shall always assume (unless otherwise stated) that the specific heat of a perfect gas is independent of temperature.

The equation of state for a perfect gas is

$$pV = p/\rho = RT/\mu, \tag{80.8}$$

where $R = 8.314 \times 10^7$ erg/deg is the gas constant, and μ the molecular weight of the gas. The velocity of sound in a perfect gas is, as shown in §63, given by

$$c^2 = \gamma RT/\mu = \gamma p/\rho, \tag{80.9}$$

where we have introduced the constant ratio of specific heats $\gamma = c_p/c_v$, which always exceeds unity; for monatomic gases $\gamma = 5/3$, and for diatomic gases $\gamma = 7/5$, at ordinary temperatures.

The internal energy of a perfect gas is, apart from an unimportant additive constant,

$$\epsilon = c_v T = pV/(\gamma - 1) = c^2/\gamma(\gamma - 1). \tag{80.10}$$

For the heat function we have the analogous formulae

$$w = c_p T = \gamma pV/(\gamma - 1) = c^2/(\gamma - 1). \tag{80.11}$$

Here we have used the well-known relation $c_p - c_v = R/\mu$. Finally, the entropy of the gas is

$$s = c_v \log(p/\rho^\gamma) = c_p \log(p^{1/\gamma}/\rho). \tag{80.12}$$

Let us now investigate steady flow, applying the general relations previously obtained to the case of a perfect gas. Substituting (80.11) in (80.3), we find that the maximum velocity of steady flow is

$$v_{\max} = c_0 \sqrt{[2/(\gamma - 1)]}. \tag{80.13}$$

For the critical velocity we obtain from the second equation (80.7)

$$\frac{c_*^2}{\gamma-1} + \tfrac{1}{2}c_*^2 = w_0 = \frac{c_0^2}{\gamma-1},$$

whence†

$$c_* = c_0\sqrt{[2/(\gamma+1)]}. \tag{80.14}$$

Bernoulli's equation (80.1), after substitution of the expression (80.11) for the heat function, gives the relation between the temperature and the velocity at any point on the streamline; similar relations for the pressure and density can then be obtained directly by means of Poisson's adiabatic equation:

$$\rho = \rho_0(T/T_0)^{1/(\gamma-1)}, \qquad p = p_0(\rho/\rho_0)^\gamma. \tag{80.15}$$

Thus we obtain the important results

$$\left.\begin{array}{l} T = T_0\left[1-\tfrac{1}{2}(\gamma-1)\dfrac{v^2}{c_0^2}\right] = T_0\left(1 - \dfrac{\gamma-1}{\gamma+1}\dfrac{v^2}{c_*^2}\right), \\[2mm] \rho = \rho_0\left[1-\tfrac{1}{2}(\gamma-1)\dfrac{v^2}{c_0^2}\right]^{1/(\gamma-1)} = \rho_0\left(1 - \dfrac{\gamma-1}{\gamma+1}\dfrac{v^2}{c_*^2}\right)^{1/(\gamma-1)}, \\[2mm] p = p_0\left[1-\tfrac{1}{2}(\gamma-1)\dfrac{v^2}{c_0^2}\right]^{\gamma/(\gamma-1)} = p_0\left(1 - \dfrac{\gamma-1}{\gamma+1}\dfrac{v^2}{c_*^2}\right)^{\gamma/(\gamma-1)}. \end{array}\right\} \tag{80.16}$$

It is sometimes convenient to use these relations in a form which gives the velocity in terms of other quantities:

$$v^2 = \frac{2\gamma}{\gamma-1}\frac{p_0}{\rho_0}\left[1-\left(\frac{p}{p_0}\right)^{(\gamma-1)/\gamma}\right] = \frac{2\gamma}{\gamma-1}\frac{p_0}{\rho_0}\left[1-\left(\frac{\rho}{\rho_0}\right)^{\gamma-1}\right]. \tag{80.17}$$

We may also give the relation between the velocity of sound and the velocity v:

$$c^2 = c_0^2 - \tfrac{1}{2}(\gamma-1)v^2 = \tfrac{1}{2}(\gamma+1)c_*^2 - \tfrac{1}{2}(\gamma-1)v^2. \tag{80.18}$$

Hence we find that the numbers M and M_* are related by

$$M_*^2 = \frac{\gamma+1}{\gamma-1+2/M^2}; \tag{80.19}$$

when M varies from 0 to ∞, M_*^2 varies from 0 to $(\gamma+1)/(\gamma-1)$.

Finally, we may give expressions for the critical temperature, pressure and density: they are obtained by putting $v = c_*$ in formulae (80.16)‡:

---

† Fig. 42 shows the ratio j/j_* as a function of v/c_* for air ($\gamma = 1.4$, $v_{max} = 2.45c_*$).
‡ For air, e.g., ($\gamma = 1.4$)

$$c_* = 0.913c_0, \quad p_* = 0.528p_0, \quad \rho_* = 0.634\rho_0, \quad T_* = 0.833T_0.$$

$$T_* = 2T_0/(\gamma+1),$$

$$p_* = p_0\left(\frac{2}{\gamma+1}\right)^{\gamma/(\gamma-1)},$$

$$\rho_* = \rho_0\left(\frac{2}{\gamma+1}\right)^{1/(\gamma-1)}.$$

$$(80.20)$$

In conclusion, it should be emphasised that the results derived above are valid only for flow in which shock waves do not occur. When shock waves are present, equation (80.2) does not hold; the entropy of the gas increases when a streamline passes through a shock wave. We shall see, however, that Bernoulli's equation (80.1) remains valid even when there are shock waves, since $w+\frac{1}{2}v^2$ is a quantity which is conserved across a surface of discontinuity (§82); formula (80.14), for example, therefore remains valid also.

PROBLEM

Express the temperature, pressure and density along a streamline in terms of the Mach number.

SOLUTION. Using the formulae obtained above, we find

$$T_0/T = 1+\tfrac{1}{2}(\gamma-1)\mathrm{M}^2, \qquad p_0/p = [1+\tfrac{1}{2}(\gamma-1)\mathrm{M}^2]^{\gamma/(\gamma-1)},$$

$$\rho_0/\rho = [1+\tfrac{1}{2}(\gamma-1)\mathrm{M}^2]^{1/(\gamma-1)}.$$

§81. Surfaces of discontinuity

In the preceding chapters we have considered only flows such that all quantities (velocity, pressure, density, etc.) vary continuously. Flows are also possible, however, for which discontinuities in the distribution of these quantities occur.

A discontinuity in a gas flow occurs over one or more surfaces; the quantities concerned change discontinuously as we cross such a surface, which is called a *surface of discontinuity*. In non-steady gas flow the surfaces of discontinuity do not in general remain fixed; here it should be emphasised, however, that the rate of motion of these surfaces bears no relation to the velocity of the gas flow itself. The gas particles in their motion may cross a surface of discontinuity.

Certain boundary conditions must be satisfied on surfaces of discontinuity. To formulate these conditions, we consider an element of the surface and use a co-ordinate system fixed to this element, with the x-axis along the normal.†

Firstly, the mass flux must be continuous: the mass of gas coming from one side must equal the mass leaving the other side. The mass flux through the surface element considered is ρv_x per unit area. Hence we must have $\rho_1 v_{1x} = \rho_2 v_{2x}$, where the suffixes 1 and 2 refer to the two sides of the surface of discontinuity.

---

† If the flow is not steady, we consider an element of the surface during a short interval of time.

The difference between the values of any quantity on the two sides of the surface will be denoted by enclosing it in square brackets; for example, $[\rho v_x] \equiv \rho_1 v_{1x} - \rho_2 v_{2x}$, and the condition just derived can be written

$$[\rho v_x] = 0. \tag{81.1}$$

Next, the energy flux must be continuous. The energy flux is given by (6.3). We therefore obtain the condition

$$[\rho v_x(\tfrac{1}{2}v^2 + w)] = 0. \tag{81.2}$$

Finally, the momentum flux must be continuous, i.e. the forces exerted on each other by the gases on the two sides of the surface of discontinuity must be equal. The momentum flux per unit area is (see §7) $p n_i + \rho v_i v_k n_k$. The normal vector $\mathbf{n}$ is along the x-axis. The continuity of the x-component of the momentum flux therefore gives the condition

$$[p + \rho v_x^2] = 0, \tag{81.3}$$

while that of the y and z components gives

$$[\rho v_x v_y] = 0, \qquad [\rho v_x v_z] = 0. \tag{81.4}$$

Equations (81.1)–(81.4) form a complete system of boundary conditions at a surface of discontinuity. From them we can immediately deduce the possibility of two types of surface of discontinuity.

In the first type, there is no mass flux through the surface. This means that $\rho_1 v_{1x} = \rho_2 v_{2x} = 0$. Since ρ_1 and ρ_2 are not zero, it follows that $v_{1x} = v_{2x} = 0$. The conditions (81.2) and (81.4) are then satisfied, and the condition (81.3) gives $p_1 = p_2$. Thus the normal velocity component and the gas pressure are continuous at the surface of discontinuity:

$$v_{1x} = v_{2x} = 0, \quad [p] = 0, \tag{81.5}$$

while the tangential velocities v_y, v_z and the density (as well as the other thermodynamic quantities except the pressure) may be discontinuous by any amount. We call this a *tangential discontinuity*.

In the second type, the mass flux is not zero, and v_{1x} and v_{2x} are therefore also not zero. We then have from (81.1) and (81.4)

$$[v_y] = 0, \quad [v_z] = 0, \tag{81.6}$$

i.e. the tangential velocity is continuous at the surface of discontinuity. The pressure, the density (and the other thermodynamic quantities) and the normal velocity, however, are discontinuous, their discontinuities being related by (81.1)–(81.3). In the condition (81.2) we can cancel ρv_x by (81.1), and replace v^2 by v_x^2 since v_y and v_z are continuous. Thus the following conditions must hold at the surface of discontinuity in this case:

$$\left.\begin{array}{c} [\rho v_x] = 0, \\[4pt] [\tfrac{1}{2}v_x^2 + w] = 0, \\[4pt] [p + \rho v_x^2] = 0. \end{array}\right\} \tag{81.7}$$

A discontinuity of this kind is called a *shock wave*, or simply a *shock*.

If we now return to the fixed co-ordinate system, we must everywhere replace v_x by the difference between the gas velocity component v_n normal to the surface of discontinuity and the velocity u of the surface itself, which is defined to be normal to the surface:

$$v_x = v_n - u. \tag{81.8}$$

The velocities v and u are taken in the fixed system. The velocity v_x is the velocity of the gas relative to the surface of discontinuity; we can also say that $-v_x = u - v_n$ is the rate of propagation of the surface relative to the gas. It should be noticed that, if v_x is discontinuous, this velocity has different values relative to the gas on the two sides of the surface.

We have already discussed (in §30) tangential discontinuities, at which the tangential velocity component is discontinuous, and we showed that, in an incompressible fluid, such discontinuities are absolutely unstable and must result in a turbulent region. A similar investigation for a compressible fluid shows that the same instability occurs, for any velocities.

A particular "degenerate" case of tangential discontinuity is that where the velocity is continuous, but not the density (and therefore the other thermodynamic quantities, except the pressure). The above remarks on instability do not relate to discontinuities of this kind.

§82. The shock adiabatic

Let us now investigate shock waves in detail. We have seen that, in this type of discontinuity, the tangential component of the gas velocity is continuous. We can therefore take a co-ordinate system in which the surface element considered is at rest, and the tangential component of the gas velocity is zero on both sides.† Then we can write the normal component v_x as v simply, and the conditions (81.7) take the form

$$\rho_1 v_1 = \rho_2 v_2 \equiv j, \tag{82.1}$$

$$p_1 + \rho_1 v_1{}^2 = p_2 + \rho_2 v_2{}^2, \tag{82.2}$$

$$w_1 + \tfrac{1}{2} v_1{}^2 = w_2 + \tfrac{1}{2} v_2{}^2, \tag{82.3}$$

where j denotes the mass flux density at the surface of discontinuity. In what follows we shall always take j positive, with the gas going from side 1 to side 2. That is, we call gas 1 the one into which the shock wave moves, and gas 2 that which remains behind the shock. We call the side of the shock wave towards gas 1 the *front* of the shock, and that towards gas 2 the *back*.

We shall derive a series of relations which follow from the above conditions. Using the specific volumes $V_1 = 1/\rho_1$, $V_2 = 1/\rho_2$, we obtain from (82.1)

$$v_1 = jV_1, \qquad v_2 = jV_2 \tag{82.4}$$

---

† This co-ordinate system is used everywhere in §§82–85, 87, 88.
   A shock wave at rest is called a *compression discontinuity*. If the shock is perpendicular to the direction of flow, we have a *normal shock*, otherwise an *oblique shock*.

and, substituting in (82.2),

$$p_1 + j^2 V_1 = p_2 + j^2 V_2,$$ (82.5)

or

$$j^2 = (p_2 - p_1)/(V_1 - V_2).$$ (82.6)

This formula, together with (82.4), relates the rate of propagation of a shock wave to the pressures and densities of the gas on the two sides of the surface.

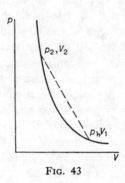

FIG. 43

Since j^2 is positive, we see that either $p_2 > p_1$, $V_1 > V_2$, or $p_2 < p_1$, $V_1 < V_2$; we shall see below that only the former case can actually occur.

We may note the following useful formula for the velocity difference $v_1 - v_2$. Substituting (82.6) in $v_1 - v_2 = j(V_1 - V_2)$, we obtain†

$$v_1 - v_2 = \sqrt{[(p_2 - p_1)(V_1 - V_2)]}.$$ (82.7)

Next, we write (82.3) in the form

$$w_1 + \tfrac{1}{2} j^2 V_1{}^2 = w_2 + \tfrac{1}{2} j^2 V_2{}^2$$ (82.8)

and, substituting j^2 from (82.6), obtain

$$w_1 - w_2 + \tfrac{1}{2}(V_1 + V_2)(p_2 - p_1) = 0.$$ (82.9)

If we replace the heat function w by $\epsilon + pV$, where ϵ is the internal energy, we can write this relation as

$$\epsilon_1 - \epsilon_2 + \tfrac{1}{2}(V_1 - V_2)(p_1 + p_2) = 0.$$ (82.10)

These relations hold between the thermodynamic quantities on the two sides of the surface of discontinuity.

For given p_1, V_1, equation (82.9) or (82.10) gives the relation between p_2 and V_2. This relation is called the *shock adiabatic* or the *Hugoniot adiabatic* (W. J. M. RANKINE, 1870; H. HUGONIOT, 1889). It is represented graphically in the pV-plane (Fig. 43) by a curve passing through the given point (p_1, V_1)

---

† Here we write the positive square root, since, as we shall see later (§84) we must have $v_1 - v_2 > 0$.

(for $p_1 = p_2$, $V_1 = V_2$ we have also $\epsilon_1 = \epsilon_2$, so that (82.10) is satisfied identically). It should be noted that the shock adiabatic cannot intersect the vertical
line $V = V_1$ except at (p_1, V_1). For the existence of another intersection
would mean that two different pressures satisfying (82.10) correspond to the
same volume. For $V_1 = V_2$, however, we have from (82.10) also $\epsilon_1 = \epsilon_2$,
and when the volumes and energies are the same the pressures must be the
same. Thus the line $V = V_1$ divides the shock adiabatic into two parts,
each of which lies entirely on one side of the line. Similarly, the shock
adiabatic meets the horizontal line $p = p_1$ only at the point (p_2, V_1).

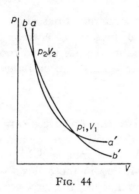

FIG. 44

Let aa' (Fig. 44) be the shock adiabatic through the point (p_1, V_1) as a
state of gas 1. We take any point (p_2, V_2) on it and draw through that point
another adiabatic bb', for which (p_2, V_2) is a state of gas 1. It is evident that
the pair of values (p_1, V_1) satisfies the equation of this adiabatic also. The
adiabatics aa' and bb' therefore intersect at the two points (p_1, V_1) and $(p_2,
V_2)$. It must be emphasised that the adiabatics are not identical, as would
happen for Poisson adiabatics through a given point. This is a consequence
of the fact that the equation of the shock adiabatic cannot be written in the
form $f(p, V) = $ constant, where f is some function, whereas the Poisson
adiabatic, for example, can be written $s(p, V) = $ constant. The Poisson
adiabatics for a given gas form a one-parameter family of curves, but the
shock adiabatic is determined by two parameters, the initial values p_1 and
V_1. This has also the following important result: if two (or more) successive
shock waves take a gas from state 1 to state 2 and from there to state 3, the
transition from state 1 to state 3 cannot in general be effected by the passage
of any one shock wave.

For a given initial thermodynamic state of the gas (i.e. for given p_1 and
V_1), the shock wave is defined by only one parameter; for instance, if the
pressure p_2 behind the shock is given, then V_2 is determined by the Hugoniot
adiabatic, and the flux density j and the velocities v_1 and v_2 are then given by
formulae (82.4) and (82.6). It should be mentioned, however, that we are

here considering the shock wave in a co-ordinate system in which the gas is moving normal to the surface. If the shock wave may be situated obliquely to the direction of flow, another parameter is needed; for example, the value of the velocity component tangential to the surface.

The following convenient graphical interpretation of formula (82.6) may be mentioned. If the point (p_1, V_1) on the shock adiabatic (Fig. 43) is joined by a chord to any other point (p_2, V_2) on it, then $(p_2 - p_1)/(V_2 - V_1) = -j^2$ is just the slope of this chord relative to the axis of abscissae. Thus j, and therefore the velocity of the shock wave, are determined at each point of the shock adiabatic by the slope of the chord joining that point to the point (p_1, V_1).

Like the other thermodynamic quantities, the entropy is discontinuous at a shock wave. By the law of increase of entropy, the entropy of a gas can only increase during its motion. Hence the entropy s_2 of the gas which has passed through the shock wave must exceed its initial entropy s_1:

$$s_2 > s_1. \tag{82.11}$$

We shall see below that this condition places very important restrictions on the manner of variation of all quantities in a shock wave.

The following fact should be emphasised. The presence of shock waves results in an increase in entropy in those flows which can be regarded as motions of an ideal fluid in all space, the viscosity and thermal conductivity being zero. The increase in entropy signifies that the motion is irreversible, i.e. energy is dissipated. Thus the discontinuities are a means by which energy can be dissipated in the motion of an ideal fluid. It follows that d'Alembert's paradox (§11) does not arise when bodies move in an ideal fluid in such a way as to cause shock waves. In such cases there is a drag force.

The true mechanism by which the entropy increases in shock waves lies, of course, in dissipative processes occurring in the very thin layers which actual shock waves are (see §87). It should be noticed, however, that the amount of this dissipation is entirely determined by the laws of conservation of mass, energy and momentum, when they are applied to the two sides of such layers; the width of the layers is just such as to give the increase in entropy required by these conservation laws.

The increase in entropy in a shock wave has another important effect on the motion: even if we have potential flow in front of the shock wave, the flow behind it is in general rotational. We shall return to this matter in §106.

§83. Weak shock waves

Let us consider a shock wave in which the discontinuity in every quantity is small; we call this a *weak shock wave*. We transform the relation (82.9) by expanding in powers of the small differences $s_2 - s_1$ and $p_2 - p_1$. We shall see that the first- and second-order terms in $p_2 - p_1$ then cancel; we must therefore carry the expansion with respect to $p_2 - p_1$ as far as the third order. In the expansion with respect to $s_2 - s_1$, only the first-order terms need be retained. We have

$$w_2 - w_1 = (\partial w/\partial s_1)_p (s_2 - s_1) + (\partial w/\partial p_1)_s (p_2 - p_1) +$$
$$+ \tfrac{1}{2}(\partial^2 w/\partial p_1^2)_s (p_2 - p_1)^2 + \tfrac{1}{6}(\partial^3 w/\partial p_1^3)_s (p_2 - p_1)^3.$$

By the thermodynamic identity $dw = T\, ds + V\, dp$ we have for the derivatives

$$(\partial w/\partial s)_p = T, \qquad (\partial w/\partial p)_s = V.$$

Hence

$$w_2 - w_1 = T_1 (s_2 - s_1) + V_1 (p_2 - p_1) +$$
$$+ \tfrac{1}{2}(\partial V/\partial p_1)_s (p_2 - p_1)^2 + \tfrac{1}{6}(\partial^2 V/\partial p_1^2)_s (p_2 - p_1)^3.$$

The volume V_2 need be expanded only with respect to $p_2 - p_1$, since the second term of equation (82.9) already contains the small difference $p_2 - p_1$, and an expansion with respect to $s_2 - s_1$ would give a term of the form $(s_2 - s_1)(p_2 - p_1)$, which is of no interest. Thus

$$V_2 - V_1 = (\partial V/\partial p_1)_s (p_2 - p_1) + \tfrac{1}{2}(\partial^2 V/\partial p_1^2)_s (p_2 - p_1)^2.$$

Substituting this expansion in (82.9), we obtain

$$s_2 - s_1 = \frac{1}{12 T_1} \left(\frac{\partial^2 V}{\partial p_1^2} \right)_s (p_2 - p_1)^3. \tag{83.1}$$

Thus the discontinuity of entropy in a weak shock wave is of the third order of smallness relative to the discontinuity of pressure.

In all cases that have been investigated, the compressibility $-(\partial V/\partial p)_s$ decreases with increasing pressure, i.e. the second derivative

$$(\partial^2 V/\partial p^2)_s > 0. \tag{83.2}$$

It should be emphasised, however, that this is not a thermodynamic relation, and cannot be derived by thermodynamic arguments. It is therefore possible in principle that the derivative might be negative. We shall find several times in what follows that the sign of the derivative $(\partial^2 V/\partial p^2)_s$ is very important in gas dynamics. In future we shall assume it to be positive.†

Let us draw through the point 1 (p_1, V_1) in the pV-plane two curves, the shock adiabatic and the Poisson adiabatic. The equation of the latter is $s_2 - s_1 = 0$. By comparing this with the equation (83.1) of the shock adiabatic near the point 1, we see that the two curves have contact of the second order at this point, both the first and the second derivatives being equal. In order to decide the relative position of the two curves near the point 1, we use the fact that, according to (83.1) and (83.2), we must have $s_2 > s_1$ on the shock adiabatic for $p_2 > p_1$, while on the Poisson adiabatic $s_2 = s_1$. The abscissa of a point on the shock adiabatic must therefore exceed that of a point on

---

† For a perfect gas $(\partial^2 V/\partial p^2)_s = (\gamma+1)V/\gamma^2 p^2$. This expression can be most simply obtained by differentiating Poisson's adiabatic equation $pV^\gamma = \text{constant}$.

the Poisson adiabatic having the same ordinate p_2. This follows at once from the fact that, by the well-known thermodynamic formula $(\partial V/\partial s)_p = (T/c_p)(\partial V/\partial T)_p$, the entropy increases with the volume at constant pressure for all substances which expand on heating, i.e. which have $(\partial V/\partial T)_p$ positive. We can similarly deduce that, for $p_2 < p_1$, the abscissa of a point on the Poisson adiabatic exceeds that of the corresponding point on the shock adiabatic. Thus, near the point of contact, the two curves lie as shown in Fig. 45 (HH' being the shock adiabatic and PP' the Poisson adiabatic)†, both being concave upwards, by (83.2).

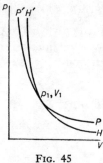

FIG. 45

For small $p_2 - p_1$ and $V_2 - V_1$, formula (82.6) can be written, in the first approximation, as $j^2 = -(\partial p/\partial V)_s$ (we take the derivative for constant entropy, since the tangents to the two adiabatics at the point 1 coincide). The velocities v_1 and v_2 are, in the same approximation, equal:

$$v_1 = v_2 = v = jV = \sqrt{[-V^2(\partial p/\partial V)_s]} = \sqrt{(\partial p/\partial \rho)_s}.$$

This is just the velocity of sound c. Thus the rate of propagation of weak shock waves is, in the first approximation, the velocity of sound:

$$v = c. \tag{83.3}$$

From the properties of the shock adiabatic near the point 1 derived above we can deduce a number of important consequences. Since we must have $s_2 > s_1$ in a shock wave, it follows that $p_2 > p_1$, i.e. the point 2 (p_2, V_2) must lie above the point 1. Moreover, since the chord 12 has a greater slope than the tangent to the adiabatic at the point 1 (Fig. 43), and the slope of the tangent is equal to the derivative $(\partial p/\partial V_1)_{s_1}$, we have $j^2 > -(\partial p/\partial V_1)_{s_1}$. Multiplying both sides of this inequality by V_1^2, we find

$$j^2 V_1^2 = v_1^2 > -V_1^2(\partial p/\partial V_1)_{s_1} = (\partial p/\partial \rho_1)_{s_1} = c_1^2,$$

where c_1 is the velocity of sound corresponding to the point 1. Thus $v_1 > c_1$.

---

† If $(\partial V/\partial T)_p$ is negative, the relative position is reversed.

Finally, from the fact that the chord 12 has a smaller slope than the tangent at the point 2, it follows in like manner that $v_2 < c_2$.†

§84. The direction of variation of quantities in a shock wave

The results of §83 show that, if the derivative $(\partial^2 V/\partial p^2)_s$ is assumed positive, it can be demonstrated very simply that for weak shocks the condition of increasing entropy ($s_2 > s_1$) necessarily means that

$$p_2 > p_1, \tag{84.1}$$

$$v_1 > c_1, \qquad v_2 < c_2. \tag{84.2}$$

From the remark made concerning (82.6) it follows that, if $p_2 > p_1$, then

$$V_1 > V_2, \tag{84.3}$$

and, since $v_1/V_1 = v_2/V_2 = j$, also

$$v_1 > v_2. \tag{84.4}$$

We shall now show that all these inequalities actually hold (still on the assumption that $(\partial^2 V/\partial p^2)_s$ is positive) for shock waves of any intensity. We shall therefore conclude, in particular, that, when gas passes through a shock wave, it is compressed, the pressure and density increasing (E. JOUGUET, 1904; G. ZEMPLEN, 1905).‡ This means, graphically, that only the upper branch of the shock adiabatic (above the point 1) has any real significance; shock waves corresponding to points on the lower branch cannot exist. We may also mention the following important result which can be derived from the inequalities (84.2). Since a shock wave moves relative to the gas in front of it with a velocity $v_1 > c_1$, it is clear that no perturbation starting from the shock wave can penetrate into that gas. In other words, the presence of the shock has no effect on the state of the gas in front of it.

We shall now prove these statements, beginning with a preliminary calculation. We differentiate the relations (82.5) and (82.8) with respect to the quantities pertaining to gas 2, assuming the state of gas 1 to be unchanged. This means that p_1, V_1 and w_1 are regarded as constants, while p_2, V_2, w_2 and also j (which depends on p_2 and V_2) are differentiated. From (82.5) we obtain

$$V_1 d(j^2) = dp_2 + j^2 dV_2 + V_2 d(j^2),$$

or

$$dp_2 + j^2 dV_2 = (V_1 - V_2)d(j^2), \tag{84.5}$$

---

† It can easily be shown in the same way that, when the derivative $(\partial^2 V/\partial p^2)_s$ is negative, the condition $s_2 > s_1$ for weak shock waves implies that $p_2 < p_1$, while the velocities again satisfy $v_1 > c_1$, $v_2 < c_2$.

‡ If we change to a co-ordinate system in which gas 1 (in front of the shock wave) is at rest, and the shock is moving, then the inequality $v_1 > v_2$ means that the gas behind the shock wave moves (with velocity $v_1 - v_2$) in the same direction as the shock itself.

and from (82.8)

$$dw_2 + j^2 V_2 dV_2 = \tfrac{1}{2}(V_1{}^2 - V_2{}^2)d(j^2),$$

or, expanding the differential dw_2,

$$T_2 ds_2 + V_2 dp_2 + j^2 V_2 dV_2 = \tfrac{1}{2}(V_1{}^2 - V_2{}^2)d(j^2).$$

Substituting this equation in (84.5), we obtain

$$T_2 ds_2 = \tfrac{1}{2}(V_1 - V_2)^2 d(j^2). \tag{84.6}$$

Hence we see that

$$d(j^2)/ds_2 > 0, \tag{84.7}$$

i.e. j^2 increases with s_2.

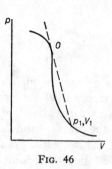

Fig. 46

We now show that there can be no point on the shock adiabatic at which it touches any line drawn from the point 1 (such as the point O is in Fig. 46). At such a point the slope of the chord from the point 1 is a minimum, and j^2 has a corresponding maximum, so that $d(j^2)/dp_2 = 0$. We see from (84.6) that in this case we also have $ds_2/dp_2 = 0$. Next, substituting in (84.5) the differential dV_2 in the form $dV_2 = (\partial V_2/\partial p_2)_{s_2} dp_2 + (\partial V_2/\partial s_2)_{p_2} ds_2$ and ds_2 in the form given by (84.6), and dividing by dp_2, we obtain

$$1 + j^2\left(\frac{\partial V_2}{\partial p_2}\right)_{s_2} = (V_1 - V_2)\left\{1 - \frac{j^2(V_1 - V_2)}{T_2}\left(\frac{\partial V_2}{\partial s_2}\right)_{p_2}\right\}\frac{d(j^2)}{dp_2}.$$

Hence it follows that, for $d(j^2)/dp_2 = 0$, we must have

$$1 + j^2(\partial V_2/\partial p_2)_{s_2} = 1 - v_2{}^2/c_2{}^2 = 0,$$

i.e. $v_2 = c_2$; conversely, if $v_2 = c_2$, it follows that $d(j^2)/dp_2 = 0$.†
Thus, of the three equations

$$d(j^2)/dp_2 = 0, \qquad ds_2/dp_2 = 0, \qquad v_2 = c_2, \tag{84.8}$$

each implies the other two and all three would hold at the point O (Fig. 46).

---

† The expression in the braces can vanish only by chance, and this possibility is therefore unlikely.

Finally, we have for the derivative of $j^2(\partial V_2/\partial p_2)_{s_2} = -v_2^2/c_2^2$ at the point O

$$\frac{\mathrm{d}}{\mathrm{d}p_2}\left(\frac{v_2^2}{c_2^2}\right) = -j^2\left(\frac{\partial^2 V_2}{\partial p_2^2}\right)_{s_2}.$$

On account of the assumption that $(\partial^2 V/\partial p^2)_s$ is positive, we therefore have at O

$$\mathrm{d}(v_2/c_2)/\mathrm{d}p_2 < 0. \tag{84.9}$$

It is now easy to show that such a point cannot exist on the shock adiabatic. At points just above the point 1, $v_2/c_2 < 1$ (see the end of §83). The equation $v_2/c_2 = 1$ can therefore be satisfied only by an increase in v_2/c_2; that is, at O we should necessarily have $\mathrm{d}(v_2/c_2)/\mathrm{d}p_2 > 0$, whereas by (84.9) the converse is true. In an entirely similar manner, we can show that the ratio v_2/c_2 also cannot become equal to unity on the part of the shock adiabatic below the point 1.

From the impossibility of the existence of a point such as O, which has just been demonstrated, we can at once deduce from the graph of the shock adiabatic that the slope of the chord from the point 1 (p_1, V_1) to the point 2 (p_2, V_2) decreases as we move up the curve, and j^2 correspondingly increases. From this property of the shock adiabatic and the inequality (84.7) it follows immediately that the necessary condition $s_2 > s_1$ implies that $p_2 > p_1$ also.

It is also easy to see that, on the upper part of the shock adiabatic, the inequalities $v_2 < c_2$, $v_1 > c_1$ hold. The former follows at once from the fact that it holds near the point 1, and the ratio v_2/c_2 can never become equal to unity. The second inequality follows from the fact that every chord from the point 1 to a point 2 above it is steeper than the tangent to the adiabatic at the point 1, since the curve cannot behave as shown in Fig. 46.

The condition $s_2 > s_1$ and all three inequalities (84.1), (84.2) are therefore satisfied on the upper part of the shock adiabatic. On the lower part, however, none of these conditions holds. They are consequently equivalent, and if one is satisfied so are all the others.

In the preceding discussion we have everywhere assumed that the derivative $(\partial^2 V/\partial p^2)_s$ is positive. If this derivative could change sign, it would no longer be possible to draw from the necessity of $s_2 > s_1$ any general conclusions concerning inequalities for the other quantities. It is important, however, that the inequalities (84.2) for the velocities can be obtained by quite different arguments, which show that shock waves in which those inequalities do not hold cannot exist, even if their existence would not be disproved by the purely thermodynamic arguments given above.

The reason is that we have still to discuss the subject of he stability of shock waves. Let us suppose that a shock wave at rest is subjected to an infinitesimal displacement in a direction (say) perpendicular to its plane. It can be shown that the result of such a displacement is that the shock wave is continually accelerated in some direction, and it is clear that this demonstrates the absolute instability of such a wave and the impossibility of its existence.

The displacement of the shock wave is accompanied by infinitesimal perturbations in the gas pressure, velocity, etc. on both sides of the surface of discontinuity. These perturbations near the shock are then propagated away from it with the velocity of sound (relative to the gas); this, however, does not apply to the perturbation in the entropy, which is transmitted only with the gas itself. Thus an arbitrary perturbation of the type in question can be regarded as consisting of sound disturbances propagated in gases 1 and 2 on both sides of the shock wave, and a perturbation of the entropy; the latter, which moves with the gas, will evidently occur only in gas 2 behind the shock. In each of the sound disturbances, the changes in the various quantities are related by certain formulae which follow from the equations of motion (as in any sound wave, §63), and therefore any such disturbance is specified by only one parameter.

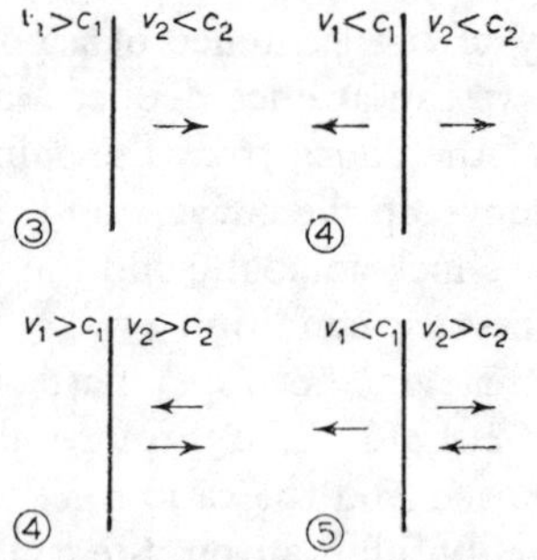

Fig. 47

Let us now compute the number of possible sound disturbances. It depends on the relative magnitudes of the gas velocities v_1, v_2 and the sound velocities c_1, c_2. We take the direction of motion of the gas (from 1 to 2) as the positive direction of the x-axis. The rate of propagation of the disturbance in gas 1 relative to the stationary shock wave is $u_1 = v_1 \pm c_1$, and in gas 2 it is $u_2 = v_2 \pm c_2$. Since these disturbances must be propagated away from the shock wave, it follows that $u_1 < 0$, $u_2 > 0$.

Let us suppose that $v_1 > c_1$, $v_2 < c_2$. Then it is clear that both values $u_1 = v_1 \pm c_1$ are positive, while only $v_2 + c_2$ of the two values of u_2 is positive. This means that the sound disturbances in which we are interested cannot exist in gas 1, while in gas 2 there can be only one, which is propagated relative to the gas with velocity c_2. The calculation in other cases is similar.

The result is shown in Fig. 47, where each arrow corresponds to one sound disturbance, propagated relative to the gas in the direction shown by the arrow. Each sound disturbance is defined, as stated above, by one parameter. Furthermore, in all four cases there are two other parameters, one determining the entropy perturbation propagated in gas 2 and one determining the displacement of the shock wave. For each of the four cases in Fig. 47, the number in a circle shows the total number of parameters, thus obtained, which define an arbitrary perturbation arising from the displacement of the shock wave.

The number of boundary conditions which must be satisfied by a perturbation on the surface of discontinuity is three (the continuity of the mass, energy and momentum fluxes). The solution of the stability problem is effected by prescribing the displacement of the shock wave (and therefore the perturbations in all the other quantities) in a form proportional to $e^{\Omega t}$, and determining the possible values of Ω by means of the boundary conditions; the existence of real positive values of Ω indicates absolute instability. In all except the first of the cases shown in Fig. 47, the number of parameters available exceeds the number of equations given by the boundary conditions at the discontinuity. In these cases, therefore, the boundary conditions admit any (and therefore any positive) value of Ω, and the shock wave is absolutely unstable. In the one case $v_1 > c_1$, $v_2 < c_2$, however, the number of parameters just equals the number of equations, and these therefore give a definite value of Ω. It is evident, without writing down the equations, that this value must be $\Omega = 0$, since the problem contains no parameter of the dimensions $\sec^{-1}$ which could determine a value of Ω different from zero but not arbitrary. There is therefore no such instability in this case.

Thus we see that the inequalities (84.2) for the velocity of the shock wave are necessary for the shock to exist, whatever the thermodynamic properties of the gas.

In order to decide the stability of shock waves for which the condition (84.2) is satisfied, we should have to investigate also the other possible modes of instability. One of these is instability with respect to perturbations of the kind considered in §30 (characterised by periodicity in the direction parallel to the surface of discontinuity and forming "ripples" on this surface). We shall not perform the calculations here, but merely mention that shock waves are almost always stable with respect to such perturbations. Instability can occur only for certain very special forms of the shock adiabatic, which seem hardly ever to occur in Nature; they all require that the derivative $(\partial^2 V/\partial p^2)_s$ should be of variable sign.[†]

A shock wave might also, in principle, be unstable with respect to breakup into more than one surface of discontinuity. This problem has not been adequately investigated, but such instabilities may likewise occur only for certain very special types of shock adiabatic.

§85. Shock waves in a perfect gas

Let us apply the general relations obtained in the previous sections to shock waves in a perfect gas. The heat function of a perfect gas is given by the simple formula $w = \gamma p V/(\gamma - 1)$. Substituting this expression in (82.9),

---

[†] See S. P. D'YAKOV, *Zhurnal éksperimental'noĭ i teoreticheskoĭ fiziki* **27**, 288, 1954; V. M. KONTOROVICH, *ibid.* **33**, 1525, 1957; *Soviet Physics JETP* **6** (33), 1179, 1958.

we have after a simple transformation

$$\frac{V_2}{V_1} = \frac{(\gamma+1)p_1+(\gamma-1)p_2}{(\gamma-1)p_1+(\gamma+1)p_2}. \tag{85.1}$$

Using this formula, we can determine any of the quantities p_1, V_1, p_2, V_2 from the other three. The ratio V_2/V_1 is a monotonically decreasing function of the ratio p_2/p_1, tending to the finite limit $(\gamma-1)/(\gamma+1)$. The curve showing p_2 as a function of V_2 for given p_1, V_1 (the shock adiabatic) is represented in Fig. 48. It is a rectangular hyperbola with asymptotes $V_2/V_1 = (\gamma-1)/(\gamma+1)$, $p_2/p_1 = -(\gamma-1)/(\gamma+1)$. As we know, only the upper part of the curve, above the point $V_2/V_1 = p_2/p_1 = 1$, has any real significance; it is shown in Fig. 48 (for $\gamma = 1\cdot4$) by a continuous line.

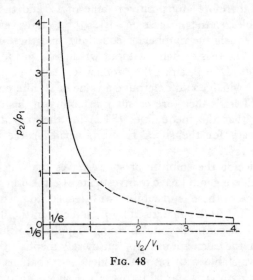

FIG. 48

For the ratio of the temperatures on the two sides of the discontinuity we find, from the equation of state for a perfect gas $T_2/T_1 = p_2V_2/p_1V_1$, that

$$\frac{T_2}{T_1} = \frac{p_2}{p_1}\frac{(\gamma+1)p_1+(\gamma-1)p_2}{(\gamma-1)p_1+(\gamma+1)p_2}. \tag{85.2}$$

For the flux density j we obtain from (82.6) and (85.1)

$$j^2 = \{(\gamma-1)p_1+(\gamma+1)p_2\}/2V_1, \tag{85.3}$$

and then for the velocities of propagation of the shock wave relative to the gas before and behind it

$$\begin{aligned}
v_1^2 &= \tfrac{1}{2}V_1\{(\gamma-1)p_1+(\gamma+1)p_2\},\\
v_2^2 &= \tfrac{1}{2}V\{(\gamma+1)p_1+(\gamma-1)p_2\}^2/\{(\gamma-1)p_1+(\gamma+1)p_2\}.
\end{aligned} \tag{85.4}$$

We may derive limiting results for very strong shock waves, in which p_2 is very large compared with p_1.† From (85.1) and (85.2) we have

$$V_2/V_1 = \rho_1/\rho_2 = (\gamma-1)/(\gamma+1), \quad T_2/T_1 = (\gamma-1)p_2/(\gamma+1)p_1. \quad (85.5)$$

The ratio T_2/T_1 increases to infinity with p_2/p_1, i.e. the temperature discontinuity in a shock wave, like the pressure discontinuity, can be arbitrarily great. The density ratio, however, tends to a constant limit; e.g., for a monatomic gas the limit is $\rho_2 = 4\rho_1$, and for a diatomic gas $\rho_2 = 6\rho_1$. The velocities of propagation of a strong shock wave are

$$v_1 = \sqrt{\{\tfrac{1}{2}(\gamma+1)p_2V_1\}}, \quad v_2 = \sqrt{\{\tfrac{1}{2}(\gamma-1)^2p_2V_1/(\gamma+1)\}}. \quad (85.6)$$

They increase as the square root of the pressure p_2.

Finally, we may give some formulae useful in applications, which express the ratios of densities, pressures and temperatures in a shock wave in terms of the Mach number $M_1 = v_1/c_1$. These formulae are easily derived from the foregoing results:

$$\rho_2/\rho_1 = v_1/v_2 = (\gamma+1)M_1^2/\{(\gamma-1)M_1^2+2\}, \quad (85.7)$$

$$p_2/p_1 = 2\gamma M_1^2/(\gamma+1)-(\gamma-1)/(\gamma+1), \quad (85.8)$$

$$T_2/T_1 = \{2\gamma M_1^2-(\gamma-1)\}\{(\gamma-1)M_1^2+2\}/(\gamma+1)^2M_1. \quad (85.9)$$

The Mach number M_2 is given in terms of M_1 by

$$M_2^2 = \{2+(\gamma-1)M_1^2\}/\{2\gamma M_1^2-(\gamma-1)\}. \quad (85.10)$$

PROBLEMS

PROBLEM 1. Derive the formula $v_1v_2 = c_*^2$, where c_* is the critical velocity.

SOLUTION. Since $w+\tfrac{1}{2}v^2$ is continuous at a shock wave, we can define a critical velocity which is the same for gases 1 and 2 by

$$\frac{\gamma p_1}{(\gamma-1)\rho_1} + \tfrac{1}{2}v_1^2 = \frac{\gamma p_2}{(\gamma-1)\rho_2} + \tfrac{1}{2}v_2^2 = \frac{\gamma+1}{2(\gamma-1)}c_*^2;$$

cf. (80.7). Determining p_2/ρ_2 and p_1/ρ_1 from these equations and substituting in

$$v_1-v_2 = \frac{p_2}{\rho_2 v_2} - \frac{p_1}{\rho_1 v_1}$$

(obtained by combining (82.1) and (82.2)), we obtain

$$\frac{\gamma+1}{2\gamma}(v_1-v_2)\left(1 - \frac{c_*^2}{v_1v_2}\right) = 0.$$

Since $v_1 \neq v_2$, this gives the required relation.

PROBLEM 2. Determine the value of the ratio p_2/p_1, for given temperatures T_1, T_2 at a discontinuity in a perfect gas with a variable specific heat.

---

† It is necessary that not only $p_2 \gg p_1$ but $p_2 \gg (\gamma+1)p_1/(\gamma-1)$.

SOLUTION. In the general case of a perfect gas with variable specific heat, we can say only that w (like ϵ) is a function of temperature alone, and that p, V and T are related by the equation of state $pV = RT/\mu$. Solving equation (82.9) for p_2/p_1, we obtain

$$\frac{p_2}{p_1} = \frac{\mu}{RT}(w_2 - w_1) - \frac{T_2 - T_1}{2T_1} + \sqrt{\left\{\left[\frac{\mu(w_2 - w_1)}{RT_1} - \frac{T_2 - T_1}{2T_1}\right]^2 + \frac{T_2}{T_1}\right\}},$$

where $w_1 = w(T_1)$, $w_2 = w(T_2)$.

PROBLEM 3. A plane sound wave meets normally a shock wave in a perfect gas. Determine the intensity of sound transmitted by the shock wave (D. I. BLOKHINTSEV, 1945).[†]

SOLUTION. Since a shock wave is propagated with supersonic velocity relative to the gas in front of it, no sound wave can be reflected from it. In gas 2, behind the discontinuity, an ordinary isentropic transmitted sound wave is propagated, and also a perturbation of the entropy (at constant pressure), which is propagated with the moving gas itself.

We consider the process in a co-ordinate system in which the shock wave is at rest, and the gas moves through it in the positive direction of the x-axis, the incident sound wave being propagated in this direction also. The perturbations on the two sides of the discontinuity are related by conditions obtained by varying the boundary conditions (82.1)–(82.3). As a result of the sound disturbance, the shock wave also begins to oscillate; denoting its oscillatory velocity by δu, we must write the change in the velocities v_1, v_2 in the boundary conditions as $\delta v_1 - \delta u$, $\delta v_2 - \delta u$. Thus[‡]

$$v_1 \delta \rho_1 + \rho_1(\delta v_1 - \delta u) = v_2 \delta \rho_2 + \rho_2(\delta v_2 - \delta u),$$

$$\delta p_1 + v_1^2 \delta \rho_1 + 2\rho_1 v_1(\delta v_1 - \delta u) = \delta p_2 + v_2^2 \delta \rho_2 + 2\rho_2 v_2(\delta v_2 - \delta u),$$

$$\delta w_1 + v_1(\delta v_1 - \delta u) = \delta w_2 + v_2(\delta v_2 - \delta u).$$

In the incident sound wave we have

$$\delta s_1 = 0, \qquad \delta v_1 = (c_1/\rho_1)\delta \rho_1 = \delta p_1/c_1\rho_1, \qquad \delta w_1 = \delta p_1/\rho_1.$$

The perturbation in medium 2 is composed of the sound wave and the "entropy wave", which we denote by one and two primes respectively:

$$\delta s_2' = 0, \qquad \delta v_2' = (c_2/\rho_2)\delta \rho_2' = \delta p_2'/c_2\rho_2, \qquad \delta w_2 = \delta p_2'/\rho_2,$$

$$\delta p_2'' = 0, \qquad \delta v_2'' = 0, \qquad \delta w_2'' = T_2\delta s_2'' = -c_2^2\delta \rho_2''/\rho_2(\gamma - 1)$$

(for a perfect gas $(\partial s/\partial \rho)_p = -c_p/\rho$).

These relations enable us to express all quantities in the transmitted waves in terms of the corresponding quantities in the incident wave. The ratio of pressures in the sound waves is found to be

$$\frac{\delta p_2'}{\delta p_1} = \frac{M_1 + 1}{M_2 + 1}\left\{\frac{2(\gamma - 1)M_1 M_2^2(M_1^2 - 1) - (M_1 + 1)[(\gamma - 1)M_1^2 + 2]}{2(\gamma - 1)M_2^2(M_1^2 - 1) - (M_2 + 1)[(\gamma - 1)M_1^2 + 2]}\right\}.$$

For a weak shock wave ($p_2 - p_1 \ll p_1$) we find

$$\frac{\delta p_2'}{\delta p_1} \approx 1 + \frac{\gamma + 1}{2\gamma} \frac{p_2 - p_1}{p_1},$$

[†] The solution of the more general problem of oblique incidence of sound on a shock wave in an arbitrary medium is given by V. M. KONTOROVICH, *Zhurnal éksperimental'noĭ i teoreticheskoĭ fiziki* 33, 1527, 1957; *Soviet Physics JETP* 6 (33), 1180, 1958.

[‡] We here denote the variable parts of quantities by δ instead of the usual prime.

and in the opposite limiting case of a strong shock wave

$$\frac{\delta p_2'}{\delta p_1} \approx \frac{1}{\gamma + \sqrt{[2\gamma(\gamma-1)]}} \frac{p_2}{p_1}.$$

In both cases the pressure amplitude in the transmitted wave is greater than that in the incident wave.

§86. Oblique shock waves

Let us consider a steady shock wave, and abandon the system of co-ordinates used hitherto, in which the gas velocity is perpendicular to the shock surface element considered. The streamlines can intersect the surface of such a shock wave at any angle,† and in doing so are "refracted": the tangential component of the gas velocity is unchanged, while the normal component is, according to (84.4), diminished: $v_{1t} = v_{2t}$, $v_{1n} > v_{2n}$. It is therefore clear that the streamlines "approach" the shock wave as they pass through it (cf. Fig. 49). Thus the streamlines are always refracted in a definite direction in passing through a shock wave.

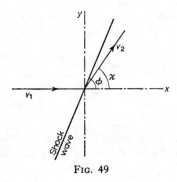

Fig. 49

The motion behind a shock wave may be either subsonic or supersonic (only the normal velocity component need be less than the velocity of sound c_2); the motion in front of it is necessarily supersonic. If the gas flow on both sides is supersonic, every disturbance must be propagated along the surface in the direction of the tangential component of the gas velocity. In this sense we can speak of the "direction" of a shock wave, and distinguish shock waves leaving and reaching any point (as we did for characteristics, the motion near which is always supersonic; see §79). If the motion behind the shock is subsonic, there is strictly no meaning in speaking of its "direction", since perturbations can be propagated in all directions on its surface.

We shall derive a relation between the two components of the gas velocity after it has passed through an oblique shock wave, supposing that we have a perfect gas. We take the direction of the gas velocity v_1 in front of the shock as the x-axis; let ϕ be the angle between the shock and the x-axis (Fig. 49).

† The only restriction is that the normal velocity component v_{1n} exceeds c_1.

The continuity of the velocity component tangential to the shock means that $v_1 \cos \phi = v_{2x} \cos \phi + v_{2y} \sin \phi$, or

$$\tan \phi = (v_1 - v_{2x})/v_{2y}. \tag{86.1}$$

Next we use formula (85.7), in which v_1 and v_2 denote the velocity components normal to the plane of the shock wave and must be replaced by $v_1 \sin \phi$ and $v_{2x} \sin \phi - v_{2y} \cos \phi$, so that

$$\frac{v_{2x} \sin \phi - v_{2y} \cos \phi}{v_1 \sin \phi} = \frac{\gamma - 1}{\gamma + 1} + \frac{2c_1^2}{(\gamma + 1)v_1^2 \sin^2 \phi}. \tag{86.2}$$

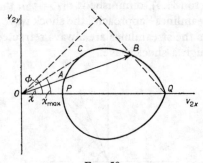

FIG. 50

We can eliminate the angle ϕ from these two relations. After some simple transformations, we obtain the following formula which determines the relation between v_{2x} and v_{2y} (for given v_1 and c_1):

$$v_{2y}^2 = (v_1 - v_{2x})^2 \frac{2(v_1 - c_1^2/v_1)/(\gamma + 1) - (v_1 - v_{2x})}{v_1 - v_{2x} + 2c_1^2/(\gamma + 1)v_1}. \tag{86.3}$$

This formula can be more intelligibly written by introducing the critical velocity. According to Bernoulli's equation and the definition of the critical velocity, we have $w_1 + \frac{1}{2}v_1^2 = \frac{1}{2}v_1^2 + c_1^2/(\gamma - 1) = (\gamma + 1)c_*^2/2(\gamma - 1)$ (see §85, Problem 1), whence

$$c_*^2 = [(\gamma - 1)v_1^2 + 2c_1^2]/(\gamma + 1). \tag{86.4}$$

Using this in (86.3), we obtain

$$v_{2y}^2 = (v_1 - v_{2x})^2 \frac{v_1 v_{2x} - c_*^2}{2v_1^2/(\gamma + 1) - v_1 v_{2x} + c_*^2}. \tag{86.5}$$

Equation (86.5) is called the equation of the *shock polar*. Fig. 50 shows a praph of the function $v_{2y}(v_{2x})$; it is a cubic curve, called a strophoid. It crosses the axis of abscissae at the points P and Q, corresponding to v_{2x}

$= c_*^2/v_1$ and $v_{2x} = v_1$.† A line (OB in Fig. 50) drawn from the origin at an angle χ to the axis of abscissae gives, by the length of the segment between O and the point where it intersects the shock polar, the gas velocity behind a discontinuity which turns the stream through an angle χ. There are two such intersections (A, B), i.e. two different shock waves correspond to a given value of χ. The direction of the shock wave also can be immediately determined from the shock polar: it is given by the direction of the perpendicular from the origin to the line QB or QA (Fig. 50 shows the angle for a shock corresponding to the point B). As χ decreases, the point A approaches P, corresponding to a normal shock ($\phi = \frac{1}{2}\pi$) with $v_2 = c_*^2/v_1$. The point B approaches Q; the intensity of the shock (velocity discontinuity) tends to zero, and the angle ϕ tends, as it should, to the Mach angle $\alpha = \sin^{-1}(c_1/v_1)$; the tangent to the shock polar at Q makes an angle $\frac{1}{2}\pi + \alpha$ with the axis of abscissae.

From the shock polar we can immediately derive the important result that the angle of deviation χ of the stream at the shock wave cannot exceed a certain maximum value χ_{max}, corresponding to the tangent from O to the curve. This quantity is, of course, a function of the Mach number $M_1 = v_1/c_1$, but we shall not give the expression for it, which is very cumbersome. For $M_1 = 1$, $\chi_{max} = 0$; as M_1 increases, χ_{max} increases monotonically, and tends to a finite limit as $M_1 \to \infty$. It is easy to discuss the two limiting cases. If the velocity v_1 is near to c_*, then v_2 is so also, and the angle χ is small; the equation (86.5) of the shock polar can then be written in the approximate form‡

$$\chi^2 = (\gamma+1)(v_1-v_2)^2(v_1+v_2-2c_*)/2c_*^3, \qquad (86.6)$$

where we have put $v_{2x} \approx v_2$, $v_{2y} \approx c_*\chi$ in view of the smallness of χ. Hence we easily find††

$$\chi_{max} = \frac{4\sqrt{(\gamma+1)}}{3\sqrt{3}}\left(\frac{v_1}{c_*}-1\right)^{\frac{3}{2}} = \frac{8\sqrt{2}}{3\sqrt{3}(\gamma+1)}(M_1-1)^{\frac{3}{2}}. \qquad (86.7)$$

In the opposite limiting case $M_1 = \infty$ (i.e. $M_{1*} = \sqrt{[(\gamma+1)/(\gamma-1)]}$), the shock polar degenerates to a circle which meets the axis of abscissae at the points $c_*\sqrt{[(\gamma-1)/(\gamma+1)]}$ and $c_*\sqrt{[(\gamma+1)/(\gamma-1)]}$. It is easy to see that we then have

$$\chi_{max} = \sin^{-1}(1/\gamma); \qquad (86.8)$$

---

† The strophoid actually continues in two branches from the point $v_{2x} = v_1$ (which is a double point) to infinite v_{2y}; these are not shown in Fig. 50. They have a common asymptote

$$v_{2x} = c_*^2/v_1 + 2v_1/(\gamma+1).$$

The points on these branches have no physical significance; they would give values for v_{2x} and v_{2y} such that $v_{2n}/v_{1n} > 1$, which is impossible.

‡ It is easily seen that equation (86.6) holds also for any (non-perfect) gas, provided that $(\gamma+1)$ is replaced by $2\alpha_*$ (95.2).

†† It may be noted that this dependence of χ_{max} on $M_1 - 1$ is in agreement with the general similarity law (118.7) for transonic flow.

for air this is 45·6°. Figure 51 shows a graph of χ_{max} as a function of M_1 for air; the upper curve is a similar graph for flow past a cone (see §105).

The circle $v_2 = c_*$ cuts the axis of abscissae between the points P and Q (Fig. 50), and therefore divides the shock polar into two parts corresponding to subsonic and supersonic gas velocities behind the discontinuity. The point where this circle crosses the polar lies to the right of, but very close to, the point C; the whole segment PC therefore corresponds to transitions to subsonic velocities, while CQ (except for a very small segment near C) corresponds to transitions to supersonic velocities.

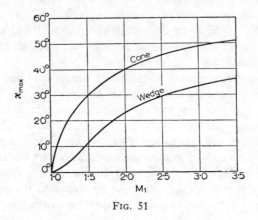

FIG. 51

For given M_1 and ϕ, the pressure change in the shock wave is given by

$$\frac{p_2}{p_1} = \frac{2\gamma M_1^2 \sin^2\phi - (\gamma - 1)}{(\gamma + 1)};$$ (86.9)

this is formula (85.8) with $M_1 \sin\phi$ in place of M_1. This ratio increases monotonically when the angle ϕ increases from its smallest value $\sin^{-1}(1/M_1)$ (when $p_2/p_1 = 1$) to $\frac{1}{2}\pi$, i.e. as we move along the shock polar from Q to P.

The two shock waves determined by the shock polar for a given deviation angle χ are often said to belong to the *weak* and *strong* families. A shock wave of the strong family (the segment PC of the polar) is strong (the ratio p_2/p_1 is large), makes a large angle ϕ with the direction of the velocity v_1, and converts the flow from supersonic to subsonic. A shock wave of the weak· family (the segment QC) is weak, is inclined at a smaller angle to the stream, and almost always leaves the flow supersonic.

PROBLEMS

PROBLEM 1. Derive the formula giving the angle of deviation χ of the velocity in an oblique shock wave (in a perfect gas) in terms of $M_1 = v_1/c_1$ and the angle ϕ between the shock wave and the direction of the velocity v_1 (Fig. 49):

$$\cot\chi = \tan\phi \left[\frac{(\gamma + 1)M_1^2}{2(M_1^2 \sin^2\phi - 1)} - 1 \right].$$

PROBLEM 2. Derive the formula giving the number $M_2 = v_2/c_2$ in terms of M_1 and ϕ:

$$M_2{}^2 = \frac{2+(\gamma-1)M_1{}^2}{2\gamma M_1{}^2 \sin^2\phi - (\gamma-1)} + \frac{2M_1{}^2 \cos^2\phi}{2+(\gamma-1)M_1{}^2 \sin^2\phi}.$$

§87. The thickness of shock waves

Hitherto we have regarded shock waves as geometrical surfaces of zero thickness. We shall now consider the structure of actual surfaces of discontinuity, and we shall see that shock waves in which the discontinuities are small are in reality transition layers of finite thickness, the thickness diminishing as the magnitude of the discontinuities increases. If the discontinuities are not small, the change occurs so sharply that the concept of thickness is meaningless.

To determine the structure and thickness of the transition layer we must take account of the viscosity and thermal conductivity of the gas, which we have hitherto neglected.

The relations (82.1)–(82.3) for a shock wave were obtained from the constancy of the fluxes of mass, momentum and energy. If we consider a surface of discontinuity as a layer of finite thickness, these conditions must be written, not as the equality of the quantities concerned on the two sides of the discontinuity, but as their constancy throughout the thickness of the layer. The first condition, (82.1), is unchanged:

$$\rho v \equiv j = \text{constant}. \tag{87.1}$$

In the other two conditions additional fluxes of momentum and energy, due to internal friction and thermal conduction, must be taken into account.

The momentum flux density (in the x-direction) due to internal friction is given by the component $-\sigma'_{xx}$ of the viscosity stress tensor; according to the general expression (15.3) for this tensor, we have $\sigma'_{xx} = (\tfrac{4}{3}\eta + \zeta)dv/dx$. The condition (82.2) then becomes

$$p+\rho v^2 - (\tfrac{4}{3}\eta+\zeta)dv/dx = \text{constant}.$$

As in §82, we introduce the specific volume V in place of the velocity $v = jV$. Since $j = \text{constant}$, $dv/dx = jdV/dx$, so that

$$p+j^2 V - (\tfrac{4}{3}\eta+\zeta)j\,dV/dx = \text{constant}.$$

At great distances from the shock wave, the thermodynamic quantities are constants, i.e. they are independent of x; in particular, $dV/dx = 0$. We denote by a suffix 1 the values of quantities far in front of the shock wave. Then we can put the constant equal to $p_1+j^2V_1$, obtaining

$$p-p_1+j^2(V-V_1)-(\tfrac{4}{3}\eta+\zeta)j\,dV/dx = 0. \tag{87.2}$$

Next, the energy flux density due to thermal conduction is $-\kappa\,dT/dx$. That due to internal friction is $-\sigma'_{xi}v_i$ or, since the velocity is along the x-axis, $-\sigma'_{xx}v = -(\tfrac{4}{3}\eta+\zeta)v\,dv/dx$. Thus the condition (82.3) can be written

$$\rho v(w+\tfrac{1}{2}v^2)-(\tfrac{4}{3}\eta+\zeta)v\,dv/dx-\kappa\,dT/dx = \text{constant}.$$

Again putting $v = jV$, we can obtain the final form

$$w + \tfrac{1}{2}j^2V^2 - j(\tfrac{4}{3}\eta + \zeta)V\,dV/dx - (\kappa/j)dT/dx = w_1 + \tfrac{1}{2}j^2V_1^2. \quad (87.3)$$

We shall here consider shock waves in which all the discontinuities are small. Then all the differences $V - V_1$, $p - p_1$, etc. between the values inside and outside the transition layer are also small. In (87.2) we expand $V - V_1$ in powers of $p - p_1$ and $s - s_1$, taking the pressure and entropy as the independent variables. It is seen from the relations obtained below that $1/\delta$ (where δ is the thickness of the discontinuity) is of the first order in $p - p_1$, and the difference $s - s_1$ is of the second order.† Hence we can write, neglecting quantities of the third order,

$$V - V_1 = (\partial V/\partial p)_s(p - p_1) + \tfrac{1}{2}(\partial^2 V/\partial p^2)_s(p - p_1)^2 + (\partial V/\partial s)_p(s - s_1).$$

The values of all the coefficients are, of course, taken outside the transition layer (i.e. for $p = p_1$, $s = s_1$). Substituting this expansion in (87.2), we obtain

$$[1 + (\partial V/\partial p)_s j^2](p - p_1) + \tfrac{1}{2}j^2(\partial^2 V/\partial p^2)_s(p - p_1)^2 + (\partial V/\partial s)_p(s - s_1)j^2$$
$$= (\tfrac{4}{3}\eta + \zeta)j\,dV/dx.$$

The derivative dV/dx can be written

$$\frac{dV}{dx} = \left(\frac{\partial V}{\partial p}\right)_s \frac{dp}{dx} + \left(\frac{\partial V}{\partial s}\right)_p \frac{ds}{dx}.$$

Differentiation with respect to x increases the order of smallness by one, since $1/\delta$ is of the first order; the derivative dp/dx is therefore of the second order, and ds/dx of the third order. The term in ds/dx can therefore be omitted. Thus the condition (87.2) becomes

$$[1 + (\partial V/\partial p)_s j^2](p - p_1) + \tfrac{1}{2}j^2(\partial^2 V/\partial p^2)_s(p - p_1)^2 + (\partial V/\partial s)_p(s - s_1)j^2$$
$$= (\tfrac{4}{3}\eta + \zeta)(\partial V/\partial p)_s(dp/dx)j. \quad (87.4)$$

Next, we multiply each term of (87.2) by $\tfrac{1}{2}(V + V_1)$ and subtract from equation (87.3). The result is

$$(w - w_1) - \tfrac{1}{2}(p - p_1)(V + V_1) - \tfrac{1}{2}j(\tfrac{4}{3}\eta + \zeta)(V - V_1)\frac{dV}{dx} - \frac{\kappa}{j}\frac{dT}{dx} = 0.$$

The third term, which contains the product $(V - V_1)\,dV/dx$, is of the third order, and may be omitted:

$$(w - w_1) - \tfrac{1}{2}(p - p_1)(V + V_1) - (\kappa/j)\,dT/dx = 0.$$

The first two terms are just the expression which we expanded in powers of $p - p_1$ and $s - s_1$ in deriving formula (83.1). The first- and second-order terms in $p - p_1$ in this expansion are zero, and we have as far as terms of the second

---

† The total entropy discontinuity $s_2 - s_1$ is, as we have seen in §83, of the third order relative to the pressure discontinuity $p_2 - p_1$, whereas $s - s_1$ is of only the second order in $p - p_1$. The reason is that, as we shall show below, the pressure in the transition layer varies monotonically from p_1 to p_2, whereas the entropy does not vary monotonically; it has a maximum within the layer.

order just $T(s-s_1)$. The derivative dT/dx can be written

$$\frac{dT}{dx} = \left(\frac{\partial T}{\partial p}\right)_s \frac{dp}{dx} + \left(\frac{\partial T}{\partial s}\right)_p \frac{ds}{dx} \approx \left(\frac{\partial T}{\partial p}\right)_s \frac{dp}{dx}.$$

The result is

$$T(s-s_1) = \frac{\kappa}{j}\left(\frac{\partial T}{\partial p}\right)_s \frac{dp}{dx}. \tag{87.5}$$

Substituting this expression for $s-s_1$ in (87.4), we find

$$\tfrac{1}{2}j^2\left(\frac{\partial^2 V}{\partial p^2}\right)_s (p-p_1)^2 + \left[1+\left(\frac{\partial V}{\partial p}\right)_s j^2\right](p-p_1)$$

$$= \left\{-\frac{\kappa}{T}\left(\frac{\partial V}{\partial s}\right)_p \left(\frac{\partial T}{\partial p}\right)_s + (\tfrac{4}{3}\eta+\zeta)\left(\frac{\partial V}{\partial p}\right)_s\right\}\frac{dp}{dx}j. \tag{87.6}$$

The flux j is, in the first approximation, $j = v/V \approx c/V$ (see (83.3)). This expression can be substituted on the right-hand side of (87.6), but it will not serve on the left-hand side; further terms have to be included in j^2. These terms could be obtained from (87.2), for instance. It is simpler, however, to argue as follows. At great distances on both sides of the surface of discontinuity, the right-hand side of (87.6) is zero, since dp/dx is zero. At such distances the pressure is p_1 or p_2. That is, we can say that the quadratic in p on the left of (87.6) has the zeros p_1 and p_2. By a well-known theorem of algebra, it can therefore be written as the product $(p-p_1)(p-p_2)$ multiplied by the coefficient of p^2, $\tfrac{1}{2}j^2(\partial^2 V/\partial p^2)_s$.

Thus we have the following differential equation for the function $p(x)$:[†]

$$\frac{1}{2}\left(\frac{\partial^2 V}{\partial p^2}\right)_s (p-p_1)(p-p_2) = -\frac{V^3}{c^3}\left\{(\tfrac{4}{3}\eta+\zeta) + \frac{\kappa}{T}\left(\frac{\partial T}{\partial p}\right)_s \left(\frac{\partial V}{\partial s}\right)_p c^2 p^2\right\}\frac{dp}{dx}.$$

From the thermodynamic formulae for derivatives, $(\partial V/\partial s)_p = (\partial T/\partial p)_s$; it is easy to see that the coefficient of $-dp/dx$ on the right-hand side of the above equation is $2V^2a$, where a is related to the sound-absorption coefficient γ (77.6) by $\gamma = a\omega^2$. Thus

$$\frac{dp}{dx} = -\frac{1}{4V^2a}\left(\frac{\partial^2 V}{\partial p^2}\right)_s (p-p_1)(p-p_2). \tag{87.7}$$

Integration gives

$$x = -\frac{4V^2a}{(\partial^2 V/\partial p^2)_s}\int \frac{dp}{(p-p_1)(p-p_2)}$$

$$= \frac{4aV^2}{\tfrac{1}{2}(p_2-p_1)(\partial^2 V/\partial p^2)_s}\tanh^{-1}\frac{p-\tfrac{1}{2}(p_2+p_1)}{\tfrac{1}{2}(p_2-p_1)} + \text{constant}.$$

[†] In considering a weak shock wave we can regard the viscosity and the thermal conductivity as constants.

Putting the constant equal to zero, we have

$$p - \tfrac{1}{2}(p_2 + p_1) = \tfrac{1}{2}(p_2 - p_1)\tanh(x/\delta), \tag{87.8}$$

where

$$\delta = 8aV^2/(p_2 - p_1)(\partial^2 V/\partial p^2)_s. \tag{87.9}$$

This gives the manner of variation of the pressure between the values p_1 and p_2 which it takes at great distances on the two sides of the shock wave. The point $x = 0$ corresponds to the median value of the pressure, $\tfrac{1}{2}(p_1 + p_2)$. For $x \to \pm\infty$, the pressure tends asymptotically to p_1 and p_2. Almost the whole change from p_1 to p_2 occurs over a distance of the order of δ, which may be called the *thickness* of the shock wave. We see that this is the less, the stronger the shock, i.e. the greater the pressure discontinuity.

The variation of the entropy across the discontinuity is obtained from (87.5) and (87.8)

$$s - s_1 = \frac{\kappa}{16caVT}\left(\frac{\partial T}{\partial p}\right)_s\left(\frac{\partial^2 V}{\partial p^2}\right)_s (p_2 - p_1)^2\frac{1}{\cosh^2(x/\delta)}. \tag{87.10}$$

From this we see that the entropy does not vary monotonically, but has a maximum inside the shock, at $x = 0$. For $x = \pm\infty$ this formula gives $s = s_1$ in either case; this is because the total entropy change $s_2 - s_1$ is of the third order in $p_2 - p_1$ (cf. (83.1)), whereas $s - s_1$ is of the second order.

Formula (87.8) is quantitatively valid only for sufficiently small differences $p_2 - p_1$. We can, however, use (87.9) qualitatively to determine the order of magnitude of the thickness in cases where the difference $p_2 - p_1$ is of the same order of magnitude as p_1 and p_2 themselves. The velocity of sound in the gas is of the same order as the thermal velocity v of the molecules. The kinematic viscosity is, as we know from the kinetic theory of gases, $\nu \sim lv \sim lc$, where l is the mean free path of the molecules. Hence $a \sim l/c^2$; an estimate of the thermal-conduction term gives the same result. Finally, $(\partial^2 V/\partial p^2)_s \sim V/p^2$, and $pV \sim c^2$. Using these relations in (87.9), we obtain

$$\delta \sim l. \tag{87.11}$$

Thus the thickness of a strong shock is of the same order of magnitude as the mean free path of the gas molecules.† In macroscopic gas dynamics, however, where the gas is treated as a continuous medium, the mean free path must be taken as zero. It follows that the methods of gas dynamics cannot strictly be used alone to investigate the internal structure of strong shock waves.

A considerable increase in the thickness of a shock wave may be caused by the presence in the gas of comparatively slow relaxation processes (slow chemical reactions, a slow energy transfer between different degrees of freedom of the molecule, and so on). This topic has been discussed by YA. B. ZEL'DOVICH (1946).

---

† A strong shock wave causes a considerable increase in temperature; l denotes the mean free path for some mean temperature of the gas in the shock.

Let τ be of the order of magnitude of the relaxation time. Both the initial and the final states of the gas must be states of complete equilibrium; it is therefore immediately clear that the total thickness of the shock wave will be of the order of τv_1, the distance traversed by the gas in the time τ. It is also found that, if the shock strength is above a certain limit, its structure becomes more complex; this may be seen as follows.

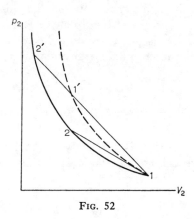

FIG. 52

In Fig. 52 the continuous curve shows the shock adiabatic drawn through a given initial point 1, on the assumption that the final states of the gas are states of complete equilibrium; the slope of the tangent at the point 1 gives the "equilibrium" velocity of sound, denoted in §78 by c_0. The dashed curve shows the shock adiabatic through the same point 1, on the assumption that the relaxation processes are "frozen" and do not occur. The slope of the tangent to this curve at the point 1 gives the velocity of sound denoted in §78 by c_∞.

If the velocity of the shock wave is such that $c_0 < v_1 < c_\infty$, the chord 12 lies as shown in Fig. 52 (the lower chord). In this case we have a simple increase in the shock thickness, all intermediate states between the initial state 1 and the final state 2 being represented in the pV-plane by points on the segment 12.†

If, however, $v_1 > c_\infty$, the chord takes the position $11'2'$. No point lying between 1 and $1'$ corresponds to any actual state of the gas; the first real point (after 1) is $1'$, which corresponds to a state in which the relaxation equilibrium is no different from that in state 1. The compression of the gas from state 1 to state $1'$ occurs discontinuously, and afterwards (over distances $\sim v_1\tau$) it is gradually compressed to the final state $2'$

---

† This follows from the fact that (neglecting ordinary viscosity and thermal conduction) all the states through which the gas passes satisfy the equations of conservation of mass, $\rho v = j =$ constant, and of momentum, $p+j^2V =$ constant (cf. the similar but more detailed discussion in §121).

§88. The isothermal discontinuity

The discussion of the structure of a shock wave in §87 involves the assumption that the viscosity and thermal conductivity are of the same order of magnitude, as is usually the case. The case where $\chi \gg \nu$ is also possible, however. If the temperature is sufficiently high, additional heat is transferred by thermal radiation in equilibrium with the matter. Radiation has a much smaller effect on the viscosity (i.e. the momentum transfer), and so ν may be small compared with χ. We shall now see that this inequality leads to a very important difference in the structure of the shock wave.

Neglecting terms in the viscosity, we can write equations (87.2) and (87.3), which determine the structure of the transition layer, as

$$p + j^2 V = p_1 + j^2 V_1, \tag{88.1}$$

$$\frac{\kappa}{j} \frac{\mathrm{d}T}{\mathrm{d}x} = w + \tfrac{1}{2} j^2 V^2 - w_1 - \tfrac{1}{2} j^2 V_1^2. \tag{88.2}$$

The right-hand side of (88.2) is zero only at the boundaries of the layer. Since the temperature behind the shock wave must be higher than that in front of it, it follows that we have

$$\mathrm{d}T/\mathrm{d}x > 0 \tag{88.3}$$

everywhere in the transition layer, i.e. the temperature increases monotonically.

All quantities in the layer are functions of a single variable, the co-ordinate x, and therefore are functions of one another. Differentiating (88.1) with respect to V, we obtain

$$\left(\frac{\partial p}{\partial T} \right)_V \frac{\mathrm{d}T}{\mathrm{d}V} + \left(\frac{\partial p}{\partial V} \right)_T + j^2 = 0.$$

The derivative $(\partial p/\partial T)_V$ is always positive in gases. The sign of the derivative $\mathrm{d}T/\mathrm{d}V$ is therefore the reverse of that of the sum $(\partial p/\partial V)_T + j^2$. In state 1 we have $j^2 < -(\partial p_1 \partial V_1/)_s$ (since $v_1 > c_1$), and, since the adiabatic compressibility is always less than the isothermal compressibility, $j^2 > -(\partial p_1/\partial V_1)_T$. On side 1, therefore, $\mathrm{d}T_1/\mathrm{d}V_1 < 0$. If this derivative remains negative everywhere in the transition layer, then, as the gas is compressed (V decreasing), the temperature increases monotonically, in accordance with (88.3), from side 1 to side 2. In other words, we have a shock wave whose thickness is much increased by the high thermal conductivity (possibly to such an extent that even to call it a shock wave is mere convention).

If, however, the shock is so strong that

$$j^2 < -(\partial p_2/\partial V_2)_T, \tag{88.4}$$

then we have in state 2 $\mathrm{d}T_2/\mathrm{d}V_2 > 0$, so that the function $T(V)$ has a maximum somewhere between V_1 and V_2 (Fig. 53). It is clear that the transition from state 1 to state 2, with V changing continuously, then becomes impossible, since the inequality (88.3) can not be satisfied everywhere.

Consequently, we have the following pattern of transition from the initial state 1 to the final state 2. First comes a region where the gas is gradually compressed from the specific volume V_1 to some V' (the value for which $T(V') = T_2$ for the first time; see Fig. 53); the thickness of this region is determined by the thermal conductivity, and may be considerable. The compression from V' to V_2 then occurs discontinuously, the temperature remaining constant at T_2. This may be called an *isothermal discontinuity*.

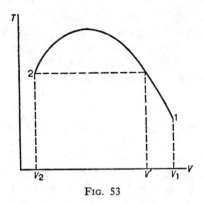

Fig. 53

Let us determine the variation of the pressure and density in an isothermal discontinuity, assuming that we have a perfect gas. The condition of continuity of momentum flux (88.1), applied to the two sides of the discontinuity, gives $p' + j^2 V' = p_2 + j^2 V_2$. For a perfect gas $V = RT/\mu p$; since $T' = T_2$, we have

$$p' + \frac{j^2 R T_2}{\mu p'} = p_2 + \frac{j^2 R T_2}{\mu p_2}.$$

This quadratic equation for p' has the solutions $p' = p_2$ (trivial) and

$$p' = j^2 R T_2/\mu p_2 = j^2 V_2. \tag{88.5}$$

We can express j^2 in the form (82.6), obtaining $p' = (p_2 - p_1) V_2/(V_1 - V_2)$, and, substituting V_2/V_1 from (85.1), we have

$$p' = \tfrac{1}{2}[(\gamma + 1) p_1 + (\gamma - 1) p_2]. \tag{88.6}$$

Since we must have $p_2 > p'$, we find that an isothermal discontinuity occurs only when the ratio of the pressures p_2 and p_1 satisfies

$$p_2/p_1 > (\gamma + 1)/(3 - \gamma) \tag{88.7}$$

(RAYLEIGH 1910). This condition can, of course, be obtained directly from (88.4).

Since, for a given temperature, the gas density is proportional to the pressure, the density ratio in an isothermal discontinuity is equal to the pressure ratio;

$$\rho'/\rho_2 = V_2/V' = p'/p_2. \tag{88.8}$$

§89. Weak discontinuities

Besides surface discontinuities, at which the quantities ρ, p, $\mathbf{v}$ etc. are discontinuous, we can also have surfaces at which these quantities, though remaining continuous, are not regular functions of the co-ordinates. The irregularity may be of various kinds. For example, the first spatial derivatives of ρ, p, $\mathbf{v}$ etc. may be discontinuous on a surface, or these derivatives may become infinite; or higher derivatives may behave in the same manner. We call such surfaces *weak discontinuities*, in contrast to the strong discontinuities (shock waves and tangential discontinuities), in which the quantities ρ, p, $\mathbf{v}$, ... themselves are discontinuous.

It is easy to see from simple considerations that weak discontinuities are propagated relative to the gas (on either side of the surface) with the velocity of sound. For, since the functions ρ, p, $\mathbf{v}$, ... themselves are continuous, they can be "smoothed" by modifying them only near the surface of discontinuity, and only by arbitrarily small amounts, in such a way that the smoothed functions have no singularity. The true distribution of the pressure, say, can thus be represented as a superposition of a perfectly smooth function p_0, free from all singularities, and a very small perturbation p' of this distribution near the surface of discontinuity; and the latter, like any small perturbation, is propagated, relative to the gas, with the velocity of sound.

It must be emphasised that, for a shock wave, the smoothed functions would differ from the true ones by quantities which in general are not small, and the foregoing arguments are therefore invalid. If, however, the discontinuities in the shock wave are sufficiently small, those arguments are again applicable, and such a shock wave is propagated with the velocity of sound, a result which was obtained by another method in §83.

If the flow is steady in a given co-ordinate system, then the surface of discontinuity is at rest in that system, and the gas flows through it. The gas velocity component normal to the surface must equal the velocity of sound. If we denote by α the angle between the direction of the gas velocity and the tangent plane to the surface, then $v_n = v \sin \alpha = c$, or $\sin \alpha = c/v$, i.e. a surface of weak discontinuity intersects the streamlines at the Mach angle. In other words, a surface of weak discontinuity is one of the characteristic surfaces, a result which is entirely reasonable if we recall the physical significance of the latter: they are surfaces along which small perturbations are propagated (see §79). It is clear that, in steady flow of a gas, weak discontinuities can occur only at velocities not less than that of sound.

Weak discontinuities differ fundamentally from strong ones in the manner
of their occurrence. We shall see that shock waves can be formed as a direct
result of the gas flow, the boundary conditions being continuous (for instance,
the formation of shock waves in a sound wave, §95). In contrast to this, weak
discontinuities cannot occur spontaneously; they are always the result of
some singularity of the initial or boundary conditions of the flow. These
singularities may be of various kinds, like the weak discontinuities themselves.
For example, a weak discontinuity may occur on account of the presence of
angles on the surface of a body past which the flow takes place; in this case
the first spatial derivatives of the velocity are discontinuous. A weak discontinuity is also formed when the curvature of the surface of the body is
discontinuous, without there being an angle; in this case the second spatial
derivatives of the velocity are discontinuous, and so on. Finally, any singularity in the time variation of the flow results in a non-steady weak discontinuity.

The gas velocity component tangential to the surface of a weak discontinuity is always directed away from the point (e.g. an angle on the surface
of a body) from which the perturbation begins which causes the discontinuity;
we shall say that the discontinuity begins from this point. This is an example
of the fact that, in a supersonic flow, perturbations are propagated downstream.

The presence of viscosity and thermal conduction results in a finite
thickness of a weak discontinuity, which is therefore in reality a transition
layer, like a shock wave. The thickness of the latter, however, depends only
on its strength and is constant in time, whereas the thickness of a weak
discontinuity increases with time after its formation. It is easy to determine
the law governing this increase. To do so, we again use the remark made at
the beginning of this section, that the motion of any part of the surface of a
weak discontinuity follows the same equations as the propagation of any weak
perturbation in the gas. When viscosity and thermal conduction occur, a
perturbation which is initially concentrated in a small volume (a "wave
packet") expands as it moves in the course of time; the manner of this expansion has been determined in §77. We can therefore conclude that the thickness δ of a weak discontinuity is of the order of

$$\delta \sim \sqrt{(act)}, \tag{89.1}$$

where t is the time from the formation of the discontinuity and a the coefficient of the squared frequency in the sound absorption coefficient. If
the discontinuity is at rest, then the time t must be replaced by l/c, where
l is the distance from the point where the discontinuity starts (e.g. for a weak
discontinuity starting from an angle on the surface of a body, l is the distance
from the vertex of the angle); consequently $\delta \sim \sqrt{(al)}$. Thus the thickness
of a weak discontinuity increases as the square root of the time from its
formation or of the distance from its starting-point.

To conclude this section, we should make the following remark, analogous to the one at the end of §79. We stated there that, among the various per-turbations of the state of a gas in motion, perturbations of entropy (at con-stant pressure) and vorticity are distinct in their properties. Such pertur-bations do not move relative to the gas, and are not propagated with the velocity of sound. Hence the surfaces at which the entropy and vorticity† are weakly discontinuous are at rest relative to the gas, and move with it relative to a fixed system of co-ordinates. Such discontinuities may be called *weak tangential discontinuities*; they pass through streamlines, and are in this respect entirely analogous to the strong tangential discontinuities.

---

† A weak discontinuity of the vorticity implies a weak discontinuity of the velocity component tangential to the surface of discontinuity; for example, the normal derivatives of the velocity may be discontinuous.

CHAPTER X

ONE-DIMENSIONAL GAS FLOW

§90. Flow of gas through a nozzle

LET us consider steady flow of a gas out of a large vessel through a tube of variable cross-section (a *nozzle*). We shall suppose that the gas flow is uniform over the cross-section at every point in the tube, and that the velocity is along the axis of the tube. For this to be so, the tube must not be too wide, and its cross-sectional area S must vary fairly slowly along its length. Thus all quantities characterising the flow will be functions only of the co-ordinate along the axis of the tube. Under these conditions we can apply the relations obtained in §80, which are valid along streamlines, directly to the variation of quantities along the axis.

The mass of gas passing through a cross-section of the tube in unit time (the *discharge*) is $Q = \rho v S$; this must evidently be constant along the tube:

$$Q = S\rho v = \text{constant}. \tag{90.1}$$

The linear dimensions of the vessel are supposed very large in comparison with the diameter of the tube. The velocity of the gas in the vessel may therefore be taken as zero, and accordingly all quantities with the suffix 0 in the formulae of §80 will be the values of those quantities in the vessel.

We have seen that the flux density $j = \rho v$ cannot exceed a certain limiting value j_*. It is therefore clear that the possible values of the total discharge Q have (for a given tube and a given state of the gas in the vessel) an upper limit Q_{max}, which is easily determined. If the value j_* of the flux density were reached anywhere except at the narrowest point of the tube, we should have $j > j_*$ for cross-sections with smaller S, which is impossible. The value $j = j_*$ can therefore be attained only at the narrowest point of the tube; let the cross-sectional area there be S_{min}. Then the upper limit to the total discharge is

$$Q_{\text{max}} = \rho_* v_* S_{\text{min}} = \sqrt{(\gamma p_0 \rho_0)}[2/(\gamma+1)]^{(1+\gamma)/2(\gamma-1)} S_{\text{min}}. \tag{90.2}$$

Let us first consider a nozzle which narrows continually towards its outer end, so that the minimum cross-sectional area is at that end (Fig. 54). By (90.1), the flux density j increases monotonically along the tube. The same is true of the gas velocity v, and the pressure accordingly falls monotonically. The greatest possible value of j is reached if v attains the value c just at the outer end of the tube, i.e. if $v_1 = c_1 = v_*$ (the suffix 1 denotes quantities pertaining to the outer end). At the same time, $p_1 = p_*$.

Let us now follow the change in the manner of outflow of the gas when the external pressure p_e diminishes. When this pressure decreases from p_0, the pressure inside the vessel, to p_*, the pressure p_1 at the outer end of the tube decreases also, and the two pressures p_1 and p_e remain equal; that is, the whole of the pressure drop from p_0 to p_e occurs in the nozzle. The velocity v_1 with which the gas leaves the tube, and the total discharge $Q = j_1 S_{\min}$, increase monotonically, however. For $p_e = p_*$ the velocity becomes equal to the local velocity of sound, and the discharge reaches the value $Q_{\max}$. When the external pressure decreases further, the pressure p_1 remains constant at p_*, and the fall of pressure from p_* to p_e occurs outside the tube, in the surrounding medium. In other words, the pressure drop along the tube cannot be greater than from p_0 to p_*, whatever the external pressure. For air ($p_* = 0.53 p_0$), the maximum pressure drop is $0.47 p_0$. The velocity at the end of the tube and the discharge also remain constant for $p_e < p_*$. Thus the gas cannot acquire a supersonic velocity in flowing through a nozzle of this kind.

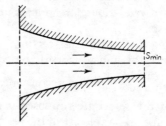

FIG. 54

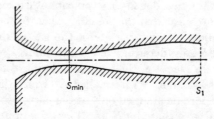

FIG. 55

If we consider only the flow in the immediate neighbourhood of the end of the tube, the motion of the gas after leaving the tube is essentially flow round an angle, the vertex of which is the edge of the tube mouth; we shall discuss this flow in detail in §104.

The impossibility of achieving supersonic velocities by flow through a continually narrowing nozzle is due to the fact that a velocity equal to the local velocity of sound can be reached only at the very end of such a tube. It is clear that a supersonic velocity can be attained by means of a nozzle which first narrows and then widens again (Fig. 55). This is called a *de Laval nozzle*.

The maximum flux density j_*, if reached, can again occur only at the narrowest cross-section, so that the discharge cannot exceed $S_{\min}j_*$. In the narrowing part of the nozzle, the flux density increases (and the pressure falls); the curve in Fig. 56 shows j as a function† of p, and the variation just described corresponds to the interval from c to b. If the maximum flux density is reached at the cross-section $S_{\min}$ (the point b in Fig. 56), the pressure continues to diminish in the widening part of the nozzle, while j begins to decrease also, corresponding to the segment ba of the curve. At the outer end of the tube j takes a definite value, $j_{1,\max} = j_* S_{\min}/S_1$, and the pressure has the corresponding value, denoted in Fig. 56 by p_1', at some point d on the curve. If, however, only some point e is reached at the cross-section $S_{\min}$, the pressure increases in the widening part of the nozzle, corresponding to a return down the curve from e towards c. At first sight it might appear that we might pass discontinuously from cb to ab, without going through the point b, by the formation of a shock wave. This, however, is impossible, since the gas "entering" the shock wave cannot have a subsonic velocity.

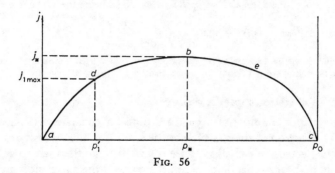

FIG. 56

Bearing in mind these results, let us now investigate the manner of variation in the outflow when the external pressure p_e is gradually increased. For small pressures, from zero to p_1', the pressure p_* and velocity $v_* = c_*$ are reached at the cross-section $S_{\min}$. In the widening part of the nozzle the velocity continues to increase, so that there results a supersonic flow of the gas, and the pressure accordingly continues decreasing, reaching the value p_1' at the outer end of the tube, whatever the pressure p_e. The pressure falls from p_1' to p_e outside the nozzle, in the rarefaction wave which leaves the edge of the tube mouth (see §104).

When p_e exceeds p_1', an oblique shock wave leaves the edge of the tube mouth, compressing the gas from p_1' to p_e (§104). We shall see, however, that a steady shock wave can leave a solid surface only if its intensity is not too

† According to formulae (80.15-80.17), the dependence is

$$j = \left(\frac{p}{p_0}\right)^{1/\gamma} \left\{\frac{2\gamma}{\gamma-1} p_0\rho_0 \left[1 - \left(\frac{p}{p_0}\right)^{(\gamma-1)/\gamma}\right]\right\}^{\frac{1}{2}}.$$

great (§103). Hence, when the external pressure increases further, the shock wave soon begins to move into the nozzle, with separation occurring in front of it on the inner surface of the tube. For some value of p_e the shock wave reaches the narrowest cross-section and then disappears; the flow becomes everywhere subsonic, with separation on the walls of the widening part of the nozzle. All these complex phenomena are, of course, three-dimensional.

PROBLEM

A small amount of heat is supplied over a short segment of a tube to a perfect gas in steady flow in the tube. Determine the change in the gas velocity when it passes through this segment.

SOLUTION. Let Sq be the amount of heat supplied per unit time, S being the cross-sectional area of the tube at the segment concerned. The mass flux density $j = \rho v$ and the momentum flux density $p+jv$ are the same on both sides of the heated segment; hence $\Delta p = -j\Delta v$, where Δ denotes the change in a quantity in passing through the segment. The difference in the energy flux density $(w+\frac{1}{2}v^2)j$ is q. Writing $w = \gamma p/(\gamma-1)\rho = \gamma p v/(\gamma-1)j$, we obtain (supposing Δv and Δp small) $vj\Delta v+\gamma(p\Delta v+v\Delta p)/(\gamma-1) = q$. Eliminating Δp, we find $\Delta v = (\gamma-1)q/\rho(c^2-v^2)$. We see that, in subsonic flow, the supply of heat accelerates the flow ($\Delta v > 0$), while in supersonic flow it retards it.

Writing the gas temperature as $T = \mu p/R\rho = \mu p v/Rj$ (R being the gas constant), we find

$$\Delta T = \frac{\mu}{Rj}(v\Delta p+p\Delta v) = \frac{\mu(\gamma-1)q}{Rj(c^2-v^2)}\left(\frac{c^2}{\gamma}-v^2\right).$$

For supersonic flow, this expression is always positive, and the gas temperature is increased; for subsonic flow, however, ΔT may be either positive or negative.

§91. Flow of a viscous gas in a pipe

Let us consider the flow of a gas in a pipe (of constant cross-section) so long that the friction of the gas against the walls, i.e. the viscosity of the gas, cannot be neglected. We shall suppose the walls to be thermally insulated, so that there is no heat exchange between the gas and the surrounding medium.

For gas velocities of the order of or exceeding the velocity of sound (the only case we shall discuss here), the gas flow in the pipe is, of course, turbulent if the radius of the pipe is not small. The turbulence of the flow is important, as regards our problem, only in one respect: we have seen in §43 that, in turbulent flow, the (mean) velocity is practically the same almost everywhere in the cross-section of the pipe, and falls rapidly to zero very close to the walls. We shall therefore suppose that the gas velocity v is a constant over the cross-section, and define it so that the product $S\rho v$ (S being the cross-sectional area) is equal to the total discharge through the cross-section.

Since the total discharge $S\rho v$ is constant along the pipe, and S is assumed constant, the mass flux density must also be constant:

$$j = \rho v = \text{constant}. \tag{91.1}$$

Next, since the pipe is thermally insulated, the total energy flux carried by the gas through any cross-section must also be constant. This flux is $S\rho v(w+\frac{1}{2}v^2)$, and by (91.1) we have

$$w+\tfrac{1}{2}v^2 = w+\tfrac{1}{2}j^2 V^2 = \text{constant}. \tag{91.2}$$

The entropy s of the gas does not, of course, remain constant, but increases as the gas moves along the pipe, because of the internal friction. If x is the co-ordinate along the pipe, with x increasing downstream, we can write

$$\mathrm{d}s/\mathrm{d}x > 0. \tag{91.3}$$

We now differentiate (91.2) with respect to x. Since $\mathrm{d}w = T\mathrm{d}s + V\mathrm{d}p$, we have

$$T\frac{\mathrm{d}s}{\mathrm{d}x} + V\frac{\mathrm{d}p}{\mathrm{d}x} + j^2 V\frac{\mathrm{d}V}{\mathrm{d}x} = 0.$$

Next, substituting

$$\frac{\mathrm{d}V}{\mathrm{d}x} = \left(\frac{\partial V}{\partial p}\right)_s \frac{\mathrm{d}p}{\mathrm{d}x} + \left(\frac{\partial V}{\partial s}\right)_p \frac{\mathrm{d}s}{\mathrm{d}x}, \tag{91.4}$$

we obtain

$$\left[T + j^2 V\left(\frac{\partial V}{\partial s}\right)_p\right]\frac{\mathrm{d}s}{\mathrm{d}x} = -V\left[1 + j^2\left(\frac{\partial V}{\partial p}\right)_s\right]\frac{\mathrm{d}p}{\mathrm{d}x}. \tag{91.5}$$

By a well-known formula of thermodynamics, $(\partial V/\partial s)_p = (T/c_p)(\partial V/\partial T)_p$. The coefficient of thermal expansion is positive for gases. We therefore conclude, using (91.3), that the left-hand side of (91.5) is positive. The sign of the derivative $\mathrm{d}p/\mathrm{d}x$ is therefore that of $-[1 + j^2(\partial V/\partial p)_s] = (v/c)^2 - 1$. We see that

$$\mathrm{d}p/\mathrm{d}x \lessgtr 0 \quad \text{for} \quad v \lessgtr c. \tag{91.6}$$

Thus, in subsonic flow, the pressure decreases downstream, as for an incompressible fluid. For supersonic flow, however, it increases.

We can similarly determine the sign of the derivative $\mathrm{d}v/\mathrm{d}x$. Since $j = v/V = $ constant, the sign of $\mathrm{d}v/\mathrm{d}x$ is the same as that of $\mathrm{d}V/\mathrm{d}x$. The latter can be expressed in terms of the positive derivative $\mathrm{d}s/\mathrm{d}x$ by means of (91.4) and (91.5). The result is that

$$\mathrm{d}v/\mathrm{d}x \gtrless 0 \quad \text{for} \quad v \lessgtr c, \tag{91.7}$$

i.e. the velocity increases downstream for subsonic flow and decreases for supersonic flow.

Any two thermodynamic quantities for a gas flowing in a pipe are functions of one another, independent of (*inter alia*) the resistance law for the pipe. These functions depend on the constant j as a parameter, and are given by the equation $w + \frac{1}{2}j^2 V^2 = $ constant, which is obtained by eliminating the velocity from the equations of conservation of mass and energy for the gas.

Let us ascertain the nature of the curves giving, for example, the entropy as a function of pressure. Rewriting (91.5) in the form

$$\frac{\mathrm{d}s}{\mathrm{d}p} = V\frac{(v/c)^2 - 1}{T + j^2 V(\partial V/\partial s)_p},$$

we see that, at the point where $v = c$, the entropy has an extremum. It is easy to see that s has a maximum. For the second derivative of s with respect to p at this point is

$$\left[\frac{d^2s}{dp^2}\right]_{v=c} = -\frac{j^2V(\partial^2V/\partial p^2)_s}{T+j^2V(\partial V/\partial s)_p} < 0;$$

we assume, as usual, that the derivative $(\partial^2V/\partial p^2)_s$ is positive.

The curves giving s as a function of p (called *Fanno lines*) are therefore as shown in Fig. 57. The region of subsonic velocities lies to the right of the maximum, and that of subsonic velocities to the left. When the parameter j increases, we go to lower curves. For, differentiating equation (91.2) with respect to j for constant p, we have

$$\frac{ds}{dj} = -\frac{jV^2}{T+j^2V(\partial V/\partial s)_p} < 0.$$

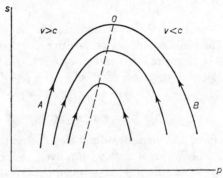

FIG. 57

We can draw an interesting conclusion from the above results. Let the gas velocity at the entrance to the pipe be less than that of sound. The entropy increases downstream, and the pressure decreases; this corresponds to a movement along the right-hand branch of the curve $s = s(p)$, from B towards O (Fig. 57). This can, however, continue only until the entropy reaches its maximum value. A further movement along the curve beyond O (i.e. into the region of supersonic velocities) is not possible, since the entropy of the gas would have to decrease as it moved along the pipe. The transition between the branches BO and OA cannot even be effected by a shock wave, since the gas entering a shock wave cannot move with subsonic velocity.

Thus we conclude that, if the gas velocity at the entrance to the pipe is less than that of sound, the flow remains subsonic everywhere in the pipe. The gas velocity becomes equal to the local velocity of sound only at the other end of the pipe, if at all (it does so if the pressure of the external medium into which the gas issues is sufficiently low).

In order that the gas should have supersonic velocities in the pipe, its velocity at the entrance must be supersonic. By the general properties of supersonic flow (the impossibility of propagating disturbances upstream), the flow will then be entirely independent of the conditions at the outlet of the pipe. In particular, the entropy will increase along the pipe in a quite definite manner, and its maximum value will be attained at a definite distance $x = l_k$ from the entrance. If the total length l of the pipe is less than l_k, the flow is supersonic throughout the pipe (corresponding to movement on the branch AO from A towards O). If, on the other hand, $l > l_k$, the flow cannot be supersonic throughout the pipe, nor can there be a smooth transition to subsonic flow, since we can move along the branch OB only in the direction shown by the arrow. In this case, therefore, a shock wave must necessarily be formed, which discontinuously changes the flow from supersonic to subsonic. The pressure is thereby increased, and we pass from the branch AO to BO without going through the point O. The flow is entirely subsonic beyond the discontinuity.

§92. One-dimensional similarity flow

An important class of one-dimensional non-steady gas flows is formed by flows occurring in conditions where there are characteristic velocities but not characteristic lengths. The simplest example of such a flow is given by gas flow in a semi-infinite cylindrical pipe terminated by a piston, when the piston begins to move with constant velocity.

Such a flow is defined by the velocity parameter and by parameters which give, say, the gas pressure and density at the initial instant. We can, however, form no combination of these parameters which has the dimensions of length or time. It therefore follows that the distributions of all quantities can depend on the co-ordinate x and the time t only through the ratio x/t, which has the dimensions of velocity. In other words, these distributions at various instants will be similar, differing only in the scale along the x-axis, which increases proportionally to the time. We can say that, if lengths are measured in a unit which increases proportionally to t, then the flow pattern does not change. When the flow pattern is unchanged with time if the scale of length varies appropriately, we speak of a *similarity flow*.

The equation of conservation of entropy for a flow which depends on only one co-ordinate, x, is $\partial s/\partial t + v_x \, \partial s/\partial x = 0$. Assuming that all quantities depend only on $\xi = x/t$, and noticing that in this case $\partial/\partial x = (1/t)\mathrm{d}/\mathrm{d}\xi$, $\partial/\partial t = -(\xi/t)\mathrm{d}/\mathrm{d}\xi$, we obtain $(v_x - \xi) \, s' = 0$ (the prime denoting differentiation with respect to ξ). Hence $s' = 0$, i.e. $s = $ constant†; thus similarity flow in one dimension must be isentropic. Likewise, from the y and z components of Euler's equation: $\partial v_y/\partial t + v_x \, \partial v_y/\partial x = 0$, $\partial v_z/\partial t + v_x \, \partial v_z/\partial x = 0$, we find that v_y and v_z are constants, which we can take as zero without loss of generality.

---

† The assumption that $v_x - \xi = 0$ would contradict the other equations of motion; from (92.3) we should have $v_x = $ constant, contrary to hypothesis.

Next, the equation of continuity and the x-component of Euler's equation are

$$\frac{\partial \rho}{\partial t} + \rho \frac{\partial v}{\partial x} + v \frac{\partial \rho}{\partial x} = 0, \tag{92.1}$$

$$\frac{\partial v}{\partial t} + v \frac{\partial v}{\partial x} = -\frac{1}{\rho} \frac{\partial p}{\partial x}; \tag{92.2}$$

here and henceforward we write v_x as v simply. In terms of the variable ξ, these equations become

$$(v - \xi)\rho' + \rho v' = 0, \tag{92.3}$$

$$(v - \xi)v' = -p'/\rho = -c^2 \rho'/\rho. \tag{92.4}$$

In the second equation we have put $p' = (\partial p/\partial \rho)_s \rho' = c^2 \rho'$, since the entropy is constant.

These equations have, first of all, the trivial solution $v = \text{constant}$, $\rho = \text{constant}$, i.e. a uniform flow of constant velocity. To find a non-trivial solution, we eliminate ρ' and v' from the equations, obtaining $(v - \xi)^2 = c^2$, whence $\xi = v \pm c$. We shall take the plus sign:

$$x/t = v + c; \tag{92.5}$$

this choice of sign means that we take the positive x-axis in a definite direction, selected in a manner shown later. Finally, putting $v - \xi = -c$ in (92.3), we obtain $c\rho' = \rho v'$, or $\rho \, dv = c \, d\rho$. The velocity of sound is a function of the thermodynamic state of the gas; taking as the fundamental thermodynamic quantities the entropy s and the density ρ, we can represent the velocity of sound as a function $c(\rho)$ of the density, for any given value of the constant entropy. With c understood as such a function, we can write

$$v = \int c \, d\rho/\rho = \int dp/c\rho. \tag{92.6}$$

This formula can also be written

$$v = \int \sqrt{(-dp \, dV)}, \tag{92.7}$$

in which the choice of dependent variable remains open.

Formulae (92.5) and (92.6) give the required solution of the equations of motion. If the function $c(\rho)$ is known, then the velocity v can be calculated as a function of density from (92.6). Equation (92.5) then determines the density as an implicit function of x/t, and so the dependence of all the other quantities on x/t is determined also.

We can derive some general properties of the solution thus obtained. Differentiating equation (92.5) with respect to x, we have

$$t \frac{\partial \rho}{\partial x} \frac{d(v+c)}{d\rho} = 1. \tag{92.8}$$

For the derivative of $v+c$ we have, by (92.6),

$$\frac{d(v+c)}{d\rho} = \frac{c}{\rho} + \frac{dc}{d\rho} = \frac{1}{\rho}\frac{d(\rho c)}{d\rho}.$$

But

$$\rho c = \rho\sqrt{(\partial p/\partial\rho)} = 1/\sqrt{(-\partial p/\partial V)};$$

differentiating, we have

$$d(\rho c)/d\rho = c^2 d(\rho c)/dp = \tfrac{1}{2}\rho^3 c^5(\partial^2 V/\partial p^2)_s. \tag{92.9}$$

Thus

$$d(v+c)/d\rho = \tfrac{1}{2}\rho^2 c^5(\partial^2 V/\partial p^2)_s > 0. \tag{92.10}$$

It therefore follows from (92.8) that $\partial\rho/\partial x > 0$ for $t > 0$.† Since $\partial p/\partial x = c^2\,\partial\rho/\partial x$, we conclude that $\partial p/\partial x > 0$ also. Finally, we have $\partial v/\partial x = (c/\rho)\partial\rho/\partial x$, so that $\partial v/\partial x > 0$. The inequalities

$$\partial\rho/\partial x > 0, \qquad \partial p/\partial x > 0, \qquad \partial v/\partial x > 0 \tag{92.11}$$

therefore hold.

The meaning of these inequalities becomes clearer if we follow the variation of quantities, not along the x-axis for given t, but with time for a given gas element as it moves about. This variation is given by the total time derivative; for the density, for example, we have, using the equation of continuity, $d\rho/dt = \partial\rho/\partial t + v\,\partial\rho/\partial x = -\rho\,\partial v/\partial x$. By the third inequality (92.11), this quantity is negative, and therefore so is dp/dt:

$$d\rho/dt < 0, \qquad dp/dt < 0. \tag{92.12}$$

Similarly (using Euler's equation (92.2)) we can see that $dv/dt < 0$; this, however, does not mean that the magnitude of the velocity diminishes with time, since v may be negative.

The inequalities (92.12) show that the density and pressure of any gas element decrease as it moves. In other words, the gas is continually rarefied as it moves. Such a flow may therefore be called a *non-steady rarefaction wave*.

A rarefaction wave can be propagated only a finite distance along the x-axis; this is seen from the fact that formula (92.5) would give an infinite velocity for $x \rightarrow \pm\infty$, which is impossible.

Let us apply formula (92.5) to a plane bounding the region of space occupied by the rarefaction wave. Here x/t is the velocity of this boundary relative to the fixed co-ordinate system chosen. Its velocity relative to the gas itself is $(x/t)-v$ and is, by (92.5), equal to the local velocity of sound. This means that the boundaries of a rarefaction wave are weak discontinuities. The

---

† There is no meaning for times $t < 0$ in the similarity flow here considered. Such a flow can occur only because of some singularity in the initial conditions ($t = 0$) of the flow at the point $x = 0$, and therefore takes place only for $t > 0$ (in our example, the piston velocity changes discontinuously at $t = 0$. See also §93).

similarity flow in different cases is therefore made up of rarefaction waves and regions of constant flow, separated by surfaces of weak discontinuity.†

The choice of sign in (92.5) is now seen to correspond to the fact that these weak discontinuities are assumed to move in the positive x-direction relative to the gas. The inequalities (92.11) arise from this choice, but the inequalities (92.12), of course, do not depend on the direction of the x-axis.

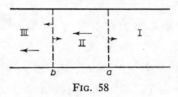

FIG. 58

We are usually concerned, in actual problems, with a rarefaction wave bounded on one side by a region where the gas is at rest. Let this region (I in Fig. 58) be to the right of the rarefaction wave. Region II is the rarefaction wave, and region III contains gas moving with constant velocity. The arrows in the figure show the direction of motion of the gas, and of the weak discontinuities bounding the rarefaction wave; the discontinuity a always moves into the gas at rest, but the discontinuity b may move in either direction, depending on the velocity reached in the rarefaction wave (see Problem 2). We may give explicitly the relations between the various quantities in such a rarefaction wave, assuming that we have a perfect gas. For an adiabatic process $\rho T^{1/(1-\gamma)}$ = constant. Since the velocity of sound is proportional to $\sqrt{T}$, we can write this relation as

$$\rho = \rho_0 (c/c_0)^{2/(\gamma-1)}. \tag{92.13}$$

Substituting this expression in the integral (92.6), we obtain

$$v = \frac{2}{\gamma-1} \int dc = \frac{2}{\gamma-1}(c-c_0);$$

the constant of integration is chosen so that $c = c_0$ for $v = 0$ (we use the suffix 0 to refer to the point where the gas is at rest). We shall express all quantities in terms of v, bearing in mind that, with the above situation of the various regions, the gas velocity is in the negative x-direction, i.e. $v < 0$. Thus

$$c = c_0 - \tfrac{1}{2}(\gamma-1)|v|, \tag{92.14}$$

which determines the local velocity of sound in terms of the gas velocity. Substituting in (92.13), we find the density to be

$$\rho = \rho_0[1 - \tfrac{1}{2}(\gamma-1)|v|/c_0]^{2/(\gamma-1)}, \tag{92.15}$$

† There may also, of course, be regions of constant flow separated by shock waves.

and similarly the pressure is

$$p = p_0[1 - \tfrac{1}{2}(\gamma - 1)|v|/c_0]^{2\gamma/(\gamma-1)}. \tag{92.16}$$

Finally, substituting (92.14) in formula (92.5), we obtain

$$|v| = \frac{2}{\gamma+1}\left(c_0 - \frac{x}{t}\right), \tag{92.17}$$

which gives v as a function of x and t.

The quantity c cannot be negative, by definition. We can therefore draw from (92.14) the important conclusion that the velocity must satisfy the inequality

$$|v| \leqslant 2c_0/(\gamma - 1); \tag{92.18}$$

when the velocity reaches this limiting value, the gas density (and also p and c) becomes zero. Thus a gas originally at rest and expanding non-steadily in a rarefaction wave can be accelerated only to velocities not exceeding $2c_0/(\gamma - 1)$.

We have already mentioned, at the beginning of this section, a simple example of similarity flow, namely that which occurs in a cylindrical pipe in which a piston begins to move with constant velocity. If the piston moves out of the pipe, it creates a rarefaction, and a rarefaction wave of the kind described above is formed. If, however, the piston moves inwards, it compresses the gas in front of it, and the transition to the original lower pressure can occur only in a shock wave, which is in fact formed in front of a piston moving forward in a pipe (see the following Problems).†

PROBLEMS

PROBLEM 1. A perfect gas occupies a semi-infinite cylindrical pipe terminated by a piston. At an initial instant the piston begins to move into the pipe with constant velocity U. Determine the resulting flow.

SOLUTION. A shock wave is formed in front of the piston, and moves along the pipe. At the initial instant this shock and the piston are coincident, but at subsequent instants the shock is ahead of the piston, and a region of gas lies between them (region 2). In front of the shock wave (region 1), the gas pressure is equal to its initial value p_1, and its velocity relative to the pipe is zero. In region 2, the gas moves with constant velocity, equal to the velocity U of the piston (Fig. 59). The difference in velocity between regions 1 and 2 is therefore also U, and, by formulae (82.7) and (85.1), we can write

$$U = \sqrt{[(p_2 - p_1)(V_1 - V_2)]}$$
$$= (p_2 - p_1)\sqrt{\{2V_1/[(\gamma - 1)p_1 + (\gamma + 1)p_2]\}}.$$

---

† We may mention also an analogous similarity flow in three dimensions: the centrally symmetrical gas flow caused by a uniformly expanding sphere. A spherical shock wave, expanding with constant velocity, is formed in front of the sphere. Unlike what happens in the one-dimensional case, the velocity of the gas between the sphere and the shock is not constant; the equation which determines it as a function of the ratio r/t (and therefore the rate of propagation of the shock wave) cannot be integrated analytically.

This problem has been discussed by L. I. SEDOV (1945; see his book *Similarity and Dimensional Methods in Mechanics*, Cleaver-Hume Press, London 1959) and by G. I. TAYLOR, *Proceedings of the Royal Society*, A186, 273, 1946.

Hence we find the gas pressure p_2 between the piston and the shock wave to be given by

$$\frac{p_2}{p_1} = 1 + \frac{\gamma(\gamma+1)U^2}{4c_1^2} + \frac{\gamma U}{c_1}\sqrt{\left[1 + \frac{(\gamma+1)^2 U^2}{16c_1^2}\right]}.$$

Knowing p_2, we can calculate, from formulae (85.4), the velocity of the shock wave relative to the gas on each side of it. Since gas 1 is at rest, the velocity of the shock relative to it is equal to the rate of propagation of the shock in the pipe. If the x co-ordinate (along the pipe) is measured from the initial position of the piston (the gas being on the side $x > 0$), we find the position of the shock wave at time t to be

$$x = t\{\tfrac{1}{4}(\gamma+1)U + \sqrt{[\tfrac{1}{16}(\gamma+1)^2 U^2 + c_1^2]}\},$$

while the position of the piston is $x = Ut$.

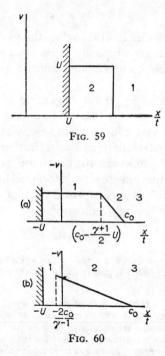

Fig. 59

Fig. 60

PROBLEM 2. The same as Problem 1, but for the case where the piston moves out of the pipe with velocity U.

SOLUTION. The piston adjoins a region of gas (region 1 in Fig. 60a) which moves in the negative x-direction with constant velocity $-U$, equal to the velocity of the piston. Then follows a rarefaction wave (2), in which the gas moves in the negative x-direction, its velocity varying linearly from $-U$ to zero according to (92.17). The pressure varies according to (92.16) from $p_1 = p_0[1-\tfrac{1}{2}(\gamma-1)U/c_0]^{2\gamma/(\gamma-1)}$ in gas 1 to p_0 in the gas 3, which is at rest. The boundary of regions 1 and 2 is given by the condition $v = -U$; according to (92.17), we have $x = [c_0 - \tfrac{1}{2}(\gamma+1)U]t = (c-U)t$, where c is the velocity of sound in gas 1. At the boundary of regions 2 and 3, $v = 0$, whence $x = c_0 t$. Both boundaries are weak discontinuities; the second is always propagated to the right (i.e. away from the piston), but the first may be propagated either to the right (as shown in Fig. 60a) or to the left (if the piston velocity $U > 2c_0/(\gamma+1)$).

The flow pattern just described can occur only if $U < 2c_0/(\gamma-1)$. If $U > 2c_0/(\gamma-1)$, a vacuum is formed in front of the piston (the gas cannot follow the piston), which extends from the piston to the point $x = -2c_0t/(\gamma-1)$ (region 1 in Fig. 60b). At this point, $v = -2c_0/(\gamma-1)$; then follow region 2, in which the velocity decreases to zero at the point $x = c_0t$, and region 3, where the gas is at rest.

PROBLEM 3. A gas occupies a semi-infinite cylindrical pipe $(x > 0)$ terminated by a valve. At time $t = 0$, the valve is opened, and the gas flows into the external medium, the pressure p_e in which is less than the initial pressure p_0 in the pipe. Determine the resulting flow.

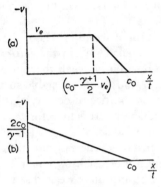

FIG. 61

SOLUTION. Let $-v_e$ be the gas velocity which corresponds to the external pressure p_e according to formula (92.16); for $x = 0$ and $t > 0$, we must have $v = -v_e$. If $v_e < 2c_0/(\gamma+1)$, the velocity distribution shown in Fig. 61a results. For $v_e = 2c_0/(\gamma+1)$ (corresponding to a rate of outflow equal to the local velocity of sound at the end of the pipe: this is easily seen by putting $v = c$ in formula (92.14)), the region of constant velocity vanishes and the pattern shown in Fig. 61b is obtained. The quantity $2c_0/(\gamma+1)$ is the greatest possible rate of outflow from the pipe in the conditions stated. If the external pressure p_e is such that

$$p_e < p_0[2/(\gamma+1)]^{2\gamma/(\gamma-1)}, \tag{1}$$

the corresponding velocity exceeds $2c_0/(\gamma+1)$. In reality, the pressure at the pipe outlet would still be equal to the limiting value (the right-hand side of (1)), and the rate of outflow would be $2c_0/(\gamma+1)$; the remaining pressure drop (to p_e) occurs in the external medium.

PROBLEM 4. An infinite pipe is divided by a piston, on one side of which $(x < 0)$ there is, at the initial instant, gas at pressure p_0, and on the other side a vacuum. Determine the motion of the piston as the gas expands.

SOLUTION. A rarefaction wave is formed in the gas; one of its boundaries moves to the right with the piston, and the other moves to the left. The equation of motion of the piston is

$$m\,dU/dt = p_0[1-\tfrac{1}{2}(\gamma-1)U/c_0]^{2\gamma/(\gamma-1)},$$

where U is the velocity of the piston and m its mass per unit area. Integrating, we obtain

$$U(t) = \frac{2c_0}{\gamma-1}\left\{1-\left[1+\frac{(\gamma+1)p_0}{2mc_0}\right]^{-(\gamma-1)/(\gamma+1)}\right\}.$$

PROBLEM 5. Determine the flow in an isothermal similarity rarefaction wave.

SOLUTION. The isothermal velocity of sound is $c_T = \sqrt{(\partial p/\partial \rho)_T} = \sqrt{(RT/\mu)}$, and for constant temperature $c_T = \text{constant} = c_{T0}$. According to (92.5) and (92.6), we therefore have

$$v = c_{T_0}\log(\rho/\rho_0) = c_{T_0}\log(p/p_0) = (x/t)-c_{T_0}.$$

§93. Discontinuities in the initial conditions

One of the most important reasons for the occurrence of surfaces of discontinuity in a gas is the possibility of discontinuities in the initial conditions. These conditions (i.e. the initial distributions of velocity, pressure, etc.) may in general be prescribed arbitrarily. In particular, they need not be everywhere continuous, but may be discontinuous on various surfaces. For example, if two masses of gas at different pressures are brought together at some instant, their surface of contact will be a surface of discontinuity of the initial pressure distribution.

It is of importance that the discontinuities of the various quantities in the initial conditions (or, as we shall say, in the *initial discontinuities*) can have any values whatever; no relation between them need exist. We know, however, that certain conditions must hold on stable surfaces of discontinuity in a gas; for instance, the discontinuities of density and pressure in a shock wave are related by the shock adiabatic. It is therefore clear that, if these conditions are not satisfied in the initial discontinuity, it cannot continue to be a discontinuity at subsequent instants. Instead, the initial discontinuity in general splits into several discontinuities, each of which is one of the possible types (shock wave, tangential discontinuity, weak discontinuity); in the course of time, these discontinuities move apart. A general discussion of the behaviour of an arbitrary discontinuity has been given by N. E. KOCHIN (1926).

During a short interval of time after the initial instant $t = 0$, the discontinuities formed from the initial discontinuity do not move apart to great distances, and the flow under consideration therefore takes place in a relatively small volume adjoining the surface of initial discontinuity. As usual, it suffices to consider separate portions of this surface, each of which may be regarded as plane. We need therefore consider only a plane surface of discontinuity, which we take as the yz-plane. It is evident from symmetry that the discontinuities formed from the initial discontinuity will also be plane, and perpendicular to the x-axis. The flow pattern will depend on the coordinate x only (and on the time), so that the problem is one-dimensional. There being no characteristic parameters of length and time, we have a similarity problem, and the results obtained in §92 can be used.

The discontinuities formed from the initial discontinuity must evidently move away from their point of formation, i.e. away from the position of the initial discontinuity. It is easy to see that either one shock wave, or one pair of weak discontinuities bounding a rarefaction wave, can move in each direction (the positive and negative x-direction). For, if there were, say, two shock waves formed at the same point at time $t = 0$ and both propagated in the positive x-direction, the leading one would have to move more rapidly than the other. According to the general properties of shock waves, however, the leading shock wave must move, relative to the gas behind it, with a velocity less than the velocity of sound c in that gas, and the following shock must move, relative to the same gas, with a velocity exceeding c (c being a constant in the region between the shock waves), i.e. it must overtake the other. For the same reason, a shock wave and a rarefaction wave cannot move in the same direction; to see this, it is sufficient to notice that weak discontinuities move with the velocity of sound relative to the gas on each side of them. Finally, two rarefaction waves formed at the same time cannot become separated, since the velocities of their backward fronts are the same.

As well as shock waves and rarefaction waves, a tangential discontinuity must in general be formed from an initial discontinuity. Such a discontinuity must occur if the transverse velocity components v_y, v_z are discontinuous in the initial discontinuity. Since these velocity components do not change in a shock or rarefaction wave, their discontinuities always occur at a tangential discontinuity, which remains at the position of the initial discontinuity; on each side of this discontinuity, v_y and v_z are constant (in reality, of course, the instability of a tangential velocity discontinuity causes its gradual smoothing into a turbulent region).

A tangential discontinuity must occur, however, even if v_y and v_z are continuous at the initial discontinuity (without loss of generality, we can, and shall, assume that they are zero). This is shown as follows. The discontinuities formed from the initial discontinuity must make it possible to go from a given state 1 of the gas on one side of the initial discontinuity to a given state 2 on the other side. The state of the gas is determined by three independent quantities, e.g. p, ρ and $v_x = v$. It is therefore necessary to have three arbitrary parameters in order to go from state 1 to an arbitrary state 2 by some choice of the discontinuities. We know, however, that a shock wave, perpendicular to the stream, propagated in a gas whose thermodynamic state is given, is completely determined by one parameter (§82). The same is true of a rarefaction wave; as we see from formulae (92.14)–(92.16), when the state of the gas entering a rarefaction wave is given, the state of the gas leaving it is completely determined by one parameter. We have seen, moreover, that at most one wave (rarefaction or shock) can move in each direction. We therefore have at our disposal only two parameters, which are not sufficient.

The tangential discontinuity formed at the position of the initial discontinuity furnishes the third parameter required. The pressure is continuous there, but the density (and therefore the temperature and entropy) is not. The tangential discontinuity is stationary with respect to the gas on both sides of it and the arguments about the "overtaking" of two waves propagated in the same direction therefore do not apply to it.

The gases on the two sides of the tangential discontinuity do not mix, since there is no motion of gas through a tangential discontinuity; in all the examples given below, these gases may be different substances.

Fig. 62 shows schematically all possible types of break-up of an initial discontinuity. The continuous line shows the variation of the pressure along the x-axis; the variation of the density would be given by a similar line, the only difference being that there would be a further jump at the tangential discontinuity. The vertical lines show the discontinuities formed, and the arrows show their direction of propagation and that of the gas flow. The co-ordinate system is always that in which the tangential discontinuity is at rest, together with the gas in the regions 3 and 3' which adjoin it. The pressures, densities and velocities of the gases in the extreme left-hand (1) and right-hand (2) regions are the values of these quantities at time $t = 0$ on each side of the initial discontinuity.

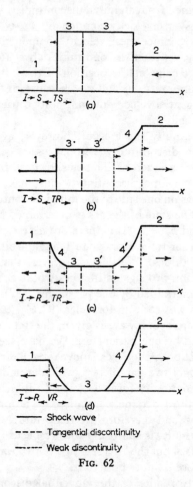

Fig. 62

In the first case, which we write $I \to S_\leftarrow TS_\to$ (Fig. 62a), the initial discontinuity I gives two shock waves S, propagated in opposite directions, and a tangential discontinuity T between them. This case occurs when two masses of gas collide with a large relative velocity.

In the case $I \to S_\leftarrow TR_\to$ (Fig. 62b), a shock wave is propagated on one side of the tangential discontinuity, and a rarefaction wave R on the other side. This case occurs, for instance, if two masses of gas at relative rest $(v_2 - v_1 = 0)$ and at different pressures are brought into contact at the initial instant. For, of all the cases shown in Fig. 62, the second is the only one in which gases 1 and 2 are moving in the same direction, and so the equation $v_1 = v_2$ is possible.

In the third case ($I \rightarrow R_{\leftarrow} TR_{\rightarrow}$, Fig. 62c), a rarefaction wave is propagated on each side of the tangential discontinuity. If gases 1 and 2 separate with a sufficiently great relative velocity $v_2 - v_1$, the pressure may decrease to zero in the rarefaction waves. We then have the pattern shown in Fig. 62d; a vacuum 3 is formed between regions 4 and 4'.

We can derive the analytical conditions which determine the manner in which the initial discontinuity breaks up, as a function of its parameters. We shall suppose in every case that $p_2 > p_1$, and take the positive x-direction from region 1 to region 2 (as in Fig. 62).

Since the gases on the two sides of the initial discontinuity may be of different substances, we shall distinguish them as gases 1 and 2.

(1) $I \rightarrow S_{\leftarrow} TS_{\rightarrow}$. If $p_3 = p_3'$, V_3 and V_3' are the pressures and specific volumes in the resulting regions 3 and 3', then we have $p_3 > p_2 > p_1$, and the volumes V_3 and V_3' are the abscissae of the points with ordinate p_3 on the shock adiabatics through (p_1, V_1) and (p_2, V_2) respectively. Since the gases in regions 3 and 3' are at rest in the co-ordinate system chosen, we can use formula (82.7) to give the velocities v_1 and v_2, which are in the positive and negative x-directions respectively:

$$v_1 = \sqrt{[(p_3 - p_1)(V_1 - V_3)]}, \qquad v_2 = -\sqrt{[(p_3 - p_2)(V_2 - V_3')]}.$$

The least value of p_3, for given p_1 and p_2, which does not contradict the initial assumption ($p_3 > p_2 > p_1$) is p_2. Since, moreover, the difference $v_1 - v_2$ is a monotonically increasing function of p_3, we find the required inequality

$$v_1 - v_2 > \sqrt{[(p_2 - p_1)(V_1 - V')]}, \qquad (93.1)$$

where V' denotes the abscissa of the point with ordinate p_2 on the shock adiabatic for gas 1 through (p_1, V_1). Calculating V' from formula (85.1) (in which V_2 is replaced by V'), we obtain the condition (93.1) for a perfect gas in the form

$$v_1 - v_2 > (p_2 - p_1)\sqrt{\{2V_1/[(\gamma_1 - 1)p_1 + (\gamma_1 + 1)p_2]\}}. \qquad (93.2)$$

It should be noted that the limits placed by (93.1) and (93.2) on the possible values of the velocity difference $v_1 - v_2$ clearly do not depend on the co-ordinate system chosen.

(2) $I \rightarrow S_{\leftarrow} TR_{\rightarrow}$. Here $p_1 < p_3 = p_3' < p_2$. For the gas velocity in region 1 we again have

$$v_1 = \sqrt{[(p_3 - p_1)(V_1 - V_3)]},$$

and the total change in velocity in the rarefaction wave 4 is, by (92.7),

$$v_2 = \int_{p_3}^{p_2} \sqrt{(-dp\, dV)}.$$

For given p_1 and p_2, p_3 can lie between them. Replacing p_3 in the difference $v_2 - v_1$ by p_1 and then by p_2, we obtain the condition

$$- \int_{p_1}^{p} \sqrt{(-dp\,dV)} < v_1 - v_2 < \sqrt{[(p_2 - p_1)(V_1 - V')]}. \qquad (93.3)$$

Here V' has the same significance as in the previous case; the upper limit of the difference $v_1 - v_2$ must be calculated for gas 1, and the lower limit for gas 2. For a perfect gas we have

$$- \frac{2c_2}{\gamma_2 - 1} \left[1 - \left(\frac{p_1}{p_2} \right)^{(\gamma_2 - 1)/2\gamma_2} \right] < v_1 - v_2$$

$$< (p_2 - p_1) \sqrt{\{2V_1 / [(\gamma_1 - 1)p_1 + (\gamma_1 + 1)p_2]\}}, \qquad (93.4)$$

where $c_2 = \sqrt{(\gamma_2 p_2 V_2)}$ is the velocity of sound in gas 2 in the state (p_2, V_2).

(3) $I \to R_\leftarrow TR_\to$. Here $p_2 > p_1 > p_3 = p_3' > 0$. By the same method we find the following condition for this case to occur:

$$- \int_0^{p_1} \sqrt{(-dp\,dV)} - \int_0^{p_2} \sqrt{(-dp\,dV)} < v_1 - v_2 < - \int_{p_1}^{p_2} \sqrt{(-dp\,dV)}. \quad (93.5)$$

The first integral in the first member is calculated for gas 1, and the others for gas 2. For a perfect gas we find

$$- \frac{2c_1}{\gamma_1 - 1} - \frac{2c_2}{\gamma_2 - 1} < v_1 - v_2 < - \frac{2c_2}{\gamma_2 - 1} \left[1 - \left(\frac{p_1}{p_2} \right)^{(\gamma_2 - 1)/2\gamma_2} \right], \qquad (93.6)$$

where $c_1 = \sqrt{(\gamma_1 p_1 V_1)}$, $c_2 = \sqrt{(\gamma_2 p_2 V_2)}$. If

$$v_1 - v_2 < - \frac{2c_1}{\gamma_1 - 1} - \frac{2c_2}{\gamma_2 - 1}, \qquad (93.7)$$

a vacuum is formed between the rarefaction waves $(I \to R_\leftarrow V R_\to)$.

The problem of a discontinuity in the initial conditions includes that of various collisions between plane surfaces of discontinuity. At the instant of collision, the two planes coincide, and form some initial discontinuity, which then leads to one of the patterns described above. The collision of two shock waves, for instance, results in two other shock waves, which move away from the tangential discontinuity remaining between them: $S_\to S_\leftarrow \to S_\leftarrow TS_\to$. When one shock wave overtakes another, there are two possibilities: $S_\to S_\to \to S_\leftarrow TS_\to$ and $S_\to S_\to \to R_\leftarrow TS_\to$. In either case a shock wave continues in the same direction.

The problem of the reflection and transmission of shock waves by a tangential discontinuity (boundary of two media) also comes under this heading. Here two cases are possible: $S_\to T \to S_\leftarrow TS_\to$ and $S_\to T \to R_\leftarrow TS_\to$. The wave transmitted into the second medium is always a shock (see also the following Problems).†

PROBLEMS

PROBLEM 1. A plane shock wave is reflected from a rigid plane surface. Determine the gas pressure behind the reflected wave (S. V. IZMAĬLOV 1935).

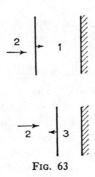

FIG. 63

SOLUTION. When a shock wave is incident on a rigid wall, a reflected shock wave is propagated away from the wall. We denote by the suffixes 1, 2 and 3 respectively quantities pertaining to the undisturbed gas in front of the incident shock, the gas behind this shock (which is also the gas in front of the reflected shock) and the gas behind the reflected shock; see Fig. 63, where the arrows indicate the direction of motion of the shock waves and of the gas itself. The gas in regions 1 and 3, which adjoin the wall, is at rest relative to the wall. The relative velocity of the gases on the two sides of the discontinuity is the same in both the incident and the reflected shock wave, and equal to the velocity of gas 2. Using formula (82.7) for the relative velocity, we therefore have $(p_2 - p_1)(V_1 - V_2) = (p_3 - p_2)(V_2 - V_3)$. The equation of the shock adiabatic (85.1) for each shock gives

$$\frac{V_2}{V_1} = \frac{(\gamma+1)p_1 + (\gamma-1)p_2}{(\gamma-1)p_1 + (\gamma+1)p_2}, \qquad \frac{V_3}{V_2} = \frac{(\gamma+1)p_2 + (\gamma-1)p_3}{(\gamma-1)p_2 + (\gamma+1)p_3}.$$

We can eliminate the specific volumes from these three equations, and the result is

$$(p_3 - p_2)^2[(\gamma+1)p_1 + (\gamma-1)p_2] = (p_2 - p_1)^2[(\gamma+1)p_3 + (\gamma-1)p_2].$$

This is a quadratic equation for p_3, which has the trivial root $p_3 = p_1$; cancelling $p_3 - p_1$, we obtain

$$\frac{p_3}{p_2} = \frac{(3\gamma-1)p_2 - (\gamma-1)p_1}{(\gamma-1)p_2 - (\gamma+1)p_1},$$

which determines p_3 from p_1 and p_2. In the limiting case of a very strong incident shock, $p_3 = (3\gamma-1)p_2/(\gamma-1)$, while for a weak shock $p_3 - p_2 = p_2 - p_1$, corresponding to the sound-wave approximation.

PROBLEM 2. Find the condition for a shock wave to be reflected from a plane boundary between two gases.

---

† For completeness we should mention that, when a shock wave collides with a weak discontinuity (a problem which is not of the similarity type considered here), the shock wave continues to be propagated in the same direction, but behind it there remain a weak discontinuity of the original kind and a weak tangential discontinuity (see the end of §89).

SOLUTION. Let $p_1 = p_{2'}$, V_1, $V_{2'}$, be the pressures and specific volumes of the two media before the incidence of the shock wave (propagated in gas 2), at their surface of separation, and p_3, V_2 the values behind the shock wave. The condition for the reflected wave to be a shock wave is given by the inequality (93.2), in which we must now put

$$v_1 - v_2 = \sqrt{[(p_2 - p_{2'})(V_{2'} - V_2)]}.$$

Expressing all quantities in terms of the ratio of pressures p_2/p_1 and the initial specific volumes V_1, $V_{2'}$, we obtain

$$\frac{V_1}{(\gamma_1 + 1)p_2/p_1 + (\gamma_1 - 1)} < \frac{V_{2'}}{(\gamma_2 + 1)p_2/p_1 + (\gamma_2 - 1)}.$$

§94. One-dimensional travelling waves

In discussing sound waves in §63, we assumed the amplitude of oscillations in the wave to be small. The result was that the equations of motion were linear and were easily solved. A particular solution of these equations is any function of $x \pm ct$ (a plane wave), corresponding to a *travelling wave* whose profile moves with velocity c, its shape remaining unchanged; by the *profile* of a wave we mean the distribution of density, velocity, etc., along the direction of propagation. Since the velocity v, the density ρ and the pressure p (and the other quantities) in such a wave are functions of the same quantity $x \pm ct$, they can be expressed as functions of one another, in which the coordinates and time do not explicitly appear ($p = p(\rho)$, $v = v(p)$, and so on).

When the wave amplitude is not necessarily small, these simple relations do not hold. It is found, however, that a general solution of the exact equations of motion can be obtained, in the form of a travelling plane wave which is a generalisation of the solution $f(x \pm ct)$ of the approximate equations valid for small amplitudes. To derive this solution, we shall begin from the requirement that, for a wave of any amplitude, the velocity can be expressed as a function of the density.

In the absence of shock waves the flow is adiabatic. If the gas is homogeneous at some initial instant (so that, in particular, $s = $ constant), then $s = $ constant at all times, and we shall assume this in what follows.

In a plane sound wave propagated in the x-direction, all quantities depend on x and t only, and for the velocity we have $v_x = v$, $v_y = v_z = 0$. The equation of continuity is $\partial\rho/\partial t + \partial(\rho v)/\partial x = 0$, and Euler's equation is

$$\frac{\partial v}{\partial t} + v\frac{\partial v}{\partial x} + \frac{1}{\rho}\frac{\partial p}{\partial x} = 0.$$

Using the fact that v is a function of ρ only, we can write these equations as

$$\frac{\partial\rho}{\partial t} + \frac{d(\rho v)}{d\rho}\frac{\partial\rho}{\partial x} = 0, \tag{94.1}$$

$$\frac{\partial v}{\partial t} + \left(v + \frac{1}{\rho}\frac{dp}{dv}\right)\frac{\partial v}{\partial x} = 0. \tag{94.2}$$

Since

$$\frac{\partial \rho / \partial t}{\partial \rho / \partial x} = -\left(\frac{\partial x}{\partial t}\right)_\rho,$$

we have from (94.1)

$$\left(\frac{\partial x}{\partial t}\right)_\rho = \frac{d(\rho v)}{d\rho} = v + \rho \frac{dv}{d\rho},$$

and similarly from (94.2)

$$\left(\frac{\partial x}{\partial t}\right)_v = v + \frac{1}{\rho}\frac{dp}{dv}. \tag{94.3}$$

Since the value of ρ uniquely determines that of v, the derivatives for constant ρ and constant v are the same, i.e. $(\partial x/\partial t)_\rho = (\partial x/\partial t)_v$, so that $\rho \, dv/d\rho = (1/\rho) \, dp/dv$. Putting $dp/dv = (dp/d\rho)(d\rho/dv) = c^2 d\rho/dv$, we obtain $dv/d\rho = \pm c/\rho$, whence

$$v = \pm \int \frac{c}{\rho} d\rho = \pm \int \frac{dp}{\rho c}. \tag{94.4}$$

This gives the general relation between the velocity and the density or pressure in the wave.†

Next, we can combine (94.3) and (94.4) to give $(\partial x/\partial t)_v = v + (1/\rho)dp/dv = v \pm c(v)$, or, integrating,

$$x = t[v \pm c(v)] + f(v), \tag{94.5}$$

where $f(v)$ is an arbitrary function of the velocity, and $c(v)$ is given by (94.4).

Formulae (94.4) and (94.5) give the required general solution (B. RIEMANN, 1860). They determine the velocity (and therefore all other quantities) as an implicit function of x and t, i.e. the wave profile at every instant. For any given value of v, we have $x = at + b$, i.e. the point where the velocity has a given value moves with constant velocity; in this sense, the solution obtained is a travelling wave. The two signs in (94.5) correspond to waves propagated (relative to the gas) in the positive and negative x-directions.

The flow described by the solution (94.4) and (94.5) is often called a *simple wave*, and we shall use this expression below. It should be noticed that the similarity flow discussed in §92 is a particular case of a simple wave, corresponding to $f(v) = 0$ in (94.5).

We can write out explicitly the relations for a simple wave in a perfect gas; for definiteness, we assume that there is a point in the wave for which $v = 0$, as usually happens in practice. Since formula (94.4) is the same as (92.6), we have by analogy with formulae (92.14)–(92.16)

$$c = c_0 \pm \tfrac{1}{2}(\gamma - 1)v, \tag{94.6}$$

$$\rho = \rho_0(1 \pm \tfrac{1}{2}(\gamma - 1)v/c_0)^{2/(\gamma - 1)},$$

$$p = p_0(1 \pm \tfrac{1}{2}(\gamma - 1)v/c_0)^{2\gamma/(\gamma - 1)}. \tag{94.7}$$

---

† In a wave of small amplitude we have $\rho = \rho_0 + \rho'$, and (94.4) gives in the first approximation $v = c_0 \rho' / \rho_0$ (where $c_0 = c(\rho_0)$), i.e. the usual formula (63.12).

Substituting (94.6) in (94.5), we obtain

$$x = t(\pm c_0 + \tfrac{1}{2}(\gamma + 1)v) + f(v). \tag{94.8}$$

It is sometimes convenient to write this solution in the form

$$v = F[x - (\pm c_0 + \tfrac{1}{2}(\gamma + 1)v)t], \tag{94.9}$$

where F is another arbitrary function.

From formulae (94.6) and (94.7) we again see (as in §92) that the velocity in a direction opposite to that of the propagation of the wave (relative to the gas itself) is of limited magnitude; for a wave propagated in the positive x-direction we have

$$-v \leqslant 2c_0/(\gamma - 1). \tag{94.10}$$

A travelling wave described by formulae (94.4) and (94.5) is essentially different from the one obtained in the limiting case of small amplitudes. The velocity of a point in the wave profile is

$$u = v \pm c; \tag{94.11}$$

it may be conveniently regarded as a superposition of the propagation of a disturbance relative to the gas with the velocity of sound and the movement of the gas itself with velocity v. The velocity u is now a function of the density, and therefore is different for different points in the profile. Thus, in the general case of a plane wave of arbitrary amplitude, there is no definite constant "wave velocity". Since the velocities of different points in the wave profile are different, the profile changes its shape in the course of time.

Let us consider a wave propagated in the positive x-direction, for which $u = v + c$. The derivative of $v + c$ with respect to the density has been calculated in §92; see (92.10). We have seen that $du/d\rho > 0$. The velocity of propagation of a given point in the wave profile is therefore the greater, the greater the density. If we denote by c_0 the velocity of sound for a density equal to the equilibrium density ρ_0, then in compressions $\rho > \rho_0$ and $c > c_0$, while in rarefactions $\rho < \rho_0$ and $c < c_0$.

The inequality of the velocity of different points in the wave profile causes its shape to change in the course of time: the points of compression move forward and those of rarefaction are left behind (Fig. 64b). Finally, the profile may become such that the function $\rho(x)$ (for given t) is no longer one-valued; three different values of ρ correspond to some x (the dashed line in Fig. 64c). This is, of course, physically impossible. In reality, discontinuities are formed where ρ is not one-valued, and ρ is consequently one-valued everywhere except at the discontinuities themselves. The wave profile then has the form shown by the continuous line in Fig. 64c. The surfaces of discontinuity are thus formed at points a wavelength apart.

When the discontinuities are formed, the wave ceases to be a simple wave. The cause of this can be briefly stated thus: when surfaces of discontinuity are present, the wave is "reflected" from them, and therefore ceases to be a wave travelling in one direction. The assumption on which the whole derivation is based, namely that there is a one-to-one relation between the various quantities, consequently ceases to be valid in general.

The presence of discontinuities (shock waves) results, as was mentioned in §82, in the dissipation of energy. The formation of discontinuities therefore leads to a marked damping of the wave. This is evident from Fig. 64. When the discontinuity is formed, the highest part of the wave profile is cut off. In the course of time, as the profile is bent over, its height becomes less, and the profile is "smoothed" to one of smaller amplitude, i.e. the wave is damped.

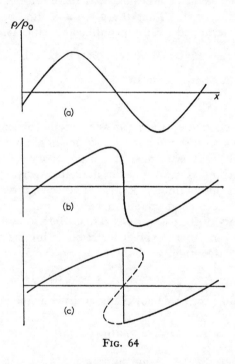

FIG. 64

It is clear from the above that discontinuities must ultimately be formed in every simple wave which contains regions where the density decreases in the direction of propagation. The only case where discontinuities do not occur is a wave in which the density increases monotonically in the direction of propagation (such, for example, is the wave formed when a piston moves out of an infinite pipe filled with gas; see the Problems at the end of this section).

Although the wave is no longer a simple one when a discontinuity has been formed, the time and place of formation of the discontinuity can be determined analytically. We have seen that the occurrence of discontinuities is mathematically due to the fact that, in a simple wave, the quantities p, ρ and v become many-valued functions of x (for given t) at times greater than a certain definite value t_0, whereas for $t < t_0$ they are one-valued functions. The time t_0 is the time of formation of the discontinuity. It is evident from geometrical considerations that, at the instant t_0, the curve giving, say, v as a function of x becomes vertical at some point $x = x_0$, which is the point where the function is subsequently many-valued. Analytically, this means that the derivative $(\partial v/\partial x)_t$ becomes infinite, and $(\partial x/\partial v)_t$ becomes zero. It is also clear that, at the instant t_0, the curve $v = v(x)$ must lie on both sides of the vertical tangent, since otherwise $v(x)$ would already be many-valued. In other words, the point $x = x_0$ must be, not an extremum of the function $x(v)$, but a point of inflexion, and therefore the second derivative $(\partial^2 x/\partial v^2)_t$ must also vanish. Thus the place and time of formation of the shock wave are determined by the simultaneous equations

$$(\partial x/\partial v) = 0, \qquad (\partial^2 x/\partial v^2)_t = 0. \tag{94.12}$$

For a perfect gas these equations are

$$t = -2f'(v)/(\gamma+1), \qquad f''(v) = 0, \tag{94.13}$$

where $f(v)$ is the function appearing in the general solution (94.8).

These conditions require modification if the simple wave adjoins a gas at rest and the shock wave is formed at the boundary. Here also the curve $v = v(x)$ must become vertical, i.e. the derivative $(\partial x/\partial v)_t$ must vanish, at the time when the discontinuity occurs. The second derivative, however, need not vanish; the second condition here is simply that the velocity is zero at the boundary of the gas at rest, so that $(\partial x/\partial v)_t = 0$ for $v = 0$. From this condition we can obtain explicit expressions for the time and place of formation of the discontinuity. Differentiating (94.5), we obtain

$$t = -f'(0)/\alpha_0, \qquad x = \pm c_0 t + f(0), \tag{94.14}$$

where α_0 is the value, for $v = 0$, of the quantity α defined by formula (95.2). For a perfect gas

$$t = -2f'(0)/(\gamma+1). \tag{94.15}$$

PROBLEMS

PROBLEM 1. A perfect gas is in a semi-infinite cylindrical pipe ($x > 0$) terminated by a piston. At time $t = 0$ the piston begins to move with a uniformly accelerated velocity $U = \pm at$. Determine the resulting flow.

SOLUTION. If the piston moves out of the pipe ($U = -at$), the result is a simple rarefaction wave, whose forward front is propagated to the right, through gas at rest, with velocity c_0; in the region $x > c_0 t$ the gas is at rest. At the surface of the piston, the gas and the piston must have the same velocity, i.e. we must have $v = -at$ for $x = -\frac{1}{2}at^2$ ($t > 0$). This condition gives for the function $f(v)$ in (94.8)

$$f(-at) = -c_0 t + \tfrac{1}{2}\gamma at^2.$$

Hence we have

$$x - [c_0 + \tfrac{1}{2}(\gamma+1)v]t = f(v)$$
$$= c_0 v/a + \tfrac{1}{2}\gamma v^2/a,$$

whence

$$-v = [c_0 + \tfrac{1}{2}(\gamma+1)at]/\gamma - \sqrt{\{[c_0 + \tfrac{1}{2}(\gamma+1)at]^2 - 2a\gamma(c_0 t - x)\}}/\gamma. \qquad (1)$$

This formula gives the change in velocity over the region between the piston and the forward front $x = c_0 t$ of the wave (Fig. 65a) during the time interval $t = 0$ to $t = 2c_0/(\gamma-1)a$. The gas velocity is everywhere to the left, like that of the piston, and decreases monotonically in magnitude in the positive x-direction; the density and pressure increase monotonically in that direction. For $t > 2c_0/(\gamma-1)a$, the inequality (94.10) does not hold for the piston velocity, and so the gas can no longer follow the piston. A vacuum is then formed in a region adjoining the piston, beyond which the gas velocity decreases from $-2c_0/(\gamma-1)$ to zero according to formula (1).

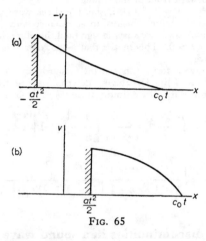

FIG. 65

If the piston moves into the pipe ($U = at$), a simple compression wave is formed; the corresponding solution is obtained by merely changing the sign of a in (1) (Fig. 65b). It is valid, however, only until a shock wave is formed; the time when this happens is determined from formula (94.15), and is

$$t = 2c_0/a(\gamma+1).$$

PROBLEM 2. The same as Problem 1, but for the case where the piston moves in any manner.

SOLUTION. Let the piston begin to move at time $t = 0$ according to the law $x = X(t)$ (with $X(0) = 0$); its velocity is $U = X'(t)$. The boundary condition on the piston ($v = U$ for $x = X$) gives $v = X'(t), f(v) = X(t) - t[c_0 + \tfrac{1}{2}(\gamma+1)X'(t)]$. If we now regard t as a parameter, these two equations determine the function $f(v)$ in parametric form. Denoting the parameter by τ, we can write the solution as

$$v = X'(\tau), \quad x = X(\tau) + (t-\tau)[c_0 + \tfrac{1}{2}(\gamma+1)X'(\tau)], \qquad (1)$$

which determines, in parametric form, the required function $v(t, x)$ in the simple wave which is caused by the motion of the piston.

PROBLEM 3. Determine the time and place of formation of the shock wave when the piston (Problem 1) moves according to the law $U = at^n$ ($n > 0$).

SOLUTION. If $a < 0$, i.e. the piston moves out of the pipe, a simple rarefaction wave results, in which no shock wave is formed. We therefore assume that $a > 0$, i.e. the piston moves into the pipe, causing a simple compression wave.

When the function $v(x, t)$ is given by the parametric formulae (1) (Problem 2), and $X = a\tau^{n+1}/n+1$, the time and place of formation of the shock wave are given by the equations

$$\left(\frac{\partial x}{\partial \tau}\right)_t = -c_0 + \tfrac{1}{2}t\tau^{n-1} an(\gamma+1) - \tfrac{1}{2}a\tau^n[\gamma - 1 + n(\gamma+1)] = 0,$$

$$\left(\frac{\partial^2 x}{\partial \tau^2}\right)_t = \tfrac{1}{2}t\tau^{n-2} an(n-1)(\gamma+1) - \tfrac{1}{2}an\tau^{n-1}[\gamma - 1 + n(\gamma+1)] = 0, \tag{1}$$

where the second equation must be replaced by $\tau = 0$ if we are concerned with the formation of a shock wave at the forward front of the simple wave.

For $n = 1$ we find $\tau = 0$, $t = 2c_0/a(\gamma+1)$, i.e. the shock wave is formed at the forward front at a finite time after the motion begins, in accordance with the results of Problem 1.

For $n < 1$, the derivative $\partial x/\partial \tau$ is of varying sign (and therefore the function $v(x)$ for given t is many-valued) for any $t > 0$. This means that a shock wave is formed at the piston as soon as it begins to move.

For $n > 1$ the shock wave is formed, not at the forward front of the simple wave, but at some intermediate point given by (1). Having determined τ and t from (1), we can then find the place of formation of the discontinuity from (1) of Problem 2. The result is

$$t = \left(\frac{2c_0}{a}\right)^{1/n} \frac{1}{\gamma+1} \left[\frac{n+1}{n-1}\gamma + 1\right]^{(n-1)/n},$$

$$x = 2c_0\left(\frac{2c_0}{a}\right)^{1/n} \left[\frac{\gamma}{\gamma+1} + \frac{n-1}{n+1}\right] \frac{1}{(n-1)^{(n-1)/n}[\gamma - 1 + n(\gamma+1)]^{1/n}}.$$

§95. Formation of discontinuities in a sound wave

A travelling plane sound wave, being an exact solution of the equations of motion, is also a simple wave. We can use the general results obtained in §94 to derive some properties of sound waves of small amplitude in the second approximation (the first approximation being that which gives the ordinary linear wave equation).

We must notice first of all that a discontinuity must ultimately appear in each wavelength of a sound wave. This leads to a very marked damping of the wave, as shown in §94. It must be remarked, however, that this happens only for a sufficiently strong sound wave; a weak sound wave is damped by the usual effects of viscosity and thermal conduction before the effects of higher order in the amplitude can develop.

The distortion of the wave profile has another effect also. If the wave is purely harmonic at some instant, it ceases to be so at later instants, on account of the change in shape of the profile. The motion, however, remains periodic, with the same period as before. When the wave is expanded in a Fourier series, terms with frequencies $n\omega$ (n being integral and ω being the fundamental frequency) appear, as well as that with frequency ω. Thus the distortion of the profile as the sound wave is propagated may be regarded as the appearance in it of higher harmonics in addition to the fundamental frequency.

The velocity u of points in the wave profile (the wave being propagated in the positive x-direction) is obtained, in the first approximation, by putting in (94.11) $v = 0$, i.e. $u = c_0$, corresponding to the propagation of the wave with no change in its profile. In the next approximation we have

$$u = c_0 + \rho' \, \partial u / \partial \rho_0 = c_0 + (\partial u / \partial \rho_0)\rho_0 v / c_0,$$

or, using the expression (92.10) for the derivative $\partial u / \partial \rho$,

$$u = c_0 + \alpha_0 v, \qquad (95.1)$$

where we have put for brevity

$$\alpha = (c^4 / 2V^3)(\partial^2 V / \partial p^2)_s. \qquad (95.2)$$

For a perfect gas, $\alpha = \frac{1}{2}(\gamma + 1)$, and formula (95.1) agrees with the exact formula (see (94.8)) for the velocity u.

In the general case of arbitrary amplitude, the wave is no longer simple after the discontinuities have appeared. A wave of small amplitude, however, is still simple in the second approximation even when discontinuities are present. This can be seen as follows. The changes in velocity, pressure and specific volume in a shock wave are related by $v_2 - v_1 = \sqrt{[(p_2 - p_1)(V_1 - V_2)]}$. The change in the velocity v over a segment of the x-axis in a simple wave is

$$v_2 - v_1 = \int_{p_1}^{p_2} \sqrt{(-\partial V / \partial p)} \, \mathrm{d}p.$$

A simple calculation, using an expansion in series, shows that these two expressions differ only by terms of the third order (it must be borne in mind that the change in entropy at a discontinuity is of the third order of smallness, while in a simple wave the entropy is constant). Hence it follows that, as far as terms of the second order, a sound wave on either side of a discontinuity in it remains simple, and the appropriate boundary condition is satisfied at the discontinuity itself. In higher approximations this is no longer true, on account of the appearance of waves reflected from the surface of discontinuity.

Let us now derive the condition which determines the location of the discontinuities in a travelling sound wave (again in the second approximation). Let u be the velocity of the discontinuity relative to a fixed co-ordinate system, and v_1, v_2 the velocities of the gases on each side of it. Then the condition that the mass flux is continuous is $\rho_1(v_1 - u) = \rho_2(v_2 - u)$, whence $u = (\rho_1 v_1 - \rho_2 v_2)/(\rho_1 - \rho_2)$. As far as the second-order terms, this is equal to the derivative $\mathrm{d}(\rho v)/\mathrm{d}\rho$ at the point where v is equal to $\frac{1}{2}(v_1 + v_2)$:

$$u = [\mathrm{d}(\rho v)/\mathrm{d}\rho]_{v = \frac{1}{2}(v_1 + v_2)}.$$

Since, in a simple wave, $\mathrm{d}(\rho v)/\mathrm{d}\rho = v + c$, we have, by (95.1),

$$u = c_0 + \frac{1}{2}\alpha(v_1 + v_2). \qquad (95.3)$$

From this we can obtain the following simple geometrical condition which determines the position of the shock wave. In Fig. 66 the curve shows the velocity profile corresponding to the simple wave; let ae be the discontinuity.

The difference of the shaded areas *abc* and *cde* is the integral

$$\int_{v_1}^{v_2} (x - x_0) dv$$

taken along the curve *abcde*. In the course of time, the wave profile moves;

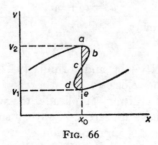

FIG. 66

let us calculate the time derivative of the above integral. Since the velocity dx/dt of points in the wave profile is given by formula (95.1), and the velocity dx_0/dt of the discontinuity by (95.3), we have

$$\frac{d}{dt} \int_{v_1}^{v_2} (x - x_0) \, dv = \alpha \{ \int_{v_1}^{v_2} v \, dv - \tfrac{1}{2}(v_1 + v_2) \int_{v_1}^{v_2} dv \} = 0;$$

in differentiating the integral, we must notice that, although the limits of integration v_1 and v_2 also vary with time, $x - x_0$ always vanishes at the limits, and so we need only differentiate the integrand.

Thus the integral $\int (x - x_0) dv$ remains constant in time. Since it is zero at the instant when the shock wave is formed (the points *a* and *e* then coinciding), it follows that we always have

$$\int_{abcde} (x - x_0) dv = 0. \qquad (95.4)$$

Geometrically this means that the areas *abc* and *cde* are equal, a condition which determines the position of the discontinuity.

Let us consider a single one-dimensional compression pulse, in which a shock wave has already been formed, and ascertain how this shock will finally be damped. By so doing, we also find the law of damping of any plane shock wave after it has been propagated for a sufficiently long time.

In the later stages of its propagation, a sound pulse containing a shock wave will have a triangular velocity profile. Let the profile be given at some instant (which we take as $t = 0$) by the triangle *ABC* (Fig. 67a). If the points in this profile move with the velocities (95.1), we obtain after time t

a profile $A'B'C'$ (Fig. 67b). In reality, the discontinuity moves to E, and the actual profile will be $A'DE$. The areas $DB'F$ and $C'FE$ are equal, by (95.4), and therefore the area $A'DE$ of the new profile is equal to the area ABC of the original profile. Let l be the length of the sound pulse at time t, and Δv the velocity discontinuity in the shock wave. During time t, the point B moves a distance $\alpha t\Delta v_0$ relative to C; the tangent of the angle $B'A'C'$ is therefore $\Delta v_0/(l_0+\alpha t\Delta v_0)$, and we obtain the condition of equal areas ABC and $A'DE$ in the form

$$l_0\Delta v_0 = l^2\,\Delta v_0/(l_0+\alpha t\,\Delta v_0),$$

whence

$$l = l_0\sqrt{[1+\alpha\Delta v_0\,t/l_0]},$$

$$\Delta v = \Delta v_0/\sqrt{[1+\alpha\Delta v_0\,t/l_0]}. \tag{95.5}$$

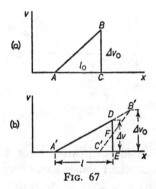

Fig. 67

For $t \to \infty$ the intensity of the shock wave diminishes asymptotically with time as $1/\sqrt{t}$ (or, what is the same thing, with distance as $1/\sqrt{x}$). The total energy of a travelling sound pulse (per unit area of its front) is

$$E = \rho_0 \int v^2\,dx = E_0/\sqrt{[1+\alpha\Delta v_0\,t/l_0]}, \tag{95.6}$$

where E_0 is the energy at time $t = 0$. For $t \to \infty$ the energy also tends to zero as $1/\sqrt{t}$.

If we have a spherical outgoing sound wave, any small section of it can be regarded as plane at sufficiently large distances r from the origin. The velocity of any point in the wave profile is then given by formula (95.1). If, however, we wish to use this formula to follow the motion of any point in the wave profile over long intervals of time, we must take into account the fact that the amplitude of a spherical wave falls off inversely as the distance r, even in the first approximation. This means that, at any given point in the profile, v is not constant, as it is for a plane wave, but decreases as $1/r$. If

v_0 is the value of v (for a given point in the profile) at a (large) distance r_0, we can put $v = v_0 r_0/r$. Thus the velocity u of points in the wave profile is $u = c_0 + \alpha v_0 r_0/r$. The first term is the ordinary velocity of sound, and corresponds to movement of the wave without change in the shape of the profile (apart from the general decrease of the amplitude as $1/r$). The second term results in a distortion of the profile. The amount δr of this additional movement of points in the profile during a time $t = (r - r_0)/c$ is obtained by multiplying by $\mathrm{d}r/c_0$ and integrating from r_0 to r; this gives

$$\delta r = (\alpha v_0 r_0/c_0)\log(r/r_0). \tag{95.7}$$

Thus the distortion of the profile of a spherical wave increases as the logarithm of the distance, i.e. much more slowly than for a plane wave, where the distortion δx increases as the distance x traversed by the wave.

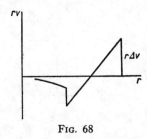

Fig. 68

The distortion of the profile ultimately leads to the formation of discontinuities in it. Let us consider shock waves formed in a single spherical sound pulse which has reached a large distance from the source (the origin). The spherical case is distinguished from the plane case primarily by the fact that the region of compression must be followed by a region of rarefaction; the excess pressure and the velocity of the gas particles in the wave must both change sign (see §69). The distortion of the profile results ultimately in the formation of two shock waves: one in the region of compression, and the other in the region of rarefaction (Fig. 68).† In the leading shock wave, the pressure increases discontinuously, then gradually decreases into a rarefaction, then again increases discontinuously in the second shock (but not to its unperturbed value, which is reached only asymptotically behind this shock).

The manner of the final damping of the shock waves with time (or, what is the same thing, with the distance r from the source) is easily found in exactly the same way as for the plane case discussed above. Using the result (95.7), we find that, at sufficiently large distances, the thickness l of the sound

---

† It should be mentioned that, since there is always ordinary damping (due to viscosity and thermal conduction) when sound is propagated in the gas, the slowness of the distortion in a spherical wave may have the result that it is damped before discontinuities can be formed.

pulse (the distance between the two discontinuities) increases as $\log^{\frac{1}{2}}(r/a)$, instead of as $\sqrt{x}$ for the plane case; a is some constant length. The intensity of the leading shock wave is damped according to $r\Delta v \sim \log^{-\frac{1}{2}}(r/a)$, or

$$\Delta v \sim 1/r \log^{\frac{1}{2}}(r/a). \qquad (95.8)$$

Finally, let us consider the cylindrical case. The general decrease in the amplitude of an outgoing sound wave occurs in inverse proportion to $\sqrt{r}$, where r is the distance from the axis. Repeating the arguments given for the spherical case, we now find the velocity u of points in the wave profile to be $u = c_0 + \alpha v_0 \sqrt{(r_0/r)}$, and so the displacement δr of points in the profile, between r_0 and r is

$$\delta r = 2\alpha(v_0/c_0)\sqrt{r_0}(\sqrt{r} - \sqrt{r_0}). \qquad (95.9)$$

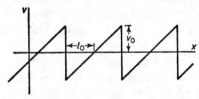

FIG. 69

The cylindrical propagation of a compression pulse must be accompanied, as in the spherical case, by a rarefaction of the gas behind the compression. Two shock waves must therefore be formed in this case also. By the same method, we find the ultimate law of increase of the thickness of the pulse: $l \sim r^{\frac{1}{4}}$, and the ultimate law of damping of the intensity of the shock wave: $\sqrt{r}\Delta v \sim r^{-\frac{3}{4}}$, or

$$\Delta v \sim r^{-\frac{5}{4}}. \qquad (95.10)$$

The formation of discontinuities in a sound wave is an example of the spontaneous occurrence of shock waves in the absence of any singularity in the external conditions of the flow. It must be emphasised that, although a shock wave can appear spontaneously at a particular instant, it cannot disappear in the same manner. Once formed, a shock wave decays only asymptotically as the time becomes infinite.

PROBLEMS

PROBLEM 1. At the initial instant, the wave profile consists of an infinite series of "teeth", as shown in Fig. 69. Determine how the profile and energy of the wave change with time.

SOLUTION. It is evident that, at subsequent instants, the wave profile will be of the same form, with l_0 unchanged but the height v_t less than v_0. Let us consider one "tooth": at time $t = 0$, the ordinate through the point where $v = v_t$ cuts off a part $v_t l_0/v_0$ of the base of the triangle. During a time t, this point moves forward a distance $\alpha v_t t$. The condition that the base of the triangle is unchanged in length is $v_t l_0/v_0 + \alpha t v_t = l_0$, whence $v_t = v_0/(1 + \alpha v_0 t/l_0)$. As $t \to \infty$, the wave amplitude diminishes as $1/t$. The energy is $E = E_0/(1 + \alpha v_0 t/l_0)^2$, i.e. it diminishes as $1/t^2$ for $t \to \infty$.

PROBLEM 2. Determine the intensity of the second harmonic formed by the distortion of the profile of a monochromatic spherical wave.

SOLUTION. Writing the wave in the form $rv = A \cos(kr - \omega t)$, we can allow for the distortion, in the first approximation, by adding δr to r on the right-hand side of this equation, and expanding in powers of δr. This gives, by (95.7),

$$rv = A \cos(kr - \omega t) - (\alpha k/2c_0)A^2 \log(r/r_0) \sin 2(kr - \omega t);$$

here r_0 must be taken as a distance at which the wave can still be regarded, with sufficient accuracy, as strictly monochromatic. The second term in this formula is the second harmonic in the spectral resolution of the wave. Its total (time average) intensity I_2 is

$$I_2 = (\alpha^2 k^2/8\pi c_0^3 \rho_0) \log^2(r/r_0) I_1^2,$$

where $I_1 = 2\pi c_0 \rho_0 A^2$ is the intensity of the first harmonic.

§96. Characteristics

The definition of characteristics, given in §79, as lines along which small disturbances are propagated (in the approximation of geometrical acoustics) is of general validity, and is not restricted to the plane steady supersonic flow discussed in §79.

For one-dimensional non-steady flow, we can introduce the characteristics as lines in the xt-plane whose slope dx/dt is equal to the velocity of propagation of small disturbances relative to a fixed co-ordinate system. Disturbances propagated relative to the gas with the velocity of sound, in the positive or negative x-direction, move relative to the fixed co-ordinate system with velocity $v \pm c$. The differential equations of the two families of characteristics, which we shall call C_+ and C_-, are accordingly

$$(dx/dt)_+ = v + c, \qquad (dx/dt)_- = v - c. \tag{96.1}$$

Disturbances transmitted with the gas are propagated in the xt-plane along characteristics belonging to a third family C_0, for which

$$(dx/dt)_0 = v. \tag{96.2}$$

These are just the "streamlines" in the xt-plane; cf. the end of §79.† It should be emphasised that, for characteristics to exist, it is no longer necessary for the gas flow to be supersonic. The "directional" propagation of disturbances, as evidenced by the characteristics, is here simply due to the causal relation between the motions at successive instants.

As an example, let us consider the characteristics of a simple wave. For a wave propagated in the positive x-direction we have, by (94.5), $x = t(v + c) + f(v)$. Differentiating this relation, we have

$$dx = (v + c)dt + [t + tc'(v) + f'(v)]dv.$$

Along a characteristic C_+, we have $dx = (v + c)dt$; comparing the two equations, we find that along such a characteristic $[t + tc'(v) + f'(v)]dv = 0$. The

---

† The same equations (96.1) and (96.2) determine the characteristics for non-steady spherically symmetrical flow, if x is replaced by the radial co-ordinate r (the characteristics now being lines in the rt-plane).

expression in brackets cannot vanish identically, and therefore $dv = 0$, i.e. v = constant. Thus we conclude that, along any characteristic C_+, the velocity is constant, and therefore so are all other quantities. The same property holds for the characteristics C_- in a wave propagated to the left. We shall see in §97 that this is no accident, but is a mathematical consequence of the nature of simple waves.

From this property of the characteristics C_+ for a simple wave, we can in turn conclude that they are a family of straight lines in the xt-plane; the velocity is constant along the lines $x = t[v+c(v)]+f(v)$ (94.5). In particular, for a similarity rarefaction wave (a simple wave with $f(v) = 0$), these lines form a pencil through the origin in the xt-plane. For this reason, a similarity simple wave is sometimes said to be *centred*.

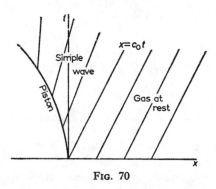

FIG. 70

Fig. 70 shows the family of characteristics C_+ for the simple rarefaction wave formed when a piston moves out of a pipe with acceleration. It is a family of diverging straight lines, which begin from the curve $x = X(t)$ giving the motion of the piston. To the right of the characteristic $x = c_0t$ lies a region of gas at rest, where the characteristics become parallel.

Fig. 71 is a similar diagram for the simple compression wave formed when a piston moves into a pipe with acceleration. In this case the characteristics are converging straight lines, which eventually intersect. Since every characteristic has a constant value of v, their intersection shows that the function $v(x, t)$ is many-valued, which is physically meaningless. This is the geometrical interpretation of the result obtained in §94: a simple compression wave cannot exist indefinitely, and a shock wave must be formed in it. The geometrical interpretation of the conditions (94.12), which determine the time and place of formation of the shock wave, is as follows. The intersecting family of rectilinear characteristics has an envelope, which, for a certain least value of t, has a cusp; this gives the instant at which many-valuedness first occurs. Every point in the region between the two branches of the envelope is on three characteristics C_+. If the equations of the characteristics are given in the parametric form $x = x(v)$, $t = t(v)$, the position of the cusp is given by equations (94.12).†

We shall now indicate briefly how the physical definition, given above, of the characteristics as lines along which disturbances are propagated corresponds to the mathematical sense of the word in the theory of partial differential equations. Let us consider a partial differential equation of the form

$$A\frac{\partial^2\phi}{\partial x^2} + 2B\frac{\partial^2\phi}{\partial x\,\partial t} + C\frac{\partial^2\phi}{\partial t^2} + D = 0, \tag{96.3}$$

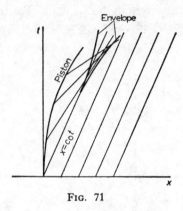

FIG. 71

which is linear in the second derivatives; the coefficients A, B, C, D can be any functions, both of the independent variables x, t and of the unknown function ϕ and its first derivatives.‡ Equation (96.3) is of the elliptic type if $B^2 - AC < 0$ everywhere, and of the hyperbolic type if $B^2 - AC > 0$. In the latter case, the equation

$$A\,dt^2 - 2B\,dx\,dt + C\,dx^2 = 0, \tag{96.4}$$

or

$$dx/dt = [B \pm \sqrt{(B^2 - AC)}]/C, \tag{96.5}$$

determines two families of curves in the xt-plane, the *characteristics* (for a given solution $\phi(x, t)$ of equation (96.3)). We may point out that, if the coefficients A, B, C are functions only of x and t, then the characteristics are independent of the particular solution ϕ.

Let a given flow correspond to some solution $\phi = \phi_0(x, t)$ of equation (96.3), and let a small perturbation ϕ_1 be applied to it. We assume that this perturbation satisfies the conditions for geometrical acoustics to be valid: it does not greatly affect the flow (ϕ_1 and its first derivatives are small), but

† The particular case where the shock wave occurs at the boundary of the gas at rest corresponds to that where one branch of the envelope is part of the characteristic $x = c_0 t$.

‡ The velocity potential satisfies an equation of this form in one-dimensional non-steady flow.

varies considerably over small distances (the second derivatives of ϕ_1 are relatively large). Putting in equation (96.3) $\phi = \phi_0 + \phi_1$, we then obtain for ϕ_1 the equation

$$A\frac{\partial^2\phi_1}{\partial x^2} + 2B\frac{\partial^2\phi_1}{\partial x\,\partial t} + C\frac{\partial^2\phi_1}{\partial t^2} = 0,$$

with $\phi = \phi_0$ in the coefficients A, B, C. Following the method used in changing from wave optics to geometrical optics, we write $\phi_1 = ae^{i\psi}$, where the function ψ (the *eikonal*) is large, and obtain

$$A\left(\frac{\partial\psi}{\partial x}\right)^2 + 2B\frac{\partial\psi}{\partial x}\frac{\partial\psi}{\partial t} + C\left(\frac{\partial\psi}{\partial t}\right)^2 = 0. \tag{96.6}$$

The equation of ray propagation in geometrical acoustics is obtained by equating dx/dt to the group velocity: $dx/dt = d\omega/dk$, where $k = \partial\psi/\partial x$, $\omega = -\partial\psi/\partial t$. Differentiating the relation $Ak^2 - 2Bk\omega + C\omega^2 = 0$, we obtain $dx/dt = (B\omega - Ak)/(C\omega - Bk)$, and, eliminating k/ω by the same relation, we again arrive at equation (96.5).

PROBLEM

Find the equation of the second family of characteristics in a centred simple wave.

Solution. In a centred simple wave propagated into gas at rest to the right of it, we have $x/t = v + c = c_0 + \frac{1}{2}(\gamma+1)v$. The characteristics C_+ form the pencil $x = \text{constant} \times t$. The characteristics C_-, on the other hand, are determined by the equation

$$\frac{dx}{dt} = v - c = \frac{3-\gamma}{\gamma+1}\frac{x}{t} - \frac{4}{\gamma+1}c_0.$$

Integrating, we find

$$x = -\frac{2}{\gamma-1}c_0 t + \frac{\gamma+1}{\gamma-1}c_0 t_0\left(\frac{t}{t_0}\right)^{(3-\gamma)/(\gamma+1)},$$

where the constant of integration has been chosen so that the characteristic C_- passes through the point $x = c_0 t_0$, $t = t_0$ on the characteristic C_+ ($x = c_0 t$) which is the boundary between the simple wave and the region at rest.

The "streamlines" in the xt-plane are given by the equation

$$\frac{dx}{dt} = v = \frac{2}{\gamma+1}\left(\frac{x}{t} - c_0\right),$$

whence

$$x = -\frac{2}{\gamma-1}c_0 t + \frac{\gamma+1}{\gamma-1}c_0 t_0\left(\frac{t}{t_0}\right)^{2/(\gamma+1)}.$$

§97. Riemann invariants

An arbitrary small disturbance is in general propagated along all three characteristics (C_+, C_-, C_0) leaving a given point in the xt-plane. However, an arbitrary disturbance can be separated into parts each of which is propagated along only one characteristic.

Let us first consider isentropic gas flow. We write the equation of continuity and Euler's equation in the form

$$\frac{\partial p}{\partial t} + v\frac{\partial p}{\partial x} + \rho c^2\frac{\partial v}{\partial x} = 0,$$

$$\frac{\partial v}{\partial t} + v\frac{\partial v}{\partial x} + \frac{1}{\rho}\frac{\partial p}{\partial x} = 0;$$

in the equation of continuity we have replaced the derivatives of the density by those of the pressure, using the formulae

$$\frac{\partial \rho}{\partial t} = \left(\frac{\partial \rho}{\partial p}\right)_s\frac{\partial p}{\partial t} = \frac{1}{c^2}\frac{\partial p}{\partial t}, \qquad \frac{\partial \rho}{\partial x} = \frac{1}{c^2}\frac{\partial p}{\partial x}.$$

Dividing the first equation by $\pm \rho c$ and adding it to the second, we obtain

$$\frac{\partial v}{\partial t} \pm \frac{1}{\rho c}\frac{\partial p}{\partial t} + \left(\frac{\partial v}{\partial x} \pm \frac{1}{\rho c}\frac{\partial p}{\partial x}\right)(v \pm c) = 0. \tag{97.1}$$

We now introduce as new unknown functions

$$J_+ = v + \int dp/\rho c, \qquad J_- = v - \int dp/\rho c, \tag{97.2}$$

which are called *Riemann invariants*. It should be remembered that, in isentropic flow, ρ and c are definite functions of p, and the integrals on the right-hand sides are therefore definite functions. For a perfect gas

$$J_+ = v + 2c/(\gamma - 1), \qquad J_- = v - 2c/(\gamma - 1). \tag{97.3}$$

In terms of these quantities, the equations of motion take the simple form

$$\left[\frac{\partial}{\partial t} + (v+c)\frac{\partial}{\partial x}\right]J_+ = 0, \qquad \left[\frac{\partial}{\partial t} + (v-c)\frac{\partial}{\partial x}\right]J_- = 0. \tag{97.4}$$

The differential operators acting on J_+ and J_- are just the operators of differentiation along the characteristics C_+ and C_- in the xt-plane. Thus we see that J_+ and J_- remain constant along each characteristic C_+ or C_- respectively. We can also say that small perturbations of J_+ are propagated only along the characteristics C_+, and those of J_- only along C_-.

In the general case of anisentropic flow, the equations (97.1) cannot be written in the form (97.4), since $dp/\rho c$ is not a perfect differential. These equations, however, still permit the separation of perturbations propagated along characteristics of only one family. For such perturbations are those of the form $\delta v \pm \delta p/\rho c$, where δv and δp are arbitrary small perturbations of the velocity and pressure. In order to obtain a complete system of equations of motion, the equations (97.1) must be supplemented by the adiabatic equation

$$\left[\frac{\partial}{\partial t} + v\frac{\partial}{\partial x}\right]s = 0, \tag{97.5}$$

which shows that perturbations δs are propagated along the characteristics C_0.

An arbitrary small perturbation can always be separated into independent parts of the three kinds mentioned.

A comparison with formula (94.4) shows that the Riemann invariants (97.2) are the quantities which, in simple waves, are constant throughout the region of the flow at all times: J_- is constant in a simple wave propagated to the right, and J_+ in one travelling to the left. Mathematically, this is the fundamental property of simple waves, from which follows, in particular, the property mentioned in §96: one family of characteristics consists of straight lines. For example, let the wave be propagated to the right. Each characteristic C_+ has a constant value of J_+ and, furthermore, a constant value of J_-, which value is the same everywhere. Since both J_+ and J_- are constant, it follows that v and p are constant (and therefore so are all the other quantities), and we obtain the property of the characteristics C_+ deduced in §96, which in turn shows that they are straight lines.

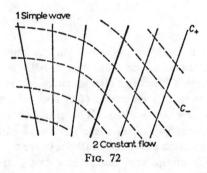

Fig. 72

If the flow in two adjoining regions of the xt-plane is described by two analytically different solutions of the equations of motion, then the boundary between the regions is a characteristic. For this boundary is a discontinuity in the derivatives of some quantity, i.e. it is a weak discontinuity, and therefore must necessarily coincide with some characteristic.

The following property of simple waves is of great importance in the theory of one-dimensional isentropic flow. The flow in a region adjoining a region of constant flow (in which $v = $ constant, $p = $ constant) must be a simple wave.

This statement is very easily proved. Let the region 1 in the xt-plane be bounded on the right by a region (2) of constant flow (Fig. 72). Both invariants J_+ and J_- are evidently constant in the latter region, and both families of characteristics are straight lines. The boundary between the two regions is a characteristic C_+, and the lines C_+ in one region do not enter the other region. The characteristics C_- pass continuously from one region to the other, and carry the constant value of J_- into region 1 from region 2. Thus J_- is constant throughout region 1 also, so that the flow in the latter is a simple wave.

The ability of characteristics to "transmit" constant values of certain quantities throws some light on the general problem of initial and boundary conditions for the equations of fluid dynamics.   In particular cases of physical interest, there is usually no doubt about the choice of these conditions, which is dictated by physical considerations.   In more complex cases, however, mathematical considerations based on the general properties of characteristics may be useful.

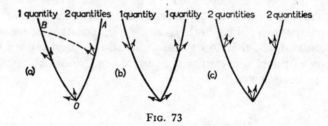

FIG. 73

For definiteness, we shall discuss a one-dimensional isentropic gas flow. Mathematically, a problem of gas dynamics usually amounts to the determination of two unknown functions (for instance, v and p) in a region of the xt-plane lying between two given curves (OA and OB in Fig. 73a), on which the boundary conditions are known. The problem is to find how many quantities can take given values on these curves. In this respect it is very important to know how each curve is situated relative to the directions (shown by arrows in Fig. 73) of the two characteristics C_+ and C_- leaving† each point of it. Two cases can occur: either both characteristics lie on the same side of the curve, or they do not. In Fig. 73a, the curve OA belongs to the first case and the curve OB to the second. It is clear that, for a complete determination of the unknown functions in the region AOB, the values of two quantities must be given on the curve OA (e.g. the two invariants J_+ and J_-), and those of only one quantity on OB. For the values of the second quantity are "transmitted" to the curve OB from the curve OA by the characteristics of the corresponding family, and therefore cannot be given arbitrarily.‡ Similarly, Figs. 73b and c show cases where one and two quantities respectively are given on each bounding curve.

It should also be mentioned that, if the bounding curve coincides with a characteristic, two independent quantities cannot be specified on it, since

---

† In the xt-plane, the characteristics leaving a given point are those which go in the direction of t increasing.

‡ An example of this case may be given as an illustration: the gas flow when a piston moves into or out of an infinite pipe. Here we are concerned with finding a solution of the equations of gas dynamics in the region of the xt-plane lying between two lines, the positive x-axis and the line $x = X(t)$ which gives the movement of the piston (Figs. 70, 71). On the first line the values of two quantities are given (the initial conditions $v = 0$, $p = p_0$ for $t = 0$), and on the second line those of one quantity ($v = u$, where $u(t)$ is the velocity of the piston).

their values are related by the condition that the corresponding Riemann invariant is constant.

The problem of specifying boundary conditions for the general case of anisentropic flow can be discussed in an entirely similar manner.

Finally, we may make the following remark. We have everywhere above spoken of the characteristics of one-dimensional flow as lines in the xt-plane. The characteristics can, however, also be defined in the plane of any two variables describing the flow. For example, we can consider the characteristics in the vc-plane. For isentropic flow, the equations of these characteristics are given simply by $J_+ = $ constant, $J_- = $ constant, with various constants on the right; we call these characteristics Γ_+ and Γ_-. For a perfect gas these are, by (97.3), two families of parallel lines (Fig. 74).

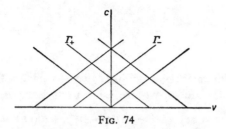

Fig. 74

It should be noted that these characteristics are entirely determined by the properties of the gas, and do not depend on any particular solution of the equations of motion. This is because the equation of isentropic flow in the variables v, c is (as we shall see in §98) a linear second-order partial differential equation with coefficients which depend only on the independent variables.

The characteristics in the xt and vc planes are transformations of one another involving the particular solution of the equations of motion. The transformation need not be one-to-one, however. In particular, only one characteristic in the vc-plane corresponds to a given simple wave, and all the characteristics in the xt-plane are transformed into it. For a wave travelling to the right (e.g.), it is one of the characteristics Γ_-; the characteristics C_- are transformed into the line Γ_-, and the characteristics C_+ into its various points.

PROBLEM

Find the general solution of the equations of one-dimensional isentropic flow of a perfect gas with $\gamma = 3$.

SOLUTION. For $\gamma = 3$ we have $J_\pm = v \pm c$, and equations (97.4) have the general integral

$$x = (v+c)t + f_1(v+c),$$

$$x = (v-c)t + f_2(v-c),$$

where f_1 and f_2 are arbitrary functions. These two equations implicitly determine the required functions $v(x, t)$ and $c(x, t)$, and therefore all other quantities. We may say that, in this case, the two quantities $v \pm c$ are propagated independently as two simple waves which do not interact.

§98. Arbitrary one-dimensional gas flow

Let us now consider the general problem of arbitrary one-dimensional isentropic gas flow (without shock waves). We shall first show that this problem can be reduced to the solution of a linear differential equation.

Any one-dimensional flow (i.e. a flow depending on only one spatial co-ordinate) must be a potential flow, since any function $v(x, t)$ can be written as a derivative: $v(x, t) = \partial\phi(x, t)/\partial x$. We can therefore use, as a first integral of Euler's equation, Bernoulli's equation (9.3): $\partial\phi/\partial t + \frac{1}{2}v^2 + w = 0$. From this, we find the differential

$$d\phi = \frac{\partial\phi}{\partial x}dx + \frac{\partial\phi}{\partial t}dt$$
$$= v\,dx - (\tfrac{1}{2}v^2 + w)dt.$$

Here the independent variables are x and t; we now change to the independent variables v and w. To do so, we use Legendre's transformation; putting

$$d\phi = d(xv) - x\,dv - d[t(w + \tfrac{1}{2}v^2)] + t\,d(w + \tfrac{1}{2}v^2)$$

and replacing ϕ by a new auxiliary function

$$\chi = \phi - xv + t(w + \tfrac{1}{2}v^2),$$

we obtain

$$d\chi = -x\,dv + t\,d(w + \tfrac{1}{2}v^2) = t\,dw + (vt - x)dv,$$

where χ is regarded as a function of v and w. Comparing this relation with the equation $d\chi = (\partial\chi/\partial w)dw + (\partial\chi/\partial v)dv$, we have $t = \partial\chi/\partial w$, $vt - x = \partial\chi/\partial v$, or

$$t = \partial\chi/\partial w, \qquad x = v\,\partial\chi/\partial w - \partial\chi/\partial v. \tag{98.1}$$

If the function $\chi(v, w)$ is known, these formulae determine v and w as functions of the co-ordinate x and the time t.

We now derive an equation for χ. To do so, we start from the equation of continuity, which has not yet been used:

$$\frac{\partial\rho}{\partial t} + \frac{\partial}{\partial x}(\rho v) \equiv \frac{\partial\rho}{\partial t} + v\frac{\partial\rho}{\partial x} + \rho\frac{\partial v}{\partial x} = 0.$$

We transform this equation to one in terms of the variables v, w. Writing the partial derivatives as Jacobians, we have

$$\frac{\partial(\rho, x)}{\partial(t, x)} + v\frac{\partial(t, \rho)}{\partial(t, x)} + \rho\frac{\partial(t, v)}{\partial(t, x)} = 0,$$

or, multiplying by $\partial(t, x)/\partial(w, v)$,

$$\frac{\partial(\rho, x)}{\partial(w, v)} + v\frac{\partial(t, \rho)}{\partial(w, v)} + \rho\frac{\partial(t, v)}{\partial(w, v)} = 0.$$

To expand these Jacobians we must use the following result. According to the equation of state of the gas, ρ is a function of any two other independent thermodynamic quantities; for example, we may regard ρ as a function of w and s. If $s = $ constant, we have simply $\rho = \rho(w)$, and the density is independent of v. Expanding the Jacobians, we therefore have

$$\frac{d\rho}{dw}\frac{\partial x}{\partial v} - v\frac{d\rho}{dw}\frac{\partial t}{\partial v} + \rho\frac{\partial t}{\partial w} = 0.$$

Substituting here the expressions (98.1) for t and x, we obtain

$$\frac{1}{\rho}\frac{d\rho}{dw}\left(\frac{\partial\chi}{\partial w} - \frac{\partial^2\chi}{\partial v^2}\right) + \frac{\partial^2\chi}{\partial w^2} = 0.$$

If $s = $ constant, we have $dw = dp/\rho$, whence $dw/dp = 1/\rho$. We can therefore write $d\rho/dw = (d\rho/dp)(dp/dw) = \rho/c^2$. We finally have for χ the equation

$$c^2\frac{\partial^2\chi}{\partial w^2} - \frac{\partial^2\chi}{\partial v^2} + \frac{\partial\chi}{\partial w} = 0; \tag{98.2}$$

here the velocity of sound c is to be regarded as a function of w. The problem of integrating the non-linear equations of motion has thus been reduced to that of solving a linear equation.

Let us apply this result to the case of a perfect gas. We have $c^2 = (\gamma - 1)w$, and the fundamental equation (98.2) becomes

$$(\gamma - 1)w\frac{\partial^2\chi}{\partial w^2} - \frac{\partial^2\chi}{\partial v^2} + \frac{\partial\chi}{\partial w} = 0. \tag{98.3}$$

This equation has an elementary general integral if $(3 - \gamma)/(\gamma - 1)$ is an even integer:

$$(3 - \gamma)/(\gamma - 1) = 2n, \quad \text{or} \quad \gamma = (3 + 2n)/(2n + 1), \quad n = 0, 1, 2, \ldots . \tag{98.4}$$

This condition is satisfied by monatomic ($\gamma = \frac{5}{3}$, $n = 1$) and diatomic ($\gamma = \frac{7}{5}$, $n = 2$) gases. Expressing γ in terms of n, we can rewrite (98.3) as

$$\frac{2}{2n + 1}w\frac{\partial^2\chi}{\partial w^2} - \frac{\partial^2\chi}{\partial v^2} + \frac{\partial\chi}{\partial w} = 0. \tag{98.5}$$

We denote by χ_n a function which satisfies this equation for a given n. For the function χ_0 we have

$$2w\frac{\partial^2\chi_0}{\partial w^2} - \frac{\partial^2\chi_0}{\partial v^2} + \frac{\partial\chi_0}{\partial w} = 0.$$

Introducing in place of w the variable $u = \sqrt{(2w)}$, we obtain

$$\frac{\partial^2 \chi_0}{\partial u^2} - \frac{\partial^2 \chi_0}{\partial v^2} = 0.$$

This is just the ordinary wave equation, whose general solution is

$$\chi_0 = f_1(u+v) + f_2(u-v),$$

f_1 and f_2 being arbitrary functions. Thus

$$\chi_0 = f_1[\sqrt{(2w)}+v] + f_2[\sqrt{(2w)}-v]. \tag{98.6}$$

We shall now show that, if the function χ_n is known, the function χ_{n+1} can be obtained by differentiation. For, differentiating equation (98.5) with respect to w, we easily find

$$\frac{2}{2n+1}w\frac{\partial^2}{\partial w^2}\left(\frac{\partial \chi_n}{\partial w}\right) + \frac{2n+3}{2n+1}\frac{\partial}{\partial w}\left(\frac{\partial \chi_n}{\partial w}\right) - \frac{\partial^2}{\partial v^2}\left(\frac{\partial \chi_n}{\partial w}\right) = 0.$$

Putting $v = v'\sqrt{[(2n+1)/(2n+3)]}$, we have for $\partial \chi_n/\partial w$ the equation

$$\frac{2}{2n+3}w\frac{\partial^2}{\partial w^2}\left(\frac{\partial \chi_n}{\partial w}\right) + \frac{\partial}{\partial w}\left(\frac{\partial \chi_n}{\partial w}\right) - \frac{\partial^2}{\partial v'^2}\left(\frac{\partial \chi_n}{\partial w}\right) = 0,$$

which is equation (98.5) for the function $\chi_{n+1}(w, v')$. Thus we conclude that

$$\chi_{n+1}(w,v') = \frac{\partial}{\partial w}\chi_n(w,v) = \frac{\partial}{\partial w}\chi_n\left(w, v'\sqrt{\frac{2n+1}{2n+3}}\right). \tag{98.7}$$

Using this formula n times and taking χ_0 from (98.6), we find that the general solution of equation (98.5) is

$$\chi = \frac{\partial^n}{\partial w^n}\{f_1[\sqrt{[2(2n+1)w]}+v] + f_2[\sqrt{[2(2n+1)w]}-v]\},$$

or

$$\chi = \frac{\partial^{n-1}}{\partial w^{n-1}}\left\{\frac{F_1[\sqrt{[2(2n+1)w]}+v] + F_2[\sqrt{[2(2n+1)w]}-v]}{\sqrt{w}}\right\}, \tag{98.8}$$

where F_1 and F_2 are again two arbitrary functions.

If we express w in terms of the velocity of sound by $w = c^2/(\gamma-1)$ $= \frac{1}{2}(2n+1)c^2$, the solution (98.8) becomes

$$\chi = \left(\frac{\partial}{c\partial c}\right)^{n-1}\left\{\frac{1}{c}F_1\left(c + \frac{v}{2n+1}\right) + \frac{1}{c}F_2\left(c - \frac{v}{2n+1}\right)\right\}. \tag{98.9}$$

The expressions $c \pm v/(2n+1) = c \pm \frac{1}{2}(\gamma-1)v$ which are the arguments of the arbitrary functions are just the Riemann invariants (97.3), which are constant along the characteristics.

In applications it is often necessary to calculate the values of the function $\chi(v, c)$ on a characteristic. The following formula[†] is useful for this purpose:

$$\left(\frac{\partial}{c\,\partial c}\right)^{n-1}\left\{\frac{1}{c}F\left(c \pm \frac{v}{2n+1}\right)\right\} = \frac{1}{2^{n-1}}\left(\frac{\partial}{\partial c}\right)^{n-1}\frac{F(2c+a)}{c^n}, \qquad (98.10)$$

with $\pm v/(2n+1) = c+a$ (a being an arbitrary constant).

Let us now ascertain the relation between the general solution just found and the solution of the equations of gas dynamics which describes a simple wave. The latter is distinguished by the property that in it v is a definite function of w: $v = v(w)$, and therefore the Jacobian $\Delta = \partial(v, w)/\partial(x, t)$ vanishes identically. In transforming to the variables v and w, however, we divided the equation of motion by this Jacobian, and the solution for which $\Delta \equiv 0$ is therefore "lost". Thus a simple wave cannot be directly obtained from the general integral of the equations of motion, but is a special integral of these equations.

To understand the nature of this special integral, we must observe that it can be obtained from the general integral by a certain passage to a limit, which is closely related to the physical significance of the characteristics as the paths of propagation of small disturbances. Let us suppose that the region of the vw-plane in which the function $\chi(v, w)$ is not zero becomes a very narrow strip along a characteristic. The derivatives of χ in the direction transverse to the characteristic then take a very wide range of values, since χ diminishes very rapidly in that direction. Such solutions $\chi(v, w)$ of the equations of motion must exist. For, regarded as a perturbation in the vw-plane, they satisfy the conditions of geometrical acoustics, and are therefore non-zero along characteristics, as such perturbations must be.

It is clear from the foregoing that, for such a function χ, the time $t = \partial\chi/\partial w$ will take an arbitrarily large range of values. The derivative of χ along the characteristic, however, is finite. Along a characteristic (for instance, a characteristic Γ_-) we have

$$\frac{dJ_-}{dv} = 1 - \frac{1}{\rho c}\frac{dp}{dw}\frac{dw}{dv} = 1 - \frac{1}{c}\frac{dw}{dv} = 0.$$

---

† It is most simply derived by using Cauchy's theorem in the theory of functions of a complex variable. For an arbitrary function $F(c+u)$ we have

$$\left(\frac{\partial}{c\,\partial c}\right)^{n-1}\frac{F(c+u)}{c} = 2^{n-1}\left(\frac{\partial}{\partial c^2}\right)^{n-1}\frac{F(c+u)}{c}$$

$$= 2^{n-1}\frac{(n-1)!}{2\pi i}\oint\frac{F(\sqrt{z}+u)}{\sqrt{z}(z-c^2)^n}dz,$$

where the integral is taken along a contour in the complex z-plane which encloses the point $z = c^2$. Putting now $u = c+a$ and substituting in the integral $\sqrt{z} = 2\zeta-c$, we obtain

$$\frac{1}{2^{n-1}}\frac{(n-1)!}{2\pi i}\oint\frac{F(2\zeta+a)}{\zeta^n(\zeta-c)^n}d\zeta,$$

where the contour of integration encloses the point $\zeta = c$; again applying Cauchy's theorem, we have the result (98.10).

The derivative of χ with respect to v along a characteristic, which we denote by $-f(v)$, is therefore

$$\frac{\mathrm{d}\chi}{\mathrm{d}v} = \frac{\partial \chi}{\partial v} + \frac{\partial \chi}{\partial w}\frac{\partial w}{\partial v} = \frac{\partial \chi}{\partial v} + c\frac{\partial \chi}{\partial w} = -f(v).$$

Expressing the partial derivatives of χ in terms of x and t by (98.1), we obtain the relation $x = (v+c)t+f(v)$, i.e. the equation (94.5) for a simple wave. The relation (94.4), which gives the relation between v and c in a simple wave, is necessarily satisfied, since J_- is constant along a characteristic Γ_-.

We have shown in §97 that, if the solution of the equations of motion reduces to constant flow in some part of the xt-plane, then there must be a simple wave in the adjoining regions. The motion described by the general solution (98.8) must therefore be separated from a region of constant flow (in particular, a region of gas at rest) by a simple wave. The boundary between the simple wave and the general solution, like any boundary between two analytically different solutions, is a characteristic. In solving particular problems, the value of the function $\chi(w, v)$ on this boundary characteristic must be determined.

The "joining" condition at the boundary between the simple wave and the general solution is obtained by substituting the expressions (98.1) for x and t in the equation of the simple wave $x = (v\pm c)t+f(v)$; this gives

$$\frac{\partial \chi}{\partial v} \pm c\frac{\partial \chi}{\partial w} + f(v) = 0.$$

Moreover, in a simple wave (and therefore on the boundary characteristic), we have $\mathrm{d}v = \pm\,\mathrm{d}p/\rho c = \pm\,\mathrm{d}w/c$, or $\pm c = \mathrm{d}w/\mathrm{d}v$. Substituting this in the above condition, we obtain

$$\frac{\partial \chi}{\partial v} + \frac{\partial \chi}{\partial w}\frac{\mathrm{d}w}{\mathrm{d}v} + f(v) = \frac{\mathrm{d}\chi}{\mathrm{d}v} + f(v) = 0,$$

or, finally,

$$\chi = -\int f(v)\,\mathrm{d}v, \tag{98.11}$$

which determines the required boundary value of χ. In particular, if the simple wave has a centre at the origin, i.e. if $f(v) \equiv 0$, then $\chi = \text{constant}$; since the function χ is defined only to within an additive constant, we can without loss of generality take $\chi = 0$ on the boundary characteristic.

PROBLEMS

PROBLEM 1. Determine the resulting flow when a centred rarefaction wave is reflected from a solid wall.

SOLUTION. Let the rarefaction wave be formed at the point $x = 0$ at time $t = 0$, and propagated in the positive x-direction; it reaches the wall after a time $t = l/c_0$, where l is the distance to the wall. Fig. 75 shows the characteristics for the reflection of the wave. In regions 1 and 1′ the gas is at rest; in region 3 it moves with a constant velocity $v = -U$.† Region 2 is the incident rarefaction wave (with rectilinear characteristics C_+), and region 5 is the reflected wave (with rectilinear characteristics C_-). Region 4 is the "region of interaction", in which the solution is required; the linear characteristics become curved on entering this region. The solution is entirely determined by the boundary conditions on the segments ab and ac. On ab (i.e. on the wall) we must have $v = 0$ for $x = l$; by (98.1), we hence obtain the condition $\partial\chi/\partial v = -l$ for $v = 0$. The boundary ac with the rarefaction wave is part of a characteristic C_-, and we therefore have $c - \frac{1}{2}(\gamma-1)v = c - v/(2n+1) = $ constant; since, at the point a, $v = 0$ and $c = c_0$, the constant is c_0. On this boundary χ must be zero, so that we have the condition $\chi = 0$ for $c - v/(2n+1) = c_0$. It is easily seen that a function of the form (98.9) which satisfies these conditions is

$$\chi = \frac{l(2n+1)}{2^n n!}\left(\frac{\partial}{c\partial c}\right)^{n-1}\left\{\frac{1}{c}\left[\left(c - \frac{v}{2n+1}\right)^2 - c_0^2\right]^n\right\},\tag{1}$$

and this gives the required solution.

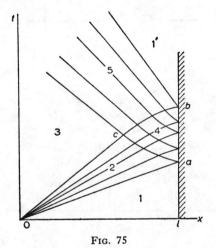

Fig. 75

The equation of the characteristic ac is (see §96, Problem)

$$x = -(2n+1)c_0 t + 2(n+1)l(tc_0/l)^{(2n+1)/2(n+1)}.$$

Its intersection with the characteristic Oc

$$x/t = c_0 - \tfrac{1}{2}(\gamma+1)U = c_0 - 2(n+1)U/(2n+1)$$

determines the time at which the incident wave disappears:

$$t_c = \frac{l(2n+1)^{n+1}c_0^n}{[(2n+1)c_0 - U]^{n+1}}.$$

In Fig. 75 it is assumed that $U < 2c_0/(\gamma+1)$; in the opposite case, the characteristic Oc is in the negative x-direction (Fig. 76). The interaction of the incident and reflected waves then lasts for an infinite time (not, as in Fig. 75, for a finite time).

---

† If the rarefaction wave is due to a piston which begins to move out of a pipe at a constant velocity, then U is the velocity of the piston.

The function (1) also describes the interaction between two equal centred rarefaction waves which leave the points $x = 0$ and $x = 2l$ at time $t = 0$ and are propagated towards each other; this is evident from symmetry (Fig. 77).

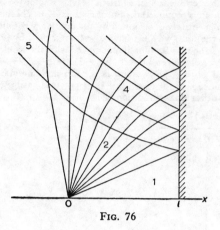

FIG. 76

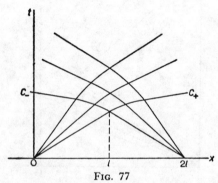

FIG. 77

PROBLEM 2. Derive the equation analogous to (98.3) for one-dimensional isothermal flow of a perfect gas.

SOLUTION. For isothermal flow, the heat function w in Bernoulli's equation is replaced by

$$\mu = \int \mathrm{d}p/\rho = c_T{}^2 \int \mathrm{d}\rho/\rho = c_T{}^2 \log \rho,$$

where $c_T{}^2 = (\partial p/\partial \rho)_T$ is the square of the isothermal velocity of sound. For a perfect gas $c_T = $ constant. Taking the quantity μ (instead of w) as an independent variable, we obtain, by the same method as in the text, the following linear equation with constant coefficients:

$$c_T{}^2 \frac{\partial^2 \chi}{\partial \mu^2} + \frac{\partial \chi}{\partial \mu} - \frac{\partial^2 \chi}{\partial v^2} = 0.$$

§99. The propagation of strong shock waves

Let us consider the propagation of a spherical shock wave of great intensity resulting from a strong explosion, i.e. from the instantaneous release of a large quantity of energy (which we denote by E) in a small volume; we suppose that the shock is propagated through a perfect gas (L. I. SEDOV, 1946).

We shall consider the wave at relatively small distances from the source, so that the amplitude is still large. These distances are, nevertheless, supposed large in comparison with the dimensions of the source; this enables us to assume that the energy E is generated at a single point (the origin).

If the shock wave is strong, the pressure discontinuity in it is very large. We shall suppose that the pressure p_2 behind the discontinuity is so large, compared with the pressure p_1 of the undisturbed gas in front of it, that $p_2/p_1 \gg (\gamma+1)/(\gamma-1)$. This means that we can everywhere neglect p_1 in comparison with p_2, and the density ratio ρ_2/ρ_1 is equal to its limiting value $(\gamma+1)/(\gamma-1)$; see §85.

Thus the gas flow pattern is entirely determined by two parameters: the initial gas density ρ_1, and the quantity of energy E generated in the explosion. From these parameters and the two independent variables (the time t and the radial co-ordinate r), we can form only one dimensionless combination, which we write as

$$\xi = r(\rho_1/Et^2)^{1/5}. \tag{99.1}$$

Consequently, we have a certain type of similarity flow.

We can say, first of all, that the position of the shock wave itself at every instant must correspond to a certain constant value ξ_0 of the dimensionless quantity ξ. This gives at once the manner in which the shock wave moves with time; denoting by r_0 the distance of the shock from the origin, we have

$$r_0 = \xi_0(Et^2/\rho_1)^{1/5}. \tag{99.2}$$

From this we find the rate of propagation of the shock wave (relative to the undisturbed gas, i.e. relative to a fixed co-ordinate system):

$$u_1 = dr_0/dt = 2r_0/5t. \tag{99.3}$$

It diminishes with time as $t^{-\frac{3}{5}}$.

The gas pressure p_2, the density ρ_2 and the velocity $v_2 = u_1 - u_2$ (relative to a fixed co-ordinate system) just behind the discontinuity can be expressed in terms of u_1 by means of the formulae derived in §85. According to (85.5) and (85.6),[†] we have

$$v_2 = 2u_1/(\gamma+1), \quad \rho_2 = \rho_1(\gamma+1)/(\gamma-1), \quad p_2 = 2\rho_1 u_1^2/(\gamma+1). \tag{99.4}$$

The density is constant in time, while v_2 and p_2 diminish as $t^{-\frac{3}{5}}$ and $t^{-\frac{6}{5}}$ respectively. We may also note that the pressure p_2 due to the shock increases with the total energy of the explosion as $E^{\frac{2}{5}}$.

Let us next determine the gas flow throughout the region behind the shock.

---

† We here denote by u_1 and u_2 the velocities of the shock wave, relative to the gas, given by formulae (85.6).

Instead of the gas velocity v, the density ρ, and the pressure p, we introduce
dimensionless variables v', ρ', p', defined by

$$v = \frac{4}{5(\gamma+1)} \frac{r}{t} v', \qquad \rho = \rho_1 \frac{\gamma+1}{\gamma-1} \rho', \qquad p = \frac{8\rho_1}{25(\gamma+1)} \frac{r^2}{t^2} p'. \quad (99.5)$$

The quantities v', ρ' and p' can be functions only of the dimensionless variable
ξ. On the surface of discontinuity (i.e. for $\xi = \xi_0$) they must have the values

$$v' = p' = \rho' = 1 \text{ for } \xi = \xi_0. \quad (99.6)$$

The equations of centrally-symmetrical adiabatic gas flow are

$$\frac{\partial v}{\partial t} + v\frac{\partial v}{\partial r} = -\frac{1}{\rho}\frac{\partial p}{\partial r}, \qquad \frac{\partial \rho}{\partial t} + \frac{\partial(\rho v)}{\partial x} + \frac{2\rho v}{r} = 0,$$

$$\quad (99.7)$$

$$\left(\frac{\partial}{\partial t} + v\frac{\partial}{\partial r}\right)\log\frac{p}{\rho^\gamma} = 0.$$

The last equation is the equation of conservation of entropy, with the expression (80.12) for the entropy of a perfect gas substituted. After substituting (99.5), we obtain a set of ordinary differential equations for the functions
v', ρ' and p'. The integration of these equations is facilitated by the fact that
one integral can be obtained immediately, using the following arguments.

The fact that we have neglected the pressure p_1 of the undisturbed gas
means that we neglect the original energy of the gas in comparison with the
energy E which it acquires as a result of the explosion. It is therefore clear
that the total energy of the gas within the sphere bounded by the shock
wave is constant and equal to E. Furthermore, since we have a similarity
flow, it is evident that the energy of the gas inside any sphere of a smaller
radius, which increases with time in such a way that $\xi =$ any constant (not
only ξ_0), must remain constant; the radial velocity of points on this sphere
is $v_n = 2r/5t$ (cf. (99.3)).

It is easy to write down the equation which expresses the constancy of this
energy. On the one hand, an amount of energy $dt \,.\, 4\pi r^2 \rho v(w + \frac{1}{2}v^2)$ leaves
the sphere (whose area is $4\pi r^2$) in time dt. On the other hand, the volume of
the sphere is increased in that time by $dt \,.\, v_n \,.\, 4\pi r^2$, and the energy of the gas
in this extra volume is $dt \,.\, 4\pi r^2 \rho v_n(\epsilon + \frac{1}{2}v^2)$. Equating the two expressions,
putting $\epsilon = p/\rho(\gamma-1)$ and $w = \gamma\epsilon$, and introducing the dimensionless functions by (99.5), we obtain

$$\frac{p'}{\rho'} = \frac{(\gamma+1-2v')v'^2}{2\gamma v' - \gamma - 1}, \quad (99.8)$$

which is the required integral. It automatically satisfies the boundary
condition (99.6) at the surface of discontinuity.

When the integral (99.8) is known, the integration of the equations is elementary though laborious. The second and third equations (99.7) give

$$\frac{dv'}{d\log\xi} + \left(v' - \frac{\gamma+1}{2}\right)\frac{d\log\rho'}{d\log\xi} = -3v',$$

$$\frac{d}{d\log\xi}\left(\log\frac{p'}{\rho'^\gamma}\right) = \frac{5(\gamma+1)-4v'}{2v'-(\gamma+1)}.$$

(99.9)

From these two equations we can express the derivatives $dv'/d\log\xi$ and $d\log\rho'/dv'$, by means of (99.8), as functions of v' only, and then an integration with the boundary conditions (99.6) gives

$$\left(\frac{\xi_0}{\xi}\right)^5 = v'^2\left[\frac{5(\gamma+1)-2(3\gamma-1)v'}{7-\gamma}\right]^{\nu_1}\left[\frac{2\gamma v'-\gamma-1}{\gamma-1}\right]^{\nu_2},$$

$$\rho' = \left[\frac{2\gamma v'-\gamma-1}{\gamma-1}\right]^{\nu_3}\left[\frac{5(\gamma+1)-2(3\gamma-1)v'}{7-\gamma}\right]^{\nu_4}\left[\frac{\gamma+1-2v'}{\gamma-1}\right]^{\nu_5},$$

$$\nu_1 = \frac{13\gamma^2-7\gamma+12}{(3\gamma-1)(2\gamma+1)}, \quad \nu_2 = -\frac{5(\gamma-1)}{2\gamma+1}, \quad \nu_3 = \frac{3}{2\gamma+1},$$

$$\nu_4 = \frac{13\gamma^2-7\gamma+12}{(2-\gamma)(3\gamma-1)(2\gamma+1)}, \quad \nu_5 = \frac{1}{\gamma-2}.$$

(99.10)

Formulae (99.8) and (99.10) give the complete solution of the problem. The constant ξ_0 is determined by the condition

$$E = \int_0^{r_0}\left(\frac{\rho v^2}{2} + \frac{p}{\gamma-1}\right)4\pi r^2\,dr,$$

which states that the total energy of the gas is equal to the energy E of the explosion. In terms of the dimensionless quantities, this condition becomes

$$\xi_0^5\frac{32\pi}{25(\gamma^2-1)}\int_0^1(\xi^4\rho'v'^2+\xi^9p')d\xi = 1.$$

(99.11)

For instance, for air ($\gamma = \frac{7}{5}$) this constant is $\xi_0 = 1\cdot033$.

The ratios v/v_2 and ρ/ρ_2 as functions of $r/r_0 = \xi/\xi_0$ are easily seen from the above formulae to tend to zero as $r/r_0 \to 0$, in the manner

$$v/v_2 \sim r/r_0, \qquad \rho/\rho_2 \sim (r/r_0)^{3/(\gamma-1)};$$

(99.12)

the ratio of pressures p/p_2, however, tends to a constant, and the ratio of temperatures therefore tends to infinity.

Fig. 78 shows the quantities v/v_2, p/p_2 and ρ/ρ_2 as functions of r/r_0 for air ($\gamma = 1\cdot4$). The very rapid decrease of the density into the sphere is noticeable: almost all the gas is in a relatively thin layer behind the shock wave. This is, of course, due to the fact that the gas on the surface of greatest radius (r_0) has a density six times the normal density.†

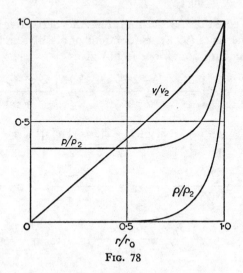

FIG. 78

§100. Shallow-water theory

There is a remarkable analogy between gas flow and the flow in a gravitational field of an incompressible fluid with a free surface, when the depth of the fluid is small (compared with the characteristic dimensions of the problem, such as the dimensions of the irregularities on the bottom of the vessel). In this case the vertical component of the fluid velocity may be neglected in comparison with its velocity parallel to the surface, and the latter may be regarded as constant throughout the depth of the fluid. In this (*hydraulic*) approximation, the fluid can be regarded as a "two-dimensional" medium having a definite velocity **v** at each point and also characterised at each point by a quantity h, the depth of the fluid.

The corresponding general equations of motion differ from those obtained in §13 only in that the changes in quantities during the motion need not be assumed small, as they were in §13 in discussing long gravity waves of small amplitude. Consequently, the second-order velocity terms in Euler's equation must be retained. In particular, for one-dimensional flow in a channel,

---

† The results of calculations for other values of γ are given by L. I. SEDOV, *Similarity and Dimensional Methods in Mechanics*, Chapter IV, §11, Cleaver-Hume Press, London 1959. The corresponding problem with cylindrical symmetry is also discussed.

depending only on one co-ordinate x (and on the time), the equations are

$$\frac{\partial h}{\partial t} + \frac{\partial(vh)}{\partial x} = 0,$$

$$\frac{\partial v}{\partial t} + v\frac{\partial v}{\partial x} = -g\frac{\partial h}{\partial x};$$

(100.1)

the depth h is here assumed constant across the channel.

Long gravity waves are, in a general sense, small perturbations of the flow now under consideration. The results of §13 show that such perturbations are propagated relative to the fluid with a finite velocity, namely

$$c = \sqrt{(gh)}.$$

(100.2)

This velocity here plays the part of the velocity of sound in gas dynamics. Just as in §79, we can conclude that, if the fluid moves with velocities $v < c$ (*streaming flow*), the effect of the perturbations is propagated both upstream and downstream. If the fluid moves with velocities $v > c$ (*shooting flow*), however, the effect of the perturbations is propagated only into certain regions downstream.

The pressure p (reckoned from the atmospheric pressure at the free surface) varies with depth in the fluid according to the hydrostatic law $p = \rho g(h-z)$, where z is the height above the bottom. It is useful to note that, if we introduce the quantities

$$\bar{\rho} = \rho h, \qquad \bar{p} = \int_0^h p\,dz = \tfrac{1}{2}\rho g h^2 = g\bar{\rho}^2/2\rho,$$

(100.3)

then equations (100.1) become

$$\frac{\partial\bar{\rho}}{\partial t} + \frac{\partial(v\bar{\rho})}{\partial x} = 0, \qquad \frac{\partial v}{\partial t} + v\frac{\partial v}{\partial x} = -\frac{1}{\rho}\frac{\partial\bar{p}}{\partial x},$$

(100.4)

which are formally identical with the equations of adiabatic flow of a perfect gas with $\gamma = 2$ ($\bar{p} \sim \bar{\rho}^2$). This enables us to apply immediately to shallow-water theory all the results of gas dynamics for flow in the absence of shock waves. If shock waves are present, however, the results of shallow-water theory differ from those of perfect-gas dynamics.

A "shock wave" in a fluid in a channel is a discontinuity in the fluid height h, and therefore in the fluid velocity v (what is called a *hydraulic jump*). The relations between the values of the quantities on the two sides of the discontinuity can be obtained from the conditions of continuity of the fluxes of mass and momentum. The mass flux density (per unit width of the channel) is $j = \rho v h$. The momentum flux density is obtained by integrating $p + \rho v^2$ over the depth of the channel, and is

$$\int_0^h (p+\rho v^2)\,dz = \tfrac{1}{2}\rho g h^2 + \rho v^2 h.$$

The conditions of continuity therefore give two equations:

$$v_1 h_1 = v_2 h_2, \tag{100.5}$$

$$v_1^2 h_1 + \tfrac{1}{2}gh_1^2 = v_2^2 h_2 + \tfrac{1}{2}gh_2^2. \tag{100.6}$$

These give the relations between the four quantities v_1, v_2, h_1, h_2, two of which can be specified arbitrarily. Expressing the velocities v_1 and v_2 in terms of the heights h_1 and h_2, we obtain

$$v_1^2 = \tfrac{1}{2}gh_2(h_1+h_2)/h_1, \qquad v_2^2 = \tfrac{1}{2}gh_1(h_1+h_2)/h_2. \tag{100.7}$$

The energy fluxes on the two sides of the discontinuity are not the same, and their difference is the amount of energy dissipated in the discontinuity per unit time. The energy flux density in the channel is

$$q = \int_0^h \left(\frac{p}{\rho} + \tfrac{1}{2}v^2\right)\rho v \, dz = \tfrac{1}{2}j(gh+v^2).$$

Using (100.7), we find the difference to be

$$q_1 - q_2 = gj(h_1^2+h_2^2)(h_2-h_1)/4h_1 h_2.$$

Let the fluid move through the discontinuity from side 1 to side 2. Then the fact that energy is dissipated means that $q_1-q_2 > 0$, and we conclude that

$$h_2 > h_1, \tag{100.8}$$

i.e. the fluid moves from the smaller to the greater height. We then can deduce from (100.7) that

$$v_1 > c_1 = \sqrt{(gh_1)}, \qquad v_2 < c_2 = \sqrt{(gh_2)}, \tag{100.9}$$

in complete analogy to the results for shock waves in gas dynamics. The inequalities (100.9) could also be derived as the necessary conditions for the discontinuity to be stable, as in §84.

CHAPTER XI

THE INTERSECTION OF
SURFACES OF DISCONTINUITY

§101. Rarefaction waves

THE line of intersection of two shock waves is, mathematically, a singular line of two functions describing the gas flow. The vertex of an acute angle on the surface of a body past which the gas flows is always such a singular line. It is found that the gas flow near the singular line can be investigated in a general manner (L. PRANDTL and T. MEYER, 1908).

In considering the region near a small segment of the singular line, we may regard the latter as a straight line, which we take as the z-axis in a system of cylindrical co-ordinates r, ϕ, z. Near the singular line, all quantities depend considerably on the angle ϕ, but their dependence on the co-ordinate r is only slight, and for sufficiently small r it can be neglected. The dependence on the co-ordinate z is also unimportant; the change in the flow pattern over a small segment of the singular line may be neglected.

Thus we have to investigate a steady flow in which all quantities are functions of ϕ only. The equation of conservation of entropy, $\mathbf{v} \cdot \mathbf{grad}\, s = 0$, gives $v_\phi\, ds/d\phi = 0$, whence $s = $ constant,† i.e. the flow is isentropic. In Euler's equation we can therefore replace $\mathbf{grad}\, p/\rho$ by $\mathbf{grad}\, w$: $(\mathbf{v} \cdot \mathbf{grad})\mathbf{v} = -\mathbf{grad}\, w$. In cylindrical co-ordinates, we have three equations:

$$\frac{v_\phi}{r}\frac{dv_r}{d\phi} - \frac{v_\phi^2}{r} = 0, \qquad \frac{v_\phi}{r}\frac{dv_\phi}{d\phi} + \frac{v_r v_\phi}{r} = -\frac{1}{r}\frac{dw}{d\phi}, \qquad v_\phi\frac{dv_z}{d\phi} = 0.$$

From the last of these we have $v_z = $ constant, and without loss of generality we can put $v_z = 0$, regarding the flow as two-dimensional; this is simply a matter of suitably defining the velocity of the co-ordinate system along the z-axis. The first two equations can be written

$$v_\phi = dv_r/d\phi, \tag{101.1}$$

$$v_\phi\left(\frac{dv_\phi}{d\phi} + v_r\right) = -\frac{1}{\rho}\frac{dp}{d\phi} = -\frac{dw}{d\phi}. \tag{101.2}$$

Substituting (101.1) in (101.2), we have

$$v_\phi\frac{dv_\phi}{d\phi} + v_r\frac{dv_r}{d\phi} = -\frac{dw}{d\phi},$$

---

† If $v_\phi = 0$, we easily deduce from the equations of motion given below that $v_r = 0$, $v_z \neq 0$. Such a flow would correspond to the intersection of surfaces of tangential discontinuity (with a discontinuous velocity v_z), and is of no interest, since such discontinuities are unstable.

or, integrating,

$$w + \tfrac{1}{2}(v_\phi^2 + v_r^2) = \text{constant.} \tag{101.3}$$

We may notice that equation (101.1) implies that **curl v** = 0, i.e. we have potential flow, as a result of which Bernoulli's equation (101.3) holds.

Next, the equation of continuity, div($\rho\mathbf{v}$) = 0, gives

$$\rho v_r + \frac{d}{d\phi}(\rho v_\phi) = \rho\left(v_r + \frac{dv_\phi}{d\phi}\right) + v_\phi\frac{d\rho}{d\phi} = 0. \tag{101.4}$$

Using (101.2), we obtain

$$\left(\frac{dv_\phi}{d\phi} + v_r\right)\left(1 - v_\phi^2\frac{d\rho}{dp}\right) = 0.$$

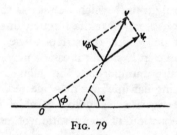

FIG. 79

The derivative $dp/d\rho$, or more correctly $(dp/d\rho)_s$, is just the square of the velocity of sound. Thus

$$\left(\frac{dv_\phi}{d\phi} + v_r\right)\left(1 - \frac{v_\phi^2}{c^2}\right) = 0. \tag{101.5}$$

This equation can be satisfied in either of two ways. Firstly, we may have $dv_\phi/d\phi + v_r = 0$. Then, from (101.2), p = constant and ρ = constant, and from (101.3) we find that $v^2 = v_r^2 + v_\phi^2$ = constant, i.e. the velocity is constant in magnitude. It is easy to see that in this case the velocity is constant in direction also. The angle χ between the velocity and some given direction in the plane of the motion is (Fig. 79)

$$\chi = \phi + \tan^{-1}(v_\phi/v_r). \tag{101.6}$$

Differentiating this expression with respect to ϕ and using formulae (101.1) and (101.2), we easily obtain

$$d\chi/d\phi = -(v_r/\rho v_\phi v^2)\,dp/d\phi. \tag{101.7}$$

Since p = constant, it follows that χ = constant. Thus, if the first factor in (101.5) is zero, we have the trivial solution of a uniform flow.

Secondly, equation (101.5) can be satisfied by putting $1 - v_\phi^2/c^2 = 0$, i.e. $v_\phi = \pm c$. The radial velocity is given by (101.3). Denoting the constant in that equation by w_0, we find that

$$v_\phi = \pm c, \qquad v_r = \pm \sqrt{[2(w_0 - w) - c^2]}.$$

In this solution, the velocity component v_ϕ perpendicular to the radius vector is equal to the local velocity of sound at every point. The total velocity $v = \sqrt{(v_\phi^2 + v_r^2)}$ therefore exceeds that of sound. Both the magnitude and the direction of the velocity are different at different points. Since the velocity of sound cannot vanish, it is clear that the function $v_\phi(\phi)$, which is continuous, must everywhere be $+c$, or else everywhere $-c$. By measuring the angle ϕ in the appropriate direction, we can take $v_\phi = c$. We shall see below that the choice of the sign of v_r follows from physical considerations, and that the plus sign must be taken. Thus

$$v_\phi = c, \qquad v_r = \sqrt{[2(w_0 - w) - c^2]}. \tag{101.8}$$

From the equation of continuity (101.4) we have $\mathrm{d}\phi = -\mathrm{d}(\rho v_\phi)/\rho v_r$. Substituting (101.8) and integrating, we have

$$\phi = -\int \frac{\mathrm{d}(\rho c)}{\rho \sqrt{[2(w_0 - w) - c^2]}}. \tag{101.9}$$

If the equation of state of the gas and the adiabatic equation are known (we recall that s is constant), this formula can be used to determine all quantities as functions of the angle ϕ. Thus formulae (101.8) and (101.9) completely determine the gas flow.

Let us now study in more detail the solution which we have obtained. First of all, we notice that the straight lines $\phi = $ constant intersect the streamlines at every point at the Mach angle (whose sine is $v_\phi/v = c/v$), i.e. they are characteristics. Thus one family of characteristics (in the xy-plane) is a pencil of straight lines through the singular point, and has an important property in this case: all quantities are constant along each characteristic. In this respect the solution concerned plays the same part in the theory of steady two-dimensional flow as does the similarity flow discussed in §92 in the theory of non-steady one-dimensional flow. We shall return to this point in §107.

It is seen from (101.9) that $(\rho c)' < 0$, the prime denoting differentiation with respect to ϕ. Putting $(\rho c)' = \rho' \mathrm{d}(\rho c)/\mathrm{d}\rho$ and noticing that the derivative $\mathrm{d}(\rho c)/\mathrm{d}\rho$ is positive (see (92.9)), we find that $\rho' < 0$, and therefore so are the derivatives $p' = c^2 \rho'$ and $w' = p'/\rho$. Next, from the fact that w' is negative it follows that the velocity $v = \sqrt{[2(w_0 - w)]}$ increases with ϕ. Finally, from (101.7), $\chi' > 0$. Thus we have

$$\mathrm{d}p/\mathrm{d}\phi < 0, \qquad \mathrm{d}\rho/\mathrm{d}\phi < 0, \qquad \mathrm{d}v/\mathrm{d}\phi > 0, \qquad \mathrm{d}\chi/\mathrm{d}\phi > 0. \tag{101.10}$$

In other words, when we go round the singular point in the direction of flow, the density and pressure decrease, while the magnitude of the velocity increases and its direction rotates in the direction of flow.

The flow just described is often called a *rarefaction wave*, and we shall use this name in what follows.

It is easy to see that a rarefaction wave cannot exist throughout the region surrounding the singular point. For, since v increases monotonically with ϕ, a complete circuit round the origin (i.e. a change of ϕ by 2π) would give a value for v different from the initial one, which is impossible. For this reason, the actual pattern of flow round the singular line must be composed of a series of sectors separated by planes $\phi =$ constant which are surfaces of discontinuity. In each of these regions we have either a rarefaction wave or a flow with constant velocity. The number and nature of these regions for various particular cases will be established in the following sections. Here we shall simply mention that the boundary between a rarefaction wave and a uniform flow must be a weak discontinuity: it cannot be a tangential discontinuity (of v_r), since the normal velocity component $v_\phi = c$ does not vanish on it. Nor can it be a shock wave, since the normal velocity component v_ϕ must be greater than the velocity of sound on one side of such a discontinuity and smaller on the other side, whereas in our problem we always have $v_\phi = c$ on one side of the boundary.

An important conclusion can be drawn from the foregoing. Disturbances which cause weak discontinuities evidently leave the singular line (the z-axis) and are propagated away from it. This means that the weak discontinuities bounding the rarefaction wave must be ones which leave this line, i.e. the velocity component v_r tangential to the weak discontinuity must be positive. This justifies the choice of the sign of v_r made in (101.8).

Let us now apply these formulae to a perfect gas. In such a gas $w = c^2/(\gamma-1)$, while the equation of the Poisson adiabatic can be written

$$\rho c^{-2/(\gamma-1)} = \text{constant}, \qquad pc^{-2\gamma/(\gamma-1)} = \text{constant}; \qquad (101.11)$$

cf. (92.13). Using these formulae, we can put the integral (101.9) in the form

$$\phi = -\sqrt{\frac{\gamma+1}{\gamma-1}} \int \frac{dc}{\sqrt{(c_*^2 - c^2)}},$$

where c_* is the critical velocity (see (80.14)). Hence

$$\phi = \sqrt{\frac{\gamma+1}{\gamma-1}} \cos^{-1}\frac{c}{c_*} + \text{constant},$$

or, if we measure ϕ in such a way that the constant is zero,

$$v_\phi = c = c_* \cos \sqrt{[(\gamma-1)/(\gamma+1)]}\phi. \qquad (101.12)$$

According to formula (101.8) we therefore have

$$v_r = \sqrt{\frac{\gamma+1}{\gamma-1}} c_* \sin \sqrt{\frac{\gamma-1}{\gamma+1}}\phi. \qquad (101.13)$$

Next, using the Poisson adiabatic equation in the form (101.11), we can find the pressure as a function of the angle ϕ:

$$p = p_* \cos^{2\gamma/(\gamma-1)} \sqrt{\frac{\gamma-1}{\gamma+1}}\phi. \tag{101.14}$$

Finally, we have for the angle χ (101.6)

$$\chi = \phi + \tan^{-1}\left(\sqrt{\frac{\gamma-1}{\gamma+1}}\cot\sqrt{\frac{\gamma-1}{\gamma+1}}\phi\right), \tag{101.15}$$

the angles χ and ϕ being measured from the same initial line.

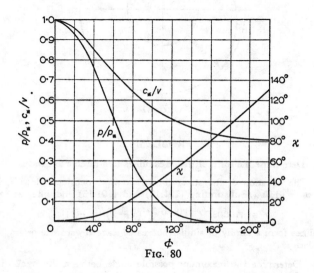

Fig. 80

Since we must have $v_r > 0$, $c > 0$, the angle ϕ in these formulae can vary only between 0 and ϕ_{max}, where

$$\phi_{max} = \tfrac{1}{2}\pi\sqrt{[(\gamma+1)/(\gamma-1)]}. \tag{101.16}$$

This means that the rarefaction wave can occupy a sector whose angle does not exceed ϕ_{max}; for a diatomic gas (air, for example), this angle is 219·3°. When ϕ varies from 0 to ϕ_{max}, the angle χ varies from $\tfrac{1}{2}\pi$ to ϕ_{max}. Thus the direction of the velocity in the rarefaction wave can turn through an angle not exceeding $\phi_{max} - \tfrac{1}{2}\pi$ ($= 129·3°$ for air).

For $\phi = \phi_{max}$ the pressure is zero. In other words, if the rarefaction wave occupies the maximum angle, the weak discontinuity on one side is a boundary with a vacuum, and is, of course, a streamline; we have $v_\phi = c = 0$, $v_r = v = \sqrt{[(\gamma+1)/(\gamma-1)]}c_* = v_{max}$, i.e. the velocity is radial and attains its limiting value v_{max} (see §80).

Fig. 80 shows graphs of p/p_*, c_*/v and χ as functions of the angle ϕ for air ($\gamma = 1·4$).

It is useful to note the form of the curve in the $v_x v_y$-plane defined by formulae (101.12) and (101.13) (called the *velocity hodograph*). It is an arc of an epicycloid between circles of radii $v = c_*$ and $v = c_* \sqrt{[(\gamma+1)/(\gamma-1)]}$ $= v_{\max}$ (Fig. 81).

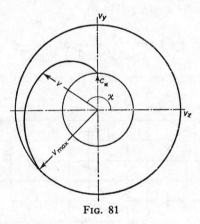

Fig. 81

PROBLEMS

PROBLEM 1. Determine the form of the streamlines in a rarefaction wave.

SOLUTION. The equation of the streamlines for two-dimensional flow is, in polar co-ordinates, $dr/v_r = r\,d\phi/v_\phi$. Substituting (101.12) and (101.13) and integrating, we obtain

$$r = r_0 \cos^{-(\gamma+1)/(\gamma-1)} \sqrt{[(\gamma-1)/(\gamma+1)]}\phi.$$

These streamlines form a family of similar curves concave toward the origin, which is the centre of similarity.

PROBLEM 2. Determine the maximum possible angle between the weak discontinuities bounding a rarefaction wave, for given values v_1, c_1 of the gas velocity and the velocity of sound at one discontinuity.

SOLUTION. The angle ϕ corresponding to the first discontinuity is, by (101.12),

$$\phi_1 = \sqrt{\frac{\gamma+1}{\gamma-1}} \cos^{-1} \frac{c_1}{c_*}.$$

The value of ϕ_2 is $\phi_{\max}$, so that the angle required is

$$\phi_2 - \phi_1 = \sqrt{\frac{\gamma+1}{\gamma-1}} \sin^{-1} \frac{c_1}{c_*}.$$

The critical velocity c_* is given in terms of v_1 and c_1 by Bernoulli's equation:

$$w_1 + \tfrac{1}{2}v_1^2 = \frac{c_1^2}{\gamma-1} + \tfrac{1}{2}v_1^2 = \frac{\gamma+1}{2(\gamma-1)}c_*^2.$$

The maximum possible angle through which the gas velocity can turn in a rarefaction wave is accordingly, by (101.15), the difference $\chi_{\max} = \chi(\phi_1) - \chi(\phi_2)$:

$$\chi_{\max} = \sqrt{\frac{\gamma+1}{\gamma-1}} \sin^{-1} \frac{c_1}{c_*} - \sin^{-1} \frac{c_1}{v_1}.$$

As a function of v_1/c_1, χ_{max} is greatest for $v_1/c_1 = 1$:

$$\chi_{max} = \tfrac{1}{2}\pi\left(\sqrt{\frac{\gamma+1}{\gamma-1}} - 1\right).$$

For $v_1/c_1 \to \infty$, χ_{max} tends to zero:

$$\chi_{max} = \frac{2}{\gamma-1}\frac{c_1}{v_1}.$$

§102. The intersection of shock waves

Shock waves can intersect along a line. In considering the flow near a small segment of this line, we can assume that it is a straight line, and that the surfaces of discontinuity are planes. It is therefore sufficient to discuss the intersection of plane shock waves.

The line of intersection of two discontinuities is, mathematically, a singular line, as has already been mentioned at the beginning of §101. The flow pattern near this line consists of a number of sectors, in each of which we have either uniform flow or a rarefaction wave of the kind described in §101. It is possible to give a general classification of the possible types of intersection of surfaces of discontinuity (L. LANDAU 1944).

First of all, we must make the following remark. If the gas flow on both sides of a shock wave is supersonic, then (as mentioned at the beginning of §86) we can speak of the "direction" of the shock wave, and accordingly distinguish shock waves leaving the line of intersection from those reaching it. In the former case, the tangential velocity component is directed away from the line of intersection, and we can say that the disturbances which cause the discontinuity leave this line. In the latter case, the perturbations leave a point not on the line of intersection.

If the flow on one side of the shock wave is subsonic, then disturbances are propagated in both directions along its surface, and the "direction" of the shock has, strictly, no meaning. In the arguments given below, however, what is important is that disturbances leaving the point of intersection can be propagated along such a discontinuity. In this sense, such shock waves play the same part in the following discussion as the purely supersonic shocks which leave the intersection, and we shall include both kinds in the term "shocks which leave the intersection".

Figs. 82–86 show the flow patterns in a plane perpendicular to the line of intersection. We can assume, without loss of generality, that the flow occurs in this plane. The velocity component parallel to the line of intersection (which lies in all the planes of discontinuity) must be the same in all regions round the line of intersection, and can therefore be made to vanish by an appropriate choice of the co-ordinate system.

It is easy to see that there can be no intersection of shock waves in which no shock reaches the intersection. For instance, in the intersection of two shock waves leaving the intersection, shown in Fig. 82a, the streamlines of the flow incident from the left would deviate in opposite directions, whereas the velocity should be constant throughout region 2, and this difficulty cannot be overcome by adding any further discontinuities in region 2.[†] Similarly, we can see that the intersection of a shock wave and a rarefaction wave both leaving the intersection, shown in Fig. 82b, is impossible; although the velocity in region 2 can be constant in direction, the pressure cannot be constant, since it increases in a shock wave but decreases in a rarefaction wave.

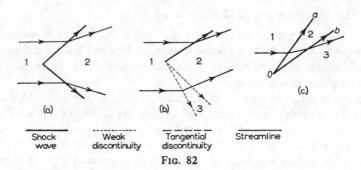

| Shock wave | Weak discontinuity | Tangential discontinuity | Streamline |

FIG. 82

Next, since the intersection cannot affect shock waves reaching it, the simultaneous intersection (along a common line) of more than two such waves, which are due to other causes, would be an improbable coincidence. Thus only one or two shock waves can reach the intersection.

The following fact is very important. The gas flowing past a point of intersection can pass through only one shock or rarefaction wave leaving this point. For example, let the gas pass through two successive shock waves leaving the point O, as shown in Fig. 82c. Since the normal velocity component v_{2n} behind the shock Oa is less than c_2, the velocity component in region 2 normal to the shock Ob must also be less than c_2, in contradiction to a fundamental property of shock waves. Similarly, we can see that the gas cannot pass through two successive rarefaction waves, or a shock wave and a rarefaction wave, leaving the point O.

These arguments evidently cannot be extended to shock waves reaching the point of intersection.

We can now proceed to enumerate the possible types of intersection. Fig. 83 shows an intersection involving one shock wave Oa reaching it and two shock waves Ob, Oc leaving it. This case may be regarded as the splitting of one shock wave into two.[‡] It is easy to see that, besides the two shock waves

---

† In order not to encumber the discussion with repetitive arguments, we shall not give similar considerations for cases where there are regions of subsonic flow and the shock leaving the intersection is actually a shock wave bounded by a subsonic region.

‡ It should be noticed that a shock wave cannot divide into a shock and a rarefaction wave; it is easily seen that the changes in the pressure and the direction of the velocities in the two waves leaving cannot be reconciled.

leaving, there must be formed a tangential discontinuity Od lying between them, which separates the gas flowing through Ob from that flowing through Oc.† For the shock Oa is due to other causes, and is therefore completely defined. This means that the thermodynamic quantities (p and ρ, say) and the velocity **v** have given values in regions 1 and 2. There remain at our disposal, therefore, only two quantities (the angles giving the directions of the discontinuities Ob and Oc) with which to satisfy, in general, four conditions (the constancy of p, ρ and two velocity components) in the region 3–4, which would have to be satisfied in the absence of the tangential discontinuity Od. The addition of the latter, however, reduces the number of conditions to two (the constancy of the pressure and of the direction of the velocity).

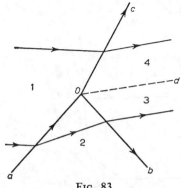

FIG. 83

An arbitrary shock wave, however, cannot divide in this manner. A shock wave reaching the intersection is defined by two parameters (for a given thermodynamic state of gas 1), say the Mach number M_1 of the incident stream and the ratio of pressures p_1/p_2. It can divide in two only in a certain region in the plane of these two parameters.‡

Intersections involving two shock waves reaching them can be regarded as "collisions" of two shocks due to other causes. Here two essentially different cases are possible, as shown in Fig. 84.

In the first case, the collision of two shock waves results in two other shock waves leaving the point of intersection. If all the necessary conditions

---

† As usual, the tangential discontinuity in reality becomes a turbulent region.

‡ The determination of this region involves very laborious algebraic calculations. The results that have been published (see, for example, R. COURANT and K. O. FRIEDRICHS, *Supersonic Flow and Shock Waves*, Interscience, New York 1948), unfortunately, are largely invalidated by the fact that they make no distinction between shock waves reaching and leaving the intersection. The ternary configurations therefore include also those where two shock waves reach the intersection and one leaves it. This, however, is the intersection of two shocks due to other causes, and therefore reaching the point of intersection with given values of all parameters. Their "fusion" into one shock is possible only when these arbitrary parameters are related in a certain way, and this would be an improbable coincidence.

are to be fulfilled, a tangential discontinuity must again be formed, and it must lie between the two resulting shock waves.

In the second case, instead of two shock waves, there are formed one shock wave and one rarefaction wave.

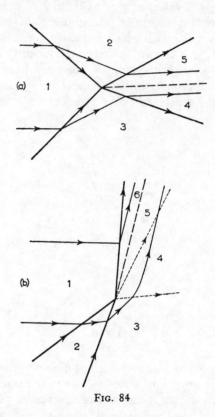

Fig. 84

Two colliding shock waves are defined by three parameters (for instance, M_1 and the ratios p_1/p_2, p_1/p_3). The types of intersection just described are possible only for certain ranges of values of these parameters. If the values of the parameters do not lie in these regions, the collision of the shock waves must be preceded by their breaking up.

Fig. 85 shows the reflection of a shock wave from the boundary between gas in motion and gas at rest. Region 5 contains gas at rest, separated from the gas in motion by a tangential discontinuity. In the two regions 1 and 4 adjoining it, the pressure must be the same and equal to p_5. Since the pressure increases in a shock wave, it is clear that the shock wave must be reflected from the tangential discontinuity as a rarefaction wave 3, which reduces the pressure to its initial value.

Finally, we may briefly discuss the intersection of a shock wave with a weak discontinuity arriving from an external source. Here two cases can occur, according as the flow behind the shock wave is supersonic or subsonic. In the former case (Fig. 86a), the weak discontinuity is "refracted" at the shock wave into the space behind the latter; the shock itself is not refracted at the intersection, but has a singularity of a higher order, like that at a weak discontinuity. Moreover, the entropy change in the shock wave must cause behind it a "weak tangential discontinuity", at which the derivatives of the entropy are discontinuous.

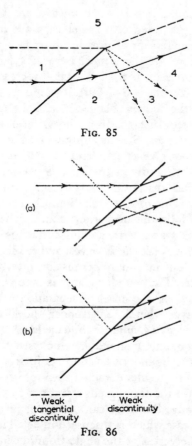

FIG. 85

FIG. 86

If, however, the flow becomes subsonic behind the shock wave, the weak discontinuity cannot penetrate into this region, and it ceases at the point of intersection (Fig. 86b). The latter is now a singular point; it can be shown that the velocity distribution behind the shock wave has a logarithmic singularity at this point. Furthermore, as in the previous case, a weak tangential discontinuity of the entropy must occur behind the shock wave.†

---

† A detailed qualitative and quantitative analysis of the possible types of intersection of shock waves with weak discontinuities is given by S.P. D'YAKOV, *Zhurnal éksperimental'noĭ i teoreticheskoĭ fiziki* **33**, 948, 962, 1957; *Soviet Physics JETP* **6** (33), 729, 739, 1958; *Doklady Akademii Nauk SSSR* **99**, 921, 1954.

§103. The intersection of shock waves with a solid surface

An important part in the phenomenon of steady intersection of shock waves with the surface of a body is played by their interaction with the boundary layer. This interaction is very complex, and has not yet been sufficiently investigated, either experimentally or theoretically. However, simple general arguments enable us to obtain some important results, which we shall now expound.†

The pressure is discontinuous in a shock wave, and increases in the direction of motion of the gas. Hence, if the shock wave intersects the surface, there must be a finite increment of pressure over a very short distance near the place of intersection, i.e. there must be a very large positive pressure gradient. We know, however, that such a rapid increase in pressure cannot occur near a solid wall (see the end of §40); it would cause separation, and the pattern of flow round the body is changed in such a way that the shock wave moves away to a sufficient distance from the surface.

These arguments, however, do not apply when the shock wave is weak. It is clear from the proof given at the end of §40 that the impossibility of a positive pressure discontinuity at the boundary layer is a consequence of the assumption that this discontinuity is large: it must exceed a certain limit depending on the value of R, which diminishes when R increases.‡

Thus we reach the following important conclusions. The steady intersection of strong shock waves with a solid surface is impossible. A solid surface can intersect only weak shock waves, and the limiting intensity is the smaller, the greater R. The maximum permissible intensity of the shock wave also depends on whether the boundary layer is laminar or turbulent. If the boundary layer is turbulent, the onset of separation is retarded (§45). In a turbulent boundary layer, therefore, stronger shock waves can leave the surface of the body than in a laminar boundary layer.††

To avoid misunderstanding, it should be emphasised that these arguments rely on the fact that the boundary layer exists in front of the shock wave (i.e. upstream of it). The results obtained therefore relate, in particular, to shock waves which leave the trailing edge, but not to those which leave the leading edge, of the body; the latter can occur, for instance in flow past an acute-angled wedge, a case which is discussed in detail in §104. In the latter case the gas reaches the vertex of the angle from outside, i.e. from a region in which there is no boundary layer. It is therefore clear that the present

---

† The boundary layer necessarily contains a subsonic part adjoining the surface, into which the shock wave cannot penetrate. In speaking of the intersection, we ignore this fact, which does not affect the following discussion.

‡ In §40, Problem, we have determined the smallest pressure change Δp over a distance Δx which can cause separation in a laminar boundary layer. In the present application, we are concerned with the pressure change over a distance of the order of the thickness δ of the boundary layer, and obtain the following law governing the decrease of Δp when the Reynolds number increases:

$$\Delta p/p \sim 1/R_x^{\frac{1}{4}} \sim 1/R_\delta^{\frac{1}{3}}.$$

†† The existing published data do not enable us to specify the maximum permissible intensity.

arguments do not deny that shock waves can occur which leave the vertex of such an angle.

In subsonic flow, separation can occur only when the pressure in the main stream increases downstream along the surface. In supersonic flow, however, it is found that separation can occur even when the pressure decreases downstream. Such a phenomenon can occur by the combination of a weak shock wave with a separation, the pressure increase necessary for separation taking place in the shock wave; the pressure may either increase or decrease downstream in the region in front of the shock wave.

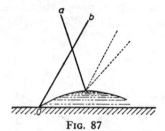

FIG. 87

The data at present existing do not enable us to give a detailed picture of the complex phenomena involved in the "reflection" of a shock wave from the subsonic part of a boundary layer (or from the turbulent region beyond the line of separation). An important part in these phenomena must be played by the fact that the disturbances due to the shock wave can be propagated both upstream and downstream through the subsonic part of the boundary layer, and can cause further discontinuities in it. In particular, the formation of another weak shock wave upstream may result in separation, which "displaces" a strong shock wave incident on the surface from outside. In Fig. 87, the line a is the incident shock wave, and b the shock wave formed upstream, which causes separation at the point O. When the incident shock is "reflected" from the subsonic part of the turbulent region, we should expect, in particular, that a rarefaction wave would be formed.

All the above discussion relates only to a steady intersection, with the shock wave and the body at relative rest. Let us now consider non-steady intersections, when a moving shock wave is incident on a solid body, so that the line of intersection moves on the surface. Such an intersection is accompanied by reflection of the shock wave: besides the incident wave, a reflected wave leaving the body is formed.

We shall examine the phenomenon in a system of co-ordinates which moves with the line of intersection; in this system the shock waves are steady. The simplest type of reflection occurs when the reflected wave leaves the line of intersection itself; this is called *regular* reflection (Fig. 88). If the angle of incidence α_1 and the intensity of the incident shock are given, the flow in region 2 is uniquely determined. The gas velocity in the reflected shock must be turned through an angle such that it is again parallel to the surface. When this angle is given, the position and intensity of the reflected shock are obtained from the equation of the shock polar. For a given angle, the shock polar determines two different shock waves, those of the weak and strong families (§86). Experimental results show that in fact the reflected shock always belongs to the weak family, and we shall assume this in what follows. It should be pointed out that, when the intensity of the incident shock tends to zero, the intensity of the reflected shock then tends to zero also, and the angle of reflection α_2 tends to the angle of incidence α_1, as we should expect in accordance with the acoustic approximation. In the limit $\alpha_1 \to 0$, the reflected shock of the weak family passes continuously into the shock obtained when a shock wave is incident "frontally" (§93. Problem 1).

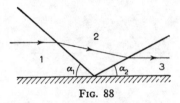

FIG. 88

The mathematical calculations for regular reflection (in a perfect gas) offer no difficulty in principle, but the algebra is extremely laborious. Here we shall give only some of the results.[†]

It is clear from the general properties of the shock polar that regular reflection is not possible for arbitrary values of the parameters of the incident wave (the angle of incidence α_1 and the ratio p_2/p_1). For a given ratio p_2/p_1 there is a maximum possible angle α_{1k},[‡] and for $\alpha_1 > \alpha_{1k}$ regular reflection is impossible. As $p_2/p_1 \to \infty$, the maximum angle tends to $\sin^{-1}(1/\gamma)$ ($= 40°$ for air). As $p_2/p_1 \to 1$, α_{1k} tends to 90°, i.e. regular reflection is possible for any angle of incidence. Fig. 89 shows α_{1k} as a function of p_1/p_2 for air.

The angle of reflection α_2 is not in general the same as the angle of incidence. There is a value α_* of the angle of incidence such that, if $\alpha_1 < \alpha_*$, the angle of reflection $\alpha_2 < \alpha_1$; if $\alpha_1 > \alpha_*$, on the other hand, $\alpha_2 > \alpha_1$. The value of α_* is $\frac{1}{2}\cos^{-1}\frac{1}{2}(\gamma-1)$ ($= 39·2°$ for air); it does not depend on the intensity of the incident wave.

† A more detailed account of the reflection of shock waves is given by R. COURANT and K. O. FRIEDRICHS, *Supersonic Flow and Shock Waves*, Interscience, New York 1948, and by W. BLEAKNEY and A. H. TAUB, *Reviews of Modern Physics* 21, 584, 1949.

The solution of complex problems concerning the regular reflection of a shock wave at almost normal incidence on the vertex of an angle close to 180°, and the diffraction of a shock wave at glancing incidence on the vertex of a similar angle, has been given by M. J. LIGHTHILL (*Proceedings of the Royal Society* A198, 454, 1949; 200, 554, 1950).

‡ This is the value of the angle of incidence for which the strong and weak reflected shocks coincide.

For $\alpha_1 > \alpha_{1k}$ regular reflection is impossible, and the incident shock wave must break up at a distance from the surface, so that we have the pattern shown in Fig. 90, with three shock waves, and a tangential discontinuity leaving the point where the incident shock wave divides.

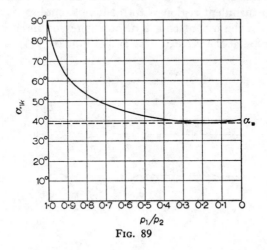

Fig. 89

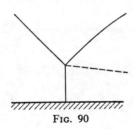

Fig. 90

§104. Supersonic flow round an angle

In investigating the flow near the vertex of an angle on the surface, it is again sufficient to consider small portions of the vertex and suppose it straight, the angle being formed by two intersecting planes. We shall speak of flow outside an angle if the angle is greater than π, and of flow inside an angle if it is less than π.

Subsonic flow past an angle is not essentially different from the flow of an incompressible fluid. Supersonic flow, however, is entirely different; an important property of it is the occurrence of discontinuities leaving the vertex of the angle.

Let us first consider the possible flow patterns when a supersonic gas stream reaches the vertex along one of the sides of the angle. In accordance with the general properties of supersonic flow, the stream remains uniform up to the vertex. The turning of the stream into the direction parallel to the other side of the angle occurs in a rarefaction wave leaving the vertex, and the flow pattern consists of three regions separated by weak discontinuities (*Oa* and *Ob* in Fig. 91): the uniform gas stream 1 moving along the side *AO* is turned into the rarefaction wave 2 and then moves, again with constant velocity, along the other side of the angle. It should be noticed that no turbulent region is formed; in a similar flow of an incompressible fluid, on the other hand, a turbulent region must be formed, with a line of separation at the vertex of the angle (Fig. 16, §35).

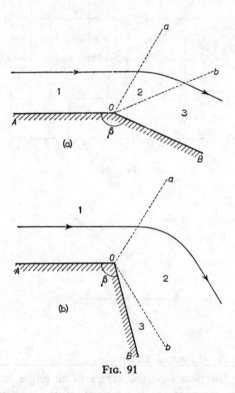

FIG. 91

Let v_1 be the velocity of the incident stream (1 in Fig. 91), and c_1 the velocity of sound in it. The position of the weak discontinuity *Oa* is determined immediately from the Mach number $M_1 = v_1/c_1$ by the condition that it intersects the streamlines at the Mach angle. The changes in velocity and pressure in the rarefaction wave are determined by formulae (101.12)–(101.15); all that is needed is the direction from which the angle ϕ in these formulae is to be measured. The straight line $\phi = 0$ corresponds to $v = c = c_*$; for $M_1 > 1$, there is in fact no such line, since $v/c > 1$ everywhere. However, if the rarefaction wave is imagined to be formally extended into the region to the left of *Oa*, we can use formula (101.12), and we find that the discontinuity Oa must correspond to a value of ϕ given by

$$\phi_1 = \sqrt{\frac{\gamma+1}{\gamma-1}}\cos^{-1}\frac{c_1}{c_*},$$

and that ϕ must increase from Oa to Ob. The position of the discontinuity Ob is determined by the fact that the direction of the velocity becomes parallel to the side OB of the angle.

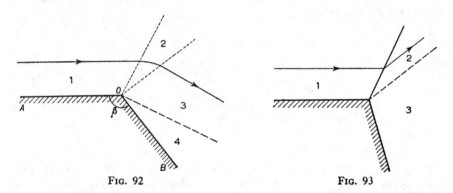

Fig. 92   Fig. 93

The angle through which the stream turns in the rarefaction wave cannot exceed the value χ_{max} determined in §101, Problem 2. If the angle β round which the flow occurs is less than $\pi-\chi_{max}$, the rarefaction wave cannot turn the stream through the necessary angle, and we have the flow pattern shown in Fig. 91b. The rarefaction in the wave 2 then proceeds to zero pressure (reached on the line Ob), so that the rarefaction wave is separated from the wall by a vacuum (region 3).

The flow pattern described above is not the only possible one, however. Figs. 92 and 93 show patterns in which a region of gas at rest adjoins the second side of the angle, this region being separated from the moving gas by a tangential discontinuity; as usual, this becomes a turbulent region, so that the case considered corresponds to the presence of separation.† The stream is turned through a certain angle in a rarefaction wave (Fig. 92) or in a shock wave (Fig. 93). The latter case, however, is possible only if the shock wave is not too strong (in accordance with the general considerations given in §103).

Which of these flow patterns will occur in any particular case depends in general on the conditions far from the angle. For instance, when gas flows out of a nozzle (the vertex of the angle being here the edge of the outlet), the relation between the pressure p_1 of the outgoing gas and the pressure p_e of the external medium is of importance. If $p_e < p_1$, the flow is of the type shown in Fig. 92; the position and angle of the rarefaction wave are then determined

---

† According to experimental results, the compressibility of the gas somewhat diminishes the angle of the turbulent region resulting from the tangential discontinuity.

by the condition that the pressure in regions 3 and 4 is equal to p_e. The smaller p_e, the greater the angle through which the stream must be turned. If, however, the angle β (Fig. 92) is large, the gas pressure cannot reach the required value p_e; the direction of the velocity becomes parallel to the side OB of the angle before the pressure falls to p_e. The flow near the edge of the outlet will then be as shown in Fig. 90. The pressure near the outer side OB of the outlet is entirely determined by the angle β, and does not depend on the pressure p_e; the final decrease of the pressure to p_e occurs only at a distance from the outlet.

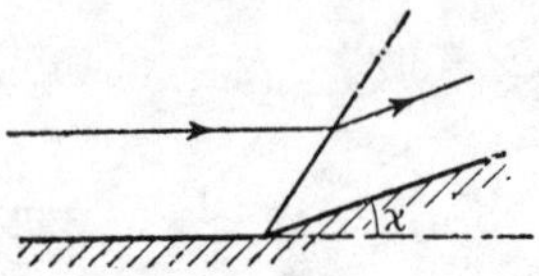

FIG. 94

If $p_e > p_1$, on the other hand, the flow round the edge of the outlet is of the type shown in Fig. 93, with a shock wave which leaves the edge and raises the pressure from p_1 to p_e. This is possible, however, only if the difference between p_e and p_1 is not too large, i.e. the shock wave is not too strong; otherwise there is separation at the inner surface of the nozzle, and the shock wave moves into the nozzle, in the manner described in §90.

Next, let us consider flow inside an angle. In the subsonic case such a flow is accompanied by separation at a point ahead of the vertex (see the end of §40). For a supersonic incident flow, however, the change in direction may be effected by a shock wave leaving the vertex (Fig. 94). Here it must again be mentioned that such a simple separationless flow pattern is possible only if the shock wave is not too strong. Its intensity increases with the angle χ through which the stream is turned, and we can therefore say that separationless flow is possible only when χ is not too large.

Let us now consider the flow pattern which results when a free supersonic stream is incident on the vertex of an angle (Fig. 95). The stream is turned into directions parallel to the sides of the angle by shock waves leaving the vertex. As has been shown in §103, this is the exceptional case where a shock wave of arbitrary intensity can leave a solid surface.

If we know the velocities v_1 and c_1 in the incident stream, we can determine the positions of the shock waves and the gas flow in the regions behind them. The direction of the velocity v_2 must be parallel to the side OA of the angle: $v_{2y}/v_{2x} = \tan\chi$. Thus v_2 and the angle ϕ giving the position of the shock wave can be determined immediately from the shock polar, using a chord through the origin at the known angle χ to the axis of abscissae (Fig. 50), as explained in §86. We have seen that, for a given χ, the shock polar gives two different shock waves, with different values of ϕ. One of these (corresponding to the point B in Fig. 50) is the weaker, and in general leaves the flow supersonic; the other, stronger, shock renders the flow subsonic. In the present case of flow past an angle on a finite solid surface,† we must always take the former, i.e. the weak shock. It should be borne in mind that this choice is really decided by the conditions of the flow far from the angle.

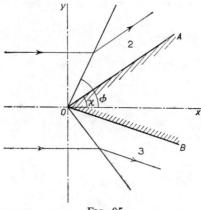

Fig. 95

In flow past a very acute angle (χ small), the resulting shock wave must obviously be very weak. It is natural to suppose that, as the angle increases, the intensity of the shock increases monotonically; this corresponds to a movement along the arc QC of the shock polar (Fig. 50), from Q towards C.

We have also seen in §86 that the angle through which the velocity vector is turned in a shock wave cannot exceed a certain value χ_{max}, which depends on M_1. The flow pattern described above is therefore impossible if either of the sides of the angle makes an angle greater than χ_{max} with the direction of the incident stream. In this case the gas flow near the angle must be subsonic; this is achieved by the appearance of a shock wave somewhere in front of the angle (see §114). Since χ_{max} increases monotonically with M_1, we can also say that, for a given value of the angle χ, M_1 for the incident stream must be greater than a certain value $M_{1\,min}$.

Finally it may be mentioned that, if the sides of the angle are situated, relative to the incident stream, as shown in Fig. 96, then a shock wave is of course formed on only one side of the angle; the stream is turned on the other side by a rarefaction wave.

PROBLEM

Determine the position and intensity of the shock wave in flow past a very small angle ($\chi \ll 1$) for very large values of M_1 ($\gg 1/\chi$).

---

† The purely formal problem of flow past a wedge formed by the intersection of two infinite planes is of no physical interest.

SOLUTION. For $\chi \ll 1$, the shock polar gives two values of ϕ, one close to zero and the other close to $\frac{1}{2}\pi$. The weak shock which we require corresponds to the former value, which is $\frac{1}{2}(\gamma+1)\chi$; see §86, Problem 1. The ratio of pressures is, by (86.9), $p_2/p_1 = \frac{1}{2}\gamma(\gamma+1)M_1^2\chi^2$. The value of M behind the shock is

$$M_2 = \frac{1}{\chi}\sqrt{\frac{2}{\gamma(\gamma-1)}},$$

i.e. it is still large compared with unity, but not large compared with $1/\chi$.

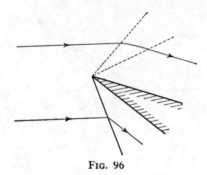

FIG. 96

§105. Flow past a conical obstacle

The problem of steady supersonic flow near a pointed projection on the surface of a body is three-dimensional, and is very much more complicated than that of flow past an angle with a line vertex. No complete general investigation of the former problem has yet been made. The only problem that has been completely solved is that of axially symmetric flow past a projecting point, and we shall discuss this case.

Near its vertex, an axially symmetric projection can be regarded as a right cone of circular cross-section, and so the problem consists in investigating the flow of a uniform stream past a cone whose axis is in the direction of incidence. The flow pattern is qualitatively as follows.

As in the analogous problem of flow past a two-dimensional angle, a shock wave must be formed (A. BUSEMANN 1929), and it is evident from symmetry that this shock is a conical surface coaxial with the cone and having the same vertex (Fig. 97 shows the cross-section of the cone by a plane through its axis). Unlike what happens in the two-dimensional case, however, the shock wave does not turn the gas velocity through the whole angle χ necessary for the gas to flow along the surface of the cone (2χ being the vertical angle of the cone). After passing through the surface of discontinuity, the streamlines are curved, and asymptotically approach the generators of the cone. This curvature is accompanied by a continuous increase in density (besides the increase which occurs at the shock itself) and by a corresponding decrease in the velocity. Immediately behind the shock wave, the velocity is in general still supersonic (as in the two-dimensional case, it is determined by the "supersonic" part of the shock polar), but on the surface of the cone it may become subsonic. As in the two-dimensional case, for every value of the Mach number $M_1 = v_1/c_1$ for the incident stream, there is a limiting value χ_{max} for the angle of the cone, above which this type of flow becomes impossible.

The conical shock wave intersects all streamlines in the incident flow at the same angle, and is therefore of constant intensity. Hence it follows (see §106) that we have isentropic potential flow behind the shock wave also.

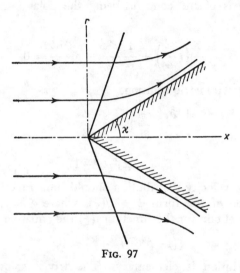

Fig. 97

From the symmetry of the problem and its similarity properties (there are no characteristic constant lengths in the conditions imposed), it is evident that the distribution of all quantities (velocity, pressure) in the flow behind the shock wave will depend only on the angle θ which the radius vector from the vertex of the cone to the point considered makes with the axis of the cone (the x-axis in Fig. 97). Accordingly, the equations of motion are ordinary differential equations; the boundary conditions on these equations at the shock wave are determined by the equation of the shock polar, while those at the surface of the cone are that the velocity should be parallel to the generators. These equations, however, cannot be integrated analytically, and have to be solved numerically. We refer the reader elsewhere[†] for the results of the calculations, and merely give the curve (Fig. 51, §86) which shows the maximum possible angle χ_{max} as a function of M_1. We may also mention that, as $M_1 \to 1$, the angle χ_{max} tends to zero:

$$\chi_{max} = \text{constant} \times \sqrt{[(M_1-1)/(\gamma+1)]}, \tag{105.1}$$

† For example, N. E. KOCHIN, I. A. KIBEL' and N. V. ROZE, *Theoretical Hydromechanics (Teoreticheskaya gidromekhanika)*, Part 2, 3rd ed., p. 193, Moscow 1948; L. HOWARTH ed., *Modern Developments in Fluid Dynamics: High Speed Flow*, vol, 1, ch. 5, Oxford 1953.

as may be deduced from the general law of transonic similarity (118.11); the constant is independent both of M_1 and of the gas involved.

An analytical solution of the problem of flow past a cone is possible only in the limit of small vertical angles. It is evident that in this case the gas velocity nowhere differs greatly from the velocity v_1 of the incident stream. Denoting by v the small difference between the gas velocity at the point considered and v_1, and using its potential ϕ, we can apply the linearised equation (106.4); if we take cylindrical co-ordinates x, r, ω with the polar axis along the axis of the cone (ω being the polar angle), this equation becomes

$$\frac{1}{r}\frac{\partial}{\partial r}\left(r\frac{\partial\phi}{\partial r}\right) + \frac{1}{r^2}\frac{\partial^2\phi}{\partial\omega^2} - \beta^2\frac{\partial^2\phi}{\partial x^2} = 0, \tag{105.2}$$

or, for an axially symmetric solution,

$$\frac{1}{r}\frac{\partial}{\partial r}\left(r\frac{\partial\phi}{\partial r}\right) - \beta^2\frac{\partial^2\phi}{\partial x^2} = 0, \tag{105.3}$$

where

$$\beta = \sqrt{(v_1^2/c_1^2 - 1)}. \tag{105.4}$$

In order that the velocity distribution should be a function of θ only, the potential must be of the form $\phi = xf(\xi)$, where $\xi = r/x = \tan\theta$. Substituting this, we obtain for the function $f(\xi)$ the equation

$$\xi(1 - \beta^2\xi^2)f'' + f' = 0,$$

of which the solution is elementary. The trivial solution $f = \text{constant}$ corresponds to a uniform flow; the other solution is

$$f = \text{constant} \times [\sqrt{(1 - \beta^2\xi^2)} - \cosh^{-1}(1/\beta\xi)].$$

The boundary condition on the surface of the cone (i.e. for $\xi = \tan\chi \approx \chi$) is

$$v_r/(v_1 + v_x) \approx (1/v_1)\,\partial\phi/\partial r = \chi \tag{105.5}$$

or $f' = v_1\chi$. Hence the constant is $v_1\chi^2$, and we have the following expression r the potential in the region $x > \beta r$:†

$$\phi = v_1\chi^2[\sqrt{(x^2 - \beta^2 r^2)} - x\cosh^{-1}(x/\beta r)]. \tag{105.6}$$

It should be noticed that ϕ has a logarithmic singularity for $r \to 0$.

We can now find the velocity components:

$$v_x = -v_1\chi^2\cosh^{-1}(x/\beta r),$$
$$v_r = (v_1\chi^2/r)\sqrt{(x^2 - \beta^2 r^2)}. \tag{105.7}$$

The pressure on the surface of the cone is calculated from formula (106.5);

---

† In this approximation, the cone $x = \beta r$ is a surface of weak discontinuity.

since ϕ has a logarithmic singularity for $r \to 0$, the velocity v_r on the surface of the cone (i.e. for small r) is large compared with v_x, and therefore we need retain only the term in v_r^2 in the formula for the pressure. The result is

$$p - p_1 = \rho_1 v_1^2 \chi^2 [\log(2/\beta\chi) - \tfrac{1}{2}]. \tag{105.8}$$

All these formulae, which have been derived by means of a linearised theory, cease to be valid for large M_1, comparable with $1/\chi$ (see §119).

The flow past a cone of arbitrary cross-section (the angle of attack being not necessarily zero) is a similarity flow, like the symmetrical flow past a circular cone. It has no characteristic length parameters, and so the velocity distribution can be a function only of the ratios y/x, z/x of the co-ordinates, i.e. it is constant along any straight line through the origin (the vertex of the cone). Such similarity flows are called *conical* flows.[†]

PROBLEM

Determine the flow past a cone of small vertical angle 2χ placed at a small angle of attack α (C. FERRARI 1937).[‡]

SOLUTION. We take the axis of the cone (not the direction of the main stream) as the x-axis; the linearised equation (105.2) for the potential is unchanged if higher-order quantities ($\sim \alpha\phi$) are neglected, and the potential determines the gas velocity as $\mathbf{v}_1 + \mathbf{grad}\ \phi$. The boundary condition on the surface of the cone is

$$\frac{v_1 \sin\alpha \cos\omega + v_r}{v_1 \cos\alpha + v_x} \approx \alpha\cos\omega + \frac{1}{v_1}\frac{\partial\phi}{\partial r} \approx \chi.$$

We seek ϕ as a sum:

$$\phi = \phi^{(1)}(x, r) + \cos\omega \cdot \phi^{(2)}(x, r), \tag{1}$$

where $\phi^{(1)}$ is the expression (105.6), and $\phi^{(2)}$ satisfies the boundary condition $\partial\phi^{(2)}/\partial r = -v_1\alpha$. The function $\phi^{(2)}$ can be written as $rf(r/x)$ and, substituting $rf \cos\omega$ in equation (105.2), we obtain for f the equation

$$\xi f''(\beta^2\xi^2 - 1) + f'(2\beta^2\xi^2 - 3) = 0.$$

The trivial solution $f = \text{constant}$ corresponds to a uniform stream incident (with velocity $v_1\alpha$) in a direction perpendicular to the axis of the cone; the other solution leads to

$$\phi^{(2)} = v_1\beta\chi^2\alpha[(x/\beta r)\sqrt{(x^2 - \beta^2 r^2)} - \beta r \cosh^{-1}(x/\beta r)].$$

The gas velocity is $\mathbf{v}_1 + \mathbf{v}^{(1)} + \mathbf{v}^{(2)}$, where $\mathbf{v}^{(2)} = \mathbf{grad}\ \phi^{(2)}$ and $\mathbf{v}^{(1)}$ is given by formulae (105.7). The pressure is calculated from the formula

$$p - p_1 = -\tfrac{1}{2}\rho_1\{(v_1\cos\alpha + \partial\phi/\partial x)^2 +$$
$$+ (v_1\sin\alpha\cos\omega + \partial\phi/\partial r)^2 + (-v_1\sin\alpha\sin\omega + \partial\phi/r\ \partial\omega)^2 - v_1^2\}$$

in which the second-order terms in α and χ must be retained. The pressure on the surface of the cone is found to be given by

$$p - p_1 = \rho_1 v_1^2\{\chi^2 \log(2/\beta\chi) - \tfrac{1}{2}(\chi^2 + \alpha^2) -$$
$$- 2\alpha\chi\cos\omega + \alpha^2\cos 2\omega\}.$$

[†] A detailed account of various problems concerning these flows is given by E. CARAFOLI, *High Speed Aerodynamics (Compressible Flow)*, Pergamon Press, London 1958.

[‡] The solution of the same problem for any thin solid of revolution is given by F. I. FRANKL' and E. A. KARPOVICH, *Gas Dynamics of Thin Bodies* §2-7, Interscience, New York 1953.

CHAPTER XII

TWO-DIMENSIONAL GAS FLOW

§106. Potential flow of a gas

IN what follows we shall meet with many important cases where the flow of a gas can be regarded as potential flow almost everywhere. Here we shall derive the general equations of potential flow and discuss the question of their validity.

After passing through a shock wave, potential flow of a gas usually becomes rotational flow. An exception, however, is formed by cases where a steady potential flow passes through a shock wave whose intensity is constant over its area; such, for example, is the case where a uniform stream passes through a shock wave intersecting every streamline at the same angle.† The flow behind the shock wave is then potential flow also. To prove this, we use Euler's equation in the form

$$\tfrac{1}{2}\mathbf{grad}\,v^2 - \mathbf{v}\times\mathbf{curl\,v} = -(1/\rho)\,\mathbf{grad}\,p$$

(cf. (2.10)), or

$$\mathbf{grad}(w+\tfrac{1}{2}v^2) - \mathbf{v}\times\mathbf{curl\,v} = T\,\mathbf{grad}\,s,$$

where we have used the thermodynamic identity $dw = T\,ds + dp/\rho$. In potential flow, however, $w + \tfrac{1}{2}v^2 = $ constant in front of the shock wave, and this quantity is continuous at the shock; it is therefore constant everywhere behind the shock wave, so that

$$\mathbf{v}\times\mathbf{curl\,v} = -T\,\mathbf{grad}\,s. \tag{106.1}$$

The potential flow in front of the shock wave is isentropic. In the general case of an arbitrary shock wave, for which the discontinuity of entropy varies over its surface, $\mathbf{grad}\,s \neq 0$ in the region behind the shock, and $\mathbf{curl\,v}$ is therefore also not zero. If, however, the shock wave is of constant intensity, then the discontinuity of entropy in it is constant, so that the flow behind the shock is also isentropic, i.e. $\mathbf{grad}\,s = 0$. From this it follows that either $\mathbf{curl\,v} = 0$ or the vectors $\mathbf{v}$ and $\mathbf{curl\,v}$ are everywhere parallel. The latter, however, is impossible; at the shock wave, $\mathbf{v}$ always has a non-zero normal component, but the normal component of $\mathbf{curl\,v}$ is always zero (since it is given by the tangential derivatives of the tangential velocity components, which are continuous).

Another important case where potential flow continues despite the shock wave is that of a weak shock. We have seen (§83) that in such a shock wave

---

† We have already met with this situation in connection with supersonic flow past a wedge or cone (§§104, 105).

the discontinuity of entropy is of the third order relative to the discontinuity of pressure or velocity. We therefore see from (106.1) that **curl v** behind the shock is also of the third order. This enables us to assume that we have potential flow behind the shock wave, the error being of a high order of smallness.

We shall now derive the general equation for the velocity potential in an arbitrary steady potential flow of a gas. To do so, we eliminate the density from the equation of continuity $\mathrm{div}(\rho\mathbf{v}) \equiv \rho\,\mathrm{div}\,\mathbf{v} + \mathbf{v}\cdot\mathbf{grad}\,\rho = 0$, using Euler's equation

$$(\mathbf{v}\cdot\mathbf{grad})\mathbf{v} = -(1/\rho)\mathbf{grad}\,p = -(c^2/\rho)\mathbf{grad}\,\rho$$

and obtaining

$$c^2\,\mathrm{div}\,\mathbf{v} - \mathbf{v}\cdot(\mathbf{v}\cdot\mathbf{grad})\mathbf{v} = 0.$$

Introducing the velocity potential by $\mathbf{v} = \mathbf{grad}\,\phi$ and expanding in components, we obtain the equation

$$\begin{aligned}(c^2 - \phi_x{}^2)\phi_{xx} + (c^2 - \phi_y{}^2)\phi_{yy} + (c^2 - \phi_z{}^2)\phi_{zz} - \\ - 2(\phi_x\phi_y\phi_{xy} + \phi_y\phi_z\phi_{yz} + \phi_z\phi_x\phi_{zx}) = 0,\end{aligned} \tag{106.2}$$

where the suffixes here denote partial derivatives. In particular, for two-dimensional flow we have

$$(c^2 - \phi_x{}^2)\phi_{xx} + (c^2 - \phi_y{}^2)\phi_{yy} - 2\phi_x\phi_y\phi_{xy} = 0. \tag{106.3}$$

In these equations, the velocity of sound must itself be expressed in terms of the velocity; this can in principle be done by means of Bernoulli's equation, $w + \tfrac{1}{2}v^2 = $ constant, and the isentropic equation, $s = $ constant. For a perfect gas, c as a function of v is given by formula (80.18).

Equation (106.2) is much simplified if the gas velocity nowhere differs greatly in magnitude or direction from that of the stream incident from infinity.† This implies that the shock waves (if any) are weak, and so the potential flow is not destroyed.

As in similar cases previously, we denote by **v** the small difference between the gas velocity at a given point and that of the main stream. Denoting the latter by $\mathbf{v}_1$, we therefore write the total velocity as $\mathbf{v}_1 + \mathbf{v}$. The potential ϕ is taken to mean that of the velocity **v**: $\mathbf{v} = \mathbf{grad}\,\phi$. The equation for this potential is obtained from (106.2) by substituting $\phi \to \phi + xv_1$; we take the x-axis in the direction of the vector $\mathbf{v}_1$. We then regard ϕ as a small quantity, and omit all terms of order higher than the first, obtaining the following linear equation:

$$(1 - M_1{}^2)\frac{\partial^2\phi}{\partial x^2} + \frac{\partial^2\phi}{\partial y^2} + \frac{\partial^2\phi}{\partial z^2} = 0, \tag{106.4}$$

where $M_1 = v_1/c_1$; the velocity of sound is, of course, given its value at infinity.

---

† One such case was discussed in §105 (flow past a narrow cone), and others will be found in connection with gas flow past arbitrary thin bodies.

The pressure at any point is determined in terms of the velocity in the same approximation, by a formula which can be obtained as follows. We regard p as a function of w (for given s), and use the fact that $(\partial w/\partial p)_s = 1/\rho$, writing $p - p_1 \approx (\partial p/\partial w)_s(w - w_1) = \rho_1(w - w_1)$. From Bernoulli's equation we have

$$w - w_1 = -\tfrac{1}{2}[(\mathbf{v}_1 + \mathbf{v})^2 - \mathbf{v}_1{}^2] \approx -\tfrac{1}{2}(v_y{}^2 + v_z{}^2) - v_1 v_x,$$

so that

$$p - p_1 = -\rho_1 v_1 v_x - \tfrac{1}{2}\rho_1(v_y{}^2 + v_z{}^2). \tag{106.5}$$

In this expression the term in the squared transverse velocity must in general be retained, since, in the region near the x-axis (and, in particular, on the surface of the body itself), the derivatives $\partial\phi/\partial y$, $\partial\phi/\partial z$ may be large compared with $\partial\phi/\partial x$.

Equation (106.4), however, is not valid if the number M_1 is very close to unity (*transonic* flow), so that the coefficient of the first term is small. It is clear that, in this case, terms of higher order in the x-derivatives of ϕ must be retained. To derive the corresponding equation, we return to the original equation (106.2); when the terms which are certainly small are neglected, this becomes

$$\left(1 - \frac{\phi_x{}^2}{c^2}\right)\phi_{xx} + \phi_{yy} + \phi_{zz} = 0. \tag{106.6}$$

In the present case, the velocity $v_x \cong v$, and the velocity of sound c is close to the critical velocity c_* ($\mathbf{v}$ now denoting the total velocity). We can therefore put $c - c_* = (v - c_*)\,(dc/dv)_{v=c_*}$, or $c - v = (c_* - v)[1 - (dc/dv)_{v=c_*}]$. Using the fact that, for $v = c = c_*$, we have by (80.4) $d\rho/dv = -\rho/c$, we put (for $v = c_*$)

$$\frac{dc}{dv} = \frac{dc}{d\rho}\frac{d\rho}{dv} = -\frac{\rho}{c}\frac{dc}{d\rho},$$

so that

$$c - v = [(c_* - v)/c]d(\rho c)/d\rho = \alpha_*(c_* - v). \tag{106.7}$$

We have here used the expression (92.9) for the derivative $d(\rho c)/d\rho$, while α_* denotes the value of α (95.2) for $v = c_*$; for a perfect gas, α is constant, so that $\alpha_* = \alpha = \tfrac{1}{2}(\gamma + 1)$. To the same accuracy, this equation can be written as

$$v/c - 1 = \alpha_*(v/c_* - 1). \tag{106.8}$$

This gives the general relation between the Mach numbers M and M_* in transonic flow.

Using this formula, we can put

$$1 - \frac{v_x{}^2}{c^2} \approx 1 - \frac{v^2}{c^2} \approx 2\left(1 - \frac{v}{c}\right) \approx 2\alpha_*\left(1 - \frac{v}{c_*}\right).$$

Finally, we introduce a new potential by the substitution $\phi \to c_*(x + \phi)$, so that

$$\frac{\partial \phi}{\partial x} = \frac{v_x}{c_*} - 1, \qquad \frac{\partial \phi}{\partial y} = \frac{v_y}{c_*}, \qquad \frac{\partial \phi}{\partial z} = \frac{v_z}{c_*}. \qquad (106.9)$$

Substituting these formulae in (106.6), we obtain the following final equation for the velocity potential in a transonic flow (with the velocity everywhere almost parallel to the x-axis):

$$2\alpha_* \frac{\partial \phi}{\partial x} \frac{\partial^2 \phi}{\partial x^2} = \frac{\partial^2 \phi}{\partial y^2} + \frac{\partial^2 \phi}{\partial z^2}. \qquad (106.10)$$

The properties of the gas appear here only through the constant α_*. We shall see later that this constant governs the entire dependence of the properties of transonic flow on the nature of the gas.

The linearised equation (106.4) becomes invalid also in another limiting case, that of very large values of M_1: however, the appearance of strong shock waves has the result that potential flow cannot actually occur for such values of M_1 (see §119).

§107. Steady simple waves

Let us determine the general form of those solutions of the equations of steady two-dimensional supersonic gas flow which describe flows in which there is a uniform plane-parallel stream at infinity, which then turns through an angle as it flows round a curved profile. We have already met a particular case of such a solution in discussing the flow near an angle; the flow considered was essentially a plane-parallel one along one side of the angle, which turned at the vertex of the angle. In this particular solution all quantities (the two velocity components, the pressure and the density) were functions of only one variable, the angle ϕ. Each of these quantities could therefore be expressed as a function of any other. Since this solution must be a particular case of the required general solution, it is natural to seek the latter on the assumption that each of the quantities p, ρ, v_x, v_y (the plane of the motion being taken as the xy-plane) can be expressed as a function of any other. This assumption is, of course, a very considerable restriction on the solution of the equations of motion, and the solution thus obtained is not the general integral of those equations. In the general case, each of the quantities p, ρ, v_x, v_y, which are functions of the two co-ordinates x, y, can be expressed as a function of any two of them.

Since we have a uniform stream at infinity, in which all quantities, and in particular the entropy s, are constants, and since in steady flow of an ideal fluid the entropy is constant along the streamlines, it is clear that $s = $ constant in all space if there are no shock waves in the gas, as we shall assume.

Euler's equations and the equation of continuity are

$$v_x \frac{\partial v_x}{\partial x} + v_y \frac{\partial v_x}{\partial y} = -\frac{1}{\rho} \frac{\partial p}{\partial x}, \qquad v_x \frac{\partial v_y}{\partial x} + v_y \frac{\partial v_y}{\partial y} = -\frac{1}{\rho} \frac{\partial p}{\partial y};$$

$$\frac{\partial}{\partial x}(\rho v_x) + \frac{\partial}{\partial y}(\rho v_y) = 0.$$

Writing the partial derivatives as Jacobians, we can convert these equations to the form

$$v_x \frac{\partial(v_x, y)}{\partial(x, y)} - v_y \frac{\partial(v_x, x)}{\partial(x, y)} = -\frac{1}{\rho} \frac{\partial(p, y)}{\partial(x, y)},$$

$$v_x \frac{\partial(v_y, y)}{\partial(x, y)} - v_y \frac{\partial(v_y, x)}{\partial(x, y)} = \frac{1}{\rho} \frac{\partial(p, x)}{\partial(x, y)};$$

$$\frac{\partial(\rho v_x, y)}{\partial(x, y)} - \frac{\partial(\rho v_y, x)}{\partial(x, y)} = 0.$$

We now take (say) x and p as independent variables. In order to effect this transformation, we need only multiply the above equations by $\partial(x, y)/\partial(x, p)$, obtaining the same equations except that $\partial(x, p)$ replaces $\partial(x, y)$ in the denominator of each Jacobian. We now expand the Jacobians, bearing in mind that all the quantities ρ, v_x, v_y are assumed to be functions of p but not of x, so that their partial derivatives with respect to x are zero. We then obtain

$$\left(v_y - v_x \frac{\partial y}{\partial x}\right) \frac{\mathrm{d}v_x}{\mathrm{d}p} = \frac{1}{\rho} \frac{\partial y}{\partial x}, \qquad \left(v_y - v_x \frac{\partial y}{\partial x}\right) \frac{\mathrm{d}v_y}{\mathrm{d}p} = -\frac{1}{\rho},$$

$$\left(v_y - v_x \frac{\partial y}{\partial x}\right) \frac{\mathrm{d}\rho}{\mathrm{d}p} + \rho\left(\frac{\mathrm{d}v_y}{\mathrm{d}p} - \frac{\partial y}{\partial x} \frac{\mathrm{d}v_x}{\mathrm{d}p}\right) = 0.$$

Here $\partial y/\partial x$ denotes $(\partial y/\partial x)_p$. All the quantities in these equations except $\partial y/\partial x$ are functions of p only, by hypothesis, and x does not appear explicitly. We can therefore conclude, first of all, that $\partial y/\partial x$ also is a function of p only: $(\partial y/\partial x)_p = f_1(p)$, whence

$$y = x f_1(p) + f_2(p), \tag{107.1}$$

where $f_2(p)$ is an arbitrary function of the pressure.

No further calculations are necessary if we use the particular solution, already known, for a rarefaction wave in flow past an angle (§§101, 104). It will be recalled that, in this solution, all quantities (including the pressure) are constants along any straight line (characteristic) through the vertex of the angle. This particular solution evidently corresponds to the case where the arbitrary function $f_2(p)$ in the general expression (107.1) is identically zero. The function $f_1(p)$ is determined by the formulae obtained in §101.

Steady simple waves

Equation (107.1) for various constant p gives a family of straight lines in the xy-plane. These lines intersect the streamlines at every point at the Mach angle. This is seen immediately from the fact that the lines $y = xf_1(p)$ in the particular solution with $f_2 \equiv 0$ have this property. Thus one of the families of characteristics (those leaving the surface of the body) consists, in the general case, of straight lines along which all quantities remain constant; these lines, however, are no longer concurrent.

The properties of the flow described above are, mathematically, entirely analogous to those of one-dimensional simple waves, in which one family of characteristics is a family of straight lines in the xt-plane (see §§94, 96, 97).

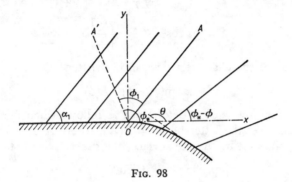

Fig. 98

Hence the class of flows under consideration occupies the same place in the theory of steady (supersonic) two-dimensional flow as do simple waves in non-steady one-dimensional flow. On account of this analogy, such flows are also called *simple waves*; in particular, the rarefaction wave which corresponds to the case $f_2 \equiv 0$ is called a *centred simple wave*.

As in the non-steady case, one of the most important properties of steady simple waves is that the flow in any region of the xy-plane bounded by a region of uniform flow is a simple wave (cf. §97).

We shall now show how the simple wave corresponding to flow round a given profile can be constructed. Fig. 98 shows the profile in question; to the left of the point O it is straight, but to the right it begins to curve. In supersonic flow the effect of the curvature is, of course, propagated only downstream of the characteristic OA which leaves the point O. Hence the flow to the left of this characteristic is uniform; we denote by the suffix 1 quantities pertaining to this region. All the characteristics there are parallel and at an angle to the x-axis which is equal to the Mach angle $\alpha_1 = \sin^{-1}(c_1/v_1)$.

In formulae (101.12)–(101.15), the angle ϕ of the characteristics is measured from the line on which $v = c = c_*$. This means (cf. §104) that the characteristic OA must have a value of ϕ given by

$$\phi_1 = \sqrt{\frac{\gamma+1}{\gamma-1}}\cos^{-1}\frac{c_1}{c_*},$$

and the angle ϕ is to be measured from OA' (Fig. 98). The angle between the characteristics and the x-axis is then $\phi_* - \phi$, where $\phi_* = \alpha_1 + \phi_1$. According to formulae (101.12)–(101.15), the velocity and pressure are given in terms of ϕ by

$$v_x = v \cos\theta, \qquad v_y = v \sin\theta, \tag{107.2}$$

$$v^2 = c_*^2\left[1 + \frac{2}{\gamma-1}\sin^2\sqrt{\frac{\gamma-1}{\gamma+1}}\phi\right], \tag{107.3}$$

$$\theta = \phi_* - \phi - \tan^{-1}\left(\sqrt{\frac{\gamma-1}{\gamma+1}}\cot\sqrt{\frac{\gamma-1}{\gamma+1}}\phi\right), \tag{107.4}$$

$$p = p_* \cos^{2\gamma/(\gamma-1)}\sqrt{\frac{\gamma-1}{\gamma+1}}\phi. \tag{107.5}$$

The equation of the characteristics can be written

$$y = x \tan(\phi_* - \phi) + F(\phi). \tag{107.6}$$

The arbitrary function $F(\phi)$ is determined as follows when the form of the profile is given. Let the latter be $Y = Y(X)$, where X and Y are the co-ordinates of points on it. At the surface, the gas velocity is tangential, i.e.

$$\tan\theta = dY/dX. \tag{107.7}$$

The equation of the line through the point (X, Y) at an angle $\phi_* - \phi$ to the x-axis is

$$y - Y = (x - X)\tan(\phi_* - \phi).$$

This equation is the same as (107.6) if we put

$$F(\phi) = Y - X \tan(\phi_* - \phi). \tag{107.8}$$

Starting from the given equation $Y = Y(X)$ and equation (107.7), we express the form of the profile in parametric equations $X = X(\theta)$, $Y = Y(\theta)$, the parameter being the inclination θ of the tangent. Substituting θ in terms of ϕ from (107.4), we obtain X and Y as functions of ϕ; finally, substituting these in (107.8), we obtain the required function $F(\phi)$.

In flow past a convex surface, the angle θ between the velocity vector and the x-axis decreases downstream (Fig. 98), and the angle $\phi_* - \phi$ between the characteristic and the x-axis therefore decreases monotonically also (we always mean the characteristic leaving the surface). For this reason, the characteristics do not intersect (in the region of flow, that is). Thus, in the region downstream of the characteristic OA (which is a weak discontinuity), we have a continuous (no shock waves) and increasingly rarefied flow.

The situation is different in flow past a concave profile. Here the inclination θ of the tangent increases downstream, and therefore so does the inclination of the characteristics. Consequently, the characteristics intersect in the region of flow. On different non-parallel characteristics, however, all quantities (velocity, pressure, etc.) have different values. Thus all these quantities become many-valued at points where characteristics intersect, which is physically impossible. We have already met a similar phenomenon in connection with a non-steady one-dimensional simple compression wave (§94). As in that case, it signifies that in reality a shock wave is formed. The position of the discontinuity cannot be completely determined from the solution under consideration, since this was derived on the assumption that there are no discontinuities. The only result that can be obtained is the place where the shock wave begins (the point O in Fig. 99, where the shock is shown by the continuous line OB). It is the point of intersection of characteristics whose streamline lies nearest to the surface of the body. On streamlines passing below O (i.e. nearer to the surface) the solution is everywhere

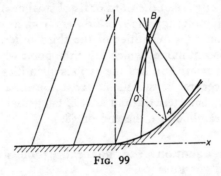

Fig. 99

one-valued; its many-valuedness "begins" at O. The equations for the co-ordinates x_0, y_0 of this point can be obtained in the same way as the corresponding equations which determine the time and place of formation of the discontinuity in a one-dimensional non-steady simple wave. If we regard the inclination of the characteristics of a function of the co-ordinates (x, y) of points through which they pass, then this function becomes many-valued when x and y exceed certain values x_0, y_0. In §94 the situation was the same in relation to the function $v(x, t)$, and so we need not repeat the arguments used there, but can write down immediately the equations

$$(\partial y/\partial\phi)_x = 0, \qquad (\partial^2 y/\partial\phi^2)_x = 0, \qquad (107.9)$$

which now determine the place of formation of the shock wave. Mathematically, this point is a cusp on the envelope of the family of straight characteristics (cf. §96).

In flow past a concave profile, the simple wave exists along streamlines passing above O as far as the points where these lines intersect the shock wave. The streamlines passing below O do not intersect the shock wave at all, but we cannot conclude from this that the solution in question is valid at all points on these streamlines. The reason is that the shock wave has a perturbing effect even on the gas which flows along these streamlines, and so alters the flow from what it would be in the absence of the shock wave. By a property of supersonic flow, however, these perturbations reach only the gas downstream of the characteristic OA (of the second family) which leaves the point where the shock wave begins. Thus the solution under consideration is valid everywhere to the left of AOB. The line OA itself is a weak discontinuity. We see that there cannot be a continuous (no shock waves) simple compression wave everywhere in flow past a concave surface, which would correspond to the simple rarefaction wave in flow past a convex surface.

The shock wave formed in flow past a concave profile is an example of a shock which "begins" at a point inside the stream, away from the solid walls. The point where the shock begins has some general properties, which may be noted here. At the point itself the intensity of the shock wave is zero, and near the point it is small. In a weak shock wave, however, the discontinuities of entropy and vorticity are of the third order of smallness, and so the change in the flow on passing through the shock differs from a continuous potential isentropic change only by quantities of the third order. Hence it follows that, in the weak discontinuities which leave the point where the shock wave begins, only the third derivatives of the various quantities can be discontinuous. There will in general be two such discontinuities: a weak discontinuity coinciding with the characteristic, and a weak tangential discontinuity coinciding with the streamline (see the end of §89).

§108. Chaplygin's equation: the general problem of steady two-dimensional gas flow

Having dealt with steady simple waves, let us now consider the general problem of an arbitrary steady plane potential flow. We assume that the flow is isentropic and contains no shock waves.

As was first shown by S. A. CHAPLYGIN in 1902, it is possible to reduce this problem to the solution of a single linear partial differential equation. This is achieved by means of a transformation to new independent variables, the velocity components v_x, v_y; this transformation is often called the *hodograph transformation*, the $v_x v_y$-plane being called the *hodograph plane* and the xy-plane the *physical plane*.

For potential flow we can replace Euler's equations by their first integral, Bernoulli's equation:

$$w + \tfrac{1}{2}v^2 = w_0. \tag{108.1}$$

The equation of continuity is

$$\frac{\partial}{\partial x}(\rho v_x) + \frac{\partial}{\partial y}(\rho v_y) = 0. \tag{108.2}$$

For the differential of the velocity potential ϕ we have $d\phi = v_x dx + v_y dy$. We transform from the independent variables x, y to the new variables v_x, v_y

by Legendre's transformation, putting

$$d\phi = d(xv_x) - xdv_x + d(yv_y) - y\,dv_y,$$

introducing the function

$$\Phi = -\phi + xv_x + yv_y, \tag{108.3}$$

and obtaining

$$d\Phi = x\,dv_x + y\,dv_y,$$

where Φ is regarded as a function of v_x and v_y. Hence

$$x = \partial\Phi/\partial v_x, \qquad y = \partial\Phi/\partial v_y. \tag{108.4}$$

It is more convenient, however, to use, not the Cartesian components of the velocity, but its magnitude v and the angle θ which it makes with the x-axis:

$$v_x = v\cos\theta, \qquad v_y = v\sin\theta. \tag{108.5}$$

The appropriate transformation of the derivatives gives, instead of (108.4),

$$x = \cos\theta\frac{\partial\Phi}{\partial v} - \frac{\sin\theta}{v}\frac{\partial\Phi}{\partial\theta}, \qquad y = \sin\theta\frac{\partial\Phi}{\partial v} + \frac{\cos\theta}{v}\frac{\partial\Phi}{\partial\theta}. \tag{108.6}$$

The relation between the potential ϕ and the function Φ is given by the simple formula

$$\phi = -\Phi + v\,\partial\Phi/\partial v. \tag{108.7}$$

Finally, in order to obtain the equation which determines the function $\Phi(v, \theta)$, we must transform the equation of continuity (108.2) to the new variables. Writing the derivatives as Jacobians:

$$\frac{\partial(\rho v_x, y)}{\partial(x, y)} - \frac{\partial(\rho v_y, x)}{\partial(x, y)} = 0,$$

multiplying by $\partial(x, y)/\partial(v, \theta)$ and substituting (108.5), we have

$$\frac{\partial(\rho v\cos\theta, y)}{\partial(v, \theta)} - \frac{\partial(\rho v\sin\theta, x)}{\partial(v, \theta)} = 0.$$

To expand these Jacobians, we must substitute (108.6) for x and y. Furthermore, since the entropy s is a given constant, if we express the density as a function of s and w and substitute $w = w_0 - \frac{1}{2}v^2$ we find that the density can be written as a function of v only: $\rho = \rho(v)$. We therefore obtain, after a simple calculation, the equation

$$\frac{d(\rho v)}{dv}\left(\frac{\partial\Phi}{\partial v} + \frac{1}{v}\frac{\partial^2\Phi}{\partial\theta^2}\right) + \rho v\frac{\partial^2\Phi}{\partial v^2} = 0.$$

According to (80.5),

$$\frac{d(\rho v)}{dv} = \rho\left(1 - \frac{v^2}{c^2}\right),$$

and so we have finally *Chaplygin's equation* for the function $\Phi(v, \theta)$:

$$\frac{\partial^2\Phi}{\partial\theta^2} + \frac{v^2}{1-v^2/c^2}\frac{\partial^2\Phi}{\partial v^2} + v\frac{\partial\Phi}{\partial v} = 0. \tag{108.8}$$

Here the velocity of sound is a known function $c(v)$, determined by the equation of state of the gas together with Bernoulli's equation.

The equation (108.8), together with the relations (108.6), is equivalent to the equations of motion. Thus the problem of solving the non-linear equations of motion is reduced to the solution of a linear equation for the function $\Phi(v, \theta)$. It is true that the boundary conditions on this equation are non-linear. These conditions are as follows. At the surface of the body, the gas velocity must be tangential. Expressing the equation of the surface in the parametric form $X = X(\theta)$, $Y = Y(\theta)$ (as in §107), and substituting X and Y in place of x and y in (108.6), we obtain two equations, which must be satisfied for all values of θ; this is not possible for every function $\Phi(v, \theta)$. The boundary condition is, in fact, that these two equations are compatible for all θ, i.e. one of them must be deducible from the other.

The satisfying of the boundary conditions, however, does not ensure that the resulting solution of Chaplygin's equation determines a flow that is actually possible everywhere in the physical plane. The following condition must also be met: the Jacobian $\Delta \equiv \partial(x, y)/\partial(\theta, v)$ must nowhere be zero, except in the trivial case when all its four component derivatives vanish. It is easy to see that, unless this condition holds, the solution becomes complex when we pass through the line (called the *limiting line*) in the xy-plane given by the equation $\Delta = 0$.† For, let $\Delta = 0$ on the line $v = v_0(\theta)$, and suppose that $(\partial y/\partial\theta)_v \neq 0$. Then we have

$$-\Delta\left(\frac{\partial\theta}{\partial y}\right)_v = \frac{\partial(x, y)}{\partial(v, \theta)}\frac{\partial(v, \theta)}{\partial(v, y)} = \frac{\partial(x, y)}{\partial(v, y)} = \left(\frac{\partial x}{\partial v}\right)_y = 0.$$

Hence we see that, near the limiting line, v is determined as a function of x (for given y) by

$$x - x_0 = \tfrac{1}{2}(\partial^2 x/\partial v^2)_y (v - v_0)^2,$$

and v becomes complex on one side or the other of the limiting line.‡

It is easy to see that a limiting line can occur only in regions of supersonic flow. A direct calculation, using the relations (108.6) and equation (108.8), gives

$$\Delta = \frac{1}{v}\left[\left(\frac{\partial^2\Phi}{\partial\theta\partial v} - \frac{1}{v}\frac{\partial\Phi}{\partial\theta}\right)^2 + \frac{v^2}{1-v^2/c^2}\left(\frac{\partial^2\Phi}{\partial v^2}\right)^2\right]. \tag{108.9}$$

---

† There is no objection to a passage through points where Δ becomes infinite. If $1/\Delta = 0$ on some line, this merely means that the correspondence between the xy and $v\theta$ planes is no longer one-to-one: in going round the xy-plane, we cover some part of the $v\theta$-plane two or three times.

‡ This result clearly remains valid even if $(\partial^2 x/\partial v^2)_y$ vanishes with Δ but $(\partial x/\partial v)_y$ again changes sign for $v = v_0$, i.e. the difference $x - x_0$ is proportional to a higher even power of $v - v_0$.

It is clear that, for $v \leqslant c$, $\Delta > 0$, and Δ can become zero only if $v > c$.

The appearance of limiting lines in the solution of Chaplygin's equation indicates that, under the given conditions, a continuous flow throughout the region is impossible, and shock waves must occur. It should be emphasised, however, that the position of these shocks is not the same as that of the limiting lines.

In §107 we discussed the particular case of steady two-dimensional super-sonic flow (a simple wave), which is characterised by the fact that the velocity in it is a function only of its direction: $v = v(\theta)$. This solution cannot be obtained from Chaplygin's equation, since $1/\Delta \equiv 0$, and the solution is "lost" when the equation of continuity is multiplied by the Jacobian Δ in the transformation to the hodograph plane. The situation is exactly analogous to that found in the theory of non-steady one-dimensional flow. The re-marks made in §98 concerning the relation between the simple wave and the general integral of equation (98.2) are wholly applicable to the relation be-tween the steady simple wave and the general integral of Chaplygin's equation.

The fact that the Jacobian Δ is positive in subsonic flow enables us to demonstrate an interesting theorem due to A. A. NIKOL'SKIĬ and G. I. TAGANOV (1946). We have identically

$$\frac{1}{\Delta} \equiv \frac{\partial(\theta, v)}{\partial(x, y)} = \frac{\partial(\theta, v)}{\partial(x, v)} \frac{\partial(x, v)}{\partial(x, y)},$$

or

$$\frac{1}{\Delta} = \left(\frac{\partial\theta}{\partial x}\right)_v \left(\frac{\partial v}{\partial y}\right)_x. \tag{108.10}$$

In a subsonic flow $\Delta > 0$, and we see that the derivatives $(\partial\theta/\partial x)_v$ and $(\partial v/\partial y)_x$ have the same sign. This has a simple geometrical significance: if we move along a line $v = \text{constant} \equiv v_0$, with the region $v < v_0$ to the right, the angle θ increases monotonically, i.e. the velocity vector turns always counterclockwise. This result holds, in particular, for the line of transition between subsonic and supersonic flow, on which $v = c = c_*$.

In conclusion, we may give Chaplygin's equation for a perfect gas, writing c explicitly in terms of v:

$$\frac{\partial^2\Phi}{\partial\theta^2} + v^2\frac{1-(\gamma-1)v^2/(\gamma+1)c_*^2}{1-v^2/c_*^2} \frac{\partial^2\Phi}{\partial v^2} + v\frac{\partial\Phi}{\partial v} = 0. \tag{108.11}$$

This equation has a family of particular integrals expressible in terms of hypergeometric functions.†

§109. Characteristics in steady two-dimensional flow

Some general properties of characteristics in steady (supersonic) two-dimensional flow have already been discussed in §79. We shall now derive

---

† See, for instance, L. I. SEDOV, *Two-dimensional Problems of Hydrodynamics and Aerodynamics* (*Ploskie zadachi gidrodinamiki i aërodinamiki*), Moscow 1950.

the equations which give the characteristics in terms of a given solution of the equations of motion.

In steady two-dimensional supersonic flow there are, in general, three families of characteristics. All small disturbances, except those of entropy and vorticity, are propagated along two of these families (which we call the characteristics C_+ and C_-); disturbances of entropy and vorticity are propagated along characteristics (C_0) of the third family, which coincide with the streamlines. For a given flow, the streamlines are known, and the problem is to determine the characteristics belonging to the first two families.

The directions of the characteristics C_+ and C_- passing through each point in the plane lie on opposite sides of the streamline through that point, and make with it an angle equal to the local value of the Mach angle α (Fig. 41, §79). We denote by m_0 the slope of the streamline at a given point, and by m_+, m_- the slopes of the characteristics C_+, C_-. Then we have

$$\frac{m_+ - m_0}{1 + m_0 m_+} = \tan \alpha, \qquad \frac{m_- - m_0}{1 + m_0 m_-} = -\tan \alpha,$$

whence

$$m_\pm = \frac{m_0 \pm \tan \alpha}{1 \mp m_0 \tan \alpha};$$

the upper signs everywhere relate to C_+ and the lower to C_-. Substituting $m_0 = v_y/v_x$, $\tan \alpha = c/\sqrt{(v^2 - c^2)}$ and simplifying, we obtain the following expression for the slopes of the characteristics:

$$m_\pm \equiv \left(\frac{\mathrm{d}y}{\mathrm{d}x} \right)_\pm = \frac{v_x v_y \pm c \sqrt{(v^2 - c^2)}}{v_x^2 - c^2}. \tag{109.1}$$

If the velocity distribution is known, this is a differential equation which determines the characteristics C_+ and C_-.†

Besides the characteristics in the xy-plane, we may consider those in the hodograph plane, which are especially useful in the discussion of isentropic potential flow; we shall take this case in what follows. Mathematically, these are the characteristics of Chaplygin's equation (108.8), which is of hyperbolic type for $v > c$. Following the general method familiar in mathematical physics (see §96), we form from the coefficients the equation of the characteristics:

$$\mathrm{d}v^2 + \mathrm{d}\theta^2 v^2/(1 - v^2/c^2) = 0,$$

or

$$\left(\frac{\mathrm{d}\theta}{\mathrm{d}v} \right)_\pm = \pm \frac{1}{v} \bigg/ \left(\frac{v^2}{c^2} - 1 \right). \tag{109.2}$$

---

† Equation (109.1) also determines the characteristics for steady axially symmetric flow if v_y and y are replaced by v_r and r, where r is the cylindrical co-ordinate (the distance from the axis of symmetry, which is the x-axis); it is clear that the derivation is unchanged if we consider, instead of the xy-plane, an xr-plane through the axis of symmetry.

The characteristics given by this equation do not depend on the particular solution of Chaplygin's equation considered, because the coefficients in that equation are independent of Φ. The characteristics in the hodograph plane are a transformation of the characteristics C_+ and C_- in the physical plane, and we call them respectively the characteristics Γ_+ and Γ_-, in accordance with the signs in (109.2).

The integration of equation (109.2) gives relations of the form $J_+(v, \theta)$ = constant, $J_-(v, \theta)$ = constant. The functions J_+ and J_- are quantities which remain constant along the characteristics C_+ and C_- (i.e. Riemann invariants). For a perfect gas, equation (109.2) can be integrated explicitly.

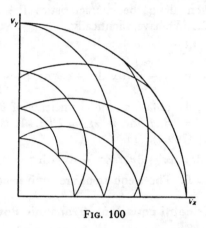

FIG. 100

There is, however, no need to go through the calculations, since the result can be seen from formulae (107.3) and (107.4). For, according to the general properties of simple waves (see §97), the dependence of v on θ for a simple wave is given by the condition that one of the Riemann invariants is constant in all space. The arbitrary constant in formulae (107.3) and (107.4) is ϕ_*; eliminating the parameter ϕ from these formulae, we obtain

$$J_\pm = \theta \pm \left\{ \sin^{-1} \sqrt{\left[\tfrac{1}{2}(\gamma+1)\left(1 - \frac{c_*^2}{v^2}\right)\right]} - \right.$$
$$\left. - \sqrt{\frac{\gamma+1}{\gamma-1}} \sin^{-1} \sqrt{\left[\tfrac{1}{2}(\gamma-1)\left(\frac{v^2}{c_*^2} - 1\right)\right]} \right\}. \tag{109.3}$$

The characteristics in the hodograph plane are a family of epicycloids, occupying the space between two circles of radii $v = c_*$ and $v = \sqrt{[(\gamma+1)/(\gamma-1)]}c_*$ (Fig. 100).

For isentropic potential flow, the characteristics Γ_+, Γ_- have the following important property: the families Γ_+, Γ_- are orthogonal to the families C_-,

C_+ respectively (it is assumed that the co-ordinate axes of x and y are transformed parallel to those of v_x and v_y).†

To prove this, we start from equation (106.3) for two-dimensional potential flow, which is of the form

$$A\frac{\partial^2\phi}{\partial x^2} + 2B\frac{\partial^2\phi}{\partial x\partial y} + C\frac{\partial^2\phi}{\partial y^2} = 0, \qquad (109.4)$$

with no free term. The slopes $m_\pm$ of the characteristics C_+ are the roots of the quadratic

$$Am^2 - 2Bm + C = 0.$$

Let us consider the expression $dv_x^+dx^- + dv_y^+dy^-$, in which the velocity differentials are taken along the characteristics Γ_+, and the co-ordinate differentials along C_-. We have, identically,

$$dv_x^+\,dx^- + dv_y^+dy^-$$

$$= \frac{\partial^2\phi}{\partial x^2}dx^+dx^- + \frac{\partial^2\phi}{\partial x\partial y}(dx^+dy^- + dx^-dy^+) + \frac{\partial^2\phi}{\partial y^2}dy^+dy^-.$$

Dividing by dx^+dx^-, we obtain as the coefficients of $\partial^2\phi/\partial x\partial y$ and $\partial^2\phi/\partial y^2$ respectively $m_+ + m_- = 2B/A$ and $m_+m_- = C/A$. It is then clear that the expression is zero, by (109.4). Thus

$$dv_x^+dx^- + dv_y^+dy^- = d\mathbf{v}^+\cdot d\mathbf{r}^- = 0.$$

Similarly, $d\mathbf{v}^-\cdot d\mathbf{r}^+ = 0$. These equations are equivalent to the result stated.

§110. The Euler–Tricomi equation. Transonic flow

The investigation of the properties of the flow resulting from the transition between subsonic and supersonic flow is of fundamental interest. Steady flows in which this transition occurs are called *mixed* or *transonic* flows, and the surface where the transition occurs is called the *transitional* or *sonic* surface.

Chaplygin's equation is particularly useful in investigating the flow near the transition, since it is much simplified there. At the boundary where the transition occurs $v = c = c_*$, and near it (in the transonic region) the differences $v-c$ and $v-c_*$ are small; they are related by (106.8):

$$(v/c) - 1 = \alpha_*[(v/c_*) - 1].$$

Let us effect the corresponding simplification in Chaplygin's equation. The third term in equation (108.8) is small compared with the second, which contains $1 - v^2/c^2$ in the denominator. In the second term we put approximately

$$\frac{v^2}{1 - v^2/c^2} = \frac{c_*^2}{2(1 - v/c)} = \frac{c_*}{2\alpha_*(1 - v/c_*)}.$$

---

† This does not apply to the characteristics of axisymmetric flow in the xr-plane.

Finally, replacing the velocity v by a new variable

$$\eta = (2\alpha_*)^{\frac{1}{3}} (v - c_*)/c_*, \tag{110.1}$$

we obtain the required equation in the form

$$\frac{\partial^2 \Phi}{\partial \eta^2} - \eta \frac{\partial^2 \Phi}{\partial \theta^2} = 0. \tag{110.2}$$

An equation of this form is called in mathematical physics the *Euler–Tricomi equation.*[†] In the half-plane $\eta > 0$ it is hyperbolic, but in $\eta < 0$ it is elliptic.

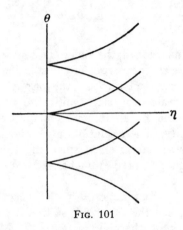

Fig. 101

We shall discuss here some mathematical properties of this equation which are important in connection with various physical problems.

The characteristics of equation (110.2) are given by the equation $\eta \, d\eta^2 - d\theta^2 = 0$, which has the general integral

$$\theta \pm \tfrac{2}{3}\eta^{\frac{3}{2}} = C, \tag{110.3}$$

where C is an arbitrary constant. This equation represents two families of curves in the $\eta\theta$-plane, which are branches of semi-cubical parabolae in the right half-plane with cusps on the θ-axis (Fig. 101).

In investigating the flow in a small region[‡] of space, where the direction of the gas velocity varies only slightly, we can always take the direction of the x-axis such that the angle θ measured from it is small throughout the region considered. The equations (108.6) which determine the co-ordinates x, y

† The application of this equation to the problem here considered is due to F. I. Frankl' (1945). The mathematical theory of equation (110.2) is given by F. Tricomi, Linear Equations of the Mixed Type (*Sulle equazioni lineari . . . di tipo misto*), *Memorie della Reale Accademia Nazionale dei Lincei, classe di scienze fisiche*, ser. 5, **14**, 133, 1922.

‡ This phrase must not be taken literally, of course. The region concerned may be the neighbourhood of the point at infinity, i.e. the region at large distances from the body.

from the function $\Phi(\eta, \theta)$ are then much simplified also:† $x = (2\alpha_*)^{\frac{1}{3}}\partial\Phi/\partial\eta$, $y = \partial\Phi/\partial\theta$. In order to avoid the appearance of the factor $(2\alpha_*)^{\frac{1}{3}}$, we shall replace the co-ordinate x by $x(2\alpha_*)^{-\frac{1}{3}}$, and call the latter quantity x. Then

$$x = \partial\Phi/\partial\eta, \qquad y = \partial\Phi/\partial\theta. \tag{110.4}$$

It is useful to note that, since it is so simply related to Φ, the function $y(\eta, \theta)$ (but not $x(\eta, \theta)$) also satisfies the Euler–Tricomi equation. Using this fact, we can write the Jacobian of the transformation from the physical plane to the hodograph plane as

$$\Delta = \frac{\partial(x, y)}{\partial(\theta, \eta)} = \Phi_{\eta\theta}^2 - \Phi_{\eta\eta}\Phi_{\theta\theta} = \left(\frac{\partial y}{\partial\eta}\right)^2 - \eta\left(\frac{\partial y}{\partial\theta}\right)^2. \tag{110.5}$$

As has already been mentioned, the Euler–Tricomi equation has usually to be applied to investigate the properties of the solution near the origin in the $\eta\theta$-plane. In cases of physical interest, the origin is a singular point of the solution. For this reason especial significance attaches to the family of particular integrals of the Euler–Tricomi equation which possess certain properties of homogeneity. These solutions are homogeneous in the variables θ^2 and η^3; such solutions must exist, since the transformation $\theta^2 \to a\theta^2$, $\eta^3 \to a\eta^3$ leaves the equation (110.2) unchanged. We shall seek these solutions in the form $\Phi = \theta^{2k} f(\xi)$, $\xi = 1 - 4\eta^3/9\theta^2$, where k is a constant, the degree of homogeneity of the function Φ with respect to the transformation mentioned. We have taken the variable ξ so that it vanishes on the characteristics which pass through the point $\eta = \theta = 0$. Making the above substitution, we obtain for the function $f(\xi)$ the equation

$$\xi(1-\xi)f'' + [\tfrac{5}{6} - 2k - \xi(\tfrac{3}{2} - 2k)]f' - k(k - \tfrac{1}{2})f = 0.$$

This is a hypergeometric equation. Using the well-known expressions for the two independent integrals of that equation, we find the required solution (for $2k + \frac{1}{6}$ not integral):

$$\Phi_k = \theta^{2k}\left[AF\left(-k, \quad -k+\tfrac{1}{2}; \quad -2k+\tfrac{5}{6}; \quad 1 - \frac{4\eta^3}{9\theta^2}\right) + \right.$$

$$\left. + B\left(1 - \frac{4\eta^3}{9\theta^2}\right)^{2k+1/6} F\left(k+\tfrac{1}{6}, \quad k+\tfrac{2}{3}; \quad 2k+\tfrac{7}{6}; \quad 1 - \frac{4\eta^3}{9\theta^2}\right)\right].$$

$$\tag{110.6}$$

Using the relations between hypergeometric functions of arguments z, $1/z$, $1 - z$, $1/(1-z)$ and $z/(1-z)$, we can also put this solution in five other forms,

---

all of which are needed in various problems. We shall give two of these:

$$\Phi_k = \theta^{2k}\left[AF\left(-k,\quad -k+\tfrac{1}{2};\quad \tfrac{2}{3};\quad \frac{4\eta^3}{9\theta^2}\right) + \right.$$

$$\left. + B\frac{\eta}{\theta^{2/3}}F\left(-k+\tfrac{1}{3},\quad -k+\tfrac{5}{6};\quad \tfrac{4}{3};\quad \frac{4\eta^3}{9\theta^2}\right)\right],\qquad (110.7)$$

$$\Phi_k = \eta^{3k}\left[AF\left(-k,\quad -k+\tfrac{1}{3};\quad \tfrac{1}{2};\quad \frac{9\theta^2}{4\eta^3}\right) + \right.$$

$$\left. + B\frac{\theta}{\eta^{3/2}}F\left(-k+\tfrac{1}{2},\quad -k+\tfrac{5}{6};\quad \tfrac{3}{2};\quad \frac{9\theta^2}{4\eta^3}\right)\right];\qquad (110.8)$$

the constants A and B in formulae (110.6)–(110.8) are not the same, of course. These expressions yield at once the following important property of the functions Φ_k, which is not evident from (110.6): the lines $\eta = 0$ and $\theta = 0$ are not singular lines (it is seen from (110.7) that, near $\eta = 0$, Φ_k can be expanded in integral powers of η, and from (110.8) the same is true of θ). It is seen from the expression (110.6) that the characteristics, on the other hand, are singular lines of the general (i.e. containing the two constants A and B) homogeneous integral Φ_k of the Euler–Tricomi equation: if $2k+\tfrac{1}{6}$ is not an integer, the factor $(9\theta^2 - 4\eta^3)^{2k+1/6}$ has branch points, while if $2k+\tfrac{1}{6}$ is an integer, one term of (110.6) is meaningless† (or degenerates to the other term if $2k+\tfrac{1}{6} = 0$), and must be replaced by the second independent solution of the hypergeometric equation, which in this case has a logarithmic singularity.

The following relations hold between the integrals Φ_k with different values of k:

$$\Phi_k = \Phi_{-k-1/6}\,(9\theta^2 - 4\eta^3)^{2k+1/6},\qquad (110.9)$$

$$\Phi_{k-1/2} = \partial\Phi_k/\partial\theta.\qquad (110.10)$$

The first of these follows immediately from (110.6), and the second from the fact that $\partial\Phi_k/\partial\theta$ satisfies the Euler–Tricomi equation, and its degree of homogeneity is that of $\Phi_{k-1/2}$. In these formulae Φ_k means, of course, the general expression, with two arbitrary constants.

In investigating the solution near the point $\eta = \theta = 0$, we have to follow its variation along a contour round this point. For example, let the function Φ_k (110.6) represent the solution at the point A near the characteristic $\theta = \tfrac{2}{3}\eta^{3/2}$ (Fig. 102), and suppose that we require the form of the solution near the characteristic $\theta = -\tfrac{2}{3}\eta^{3/2}$ (at the point B). The passage from A to B involves crossing the axis of abscissae, and $\theta = 0$ is a singular line of the hypergeometric functions in the expression (110.6), so that their argument is infinite there. In order to go from A to B, therefore, it is necessary to transform the hypergeometric functions into functions of the reciprocal argument

---

† We recall that the series $F(\alpha, \beta; \gamma; z)$ is meaningless for $\gamma = 0, -1, -2, \ldots$.

$9\theta^2/(9\theta^2 - 4\eta^3)$, for which $\theta = 0$ is not a singularity, and then change the sign of θ, finally returning to the original argument by repeating the transformation. In this way we obtain the following transformation formulae for the functions which appear in (110.6):

$$F_1 \to \frac{F_1}{2\sin(2k+\frac{1}{6})\pi} + F_2 \cdot 2^{-4k-1/3} \frac{\Gamma(-2k-\frac{1}{6})\Gamma(-2k+\frac{5}{6})}{\Gamma(-2k)\Gamma(-2k+\frac{2}{3})},$$

$$(110.11)$$

$$F_2 \to \frac{-F_2}{2\sin(2k+\frac{1}{6})\pi} + F_1 \cdot 2^{4k+1/3} \frac{\Gamma(2k+\frac{1}{6})\Gamma(2k+\frac{7}{6})}{\Gamma(2k+1)\Gamma(2k+\frac{1}{3})},$$

where F_1 and F_2 signify

$$F_1 = |\theta|^{2k} F\left(-k, \quad -k+\tfrac{1}{2}; \quad -2k+\tfrac{5}{6}; \quad 1-\frac{4\eta^3}{9\theta^2}\right),$$

$$(110.12)$$

$$F_2 = |\theta|^{2k}\left|1-\frac{4\eta^3}{9\theta^2}\right|^{2k+1/6} F\left(k+\tfrac{1}{6}, \quad k+\tfrac{2}{3}; \quad 2k+\tfrac{7}{6}; \quad 1-\frac{4\eta^3}{9\theta^2}\right),$$

in which the moduli of θ and $1-4\eta^3/9\theta^2$ are taken in the coefficients of the hypergeometric functions.

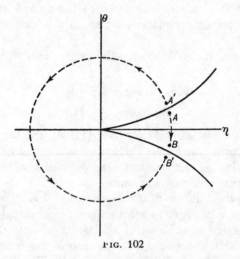

FIG. 102

We can similarly obtain transformation formulae for the passage from A' to B' (Fig. 102) round the origin in the opposite direction. The calculations are more involved, since we have to pass through three singularities of the hypergeometric function (one with $\theta = 0$ and two with $\eta = 0$; we recall that the singularities of a hypergeometric function with argument z

are $z = 1$ and $z = \infty$). The final formulae are

$$F_1 \to -\frac{\sin(4k - \tfrac{1}{6})\pi}{\sin(2k + \tfrac{1}{6})\pi} F_1 + F_2 \cdot 2^{-4k+2/3} \cos(2k + \tfrac{1}{6})\pi \frac{\Gamma(-2k - \tfrac{1}{6})\Gamma(-2k + \tfrac{5}{6})}{\Gamma(-2k)\Gamma(-2k + \tfrac{2}{3})},$$

(110.13)

$$F_2 \to \frac{\sin(4k - \tfrac{1}{6})\pi}{\sin(2k + \tfrac{1}{6})\pi} F_2 + F_1 \cdot 2^{4k+4/3} \cos(2k + \tfrac{1}{6})\pi \frac{\Gamma(2k + \tfrac{1}{6})\Gamma(2k + \tfrac{7}{6})}{\Gamma(2k + 1)\Gamma(2k + \tfrac{4}{3})}.$$

(110.14)

As well as this family of homogeneous solutions there are, of course, other families of particular integrals of the Euler–Tricomi equation. We may mention here a family which results from a Fourier expansion in terms of θ. If we seek Φ in the form

$$\Phi_\nu = g_\nu(\eta)e^{\pm i\nu\theta},$$

(110.14)

where ν is an arbitrary constant, we obtain for the function g_ν the equation $g_\nu'' + \nu^2 \eta g_\nu = 0$. This is the equation for the Airy function; its general integral is

$$g_\nu(\eta) = \sqrt{\eta} Z_{\frac{1}{3}}(\tfrac{2}{3}\nu\eta^{3/2}),$$

(110.15)

where $Z_{\frac{1}{3}}$ is an arbitrary linear combination of Bessel functions of order $\tfrac{1}{3}$.

Finally, it is useful to bear in mind that the general integral of the Euler–Tricomi equation may be written

$$\Phi = \int_{C_z} f(\zeta)dz, \qquad \zeta = z^3 - 3\eta z + 3\theta,$$

(110.16)

where $f(\zeta)$ is an arbitrary function and the integration in the complex z-plane is taken along any contour C_z at whose ends the derivative $f'(\zeta)$ has equal values. For a direct substitution of (110.16) in the Euler–Tricomi equation gives

$$\frac{\partial^2 \Phi}{\partial \eta^2} - \eta \frac{\partial^2 \Phi}{\partial \theta^2} = 9 \int_{C_z} (z^2 - \eta^2) f''(\zeta)dz = 3 \int_{C_\zeta} f''(\zeta)d\zeta$$

$$= 3[f'(\zeta)]c_\zeta = 0,$$

i.e. the equation is satisfied.

§111. Solutions of the Euler–Tricomi equation near non-singular points of the sonic surface

Let us now ascertain which solutions Φ_k correspond to cases where the gas flow has no physical singularities (weak discontinuities or shock waves) near the transition. To do this it is more convenient to start, not from the Euler–Tricomi equation itself, but from the equation for the velocity potential in the physical plane. This equation has been derived in §106;

for a two-dimensional flow, equation (106.10) becomes, with the substitution $x \to x(2\alpha_*)^{1/3}$,

$$\frac{\partial\phi}{\partial x}\frac{\partial^2\phi}{\partial x^2} = \frac{\partial^2\phi}{\partial y^2}. \tag{111.1}$$

We recall that the potential ϕ in this equation is defined so that its derivatives with respect to the co-ordinates give the velocity according to the equations

$$\partial\phi/\partial x = \eta, \qquad \partial\phi/\partial y = \theta. \tag{111.2}$$

We may also note that the Euler–Tricomi equation can be obtained directly from equation (111.1) by changing to the independent variables θ, η by Legendre's transformation, with $\Phi = -\phi + x\eta + y\theta$, or

$$\phi = -\Phi + \eta\,\partial\Phi/\partial\eta + \theta\,\partial\Phi/\partial\theta. \tag{111.3}$$

Taking the origin in the xy-plane at the point on the transition line whose neighbourhood we are investigating, we expand ϕ in powers of x and y. In the general case, the first term of an expansion which satisfies equation (111.1) is

$$\phi = xy/a. \tag{111.4}$$

Here $\theta = x/a$, $\eta = y/a$, so that

$$\Phi = a\theta\eta. \tag{111.5}$$

It is clear from the degree of homogeneity of this function that it corresponds to one of the functions $\Phi_{5/6}$; this is the second term of the expression (110.7), in which the hypergeometric function with $k = 5/6$ reduces to 1 simply: $\eta\theta F(-\tfrac{1}{2}, 0; \tfrac{4}{3}; 4\eta^3/9\theta^2) = \eta\theta$.

If we wish to find the equation of the transition line in the physical plane, the first term of the expansion does not suffice. The next term is of degree 1, i.e. it corresponds to one of the functions Φ_1, namely the first term in the expression (110.7), which reduces to a polynomial for $k = 1$:

$$\theta^2 F(-1, \ -\tfrac{1}{2}; \ \tfrac{2}{3}; \ 4\eta^3/9\theta^2) = \theta^2 + \tfrac{1}{3}\eta^3.$$

Thus the first two terms of the expansion of Φ are

$$\Phi = a\eta\theta + b(\theta^2 + \tfrac{1}{3}\eta^3). \tag{111.6}$$

Hence

$$x = a\theta + b\eta^2,$$
$$y = a\eta + 2b\theta. \tag{111.7}$$

The transition line ($\eta = 0$) is the straight line $y = 2bx/a$.

To find the equation of the characteristics in the physical plane we need only the first term of the expansion. Substituting $\theta = x/a$, $\eta = y/a$ in the equation of the hodograph characteristics $\theta = \pm\tfrac{2}{3}\eta^{3/2}$, we obtain $x = \pm\tfrac{2}{3}y^{3/2}/\sqrt{a}$, i.e. again two branches of a semi-cubical parabola with a cusp on the transition line. This property of the characteristics is evident also from the following simple argument. At points on the transition line, the Mach angle is $\frac{1}{2}\pi$. This means that the tangents to the characteristics of the two families coincide, so that there is a cusp (Fig. 103). The streamlines intersect the transition line perpendicularly to the characteristics, and do not have singularities there.

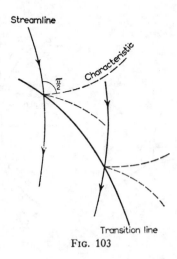

FIG. 103

The solution (111.6) is not applicable in the exceptional case where the streamline is perpendicular to the transition line at the point considered.† Near such a point the flow is evidently symmetrical about the x-axis. This case requires special consideration, which has been given by F. I. FRANKL' and S. V. FAL'KOVICH (1945).

The symmetry of the flow means that, when the sign of y is changed, the velocity v_y changes sign and v_x remains unchanged. That is, the potential ϕ must be an even function of y, and the potential Φ an even function of θ. The first terms in the expansion of ϕ in this case are therefore of the form

$$\phi = \tfrac{1}{2}ax^2 + \tfrac{1}{2}a^2xy^2 + \tfrac{1}{24}a^3y^4; \tag{111.8}$$

the relative order of smallness of x and y is not known *a priori*, so that all three terms may be of the same order. Hence we find the following formulae for the transformation from the physical plane to the hodograph plane:

$$\eta = ax + \tfrac{1}{2}a^2y^2,$$
$$\theta = a^2xy + \tfrac{1}{6}a^3y^3. \tag{111.9}$$

† This would correspond to the case $a = 0$ in (111.6); the solution then ceases to hold, because the Jacobian Δ vanishes on the line $\eta = 0$.

Without explicitly solving these equations for x and y, we can easily see that the degree of the function $y(\theta, \eta)$ is $\frac{1}{6}$. Hence the corresponding function Φ has $k = \frac{1}{6} + \frac{1}{2} = \frac{2}{3}$, i.e. it is a particular case of the general integral $\Phi_{2/3}$.

Eliminating x from equations (111.9), we obtain a cubic equation for the function $y(\theta, \eta)$:

$$(ay)^3 - 3\eta ay + 3\theta = 0. \tag{111.10}$$

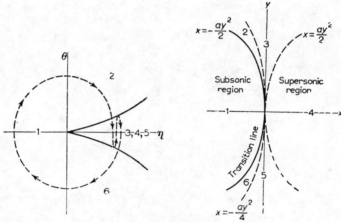

FIG. 104

For $9\theta^2 - 4\eta^3 > 0$, i.e. throughout the region to the left of the hodograph characteristics which pass through the point $\eta = \theta = 0$ (including the whole of the subsonic region $\eta < 0$; Fig. 104), this equation has only one real root, which must be the function $y(\theta, \eta)$. In the region to the right of the characteristics, all three roots are real, and we must take the one which is the continuation of the real root in the region to the left.

The characteristics in the physical plane (which pass through the origin) are obtained by substituting the expressions (111.9) in the equation $4\eta^3 = 9\theta^2$. This gives two parabolae:

the characteristics 23 and 56: $x = -\frac{1}{4}ay^2$,

the characteristics 34 and 45: $x = \frac{1}{2}ay^2$. $\qquad$ (111.11)

The numbers show which two regions in the physical plane are separated by the characteristic in question. The transition line ($\eta = 0$ in the hodograph plane) is the parabola $x = -\frac{1}{2}ay^2$ in the physical plane (Fig. 104). We may notice the following property of the point where the transition line intersects the axis of symmetry: four branches of characteristics leave this point, whereas only two leave any other point on the transition line.

Fig. 104 shows by corresponding numbers the regions of the hodograph plane which correspond to the various regions of the physical plane. This correspondence is not one-to-one;† when we go completely round the origin in the physical plane, the region between the two characteristics in the hodograph plane is covered three times, as shown by the dashed line in Fig. 104, which is twice "reflected" from the characteristics.

Since the function $y(\theta, \eta)$ itself satisfies the Euler–Tricomi equation, it must be obtainable from the general integral $\Phi_{1/6}$. Near the characteristic 23 in the physical plane, it is

$$y = \frac{1}{a}\sqrt{\frac{3\theta}{2}}F\left(-\tfrac{1}{6}, \ \tfrac{1}{3}; \ \tfrac{1}{2}; \ 1 - \frac{4\eta^3}{9\theta^2}\right); \tag{111.12}$$

the first term in (110.6) has no singularity on this characteristic. Continuing this analytically to the neighbourhood of the characteristic 56 (by a path through the subsonic region 1, i.e. by means of formulae (110.13)), we obtain the same function there. Near the characteristics 34 and 45, however, $y(\theta, \eta)$ is given by linear combinations of that function and

$$\theta^{1/3}\sqrt{\left(\frac{4\eta^3}{9\theta^2} - 1\right)}F\left(\tfrac{1}{3}, \ \tfrac{5}{6}; \ \tfrac{3}{2}; \ 1 - \frac{4\eta^3}{9\theta^2}\right), \tag{111.13}$$

i.e. the second term of (110.6). These combinations are obtained by analytical continuation, using formulae (110.11); here it must be borne in mind that the square root in (111.13) changes sign at each "reflection" from a hodograph characteristic.

Mathematically, these results show that the functions $\Phi_{1/6}$ are linear combinations of the roots of the cubic equation

$$f^3 - 3\eta f + 3\theta = 0, \tag{111.14}$$

i.e. they are algebraic functions.‡ As well as $\Phi_{1/6}$, all the Φ_k with

$$k = \tfrac{1}{6} \pm \tfrac{1}{2}n, \qquad n = 0, 1, 2, \dots \tag{111.15}$$

reduce to algebraic functions; they are obtained from $\Phi_{1/6}$, according to formulae (110.9) and (110.10), by successive differentiation, a remark due to F. I. FRANKL' (1947).

The functions Φ_k with

$$k = \pm \tfrac{1}{2}n, \qquad k = \tfrac{1}{3} \pm \tfrac{1}{2}n, \tag{111.16}$$

in which the hypergeometric function reduces to a polynomial,†† also reduce

---

† In accordance with the fact that $\Delta = \infty$ on the characteristic $x = \tfrac{1}{2}ay^2$ in the physical plane; see the first footnote to §108.

‡ It is not convenient in practice to use the explicit forms of these functions, which are obtained from (111.14) by Cardan's formula.

†† Here it must be recalled that $F(\alpha, \beta; \gamma; z)$ reduces to a polynomial if α (or β) is such that $\alpha = -n$ or $\gamma - \alpha = -n$.

to algebraic functions; e.g. for $k = \frac{1}{2}n$ we have the first term of the expression (110.6), and for $k = -\frac{1}{2}n$ the second term.

These three families of algebraic functions Φ_k include, in particular, all the functions which can be potentials Φ corresponding to flows having no singularity in the physical plane. In such flows, all the terms in the expansion of Φ near an asymmetric point on the transition line (the first two terms of which are given by formula (111.6)) must have either $k = \frac{5}{6} + \frac{1}{2}n$ or $k = 1 + \frac{1}{2}n$. The expansion of Φ near a symmetric point, however, which begins with a term with $k = \frac{2}{3}$, can also contain functions with $k = \frac{2}{3} + \frac{1}{2}n$.

§112. Flow at the velocity of sound

The simplified form of Chaplygin's equation (i.e. the Euler–Tricomi equation) is of fundamental importance in the gas dynamics of steady flow past bodies, since it must be used to investigate the basic qualitative properties of such flow. These include, in the first place, problems concerning the formation of shock waves. For example, if a shock wave is formed in subsonic flow (in a local supersonic region adjoining the surface of the body†), it must terminate at a finite distance from the body, and the question arises of the properties of its terminal point (see §113). Another similar problem is that of the properties of a shock wave just formed near its intersection with the surface. In both cases the shock wave is weak, i.e. it is in a transonic region, and so the investigation must be performed by means of the Euler–Tricomi equation.‡

We shall discuss here another problem of theoretical importance, that of the nature of steady two-dimensional flow past a body when the velocity of the incident stream is exactly equal to the velocity of sound. We shall see, in particular, that a shock wave must extend from the surface of the body to infinity. From this we can draw the important conclusion that the shock wave must first appear for a Mach number M_∞ which is certainly less than unity.

For, let us consider two-dimensional flow past a body ("wing") of infinite span and arbitrary (not necessarily symmetrical) cross-section. Here we are interested in the flow pattern at distances from the body which are large compared with its dimension. For convenience we shall first describe the results in a qualitative manner, and afterwards give a quantitative calculation.

In Fig. 105, AB and $A'B'$ are transition lines, so that the subsonic region lies to the left of them (upstream); the arrow shows the direction of the main stream, which we shall take as the x-axis, with the origin anywhere near the body. At a certain distance from the transition line we have shock waves leaving the body (EF and $E'F'$ in Fig. 105). It is found that the characteristics leaving the body (between the transition line and the shock wave) can be

---

† The smallest Mach number $M_\infty < 1$ of the main stream for which the local value of M anywhere exceeds unity is sometimes called the *critical Mach number*.

‡ It should be recalled that, in a weak shock wave, the changes in the entropy and vorticity are of a high order of smallness. In the first approximation, therefore, we can assume isentropic potential flow on both sides of the discontinuity.

divided into two groups. The characteristics in the first group meet the transition line and end there (that is to say, they are "reflected" from it as characteristics which reach the body; Fig. 105 shows one such characteristic). The characteristics in the second group end at the shock wave. The two groups are separated by limiting characteristics, the only ones which go to infinity and meet neither the transition line nor the shock wave (*CD* and *C'D'* in Fig. 105). Since disturbances (caused, for instance, by a change in the shape of the body) which are propagated from the body along characteristics of the first group reach the boundary of the subsonic region, it is clear that the part of the supersonic region which lies between the transition line and the limiting characteristic affects the subsonic region, but the flow to the right of the limiting characteristics has no effect on the flow to the left: the flow to the left is not affected by a disturbance of the flow to the right (such as a

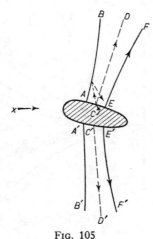

Fig. 105

change in the profile to the right of *C* or *C'*). The flow behind the shock wave has, as we know, no effect on the flow in front of it. Thus the whole flow can be divided into three parts (to the left of *DCC'D'*, between *DCC'D'* and *FEE'F'*, and to the right of *FEE'F'*), such that the flow in the second part has no effect on that in the first, and the flow in the third part has no effect on that in the second.

We shall now give a quantitative account (and verification) of the flow pattern just described.

The origin in the hodograph plane ($\theta = \eta = 0$) corresponds to an infinitely distant region of the physical plane, and the hodograph characteristics leaving the origin correspond to the limiting characteristics *CD* and *C'D'*. Fig. 106 shows the neighbourhood of the origin, the letters corresponding to those in Fig. 105. The shock wave corresponds not to one line but to two lines in the hodograph plane (corresponding to the gas flow on the two sides of the discontinuity); the region between these lines (shaded in Fig. 106) does not correspond to any part of the physical plane.

We must ascertain, first of all, which of the general integrals Φ_k corresponds to this case. If $\Phi(\theta, \eta)$ is of degree k, then the functions $x = \partial\Phi/\partial\eta$ and $y = \partial\Phi/\partial\theta$ are homogeneous and of degree $k-\frac{1}{3}$ and $k-\frac{1}{2}$ respectively. As θ and η tend to zero we must, in general, reach infinity in the physical plane (x and y tend to infinity). It is evident that, for this to be so, we must have $k < \frac{1}{3}$. The limiting characteristics in the physical plane, however, need not lie entirely at infinity, i.e. $y = \pm \infty$ need not hold everywhere on the curve $9\theta^2 = 4\eta^3$. In that case (for $2k+\frac{1}{6} < \frac{5}{6}$), the second term in the brackets in (110.6) must be zero. Thus the function $\Phi(\theta, \eta)$ must be given by the first term of (110.6):

$$\Phi = A\theta^{2k}F\left(-k, \quad -k+\tfrac{1}{2}; \quad -2k+\tfrac{5}{6}; \quad 1 - \frac{4\eta^3}{9\theta^2}\right). \qquad (112.1)$$

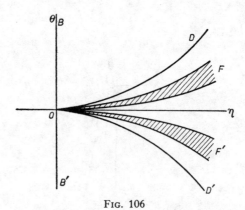

Fig. 106

The function $y(\theta, \eta)$ (which also satisfies the Euler–Tricomi equation) is of the same form, but with $k-\frac{1}{2}$ instead of k.

If the expression (112.1) is valid near (e.g.) the upper characteristic ($\theta = +\frac{2}{3}\eta^{3/2}$), however, it will not be valid near the lower characteristic also ($\theta = -\frac{2}{3}\eta^{3/2}$) for an arbitrary $k < \frac{1}{3}$. We must therefore require also that the form (112.1) of the function $\Phi(\theta, \eta)$ is maintained on going round the origin in the hodograph plane from one characteristic to the other through the half-plane $\eta < 0$ (the path $A'B'$ in Fig. 102). This path corresponds to a passage in the physical plane from distant points on one of the limiting characteristics to distant points on the other, along a path which passes through the subsonic region and therefore nowhere intersects the shock wave, at which the flow is discontinuous. The transformation of the hypergeometric function in (112.1) in going along such a path is given by the first formula (110.13), and we must require that the coefficient of F_2 in this formula is zero. This condition is fulfilled (for $k < \frac{1}{3}$) when $k = \frac{1}{8} - \frac{1}{2}n$ ($n = 0, 1, 2, ...$).

Of these values, only one can be taken, namely

$$k = -\tfrac{1}{3}; \tag{112.2}$$

it can be shown that all values of k with $n > 1$ give a transformation of the hodograph plane into the physical plane which is not one-to-one (in going once round the former we go more than once round the latter), and so the physical flow is many-valued, which is of course impossible. The value $k = \tfrac{1}{6}$, on the other hand, gives a solution in which we do not go to infinity in every direction in the physical plane when θ and η tend to zero; such a solution is, evidently, likewise physically impossible.

For $k = -\tfrac{1}{3}$ the coefficient of F_1 on the right-hand side of formula (110.13) is unity, i.e. the function Φ is unchanged when we go from one characteristic to the other. This means that Φ is an even function of θ, and the co-ordinate $y = \partial\Phi/\partial\theta$ is therefore an odd function. Physically, this means that, in the first approximation here considered, the flow pattern at large distances from the body is symmetrical about the plane $y = 0$, whatever the shape of the body, and in particular whether there is a lift force or not.

Thus we have determined the nature of the singularity of $\Phi(\eta, \theta)$ at the point $\eta = \theta = 0$. From this we can at once deduce the form of the transition line, the limiting characteristics and the shock wave at great distances from the body. Each of these lines must correspond to a definite value of the ratio θ^2/η^3 and, since Φ is of the form $\theta^{-2/3} f(\eta^3/\theta^2)$, we find from formulae (110.4) that $x \sim \theta^{-4/3}$, $y \sim \theta^{-5/3}$. Hence these lines are given by equations of the form

$$x = \text{constant} \times y^{4/5}, \tag{112.3}$$

with various values of the constant. Along these lines, θ and η decrease according to

$$\theta = \text{constant} \times y^{-3/5}, \qquad \eta = \text{constant} \times y^{-2/5}. \tag{112.4}$$

These results are due to F. I. FRANKL' (1947).

In what follows we shall, for definiteness, write the formulae with the signs appropriate to the upper half-plane $(y > 0)$.

We shall show how the coefficients in these formulae may be calculated. The value $k = -\tfrac{1}{3}$ is one of those for which the Φ_k reduce to algebraic functions (see §111). The particular integral which determines Φ in the present case can be written as $\Phi = \tfrac{1}{2}a_1 \partial f/\partial\theta$, where a_1 is an arbitrary positive constant, and f is that root of the cubic equation

$$f^3 - 3\eta f + 3\theta = 0 \tag{112.5}$$

which is the real root for $9\theta^2 - 4\eta^3 > 0$. Hence

$$\Phi = \tfrac{1}{2}a_1 \partial f/\partial\theta = -a_1/2(f^2 - \eta), \tag{112.6}$$

and we have for the co-ordinates

$$\begin{aligned} x &= \partial\Phi/\partial\eta = \tfrac{1}{2}a_1(f^2 + \eta)/(f^2 - \eta)^3, \\ y &= \partial\Phi/\partial\theta = -a_1 f/(f^2 - \eta)^3. \end{aligned} \tag{112.7}$$

These formulae can be put in a convenient parametric form by using as a parameter $s = f^2/(f^2 - \eta)$. Then

$$
\left.
\begin{aligned}
x/y^{4/5} &= a_1^{1/5}(2s-1)/2s^{2/5}, \\
\eta y^{2/5} &= a_1^{2/5} s^{1/5}(s-1), \\
\theta y^{3/5} &= \tfrac{1}{3} a_1^{3/5} s^{4/5}(3-2s),
\end{aligned}
\right\}
\qquad (112.8)
$$

which give, in parametric form, η and θ as functions of the co-ordinates. The parameter s takes positive values from zero upwards ($s = 0$ corresponding to $x = -\infty$, i.e. to the stream incident from infinity). In particular, the value $s = \tfrac{1}{2}$ corresponds to $x = 0$, i.e. it gives the velocity distribution for large y in a plane perpendicular to the x-axis and passing near the body. The value $s = 1$ corresponds to the transition line ($\eta = 0$), and $s = \tfrac{4}{3}$, as is easily seen, to the limiting characteristic. The value of the constant a_1 depends on the actual shape of the body, and can be determined only from an exact solution of the problem in all space.

Formulae (112.8) relate only to the region in front of the shock wave. The necessity for the shock to appear can be seen as follows. A simple calculation from formula (110.5) gives for the Jacobian Δ the expression $a_1^2(4f^2-\eta)/(f^2-\eta)^3$. It is easy to see that $\Delta > 0$ (and does not vanish) on the characteristics and everywhere to the left of them, corresponding to the region upstream of the limiting characteristics in the physical plane. To the right of the characteristics, however, Δ becomes zero, and so a shock wave must appear in this region.

The boundary conditions at the shock wave which must be satisfied by the solution of the Euler–Tricomi equation are as follows. Let θ_1, η_1 and θ_2, η_2 be the values of θ and η on the two sides of the discontinuity. First of all, they must correspond to the same curve in the physical plane, i.e.

$$
x(\theta_1, \eta_1) = x(\theta_2, \eta_2), \qquad y(\theta_1, \eta_1) = y(\theta_2, \eta_2). \qquad (112.9)
$$

Next, the condition that the velocity component tangential to the discontinuity is continuous (i.e. that the derivative of the potential ϕ along the discontinuity is continuous) is equivalent to the condition that the potential itself is continuous:

$$
\phi(\theta_1, \eta_1) = \phi(\theta_2, \eta_2); \qquad (112.10)
$$

the potential ϕ is determined from the function Φ by (111.3). Finally, another condition can be obtained from the limiting form of the equation (86.6) of the shock polar, which gives a relation between the velocity components on the two sides of the discontinuity. Replacing the angle χ in (86.6) by $\theta_2 - \theta_1$, and introducing η_1, η_2 in place of v_1, v_2, we obtain the relation

$$
2(\theta_2 - \theta_1)^2 = (\eta_2 - \eta_1)^2(\eta_2 + \eta_1). \qquad (112.11)
$$

In the present case, the solution of the Euler–Tricomi equation behind the shock wave (the region between OF and OF' in the hodograph plane, Fig.

106) is of the same form (112.5), (112.6), but of course with a different constant coefficient (which we call $-a_2$) in place of a_1. The four simultaneous equations (112.9)–(112.11) determine the ratio a_2/a_1 and relate the quantities $\eta_1, \theta_1, \eta_2, \theta_2$. The solution of these equations is fairly complicated; it gives the following results. The shock wave corresponds to the value $s = \frac{1}{6}(5\sqrt{3}+8) = 2\cdot78$ of the parameter s in formulae (112.8), which give the form of the shock and the velocity distribution on the forward side of the discontinuity. In the region behind (downstream of) the shock, the coefficient $-a_2$ is negative, and $f^2/(f^2-\eta)$ takes negative values. Using as the parameter the positive quantity $s = f^2/(\eta-f^2)$, we have instead of (112.8) the formulae

$$x/y^{4/5} = a_2^{1/5}(2s+1)/2s^{2/5}, \qquad \eta y^{2/5} = a_2^{2/5}s^{1/5}(s+1),$$

$$\theta y^{3/5} = -\tfrac{1}{3}a_2^{3/5}s^{4/5}(2s+3), \qquad (112.12)$$

where

$$a_2/a_1 = (9\sqrt{3}+1)/(9\sqrt{3}-1) = 1\cdot14,$$

and s takes values from $\frac{1}{6}(5\sqrt{3}-8) = 0\cdot11$ on the shock wave to zero at an infinite distance downstream.

Fig. 107 shows graphs of $\eta y^{2/5}$ and $\theta y^{3/5}$ as functions of $xy^{-4/5}$, calculated from formulae (112.8) and (112.12) (the constant a_1 being arbitrarily taken as unity).

§113. The intersection of discontinuities with the transition line

As a further example of the investigation of the properties of transonic flow by means of the Euler–Tricomi equation, let us consider the reflection of a weak discontinuity from the transition line (L. D. LANDAU and E. M. LIFSHITZ, 1954).

We shall assume that the weak discontinuity incident on the transition line (reaching the point of intersection) is of the ordinary type, formed (say) by flow past an acute angle, i.e. the first spatial derivatives of the velocity are discontinuous in it. It is "reflected" from the transition line as another weak discontinuity, the nature of which, however, is unknown *a priori* and must be determined by investigating the flow near the point of intersection. We take this point as the origin in the xy-plane, and the x-axis in the direction of the gas velocity there, so that it corresponds to the origin in the hodograph plane also.

Weak discontinuities coincide with characteristics, as we know. Let the characteristic Oa in the hodograph plane (Fig. 108a) correspond to the incident discontinuity. Since the co-ordinates x, y are continuous at the discontinuity, the first derivatives Φ_η, Φ_θ must be continuous also. The second derivatives of Φ, on the other hand, can be expressed in terms of the first spatial derivatives of the velocity, and therefore must be discontinuous.

Denoting the discontinuities of quantities by placing them in brackets, we therefore have

$$\text{on } Oa \; [\Phi_\eta] = [\Phi_\theta] = 0; \quad [\Phi_{\theta\theta}], \quad [\Phi_{\theta\eta}], \quad [\Phi_{\eta\eta}] \neq 0. \quad (113.1)$$

The functions Φ themselves in the regions 1 and 2 on each side of the characteristic Oa must not have singularities on the characteristic. Such a

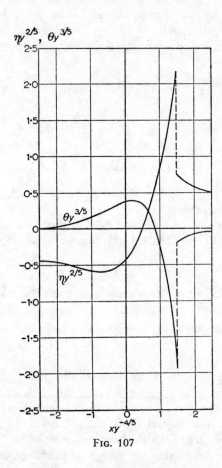

FIG. 107

solution can be constructed from the second term in (110.6) with $k = 11/12$, which is proportional to the square of the difference $1 - 4\eta^3/9\theta^2$ (the other independent solution $\Phi_{11/12}$ has a singularity on the characteristic; see below). The first derivatives of this function vanish on the characteristic, and the second derivatives are finite. Furthermore, Φ can include those particular solutions of the Euler–Tricomi equation which do not give singularities of the flow in the physical plane. The solution of this kind which is of the lowest degree in θ and η is $\eta\theta$ (§111). Thus we seek Φ near the characteristic Oa and on either side of it in the forms:

in region 1, $\Phi = -A\eta\theta - \pi B\zeta^2\theta^{11/6}\,F(\tfrac{13}{12},\ \tfrac{19}{12};\ 3;\ \zeta),$

in region 2, $\Phi = -A\eta\theta - \pi C\zeta^2\theta^{11/6}\,F(\tfrac{13}{12},\ \tfrac{19}{12};\ 3;\ \zeta),$ (113.2)

where A, B, C are constants (which we shall show to be positive), and $\zeta \equiv 1 - 4\eta^3/9\theta^2$; on the characteristics, $\zeta = 0$.

A second characteristic in the hodograph plane (Ob in Fig. 108a) corresponds to the weak discontinuity reflected from the transition line. The form of the function Φ near this characteristic is obtained by analytical continuation of the functions (113.2), using (110.11)–(110.13). For $k = 11/12$, however, the function F_1 is meaningless, and therefore we cannot use these formulae directly. Instead, we must first put $k = (11/12) + \epsilon$, and then let ϵ tend to zero. Logarithmic terms then appear, in accordance with the general theory of hypergeometric functions.

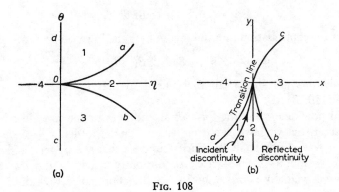

Fig. 108

The calculation with (110.13) gives the following expression for the function Φ in region 3 near the characteristic Ob (we retain terms up to the second order in ζ):

$$\Phi = -A\theta\eta + B(-\theta)^{11/6}\{\zeta^2 \log|\zeta| - 108 + 41{\cdot}1\zeta + 4{\cdot}86\zeta^2\};   (113.3)$$

in determining the nature of the singularity in the velocity distribution at the reflected discontinuity, only the logarithmic term is actually important (see below).

A similar transformation (using formula (110.11)) of the function Φ in region 1 from the neighbourhood of the characteristic Oa to that of Ob gives an expression similar to (113.3), with $\tfrac{1}{2}C$ in place of B. The condition that the co-ordinates x and y are continuous at the characteristic Ob therefore gives

$$C = 2B.   (113.4)$$

Next, we must verify that the Jacobian Δ (110.5) is positive, since it must not vanish anywhere. Near the characteristic Oa, Δ can be calculated from the functions (113.2), and is easily seen to be positive; the leading term in Δ is A^2. Near the characteristic Ob, a calculation using (113.3) gives as the leading term in Δ

$$\Delta = -16(3/2)^{1/6} AB\eta^{1/4} \log|\zeta|.$$

As we approach the characteristic, the logarithm tends to $-\infty$. The condition $\Delta > 0$ therefore gives $AB > 0$, i.e. A and B must have the same sign.

Finally, to determine the form of the transition line, we need an expression for Φ near the upper and lower halves of the axis $\eta = 0$. An expression valid near the upper half is obtained by simply transforming the hypergeometric function in Φ (113.2) into hypergeometric functions of argument $1 - \zeta = 4\eta^3/9\theta^2$, which vanishes for $\eta = 0$. On calculating the numerical values of the coefficients in this transformation and retaining only terms of the lowest degrees in η, we obtain

$$\Phi = -A\eta\theta - 18 \cdot 6B\theta^{11/6}. \tag{113.5}$$

An analytical continuation into the region near the lower half of the axis gives

$$\Phi = -A\eta\theta - 18 \cdot 6 \sqrt{3}B(-\theta)^{11/6}; \tag{113.6}$$

the calculations are similar to those used in deriving the transformation formulae (110.13).

We can now determine the form of all the lines under consideration. On the characteristics we have, omitting terms of higher order, $x = \Phi_\eta = -A\theta$, $y = \Phi_\theta = -A\eta$. We shall arbitrarily suppose that the upper characteristic ($\theta > 0$) corresponds to the weak discontinuity reaching the intersection. Since the gas velocity is in the positive x-direction, this discontinuity is the one which reaches the intersection if it lies in the half-plane $x < 0$. Hence it follows that the constant A, and therefore the constant B also, must be positive. The equation of the line of discontinuity in the physical plane is

$$x = -\tfrac{2}{3}A^{-1/2}(-y)^{3/2}. \tag{113.7}$$

The "reflected" discontinuity, which corresponds to the lower characteristic, is given by the equation

$$x = \tfrac{2}{3}A^{-1/2}(-y)^{3/2}, \tag{113.8}$$

i.e. the two discontinuities are the branches of a semi-cubical parabola with a cusp on the transition line (Fig. 108b, in which the lines and regions are marked in correspondence with those in Fig. 108a).

The equation of the transition line is obtained from the functions (113.5) and (113.6). Effecting the differentiation with respect to η and θ, and then putting $\eta = 0$, we obtain from (113.5) the equation of the part for which $\theta > 0$: $x = -A\theta$, $y = -\tfrac{11}{6} \cdot 19 \cdot 6B\theta^{5/6}$, whence

$$y = -36 \cdot 0B(-x/A)^{5/6}. \tag{113.9}$$

This is the lower part of the transition line in Fig. 108b. Similarly, we obtain from (113.6) the equation of the upper part of this line:

$$y = \sqrt{3} \cdot 36 \cdot 0 B \, (x/A)^{5/6}. \tag{113.10}$$

Thus both discontinuities and both branches of the transition line have a common tangent (the y-axis) at the point of intersection O. Near this point the two branches of the transition line are on opposite sides of the y-axis.

On the discontinuity which reaches O, the spatial derivatives of the velocity are discontinuous; as a characteristic quantity we may consider the discontinuity of the derivative $(\partial\eta/\partial x)_y$. Using the fact that

$$\left(\frac{\partial\eta}{\partial x}\right)_y = \frac{\partial(\eta,y)}{\partial(x,y)} = \frac{\partial(\eta,y)}{\partial(\eta,\theta)} \Big/ \frac{\partial(x,y)}{\partial(\eta,\theta)} = -\frac{\Phi_{\theta\theta}}{\Delta}$$

and formulae (113.2), (113.4), we obtain

$$[(\partial\eta/\partial x)_y]_2^1 = 26 \cdot 9 B\eta^{-1/4}/A^2 = 29 \cdot 9 B) - y)^{-1/4}/A^{7/4}. \tag{113.11}$$

Thus this discontinuity increases as $|y|^{-1/4}$ as we approach the point of intersection.

On the reflected weak discontinuity, the derivatives of the velocity are not discontinuous, but the velocity distribution has a very curious singularity. Calculating the co-ordinates $x = \Phi_\eta$ and $y = \Phi_\theta$ as functions of η, θ from (113.3) (keeping only the first term in the braces), we can put the dependence of η on x for given y near the reflected discontinuity in the parametric form

$$\eta = \frac{|y|}{A} + \frac{x-x_0}{2\sqrt{(A|y|)}} - \frac{1}{6A}|y|\zeta,$$

$$\tag{113.12}$$

$$x - x_0 = \frac{1}{3\sqrt{A}}|y|^{3/2}\zeta - 5 \cdot 7\frac{B|y|^{7/4}}{A^{7/4}}\zeta \log|\zeta|,$$

where ζ is the parameter and $x_0 = x_0(y)$ is the equation of the discontinuity in the physical plane.

The Euler–Tricomi equation can also be used in the problem of whether a shock wave can terminate at its intersection with the transition line (the point O in Fig. 109a, which the shock wave reaches†). Near such a point the shock wave is weak, i.e. the flow is transonic. The Euler–Tricomi equation, however, apparently has no solution which could describe such a flow and satisfy all the necessary conditions at the shock. Nor, apparently, do there exist solutions which would correspond to the termination of both the shock wave and the transition line at their point of intersection (Fig. 109b; the shock wave does not then literally reach the point O, since the flow on one side of it is subsonic). This means that the shock wave must either go to

---

† The "origin" of the shock wave may be at any point in a supersonic flow, and its properties can be investigated without especial difficulty; cf. the end of §107.

infinity or (if it was formed in a local supersonic region; cf. the beginning of §112) must curve as shown in Fig. 109c so as to be "leaving" with respect to

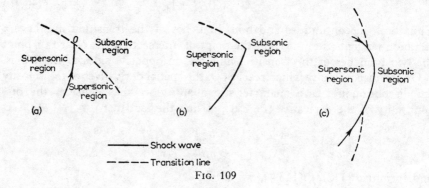

FIG. 109

both ends of itself. The transition line can terminate at a point of intersection with a shock wave, but the intensity of the latter does not vanish there.

CHAPTER XIII

FLOW PAST FINITE BODIES

§114. The formation of shock waves in supersonic flow past bodies

SIMPLE arguments show that, in supersonic flow past an arbitrary body, a shock wave must be formed in front of the body. For the disturbances in the supersonic flow caused by the presence of the body are propagated only downstream. Hence a uniform supersonic stream incident on the body would be unperturbed as far as the leading end of the body. The normal component of the gas velocity would then be non-zero at the surface there, in contradiction to the necessary boundary condition. The resolution of this difficulty can only be the occurrence of a shock wave, as a result of which the gas flow between it and the leading end of the body becomes subsonic.

Thus a shock wave is formed in front of the body when the incident flow is supersonic, and this shock does not, in general, touch the body; it is often called the *bow wave*. In front of the shock wave, the flow is uniform; behind it, the flow is modified and bends round the body (Fig. 110a). The surface

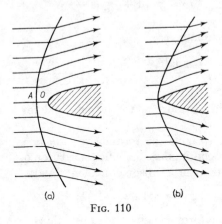

(a)   (b)

FIG. 110

of the shock wave extends to infinity, and at great distances from the body, where the shock is weak, it intersects the incident streamlines at an angle approaching the Mach angle.

The shock wave can touch the body only when the leading end of the latter is pointed. The surface of discontinuity then has a point at the same place (Fig. 110b); it must be borne in mind that, in asymmetric flow, part of this surface may be a weak discontinuity.

For a body of a given shape, however, this type of flow pattern is possible only for velocities exceeding a certain limit; at lower velocities, the shock wave is "detached" from the leading end of the body (see §§104, 105), even if the latter is pointed.

Let us consider supersonic flow past a solid of revolution (in a direction parallel to its axis), and determine the gas pressure at the rounded leading end of the body (the stagnation point O in Fig. 110a). It is evident from symmetry that the streamline which terminates at O intersects the shock wave at right angles, so that the velocity component at A normal to the surface of discontinuity is the same as the total velocity. The values of quantities in the incident stream will be denoted, as usual, by the suffix 1, and the values behind the shock wave at the point A by the suffix 2. The latter are determined at once from formulae (85.7) and (85.8):

$$p_2 = p_1[2\gamma M_1^2 - (\gamma - 1)]/(\gamma + 1),$$

$$v_2 = c_1 \frac{2 + (\gamma - 1)M_1^2}{(\gamma + 1)M_1}, \qquad \rho_2 = \rho_1 \frac{(\gamma + 1)M_1^2}{2 + (\gamma - 1)M_1^2}.$$

The pressure p_0 at the point O (where the gas velocity $v = 0$) can now be obtained by means of the formulae which give the variation of quantities along a streamline. We have (see §80, Problem)

$$p_0 = p_2 \left[1 + \frac{\gamma - 1}{2} \frac{v_2^2}{c_2^2} \right]^{\gamma/(\gamma - 1)},$$

and a simple calculation gives

$$p_0 = p_1 \left(\frac{\gamma + 1}{2} \right)^{(\gamma + 1)/(\gamma - 1)} \frac{M_1^2}{[\gamma - (\gamma - 1)/2M_1^2]^{1/(\gamma - 1)}}. \qquad (114.1)$$

This determines the pressure at the leading end for a supersonic incident flow ($M_1 > 1$).

For comparison, we give the formula for the pressure at the stagnation point obtained for a continuous adiabatic retardation of the gas, with no shock wave (as would be true for a subsonic incident flow):

$$p_0 = p_1[1 + \tfrac{1}{2}(\gamma - 1)M_1^2]^{\gamma/(\gamma - 1)}. \qquad (114.2)$$

For $M_1 = 1$, the two formulae give the same value of p_0, but for $M_1 > 1$ the pressure given by formula (114.2) is always greater than the true pressure p_0 given by formula (114.1).[†]

[†] This statement is true generally, and does not depend on the assumption of a perfect gas. For, when a shock wave is present, the entropy s_0 of the gas at the point O is greater than s_1, whereas if the shock wave were absent s_0 would be equal to s_1. The heat function is in either case $w_0 = w_1 + \tfrac{1}{2}v_1^2$, since the quantity $w + \tfrac{1}{2}v^2$ is unchanged when a streamline intersects a normal compression discontinuity. From the thermodynamic identity $dw = T ds + dp/\rho$, it therefore follows that the derivative $(\partial p/\partial s)_w = -\rho T < 0$, i.e. an increase in entropy when w remains constant involves a decrease in pressure, whence the result follows.

In the limit of very large velocities ($M_1 \gg 1$), formula (114.1) gives

$$p_0 = p_1\left(\frac{\gamma+1}{2}\right)^{(\gamma+1)/(\gamma-1)} \gamma^{-1/(\gamma-1)} M_1^2, \qquad (114.3)$$

i.e. the pressure p_0 is proportional to the square of the incident velocity. From this result we can conclude that the total drag force on the body at velocities large compared with that of sound is proportional to the square of the velocity. It should be noticed that this is the same as the law governing the drag force at velocities small compared with that of sound but yet so large that the Reynolds number is large (see §45).

No complete investigation has yet been made of the basic properties of supersonic flow past arbitrary bodies. Besides the fact that shock waves must be formed, we can also say that there must be two successive shock waves at large distances from the body (L. LANDAU 1945). For the disturbances caused by the body at large distances are small, and can therefore be regarded as a cylindrical sound wave outgoing from the x-axis (which passes through the body parallel to the direction of flow); considering the flow, as usual, in a co-ordinate system where the body is at rest, we have a wave in which the time is represented by x/v_1, and the rate of propagation by $c_1/\sqrt{(M_1^2-1)}$ (see §115). We can therefore apply immediately the results obtained in §95 for a cylindrical wave at large distances from the source. We thus arrive at the following pattern of shock waves far from the body: in the first shock, the pressure increases discontinuously, so that behind it there is a condensation; then follows a region where the pressure gradually decreases into a rarefaction, after which the pressure again increases discontinuously in the second shock. The intensity of the leading shock decreases as $r^{-3/4}$ with increasing distance from the body, and the distance between the two shocks increases as $r^{1/4}$.

Let us now examine the appearance and development of the shock waves as the number M_1 gradually increases. A supersonic region first appears for some value of M_1 less than unity, as a region adjoining the surface of the body. At least one shock wave occurs in this region. It is not known, however, whether this shock must appear as soon as the supersonic region is formed, or whether it appears for some greater value of M_1 (still less than unity). It is also unknown whether the shock leaves the surface of the body when it is first formed (and is still very weak), or begins at some distance away. At the boundary of the supersonic zone the shock wave terminates, of course; no investigation has yet been made of the properties of the shock wave near the point where it terminates (as already mentioned at the beginning of §112).

As M_1 increases, the supersonic region expands, and the length of the shocks wave increases; for $M_1 = 1$ it reaches infinity. This is the shock wave whose existence for $M_1 = 1$ has been demonstrated (for the two-dimensional case) in §112; it follows also that the shock wave must first appear for $M_1 < 1$.

As soon as M_1 exceeds unity, another shock wave appears, the bow wave, which intersects the whole of the infinitely wide incident stream of gas. For M_1 exactly unity, the flow in front of the body is entirely subsonic (§112). For $M_1 > 1$ but arbitrarily close to unity, therefore, the supersonic part of the incident stream, and consequently the bow wave, are arbitrarily far in front of the body. As M_1 increases further, the bow wave gradually approaches the body.

§115. Supersonic flow past a pointed body

The shape which a body must have in order to be streamlined in supersonic flow, i.e. to be subject to as small a drag force as possible, is quite different from the corresponding shape for subsonic flow. We may recall that, in the subsonic case, streamlined bodies are those which are elongated, rounded in front, and pointed behind. In supersonic flow past such a body, however, a strong shock wave would be formed in front of it, leading to a considerable increase in the drag. In the supersonic case, therefore, a long streamlined body must be pointed at both ends, and the angle of the point must be small; if the body is inclined to the direction of flow, the angle between them (angle of attack) must also be small.

In steady supersonic flow past a body of this shape, the gas velocity is nowhere very different in magnitude or direction from the incident velocity, even near the body, and the shock waves formed are weak; the intensity of the bow wave decreases with the angle at the front of the body. Far from the body, the gas flow consists of outgoing sound waves. The main part of the drag can be regarded as due to the conversion of kinetic energy of the moving body into the energy of the sound waves which it emits. This drag, which occurs only in supersonic flow, is called *wave drag*;[†] it can be calculated in a general form valid for any cross-section of the body (T. VON KÁRMÁN 1936).

The nature of the flow just described makes it possible to use the linearised equation (106.4) for the potential:

$$\frac{\partial^2 \phi}{\partial y^2} + \frac{\partial^2 \phi}{\partial z^2} - \beta^2 \frac{\partial^2 \phi}{\partial x^2} = 0, \tag{115.1}$$

where we have introduced for brevity the positive constant

$$\beta^2 = (v_1^2 - c_1^2)/c_1^2; \tag{115.2}$$

the x-axis is in the direction of the flow, the suffix 1 denotes quantities pertaining to the incident stream, and $1/\beta$ is just the tangent of the Mach angle.

Equation (115.1) is formally identical with the two-dimensional wave equation with x/v_1 representing the time and v_1/β the velocity of propagation of the waves. This is no accident; the physical significance is that the gas

---

† The total drag is obtained by adding to the wave drag the forces due to friction and to separation at the trailing end of the body.

flow far from the body consists, as already mentioned, of outgoing sound waves "emitted" by the body. If the gas at infinity is regarded as being at rest, and the body as being in motion, the cross-section of the body at a given point in space will vary with time, and the distance to which a disturbance is propagated at time t (i.e. the distance to the Mach cone) will increase as $v_1 t/\beta$. Thus we shall have a two-dimensional emission of sound (propagated with velocity v_1/β) by the variable profile.

Using this "sonic analogy" as a guide, we can immediately write down the required expression for the velocity potential of the gas, using formula (73.15) for the potential of cylindrical sound waves emitted from a source (at distances large compared with the dimension of the source) and replacing ct by x/β.

Let $S(x)$ be the area of the cross-section of the body in a plane perpendicular to the direction of flow (the x-axis), and l the length of the body in that direction; we take the origin at the leading end of the body. Then

$$\phi(x, r) = -\frac{v_1}{2\pi} \int_0^{x-\beta r} \frac{S'(\xi)d\xi}{\sqrt{[(x-\xi)^2 - \beta^2 r^2]}}; \tag{115.3}$$

the lower limit is taken as zero, since $S(x) \equiv 0$ for $x < 0$ (and for $x > l$).

Thus we have completely determined the gas flow at distances r from the axis which are large compared with the thickness of the body.† Disturbances leaving the body in a supersonic flow are, of course, propagated only into the region behind the cone $x - \beta r = 0$, whose vertex is at the leading end of the body; in front of this cone we have simply $\phi = 0$ (uniform flow). Between the cones $x - \beta r = 0$ and $x - \beta r = l$, the potential is determined by formula (115.3); behind the latter cone (whose vertex is at the trailing end of the body) the upper limit of the integral in (115.3) is evidently the constant l. Both these cones are weak discontinuities, in the approximation considered; in reality, they are weak shock waves.

The drag force acting on the body is just the x-component of the momentum carried away by the sound waves per unit time. We take a cylindrical surface of large radius r and axis along the x-axis. The x-component of the momentum flux density through this surface is $\Pi_{xr} = \rho v_r(v_x + v_1)$ $\simeq \rho_1(\partial\phi/\partial r)(v_1 + \partial\phi/\partial x)$. On integration over the whole surface, the first term gives zero, since the integral of ρv_r is the total mass flux through the surface, which is zero. Thus

$$F_x = -2\pi r \int_{-\infty}^{\infty} \Pi_{xr}\, dx = -2\pi r \rho_1 \int_{-\infty}^{\infty} \frac{\partial\phi}{\partial r}\frac{\partial\phi}{\partial x} dx. \tag{115.4}$$

At large distances (in the "wave region"), the derivatives of the potential can

---

† For axial flow past an axially symmetric body, formula (115.3) is valid for all r up to the surface of the body. In particular, we can derive from it formula (105.6) for flow past a narrow cone.

be calculated as in §73 (see formula (73.17)), and we have

$$\frac{\partial \phi}{\partial r} = -\frac{1}{\beta}\frac{\partial \phi}{\partial x} = \frac{v_1}{2\pi}\sqrt{\frac{\beta}{2r}}\int_0^{x-\beta r}\frac{S''(\xi)d\xi}{\sqrt{(x-\xi-\beta r)}}.$$

This expression is substituted in (115.4), and the squared integral is written as a double integral; putting for brevity $x - \beta r = X$, we obtain

$$F_x = \frac{\rho_1 v_1{}^2}{4\pi}\int_{-\infty}^{\infty}\int_0^X\int_0^X\frac{S''(\xi_1)S''(\xi_2)d\xi_1\,d\xi_2\,dX}{\sqrt{[(X-\xi_1)(X-\xi_2)]}}.$$

The integration over X can be effected; after changing the order of integration, the integral is from the greater of ξ_1 and ξ_2 to infinity. We first take as the upper limit a large but finite quantity L, which later tends to infinity. Thus

$$F_x = -\frac{\rho_1 v_1{}^2}{2\pi}\int_0^l\int_0^{\xi_2}S''(\xi_1)S''(\xi_2)[\log(\xi_2-\xi_1)-\log 4L]d\xi_1\,d\xi_2.$$

The integral of the term containing the constant factor $\log 4L$ is zero, since not only the area $S(x)$ but also its derivative $S'(x)$ vanishes at the pointed ends of the body. We therefore have

$$F_x = -\frac{\rho_1 v_1{}^2}{2\pi}\int_0^l\int_0^{\xi_2}S''(\xi_1)S''(\xi_2)\log(\xi_2-\xi_1)d\xi_1\,d\xi_2,$$

or

$$F_x = -\frac{\rho_1 v_1{}^2}{4\pi}\int_0^l\int_0^l S''(\xi_1)S''(\xi_2)\log|\xi_2-\xi_1|d\xi_1\,d\xi_2. \tag{115.5}$$

This is the required formula for the wave drag on a thin pointed body.† The order of magnitude of the integral is $(S/l^2)^2 l^2$, where S is some mean cross-sectional area of the body. Hence $F_x \sim \rho_1 v_1{}^2 S^2/l^2$. The drag coefficient for an elongated body may be conventionally defined, in terms of the square of the length, as $C_x = F_x/\tfrac{1}{2}\rho_1 v_1{}^2 l^2$. Then, in this case

$$C_x \sim S^2/l^4; \tag{115.6}$$

it is proportional to the square of the cross-sectional area.

---

† The lift (for a body not axially symmetric or a non-zero angle of attack) is zero in the approximation here considered.

We may point out the complete formal analogy between formula (115.5) and formula (47.4) for the induced drag on a thin wing; the function $\Gamma(z)$ in (47.4) is here replaced by the function $v_1 S'(x)$. On account of this analogy we can use, to calculate the integral in (115.5), the method described at the end of §47.

It should also be noticed that the wave drag given by formula (115.5) is unchanged if the direction of flow is reversed: the integral is independent of the direction in which the body extends. This property of the drag force is characteristic of the linearised theory.†

Finally, let us briefly discuss the range of applicability of this formula. This subject may be approached as follows. The amplitude of oscillation of the gas particles in the sound waves "emitted" by the body is of the order of magnitude of the thickness of the body, which we denote by δ. The velocity of the oscillations is accordingly of the order of the ratio $\delta : (l/v_1)$ of the amplitude δ to the period l/v_1 of the wave. The linear approximation for the propagation of sound waves (i.e. the linearised equation for the potential), however, always requires that the gas velocity should be small compared with the velocity of sound, i.e. we must have $v_1/\beta \gg v_1\delta/l$, or, what is in practice the same,

$$M_1 \ll l/\delta. \tag{115.7}$$

Thus the theory given above becomes inapplicable for values of M_1 comparable with the ratio of length to thickness of the body.

It is also inapplicable, of course, in the opposite limiting case where M_1 is close to unity and the linearisation of the equations is invalid.

PROBLEM

Determine the form of the elongated solid of revolution which experiences the smallest drag for a given volume V and length l.

SOLUTION. On account of the analogy mentioned in the text, we introduce a variable θ such that $x = \frac{1}{2}l(1-\cos\theta)$ ($0 \leqslant \theta \leqslant \pi$; the origin of x is at the leading end of the body); and write the function $f(x) = S'(x)$ as

$$f = -l\sum_{n=2}^{\infty} A_n \sin n\theta;$$

the condition $S = 0$ for $x = 0$ and l means that only terms with $n \geqslant 2$ can appear in the sum. The drag coefficient is then

$$C_x = \tfrac{1}{4}\pi \sum_{n=2}^{\infty} n A_n^2.$$

The area $S(x)$ and the total volume V of the body are calculated from the function $f(x)$ as

$$S = \int_0^x f(x)\,dx, \qquad V = \int_0^l S(x)\,dx.$$

---

† It also holds in the theory of the wave drag on thin wings given in §117.

A simple calculation gives $V = \pi l^3 A_2/16$, i.e. the volume is determined by the coefficient A_2 alone. The minimum F_x is therefore reached if $A_n = 0$ for $n \geqslant 3$. The result is

$$C_{x,\min} = (128/\pi)(V/l^3)^2 = (9\pi/2)(S_{\max}/l^2)^2.$$

The cross-sectional area of the body is $S = \frac{1}{4}l^2 A_2 \sin^3\theta$, and the radius as a function of x is therefore $R(x) = \sqrt{2}(8/\pi)(V/3l^4)^{1/2}[x(l-x)]^{3/4}$. The body is symmetrical about the plane $x = \frac{1}{2}l.$†

§116. Subsonic flow past a thin wing

Let us consider subsonic flow of a gas past a thin streamlined wing. As for an incompressible fluid, a wing which is streamlined for subsonic flow must be thin, pointed at the trailing edge, and rounded at the leading edge, and the angle of attack must be small. We take the direction of flow as the x-axis and the direction of the span as the z-axis.

The gas velocity nowhere‡ differs greatly from the velocity $\mathbf{v}_1$ of the incident stream, so that we can use the linearised equation (106.4) for the potential:

$$(1-M_1{}^2)\frac{\partial^2\phi}{\partial x^2} + \frac{\partial^2\phi}{\partial y^2} + \frac{\partial^2\phi}{\partial z^2} = 0. \tag{116.1}$$

At the surface of the wing (which we call C), the velocity must be tangential; introducing a unit vector $\mathbf{n}$ along the normal to the surface, we can write this condition as

$$\left(v_1 + \frac{\partial\phi}{\partial x}\right)n_x + \frac{\partial\phi}{\partial y}n_y + \frac{\partial\phi}{\partial z}n_z = 0.$$

Since the wing is flattened and the angle of attack is small, the normal $\mathbf{n}$ is almost parallel to the y-axis, so that $|n_y|$ is almost unity, while n_x and n_z are small. We can therefore neglect the second-order terms $n_x\partial\phi/\partial x$ and $n_z\partial\phi/\partial z$, and replace n_y by ± 1 ($+1$ on the upper surface of the wing and -1 on the lower surface). Thus the boundary condition on equation (116.1) is

$$v_1 n_x \pm \partial\phi/\partial y = 0. \tag{116.2}$$

Since the wing is assumed thin, $\partial\phi/\partial y$ on its surface can be taken as the limiting value for $y \to 0$.

The solution of equation (116.1) with the condition (116.2) can easily be reduced to the solution of a problem of incompressible flow. To do so, we use instead of the co-ordinates x, y, z the variables

$$x' = x, \qquad y' = y\sqrt{(1-M_1{}^2)}, \qquad z' = z\sqrt{(1-M_1{}^2)}. \tag{116.3}$$

In these variables, equation (116.1) becomes

$$\frac{\partial^2\phi}{\partial x'^2} + \frac{\partial^2\phi}{\partial y'^2} + \frac{\partial^2\phi}{\partial z'^2} = 0, \tag{116.4}$$

---

† Although $R(x)$ vanishes at the ends of the body, $R'(x)$ becomes infinite, i.e. the body is not pointed; the approximation underlying the method is therefore not strictly applicable near the ends.
‡ Except in a small region near the leading edge of the wing, where there is a stagnation line.

i.e. Laplace's equation. The surface of the body is replaced by another, C', obtained by leaving unchanged the profiles of cross-sections by planes parallel to the xy-plane, but reducing in the ratio $\sqrt{(1-M_1^2)}$ all dimensions in the direction of the span (the z-direction).

The boundary condition (116.2) then becomes

$$v_1 n_x \pm \frac{\partial\phi}{\partial y'}\sqrt{(1-M_1^2)} = 0,$$

and it can be reduced to the previous form by introducing in place of ϕ a new potential ϕ':

$$\phi' = \phi\sqrt{(1-M_1^2)}. \tag{116.5}$$

We then have for ϕ' Laplace's equation with the boundary condition

$$v_1 n_x \pm \partial\phi'/\partial y' = 0, \tag{116.6}$$

which must be satisfied for $y' = 0$.

Equation (116.4) with the boundary condition (116.6) is, however, the equation which must be satisfied by the velocity potential of an incompressible fluid flowing past the surface C'. Thus the problem of determining the velocity distribution in compressible flow past a wing with surface C is equivalent to that of finding the velocity distribution in incompressible flow past a wing with surface C'.

Next, let us consider the lift force F_y acting on the wing. First of all, we note that the derivation of Zhukovskii's formula (37.4) given in §37 is entirely valid for a compressible fluid, since the variable density ρ of the fluid can be replaced in that approximation by a constant ρ_1. Thus

$$F_y = -\rho_1 v_1 \int \Gamma \, dz, \tag{116.7}$$

where the integration is taken along the span l_z of the wing. From the relation (116.5) and the equality of the transverse profiles of the wings C and C' it follows that the velocity circulation Γ in compressible flow past the wing C is related to the circulation Γ' in incompressible flow past the wing C' by

$$\Gamma' = \Gamma\sqrt{(1-M_1^2)}. \tag{116.8}$$

Substituting this in (116.7) and changing to an integration over z', we obtain

$$F_y = -\rho_1 v_1 \int \Gamma' \, dz'/(1-M_1^2).$$

The numerator is the lift force on the wing C' in an incompressible fluid. Denoting it by F'_y, we have

$$F_y = F'_y/(1-M_1^2). \tag{116.9}$$

Introducing the lift coefficients

$$C_y = F_y/\tfrac{1}{2}\rho_1 v_1^2 l_x l_z, \qquad C'_y = F'_y/\tfrac{1}{2}\rho_1 v_1^2 l_x l_z'$$

(where l_x, l_z and l_x, $l_z' = l_z\sqrt{(1-M_1^2)}$ are the lengths of the wings C and C' in the x and z directions), we can rewrite this equation as

$$C_y = C'_y/\sqrt{(1-M_1^2)}. \tag{116.10}$$

For wings of large span (and constant profile), the lift coefficient in an incompressible fluid is proportional to the angle of attack, and does not depend on the length or width of the wing:

$$C'_y = \text{constant} \times \alpha, \tag{116.11}$$

where the constant depends only on the shape of the profile (see §46). In this case, therefore, (116.10) can be replaced by

$$C_y = C_y^{(0)}/\sqrt{(1-M_1^2)}, \tag{116.12}$$

where C_y and $C_y^{(0)}$ are the lift coefficients for the same wing in compressible and incompressible fluids respectively. Thus we have the rule that the lift force acting on a long wing in a compressible fluid is $1/\sqrt{(1-M_1^2)}$ times that on the same wing (at the same angle of attack) in an incompressible fluid (L. Prandtl 1922, H. Glauert 1928).

Similar relations can be obtained for the drag force. Together with Zhukovskiĭ's formula for the lift force, formula (47.4) for the induced drag on a wing is also entirely applicable to compressible flow. Effecting the same transformations (116.3) and (116.8), we obtain

$$F_x = F'_x/\sqrt{(1-M_1^2)}, \tag{116.13}$$

where F'_x is the drag on the wing C' in an incompressible fluid. When the span increases, the induced drag tends to a constant limit (§47). For sufficiently long wings we can therefore replace F'_x by $F_x^{(0)}$ (the drag in an incompressible fluid for the wing C). Then the drag coefficient is

$$C_x = C_x^{(0)}/(1-M_1^2). \tag{116.14}$$

Comparing this with (116.12), we see that the ratio C_y^2/C_x is the same for compressible and incompressible fluids.

All the results given here are, of course, invalid for values of M_1 close to unity, since the linearised theory then becomes inapplicable.

§117. Supersonic flow past a wing

If a wing is streamlined in a supersonic stream, it must be pointed at both ends, like the thin bodies discussed in §115.

Here we shall consider only the flow past a thin wing of very large span, the profile being constant along the span. Regarding the span as infinite, we have a two-dimensional gas flow (in the xy-plane). Instead of equation

(115.1), we now have for the potential the equation

$$\frac{\partial^2 \phi}{\partial y^2} - \beta^2 \frac{\partial^2 \phi}{\partial x^2} = 0, \qquad (117.1)$$

with the boundary condition

$$[\partial\phi/\partial y]_{y\to\pm 0} = \mp v_1 n_x, \qquad (117.2)$$

where the signs $\mp$ on the right relate to the upper and lower surfaces of the wing respectively. Equation (117.1) is a one-dimensional wave equation, and its general solution is of the form $\phi = f_1(x-\beta y) + f_2(x+\beta y)$. The fact that disturbances which affect the flow start from the body means that above

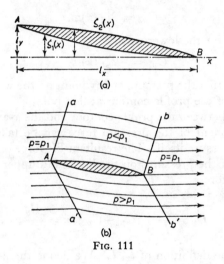

(a)

(b)

Fig. 111

the wing $(y > 0)$ we must have $f_2 \equiv 0$, so that $\phi = f_1(x-\beta y)$, and below the wing $(y < 0)$ $\phi = f_2(x+\beta y)$. For definiteness, we shall consider the region above the wing, where $\phi = f(x-\beta y)$. The function f is determined from the condition (117.2) by putting $n_x \approx -\zeta_2'(x)$, where $y = \zeta_2(x)$ is the equation of the upper part of the wing profile (Fig. 111a). We have $[\partial\phi/\partial y]_{y\to +0} = -\beta f'(x) = v_1\zeta_2'(x)$, whence $f = -v_1\zeta_2(x)/\beta$. Thus the velocity distribution for $y > 0$ is given by the potential

$$\phi(x, y) = -(v_1/\beta)\zeta_2(x-\beta y). \qquad (117.3)$$

Similarly we obtain, for $y < 0$, $\phi = (v_1/\beta)\zeta_1(x+\beta y)$, where $y = \zeta_1(x)$ is the equation of the lower part of the profile. It should be noticed that the potential, and therefore the other quantities, are constant along the straight lines $x \pm \beta y = $ constant (the characteristics), in accordance with the results of §107, of which the solution just obtained is a particular case.

The flow pattern is qualitatively as follows. Weak discontinuities (aAa' and bBb' in Fig. 111b) leave the pointed leading and trailing edges.† In the regions in front of the discontinuity aAa' and behind bBb' the flow is uniform, but between them it is turned so as to go round the surface of the wing; the flow here is a simple wave, and in the present linearised approximation the characteristics are all parallel and inclined at the Mach angle of the incident stream.

The pressure distribution is given by the formula $p - p_1 = -\rho_1 v_1\, \partial\phi/\partial x$; the term in $v_y{}^2$ in the general formula (106.5) can here be omitted, since v_x and v_y are of the same order of magnitude. Substituting (117.3) and introducing the *pressure coefficient* C_p, we obtain in the upper half-plane $C_p = (p - p_1)/\tfrac12\rho_1 v_1{}^2 = 2\zeta_2'(x - \beta y)/\beta$. In particular, the pressure coefficient on the upper surface of the wing is

$$C_{p2} = 2\zeta_2'(x)/\beta. \tag{117.4}$$

Similarly, we find for the lower surface

$$C_{p1} = -2\zeta_1'(x)/\beta. \tag{117.5}$$

It should be noted that the pressure at any point on the wing profile depends only on the slope of the profile contour at that point.

Since the angle between the profile contour and the x-axis is always small, the vertical component of the pressure force can be taken, with sufficient accuracy, as the pressure itself. The resultant lift force on the wing is equal to the difference of the pressures on the lower and upper surfaces. The lift coefficient is therefore

$$C_y = \frac{1}{l_x} \int_0^{l_x} (C_{p1} - C_{p2})\mathrm{d}x = \frac{4l_y}{\beta l_x};$$

see Fig. 111 for the definition of l_x, l_y. We define the angle of attack α as the angle between the chord AB through the ends of the profile (Fig. 111) and the x-axis: $\alpha \approx l_y/l_x$, and obtain the following simple formula:

$$C_y = 4\alpha/\sqrt{(\mathrm{M}_1{}^2 - 1)} \tag{117.6}$$

(J. ACKERET 1925). We see that the lift force is determined by the angle of attack, and does not depend on the form of the wing cross-section, unlike what happens for subsonic flow (see formula (48.7)).

Let us next determine the drag force on the wing (i.e. the wave drag, which is of the same nature as that on thin bodies; see §115). To do so, we must take the x-component of the pressure force and integrate over the profile contour.

---

† This statement is valid only in the approximation used here. In reality we have not weak discontinuities but weak shock waves or narrow centred rarefaction waves, depending on the direction in which the velocity is turned by them. For the profile shown in Fig. 111b, for example, Aa and Bb' are rarefaction waves, while Aa' and Bb are shock waves.

The streamline leaving the trailing edge (B in Fig. 111b) is actually a tangential discontinuity of the velocity (which in practice becomes a narrow turbulent wake).

The drag coefficient is then found to be

$$C_x = \frac{2}{\beta l_x} \int_0^{l_x} (\zeta_1'^2 + \zeta_2'^2)\,dx. \tag{117.7}$$

We put $\zeta_1' = \theta_1 - \alpha$, $\zeta_2' = \theta_2 - \alpha$, where $\theta_1(x)$ and $\theta_2(x)$ are the angles between the upper and lower parts of the contour and the chord AB. The integrals of θ_1 and θ_2 are evidently zero, and the result is therefore

$$C_x = [4\alpha^2 + 2(\overline{\theta_1^2} + \overline{\theta_2^2})]/\sqrt{(M_1^2 - 1)}; \tag{117.8}$$

the bar denotes an average with respect to x. For a given angle of attack, the drag coefficient is seen to be least for a wing in the form of a flat plate (for which $\theta_1 = \theta_2 = 0$). In this case $C_x = \alpha C_y$. If we apply formula (117.8) to a rough surface, we find that the roughness may result in a considerable increase in the drag, even if the height of the irregularities is small.† For the drag is independent of the height of the irregularities if the mean slope of the surface, i.e. the mean ratio of the height of the irregularities to the distance between them, remains constant.

Finally, we may make the following remark. Here, as everywhere, when we speak of a wing we imply that its edges are perpendicular to the flow. The generalisation to the case of any angle γ between the direction of flow and the edge (the *angle of yaw*) is quite obvious. It is clear that the forces on an infinite wing of constant cross-section depend only on the component of the incident velocity normal to its edges; in an ideal fluid, the velocity component parallel to the edges does not result in a force. The forces acting on a wing at an angle of yaw other than $\frac{1}{2}\pi$ in a stream with Mach number M_1 are the same as those on the same wing for $\gamma = \frac{1}{2}\pi$ in a stream with Mach number $M_1 \sin \gamma$. In particular, if $M_1 > 1$ but $M_1 \sin \gamma < 1$, the wave drag, which is peculiar to supersonic flow, will not occur.

§118. The law of transonic similarity

The theory of supersonic and subsonic flow past thin bodies developed in §§115–117 is not applicable to transonic flow, when the linearised equation for the potential becomes invalid. In this case the flow pattern in all space is given by the non-linear equation (106.10):

$$2\alpha_* \frac{\partial \phi}{\partial x} \frac{\partial^2 \phi}{\partial x^2} = \frac{\partial^2 \phi}{\partial y^2} + \frac{\partial^2 \phi}{\partial z^2} \tag{118.1}$$

(or, for two-dimensional flow, by the equivalent Euler–Tricomi equation). The solution of these equations for particular cases is very difficult, however. The similarity rules which can be established for such flows, without finding any particular solution, are therefore of great interest.

---

† But nevertheless greater than the thickness of the boundary layer.

Let us first consider two-dimensional flow, and let

$$Y = \delta f(x/l) \qquad (118.2)$$

be the equation which gives the shape of the thin contour past which the flow takes place, l being its length (in the direction of flow) and δ some characteristic thickness ($\delta \ll l$). By varying the two parameters l and δ, we obtain a family of similar contours.

The equation of motion is

$$2\alpha_* \frac{\partial \phi}{\partial x} \frac{\partial^2 \phi}{\partial x^2} = \frac{\partial^2 \phi}{\partial y^2}, \qquad (118.3)$$

with the following boundary conditions. At infinity, the velocity equals the velocity v_1 of the undisturbed stream, i.e.

$$\frac{\partial \phi}{\partial y} = 0, \qquad \frac{\partial \phi}{\partial x} = M_{1*} - 1 = (M_1 - 1)/\alpha_*; \qquad (118.4)$$

see the definition of the potential ϕ, (106.9). On the profile, the velocity must be tangential:

$$v_y/v_x \approx \partial\phi/\partial y = dY/dX = (\delta/l)f'(x/l); \qquad (118.5)$$

since the profile is thin, this condition can be imposed at $y = 0$.

We introduce dimensionless variables thus:

$$x = l\bar{x}, \qquad y = l\bar{y}/(\theta\alpha_*)^{1/3}, \qquad \phi = (l\theta^{2/3}/\alpha_*^{1/3})\bar{\phi}(\bar{x}, \bar{y}); \qquad (118.6)$$

here $\theta = \delta/l$ gives the angular thickness of the wing or angle of attack. Then

$$2\frac{\partial \bar{\phi}}{\partial \bar{x}} \frac{\partial^2 \bar{\phi}}{\partial \bar{x}^2} = \frac{\partial^2 \bar{\phi}}{\partial \bar{y}^2},$$

with the following boundary conditions:

$$\partial\bar{\phi}/\partial\bar{x} = K, \qquad \partial\bar{\phi}/\partial\bar{y} = 0 \text{ at infinity},$$

$$\partial\bar{\phi}/\partial\bar{y} = f'(\bar{x}) \text{ at } \bar{y} = 0,$$

where

$$K = (M_1 - 1)/(\alpha_*\theta)^{2/3}. \qquad (118.7)$$

These conditions contain only one parameter, K. Thus we have obtained the required similarity law: two-dimensional transonic flows with the same value of K are similar, as is shown by formulae (118.6) (S. V. FAL'KOVICH 1947).

It should be noticed that the expression (118.7) involves only a single parameter α_* which characterises the properties of the gas itself. The similarity law therefore determines also the similarity with respect to a change in the gas.

In the approximation here considered, the pressure is given by the formula $p - p_1 \approx -\rho_1 v_1 (v_x - v_1)$. A calculation using the expressions (118.6) shows that the pressure coefficient on the profile is of the form

$$C_p = \frac{p - p_1}{\frac{1}{2}\rho_1 v_1^2} = \frac{\theta^{2/3}}{\alpha_*^{1/3}} P\left(K, \frac{x}{l}\right).$$

The drag and lift coefficients are given by integrals along the contour of the profile:

$$C_x = \frac{1}{l} \oint C_p \frac{dY}{dx} dx,$$

$$C_y = \frac{1}{l} \oint C_p \, dx,$$

and are therefore of the form†

$$C_x = \frac{\theta^{5/3}}{\alpha_*^{1/3}} f_x(K), \qquad C_y = \frac{\theta^{2/3}}{\alpha_*^{1/3}} f_y(K). \tag{118.8}$$

In an entirely similar manner, we can obtain the similarity law for a three-dimensional thin body whose shape is given by equations of the form

$$Y = \delta f_1(x/l), \qquad Z = \delta f_2(x/l), \tag{118.9}$$

with the two parameters δ and l ($\delta \ll l$). There is an important difference from the two-dimensional case, because the potential has a logarithmic singularity for $y \to 0$, $z \to 0$ (see, for instance, the formulae for flow past a narrow cone in §105). Hence the boundary condition at the x-axis must determine, not the derivatives $\partial\phi/\partial y$, $\partial\phi/\partial z$ themselves, but the products $y \, \partial\phi/\partial y = Y \, dY/dx$, $z \, \partial\phi/\partial z = Z \, dZ/dx$, which remain finite. It is easy to see that in this case the similarity transformation is

$$x = l\bar{x}, \qquad y = (l/\theta\alpha_*^{\frac{1}{2}})\bar{y}, \qquad z = (l/\theta\alpha_*^{\frac{1}{2}})\bar{z}, \qquad \phi = l\theta^2\bar{\phi}, \tag{118.10}$$

the similarity parameter being

$$K = (M_1 - 1)/\theta^2\alpha_* \tag{118.11}$$

(T. von Kármán 1947). The pressure coefficient at the surface of the body is found to be of the form $C_p = \theta^2 P(K, x/l)$, and the drag coefficient is accordingly‡

$$C_x = \theta^4 f(K). \tag{118.12}$$

All these formulae hold, of course, for both small positive and small negative values of $M_1 - 1$. If $M_1 = 1$ exactly, the similarity parameter

---

† The range of validity of these formulae is given by the condition $|M_1 - 1| \ll 1$. The linearised theory, however, corresponds to large K, i.e. $|M - 1| \gg \theta^{2/3}$. In the range $1 \gg M_1 - 1 \gg \theta^{2/3}$, formulae (118.8) must therefore become the formulae (117.6)–(117.8) given by the linearised theory. This means that, for large K, the functions f_x and f_y must be proportional to $K^{-1/2}$.

‡ In the range $1 \gg M_1 - 1 \gg \theta^2$, we must obtain the formula (115.6) given by the linearised theory, according to which $C_x \sim \theta^4$; this means that the function $f(K)$ tends to a constant as K increases.

$K = 0$, and the functions in formulae (118.8) and (118.12) reduce to constants, so that these formulae completely determine C_x and C_y as functions of θ and α_*, which represents the properties of the gas.

§119. The law of hypersonic similarity

The linearised theory is invalid for supersonic flow past thin bodies for very large values of the Mach number M_1 (*hypersonic flow*), as has already been mentioned at the end of §106. A simple similarity rule which can be established for this case (H. S. TSIEN 1946) is therefore of interest.

The shock waves formed in such flow are at a small angle to the direction of flow, of the order of the ratio $\theta = \delta/l$ of thickness to length of the body. These shocks are in general curved and also strong; the velocity discontinuity in them is relatively small, but the pressure discontinuity (and therefore the entropy discontinuity) is large. The gas flow is therefore not in general potential flow.

We shall assume that the Mach number M_1 is of the order of $1/\theta$ or greater. A shock wave reduces the local value of M, but the latter always remains of the order of $1/\theta$ (see §104, Problem), so that M is large everywhere.

We use the "sonic analogy" mentioned in §115: a three-dimensional problem of steady flow past a thin body of variable cross-section $S(x)$ is equivalent to a two-dimensional problem of non-steady emission of sound waves by a contour whose area varies with time according to the law $S(v_1 t)$;[†] the velocity of sound is represented by $v_1/\sqrt{(M_1{}^2-1)}$, or, for large M_1, by c_1 simply. It should be emphasised that the only condition necessary for the two problems to be equivalent is that the ratio δ/l should be small; this enables us to regard small annular regions of the surface of the body as cylindrical. For large M_1, however, the rate of propagation of the "emitted" waves is comparable with the velocity of the gas particles in the waves (cf. the end of §115), and the problem therefore has to be solved on the basis of the exact (non-linearised) equations.

In this two-dimensional problem, the linear velocity of the source is of the order of $v_1\theta$; the only other independent parameters of the problem are the velocity of sound c_1, the dimension δ of the source, and the density ρ_1.[‡] From these we can form only one dimensionless combination,

$$K = M_1\theta, \tag{119.1}$$

which is the similarity parameter.[††] The scales of length for the co-ordinates

---

[†] For example, the problem of flow past a narrow cone is equivalent to that of the emission of cylindrical sound waves by a uniformly expanding circular cylinder.

[‡] We are considering, of course, not only the equations of motion of the gas, but also the boundary conditions on them at the surface of the body and the conditions which must be satisfied at the shock waves which are formed. We take the case of a perfect gas, so that the gas-dynamic properties depend only on the dimensionless parameter γ; the similarity rule obtained below, however, does not determine the dependence of the flow on this parameter.

[††] If M_1 is not supposed large, we obtain a similarity rule with parameter $K = \theta\sqrt{(M_1{}^2-1)}$. This is of no interest, however, since for small M_1 the linearised theory determines all quantities as functions of this parameter.

y, z and of time must be taken to have the appropriate dimensions, and be formed from the same parameters, e.g. δ and $\delta/v_1\theta = l/v_1$. Returning to the co-ordinate x, we find that v/c_1 and C_p are functions of the dimensionless variables x/l, y/δ, z/δ and of the parameter K.

The drag coefficient for a thin body is easily found to be of the form

$$C_x = \theta^4 f(K). \tag{119.2}$$

The same similarity law is evidently obtained in the two-dimensional case of flow past a thin wing of infinite span.

The drag and lift coefficients are found to be of the form

$$C_x = \theta^3 f_x(K), \qquad C_y = \theta^2 f_y(K). \tag{119.3}$$

PROBLEM

Determine the lift force on a flat wing of infinite span inclined at a small angle of attack α to the direction of flow, for large values of the Mach number M_1 ($M_1 \gtrsim 1/\alpha$).

SOLUTION. The flow pattern is as shown in Fig. 112: a shock wave and a rarefaction wave leave each of the two edges of the plate, and the stream is turned in them through an angle α in opposite directions.

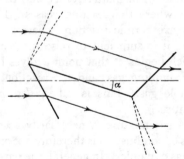

FIG. 112

According to the sonic analogy, the problem of steady flow past such a plate is equivalent to that of non-steady one-dimensional gas flow on each side of a piston moving with uniform velocity αv_1. In front of the piston a shock wave is formed, and behind it a rarefaction wave (see §92, Problems 1 and 2). Using the results there obtained, we find the required lift force as the difference of the pressures on the two sides of the plate. The lift coefficient is

$$C_y = \alpha^2 \left\{ \frac{2}{\gamma K^2} + \frac{\gamma+1}{2} + \sqrt{\left[\frac{4}{K^2} + \left(\frac{\gamma+1}{2} \right)^2 \right]} \right\} - \frac{2\alpha^2}{\gamma K^2} \left[1 - \frac{\gamma-1}{2} K \right]^{2\gamma/\gamma-1},$$

where $K = \alpha M_1$. For $K \geqslant 2/(\gamma-1)$, a vacuum is formed under the plate, and the second term must be omitted. In the range $1 \ll M_1 \ll 1/\alpha$, this formula becomes $C_y = 4\alpha/M_1$, as given by the linearised theory, in accordance with the fact that both procedures are applicable in that range.

CHAPTER XIV

FLUID DYNAMICS OF COMBUSTION

§120. Slow combustion

THE speed of a chemical reaction (measured, say, by the number of molecules reacting in unit time) depends on the temperature of the mixture of gases in which it occurs, increasing with the temperature. In many cases this dependence is very marked.† The speed of the reaction may be so small at ordinary temperatures that the reaction hardly occurs, even though the gas mixture corresponding to a state of thermodynamic (chemical) equilibrium would be one in which the reaction had occurred. When the temperature rises sufficiently, the reaction proceeds rapidly. If it is endothermic, a continuous supply of heat from an external source is necessary for the reaction to be maintained; if the temperature is merely raised at the beginning of the reaction, only a small amount of matter reacts, and thereby reduces the gas temperature to a point where the reaction ceases. The situation is quite different for a strongly exothermic reaction, where a considerable quantity of heat is evolved. Here it is sufficient to raise the temperature at a single point; the reaction which begins at that point evolves heat and so raises the temperature of the surrounding gas, and the reaction, once having begun, will extend to the whole gas. This is called *slow combustion* or simply *combustion* of a gas mixture.‡

The combustion of a gas mixture is necessarily accompanied by motion of the gas. The process of combustion is therefore not only a chemical phenomenon but also one of gas dynamics. In general, the nature of the combustion process has to be determined by a solution of simultaneous equations which include both those of chemical kinetics for the reaction and those of gas dynamics for the mixture concerned.

The situation is much simplified, however, in the very important case (the one usually encountered) where the characteristic dimension l of the problem is large (in a sense to be defined later). We shall see that, in such cases, the problems of gas dynamics and chemical kinetics can be, to a certain extent, considered separately.

The region of burnt gas (i.e. the region where the reaction is over and the

---

† The reaction rate usually depends exponentially on the temperature, being nearly proportional to a factor of the form $e^{-U/RT}$, where U is a constant for any given reaction and is called the *activation energy*. The greater U, the more strongly the reaction rate depends on the temperature.

‡ It should be borne in mind that, in a mixture capable of combustion, the spontaneous propagation of the combustion may be impossible in certain circumstances. This limitation is due to heat losses resulting from such factors as conduction through the walls of a pipe in which combustion occurs, radiation losses, etc. For this reason combustion is not possible in pipes of very small radius, for example.

gas is a mixture of combustion products) is separated from the gas where combustion has not yet begun by a transition layer, where the reaction is in progress (the *combustion zone* or *flame*); in the course of time, this layer moves forward, with a velocity which may be called the velocity of propagation of combustion in the gas. The magnitude of this velocity depends on the amount of heat transfer from the combustion zone to the cold gas mixture. The main mechanism of heat transfer is ordinary conduction. The theory of this means of propagation of combustion was first developed by V. A. MIKHEL'SON (1890).

We denote by δ the order of magnitude of the width of the combustion zone. It is determined by the mean distance over which heat evolved in the reaction is propagated during the time τ for which the reaction lasts (at the point concerned). The time τ is characteristic of the reaction, and depends only on the thermodynamic state of the gas undergoing combustion (and not on the parameter l). If χ is the thermometric conductivity of the gas, we have (see (51.7))†

$$\delta \sim \sqrt{(\chi\tau)}. \tag{120.1}$$

Let us now make more precise the above assumption: we shall assume that the characteristic dimension is large compared with the width of the combustion zone ($l \gg \delta$). When this condition holds, the problem of gas dynamics can be considered separately. In determining the gas flow, we can neglect the width of the combustion zone, regarding it as a surface which separates the combustion products from the unburnt gas. On this surface (the *flame front*) the state of the gas changes discontinuously, i.e. it is a surface of discontinuity.

The velocity v_1 of this discontinuity relative to the gas itself (in a direction normal to the front) is called the *normal velocity* of the flame. In a time τ, the combustion is propagated through a distance of the order of δ, and so the flame velocity is‡

$$v_1 \sim \delta/\tau \sim \sqrt{(\chi/\tau)}. \tag{120.2}$$

The ordinary thermometric conductivity of the gas is of the order of the mean free path of the molecules multiplied by their thermal velocity or, what is the same thing, the mean free time τ_{tr} multiplied by the square of this velocity. Since the thermal velocity of the molecules is of the same order as the velocity of sound, we have $v_1/c \sim \sqrt{(\chi/\tau c^2)} \sim \sqrt{(\tau_{tr}/\tau)}$. Not every collision between molecules results in a chemical reaction between them; on the contrary, only a very small fraction of colliding molecules react. This means that

---

† To avoid misunderstanding, it should be mentioned that, when τ depends markedly on the temperature, a fairly large coefficient should appear in formula (120.1) if τ is the value for the temperature of the combustion products. The important fact for our purposes, however, is that δ does not depend on l.

‡ As an example, it may be mentioned that the flame velocity in a mixture of methane (6 per cent) and air is only 5 cm/sec, whereas in detonating mixture ($2H_2+O_2$) it is 1000 cm/sec; the widths of the combustion zones in these two cases are about 5×10^{-2} cm and 5×10^{-4} cm respectively.

$\tau_{tr} \ll \tau$, and therefore $v_1 \ll c$. Thus the flame velocity is, in this case, small compared with the velocity of sound.†

On the surface of discontinuity which replaces the combustion zone, the fluxes of mass, momentum and energy must be continuous, as at any discontinuity. The first of these conditions, as usual, determines the ratio of the components, normal to the surface, of the gas velocities relative to the discontinuity: $\rho_1 v_1 = \rho_2 v_2$, or

$$v_1/v_2 = V_1/V_2, \tag{120.3}$$

where V_1 and V_2 are the specific volumes of the unburnt gas and the combustion products. According to the general results obtained in §81 for arbitrary discontinuities, the tangential velocity component must be continuous if the normal component is discontinuous. The streamlines are therefore "refracted" at the discontinuity.

On account of the smallness of the normal velocity of the flame relative to that of sound, the condition of continuity of the momentum flux reduces to the continuity of pressure, and that for the energy flux reduces to the continuity of the heat function:

$$p_1 = p_2, \qquad w_1 = w_2. \tag{120.4}$$

In using these conditions, it must be remembered that the gases on the two sides of the discontinuity under consideration are chemically different, and so the thermodynamic quantities are not the same functions of one another.

For a perfect gas we have $w_1 = w_{01} + c_{p1}T_1$, $w_2 = w_{02} + c_{p2}T_2$; the constant terms cannot be put equal to zero as for a single gas (by an appropriate choice of the zero of energy), since w_{01} and w_{02} are different. We put $w_{01} - w_{02} = q$; this is just the heat evolved (per unit mass) in the reaction, if the reaction occurs at a temperature of absolute zero. Then we obtain the following relations between the thermodynamic quantities for the unburnt gas (1) and the burnt gas (2):

$$p_1 = p_2, \qquad T_2 = \frac{q}{c_{p2}} + \frac{c_{p1}}{c_{p2}}T_1, \qquad V_2 = V_1\frac{\gamma_1(\gamma_2-1)}{\gamma_2(\gamma_1-1)}\left(\frac{q}{c_{p1}T_1} + 1\right). \tag{120.5}$$

Since the flame has a definite normal velocity, independent of the gas velocities themselves, the flame front has a definite form for steady combustion in a moving gas. An example is the combustion of gas leaving the end of a tube (a burner outlet). If v is the gas velocity averaged over the cross-section of the tube, it is evident that $v_1 S_1 = vS$, where S is the cross-sectional area of the tube and S_1 the total surface area of the flame front.

If this situation is to be realised, it must be stable with respect to small perturbations, and the question arises of the limits of this stability. The

---

† The diffusion of the components of the burning mixture also has a certain effect on the propagation of combustion; this, however, does not alter the orders of magnitude of the flame velocity and width.

stability of the flame front can be investigated in a similar manner to that of a tangential discontinuity in §30. Since the gas velocity is small compared with that of sound, we can regard the gas as an incompressible ideal fluid, the normal velocity of the flame front being taken as a given constant. Such an investigation (see Problem 1) leads to the result that the flame front is absolutely unstable, and the flame must therefore become turbulent (L. LANDAU 1944). In this form the investigation is valid only for large Reynolds numbers. When the viscosity of the gas is taken into account, however, it cannot here result in a very large critical Reynolds number.

The experimental data, on the other hand, show that the "self-turbulence" of the flame does not occur up to very large Reynolds numbers.[†] This means that there must be other factors which stabilise the flame front and postpone its becoming turbulent until very large Reynolds numbers are reached. It is possible that the change in the normal velocity when the front is deformed is of importance: where the front is concave, v_1 increases (since the heat transfer into the unburnt mixture in the concavity is improved), while where it is convex v_1 is reduced (YA. B. ZEL'DOVICH). This important problem has not yet been resolved.[‡]

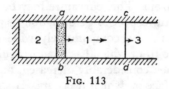

FIG. 113

A flame propagated in a mixture of combustible gases results in a motion of the surrounding gas up to a considerable distance. The fact that a motion of the gas must accompany combustion is evident from the fact that, because of the difference between the velocities v_1 and v_2, the combustion products must move with velocity $v_1 - v_2$ relative to the unburnt gas. In some cases this motion results in the formation of shock waves. These shocks bear no direct relation to the process of combustion, and their occurrence is due to the necessity of satisfying the boundary conditions. Let us consider, for example, combustion propagated from the closed end of a pipe. In Fig. 113, ab is the combustion zone. The gas in regions 1 and 3 is the original unburnt mixture, while that in region 2 consists of combustion products. The velocity v_1 with which the combustion zone moves relative to the gas 1

---

† In spherical propagation of combustion in free space the self-turbulence, if it occurs, does so for $R \sim 10^5$, the characteristic dimension being the radius of the spherical flame (YA. B. ZEL'DOVICH and A. I. ROZLOVSKIĬ 1947). When a gas burns in a pipe, the self-turbulence does not occur, being preceded by the turbulence which appears, for large R, owing to the effect of the walls on the gas flow which accompanies combustion (i.e. owing to the instability of laminar Poiseuille flow).

‡ There are special factors which stabilise the flame in the combustion of a gas evaporating from the surface of a liquid (the reaction occurring in the gas itself, and not with any external substance such as the oxygen in the air). In this case capillary forces and the gravitational field have a stabilising effect (see Problem 2).

in front of it is, from above, determined by the properties of the reaction and the conditions of heat transfer, and must be regarded as given. The velocity v_2 with which the flame moves relative to gas 2 is then determined at once by the condition (120.3). At the closed end of the pipe, the gas velocity must vanish, and so the gas in region 2 will be at rest. Gas 1, therefore, must move relative to the pipe with a constant velocity $v_2 - v_1$. In the forward part of the pipe, far from the flame, the gas is again at rest. This condition can be satisfied only by the presence of a shock wave (*cd* in Fig. 113), in which the gas velocity is discontinuous in such a way that gas 3 is at rest. From the given discontinuity of velocity we can find the discontinuities in the other quantities and the velocity of propagation of the shock itself. Thus we see that the flame front acts as a piston on the gas in front of it. The shock wave moves faster than the flame, so that the mass of gas set in motion increases in the course of time.

For sufficiently large Reynolds numbers, the gas flow which accompanies combustion in a pipe becomes turbulent, and this in turn affects the flame which causes the motion. According to K. I. Shchelkin, the structure of the combustion zone is then as follows. The turbulent eddies which are large compared with the ordinary width δ of the flame result in an irregular curvature of the flame front. This curvature may be considerable, since the stability of the front with respect to deformations is probably small, as mentioned above. The result is a comparatively wide combustion zone, consisting of a narrow flame front which has been irregularly concertina'd. The rate of combustion is then considerably increased, on account of the marked increase in the total surface on which it occurs. It should be noted that this picture is very different from what would occur if the flame were self-turbulent: in that case the combustion zone would be a homogeneous region in which turbulent eddies small compared with the pipe radius would effect thorough mixing.

PROBLEMS

PROBLEM 1. Investigate the stability of a plane flame front (propagated with a velocity small compared with that of sound) with respect to infinitesimal disturbances.

SOLUTION. We take the plane of discontinuity as the yz-plane, with the unperturbed gas velocity in the positive x-direction. On the flow with constant velocities v_1, v_2 (on the two sides of the discontinuity) we superpose a perturbation periodic in the y-direction and in time. From the equations of motion

$$\operatorname{div} \mathbf{v}' = 0, \qquad \partial \mathbf{v}'/\partial t + (\mathbf{v} \cdot \mathbf{grad}) \mathbf{v}' = -(1/\rho) \mathbf{grad} p' \tag{1}$$

($\mathbf{v}$, ρ being either $\mathbf{v}_1$, ρ_1 or $\mathbf{v}_2$, ρ_2), we obtain as in §30 the equation

$$\Delta p' = 0. \tag{2}$$

On the surface of discontinuity (i.e. for $x \approx 0$) the following conditions must be satisfied: the equation of continuity of pressure

$$p'_1 = p'_2, \tag{3}$$

the condition of continuity of the velocity component tangential to the surface

$$v'_{1y} + v_1 \, \partial \zeta / \partial y = v'_{2y} + v_2 \, \partial \zeta / \partial y \tag{4}$$

(where $\zeta(y, t)$ is the small displacement of the surface of discontinuity along the x-axis due to the disturbance), and the condition that the gas velocity normal to the surface of discontinuity is unchanged,

$$v'_{1x} - \partial\zeta/\partial t = v'_{2x} - \partial\zeta/\partial t = 0. \tag{5}$$

In the region $x < 0$ (the unburnt gas 1), the solution of equations (1) and (2) can be written

$$v'_{1x} = Ae^{iky+kx-i\omega t}, \qquad v'_{1y} = iAe^{iky+kx-i\omega t},$$

$$p'_1 = A\rho_1\left(\frac{i\omega}{k} - v_1\right)e^{iky+kx-i\omega t}. \tag{6}$$

In the region $x > 0$ (the combustion products, gas 2), besides the solution of the form constant$\times e^{iky-kx-i\omega t}$, we must take into account another particular solution of equations (1) and (2), in which the dependence on y and t is given by the same factor $e^{iky-i\omega t}$. This solution is obtained by putting $p' = 0$; then the right-hand side of Euler's equation is zero, and the resulting homogeneous equation has a solution in which v'_x and v'_y are proportional to $e^{iky-i\omega t+i\omega x/v}$. The reason why this solution need be taken into consideration only in gas 2, and not in gas 1, is that our ultimate purpose is to determine whether frequencies ω can exist having positive imaginary parts; for such ω, however, the factor $e^{i\omega x/v}$ increases without limit with $|x|$ for $x < 0$, and so such a solution is not possible in region 1. Again choosing appropriate values of the constant coefficients, we seek a solution for $x > 0$ in the form

$$\left.\begin{array}{l} v'_{2x} = Be^{iky-kx-i\omega} + Ce^{iky-i\omega t+i\omega x/v_2}, \\[2mm] v'_{2y} = -iBe^{iky-kx-i\omega} - (\omega/kv_2)Ce^{iky-i\omega t+i\omega x/v_2}, \\[2mm] p'_2 = -B\rho_2[v_2+(i\omega/k)]e^{iky-kx-i\omega t}. \end{array}\right\} \tag{7}$$

Putting also

$$\zeta = De^{iky-i\omega t}, \tag{8}$$

and substituting these expressions in the conditions (3)–(5), we obtain four homogeneous equations for the coefficients A, B, C, D. A simple calculation (using the fact that $j \equiv \rho_1 v_1 = \rho_2 v_2$) gives the following condition for these equations to be compatible:

$$\Omega^2(v_1+v_2) + 2\Omega k v_1 v_2 + k^2 v_1 v_2(v_1-v_2) = 0, \tag{9}$$

where $\Omega = -i\omega$. If $v_1 > v_2$, this equation has either two negative real roots or two complex conjugate roots with negative real parts. If $v_1 < v_2$, however, the roots are real and of opposite signs. Thus, if $v_1 > v_2$, we always have re $\Omega < 0$ and the motion is stable, but if $v_1 < v_2$ there are Ω for which re $\Omega > 0$, and the original motion is unstable. The density ρ_2 of the combustion products is actually always less than the density ρ_1 of the unburnt gas, on account of the considerable heating during combustion. Hence it follows, since $\rho_1 v_1 = \rho_2 v_2$, that $v_2 > v_1$, and we conclude that the flame front is unstable in the conditions considered.

PROBLEM 2. Combustion occurs on the surface of a liquid, the reaction taking place in vapour evaporating from the surface. Determine the stability condition in this case, taking into account the effect of the gravitational field and capillary forces (L. D. LANDAU, 1944).

SOLUTION. Let us consider the combustion zone in vapour near the liquid surface as a surface of discontinuity, but now let this surface have a surface tension α. The calculations are entirely similar to those of Problem 1, the only difference being that, instead of the boundary condition (3), we now have $p'_1 - p'_2 = -\alpha\partial^2\zeta/\partial y^2 + (\rho_1-\rho_2)g\zeta$; medium 1 is the liquid and medium 2 the burnt gas. The conditions (4) and (5) are unchanged. In place of equation (9) we obtain

$$\Omega^2(v_1+v_2) + 2\Omega k v_1 v_2 + \left[k^2(v_1-v_2) + \frac{gk(\rho_1-\rho_2)+\alpha k^3}{j}\right]v_1 v_2 = 0.$$

The stability condition in this case is that the roots of this equation should have negative real parts, i.e. the free term must be positive for all k. This requirement gives the stability condition $j^4 < 4\alpha g\rho_1^2\rho_2^2/(\rho_1-\rho_2)$. Since the density of the gaseous combustion products is small compared with that of the liquid ($\rho_1 \gg \rho_2$), the condition becomes in practice

$$j^4 < 4\alpha g\rho_1\rho_2^2.$$

PROBLEM 3. Determine the temperature distribution in the gas in front of a plane flame.

SOLUTION. In a system of co-ordinates moving with the front the temperature distribution is steady, and the gas moves with velocity $-v_1$. The equation of thermal conduction,

$$\mathbf{v \cdot grad}\, T = -v_1\, dT/dx = \chi\, d^2 T/dx^2,$$

has the solution $T = T_0 e^{-v_1 x/\chi}$, where T_0 is the temperature on the flame front, the temperature far from the front being taken as zero.

§121. Detonation

In the type of combustion (slow combustion) described above, the propagation through the gas is due to the heating which results from the direct transfer of heat from the burning gas to that which is still unburnt. Another entirely different mechanism of propagation of combustion, involving shock waves, is also possible. The shock wave heats the gas as it passes; the gas temperature behind the shock is higher than in front of it. If the shock wave is sufficiently strong, the rise in temperature which it causes may be sufficient for combustion to begin. The shock wave will then "ignite" the gas mixture as it moves, i.e. the combustion will be propagated with the velocity of the shock, or much faster than ordinary combustion. This mechanism of propagation of combustion is called *detonation*.

When the shock wave passes some point in the gas, the reaction begins at that point, and continues until all the gas there is burnt, i.e. for a time τ which characterises the kinetics of the reaction concerned. It is therefore clear that the shock wave will be followed by a layer moving with it in which combustion is occurring, and the width of this layer is equal to the speed of propagation of the shock multiplied by the time τ. It is of importance that the width does not depend on the dimensions of any bodies that are present. When the characteristic dimensions of the problem are sufficiently large, therefore, we can regard the shock wave and the combustion zone following it as a single surface of discontinuity which separates the burnt and unburnt gases. We call such a surface a *detonation wave*.

At a detonation wave the flux densities of mass, energy and momentum must be continuous, and the relations (82.1)–(82.10) derived previously, which follow from these continuity conditions alone, remain valid. In particular, the equation

$$w_1 - w_2 + \tfrac{1}{2}(V_1 + V_2)(p_2 - p_1) = 0 \tag{121.1}$$

holds; the suffix 1 always pertains to the unburnt gas and the suffix 2 to the combustion products. The curve of p_2 as a function of V_2 given by this equation is called the *detonation adiabatic*. Unlike the shock adiabatic considered earlier, this curve does not pass through the given initial point (p_1, V_1). The fact that the shock adiabatic passes through this point is due to the fact that w_1 and w_2 are the same functions of p_1, V_1 and p_2, V_2 respectively, whereas this does not now hold, on account of the chemical difference between the two gases. In Fig. 114 the continuous line shows the detonation adiabatic. The ordinary shock adiabatic for the unburnt gas mixture is drawn (dashed) through the point (p_1, V_1). The detonation adiabatic always lies above the shock adiabatic, because a high temperature is reached in combustion, and the gas pressure is therefore greater than it would be in the unburnt gas for the same specific volume.

The previous formula (82.6) holds for the mass flux density:

$$j^2 = (p_2 - p_1)/(V_1 - V_2), \tag{121.2}$$

so that graphically j^2 is again the slope of the chord from the point (p_1, V_1) to any point (p_2, V_2) on the detonation adiabatic (for instance, the chord ac in Fig. 114). It is seen at once from the diagram that j^2 cannot be less than

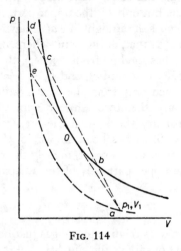

FIG. 114

the slope of the tangent aO. The flux j is just the mass of gas which is ignited per unit time per unit area of the surface of the detonation wave; we see that, in a detonation, this quantity cannot be less than a certain limiting value j_{min} (which depends on the initial state of the unburnt gas).

Formula (121.2) is a consequence only of the conditions of continuity of the fluxes of mass and momentum. It therefore holds (for a given initial state of the gas) not only for the final state of the combustion products, but also for all intermediate states, in which only part of the reaction energy has been evolved. In other words, the pressure p and specific volume V of the gas in any state obey the linear relation

$$p = p_1 + j^2(V_1 - V), \tag{121.3}$$

which is shown graphically by the chord ad. This result is of importance in the theory of detonation; it was first stated by V. A. MIKHEL'SON (1890).

Let us now use a procedure developed by YA. B. ZEL'DOVICH (1940) to investigate the variation of the state of the gas through the layer of finite width which a detonation wave actually is. The forward front of the detonation wave is a true shock wave in the unburnt gas 1. In it, the gas is compressed and heated to a state represented by the point d (Fig. 114) on the shock adiabatic of gas 1. The chemical reaction begins in the compressed gas, and as the reaction proceeds the state of the gas is represented by a point which moves down the chord da; heat is evolved, the gas expands, and its pressure decreases. This continues until combustion is complete and the whole heat of the reaction has been evolved. The corresponding point is c, which lies on the detonation adiabatic representing the final state of the combustion products. The lower point b at which the chord ad intersects the detonation adiabatic cannot be reached for a gas in which combustion is caused by compression and heating in a shock wave.†

Thus we conclude that the detonation is represented, not by the whole of the detonation adiabatic, but only by the upper part, lying above the point O where this adiabatic touches the straight line aO drawn from the initial point a.

It has been shown in §84 that, at the point where $d(j^2)/dp_2 = 0$, i.e. where the shock adiabatic touches the line from (p_1, V_1), the velocity v_2 is equal to the corresponding velocity of sound c_2, and $v_2 < c_2$ above that point. These results have been obtained only from the conservation laws for the surface of discontinuity, and are therefore entirely applicable to the detonation wave also. On the ordinary shock adiabatic for a single gas there are no points with $d(j^2)/dp_2 = 0$, as has been shown in §84. On the detonation adiabatic, however, there is such a point, namely the point O. Since detonation corresponds to the upper part only of the adiabatic, above the point O, we conclude that

$$v_2 \leqslant c_2, \tag{121.4}$$

i.e. a detonation wave moves relative to the gas just behind it with a velocity equal to or less than that of sound; the equality $v_2 = c_2$ holds for a detonation corresponding to the point O (called the *Jouguet point*).‡

The velocity of the detonation wave relative to gas 1 is always supersonic (even for the point O):

$$v_1 > c_1. \tag{121.5}$$

This is most simply seen directly from Fig. 114. The velocity of sound c_1 is given graphically by the slope of the tangent to the shock adiabatic for gas 1 (dashed curve) at the point a. The velocity v_1, on the other hand, is given by the slope of the chord ac. Since all the chords concerned are steeper than the tangent, we always have $v_1 > c_1$. Moving with supersonic velocity, the detonation wave, like a shock wave, does not affect the state of the gas in front of it. The velocity v_1 with which the detonation wave moves relative to the unburnt gas at rest is the velocity of propagation of the detonation. Since $v_1/V_1 = v_2/V_2 \equiv j$, and $V_1 > V_2$, it follows that $v_1 > v_2$. The

---

† For completeness, it should also be mentioned that a discontinuous transition from state c to state b in another shock wave is also impossible, since the gas would have to cross such a shock from high pressure to low pressure.

‡ It should be recalled that the velocities v_1, v_2 always signify the velocities normal to the surface of discontinuity.

difference $v_1 - v_2$ is evidently the velocity of the combustion products relative to the unburnt gas. This difference is positive, i.e. the combustion products move in the direction of propagation of the detonation wave.

We may note also the following. In §84 it was also shown that $\mathrm{d}s_2/\mathrm{d}(j^2) > 0$. At the point where j^2 has a minimum, s_2 therefore also has a minimum. This point is O, and we conclude that it corresponds to the least value of the entropy s_2 on the detonation adiabatic. The entropy s_2 also has an extremum at O if we consider the change in state along the line ae (since the slopes of the curve and the tangent at O are the same). This extremum, however, is a maximum. For a displacement from e to O corresponds to the change of state as the combustion reaction occurs in the compressed gas, and this is accompanied by the evolution of heat and an increase in entropy; a passage from O to a, however, would correspond to the endothermic conversion of the combustion products into the original gases, with a decrease in entropy.

If the detonation is caused by a shock wave which is produced by some external source and is then incident on the gas, any point on the upper part of the detonation adiabatic may correspond to the detonation. It is of particular interest, however, to consider a detonation which is due to the combustion process itself. We shall see in §122 that, in a number of important cases, such a detonation must correspond to the Jouguet point, so that the velocity of the detonation wave relative to the combustion products just behind it is exactly equal to the velocity of sound, while the velocity $v_1 = jV_1$ relative to the unburnt gas has its least possible value. This result was put forward as a hypothesis by D. L. CHAPMAN (1899) and E. JOUGUET (1905), but its complete theoretical justification is due to YA. B. ZEL'DOVICH (1940).

Let us now derive the relations between the various quantities in a detonation wave in a perfect gas. Substituting in the general equation (121.1) the heat function in the form

$$w = w_0 + c_p T = w_0 + \gamma p V/(\gamma - 1),$$

we obtain

$$\frac{\gamma_2 + 1}{\gamma_2 - 1} p_2 V_2 - \frac{\gamma_1 + 1}{\gamma_1 - 1} p_1 V_1 - V_1 p_2 + V_2 p_1 = 2q, \tag{121.6}$$

where $q = w_{01} - w_{02}$ again denotes the heat of the reaction, reduced to the absolute zero of temperature. The curve $p_2(V_2)$ given by this equation is a rectangular hyperbola. For $p_2/p_1 \to \infty$, the ratio of densities tends to a finite limit $\rho_2/\rho_1 = V_1/V_2 = (\gamma_2 + 1)/(\gamma_2 - 1)$; this is the greatest compression that can be achieved in a detonation wave.

The formulae are much simplified in the important case of strong detonation waves, which are obtained when the heat evolved in the reaction is large compared with the internal heat energy of the original gas, i.e. $q \gg c_{v1} T_1$. In this case we can neglect the terms containing p_1 in (121.6), obtaining

$$p_2 \left(\frac{\gamma_2 + 1}{\gamma_2 - 1} V_2 - V_1 \right) = 2q. \tag{121.7}$$

Let us consider in more detail a detonation corresponding to the Jouguet point, which is of particular interest, as we see from the above discussion. At this point $j^2 = c_2{}^2/V_2{}^2 = \gamma_2 p_2/V_2$. From this relation and (121.2) we can express p_2 and V_2 in the form

$$p_2 = (p_1 + j^2 V_1)/(\gamma_2 + 1), \qquad V_2 = \gamma_2(p_1 + j^2 V_1)/j^2(\gamma_2 + 1). \quad (121.8)$$

Substituting these expressions in equation (121.6) and replacing j by v_1/V_1, we have after a simple reduction the following quartic equation for the velocity v_1:

$$v_1{}^4 - 2v_1{}^2[(\gamma_2{}^2 - 1)q + (\gamma_2{}^2 - \gamma_1)c_{v1}T_1] + \gamma_2{}^2(\gamma_1 - 1)^2 c_{v1}{}^2 T_1{}^2 = 0,$$

where the temperature has been introduced by $T = pV/(c_p - c_v) = pV/c_v(\gamma - 1)$. Hence†

$$v_1 = \sqrt{\{\tfrac{1}{2}(\gamma_2 - 1)[(\gamma_2 + 1)q + (\gamma_1 + \gamma_2)c_{v1}T_1]\}} + \\ + \sqrt{\{\tfrac{1}{2}(\gamma_2 + 1)[(\gamma_2 - 1)q + (\gamma_2 - \gamma_1)c_{v1}T_1]\}}. \quad (121.9)$$

This formula determines the velocity of propagation of the detonation in terms of the temperature T_1 of the original gas mixture.

We can rewrite formulae (121.8) in the form

$$\frac{p_2}{p_1} = \frac{v_1{}^2 + (\gamma_1 - 1)c_{v1}T_1}{(\gamma_2 + 1)(\gamma_1 - 1)c_{v1}T_1}, \qquad \frac{V_2}{V_1} = \frac{\gamma_2[v_1{}^2 + (\gamma_1 - 1)c_{v1}T_1]}{(\gamma_2 + 1)v_1{}^2}.$$
$$(121.10)$$

Together with (121.9), they determine the ratios of pressure and density between the combustion products and the unburnt gas at temperature T_1.

The velocity v_2 is calculated as $v_2 = V_2 v_1/V_1$, using formulae (121.9) and (121.10). The result is

$$v_2 = \sqrt{\{\tfrac{1}{2}(\gamma_2 - 1)[(\gamma_2 + 1)q + (\gamma_1 + \gamma_2)c_{v1}T_1]\}} + \\ + \frac{\gamma_2 - 1}{\gamma_2 + 1}\sqrt{\{\tfrac{1}{2}(\gamma_2 + 1)[(\gamma_2 - 1)q + (\gamma_2 - \gamma_1)c_{v1}T_1]\}}. \quad (121.11)$$

The difference $v_1 - v_2$, i.e. the velocity of the combustion products relative to the unburnt gas, is

$$v_1 - v_2 = \sqrt{\{2[(\gamma_2 - 1)q + (\gamma_2 - \gamma_1)c_{v1}T_1]/(\gamma_2 + 1)\}}. \quad (121.12)$$

The temperature of the combustion products is calculated from the formula

$$c_{v2}T_2 = v_2{}^2/\gamma_2(\gamma_2 - 1) \quad (121.13)$$

(since $v_2 = c_2$).

---

† If $x^4 - 2px^2 + q = 0$, then
$$x = \sqrt{[p \pm \sqrt{(p^2 - q)}]} = \sqrt{[\tfrac{1}{2}(p + \sqrt{q})]} \pm \sqrt{[\tfrac{1}{2}(p - \sqrt{q})]}.$$

The two signs in this case correspond to the fact that two tangents can be drawn from the point a to the detonation adiabatic: one upwards, as shown in Fig. 114, and the other downwards. The upward tangent, in which we are interested, has the steeper slope, and we accordingly take the plus sign.

All these somewhat complex formulae are much simplified for strong detonation waves. In this case the velocities are given by the simple formulae

$$v_1 = \sqrt{[2(\gamma_2{}^2 - 1)q]}, \qquad v_1 - v_2 = v_1/(\gamma_2 + 1). \qquad (121.14)$$

The thermodynamic state of the combustion products is given by the formulae

$$V_2/V_1 = \gamma_2/(\gamma_2 + 1), \qquad T_2 = 2\gamma_2 q/c_{v2}(\gamma_2 + 1),$$

$$\frac{p_2}{p_1} = \frac{2(\gamma_2 - 1)}{\gamma_1 - 1} \frac{q}{c_{v1}T_1} = \frac{\gamma_1 v_1{}^2}{(\gamma_2 + 1)c_1{}^2}. \qquad (121.15)$$

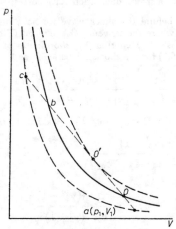

Fig. 115

If we compare formulae (121.15) with the corresponding formulae (120.5) for slow combustion, we notice that, in the limiting case $q \gg c_{v1}T_1$, the ratio of the temperatures of the combustion products after detonation and slow combustion is $T_{2,\,\mathrm{det}}/T_{2,\,\mathrm{com}} = 2\gamma_2{}^2/(\gamma_2 + 1)$. This ratio always exceeds unity (since $\gamma_2 > 1$).

In all the foregoing arguments, which were based on Fig. 114, it has been tacitly assumed that the chemical reaction of combustion is exothermic from beginning to end (i.e. at all intermediate stages between the original unburnt gas and the final combustion products). In the majority of cases this assumption is justified. Reactions are, however, possible in principle which are initially exothermic but endothermic in their final stages (YA. B. ZEL'DOVICH and S. B. RATNER, 1941). The intermediate mixture for which heat is first absorbed and not evolved then corresponds to an adiabatic lying above the detonation adiabatic which corresponds to the state of the final combustion products (Fig. 115).

Any chord along which the state of the detonating mixture varies must cross this intermediate adiabatic. The value j_{min} corresponding to the least possible value of the velocity of propagation of combustion is therefore determined by the slope of the tangent aO'. Detonation waves with $j > j_{min}$ correspond to points lying on the detonation adiabatic above the point b, and $v_2 < c_2$. If $j = j_{min}$, the state of the gas varies along the line ca from c to O' and then further downwards to O, which replaces the ordinary Jouguet point as the point corresponding to spontaneous detonation; here $v_2 > c_2$, contrary to the usual result.

PROBLEM

Determine the thermodynamic quantities for the gas immediately behind the shock wave which is the forward front of a strong detonation wave corresponding to the Jouguet point

SOLUTION. Immediately behind the shock wave we have unburnt gas, and its state is represented by the point e where the tangent aO produced (Fig. 114) intersects the shock adiabatic of gas 1, shown dashed. Denoting the co-ordinates of this point by (p_1', V_1'), we have, firstly, by equation (85.1) for the shock adiabatic of gas 1,

$$\frac{V_1'}{V_1} = \frac{(\gamma_1+1)p_1+(\gamma_1-1)p_1'}{(\gamma_1-1)p_1+(\gamma_1+1)p_1'}$$

and, secondly, $(p_1'-p_1)/(V_1-V_1') = j^2 = v_1^2/V_1^2$. Taking v_1 from (121.14), we obtain

$$p_1' = p_1\frac{4(\gamma_2^2-1)}{\gamma_1^2-1}\frac{q}{c_{v1}T_1}, \qquad V_1' = V_1\frac{\gamma_1-1}{\gamma_1+1},$$

$$T_1' = \frac{q}{c_{v1}}\frac{4(\gamma_2^2-1)}{(\gamma_1+1)^2}.$$

The ratio of the pressure p_1' to the pressure p_2 behind the detonation wave is

$$p_1'/p_2 = 2(\gamma_2+1)/(\gamma_1+1).$$

§122. The propagation of a detonation wave

Let us now consider some actual cases of the propagation of detonation waves in a gas initially at rest. We take first the case of detonation in a gas in a pipe closed at one end ($x = 0$). The boundary conditions in this case are that the gas velocity is zero both in front of the detonation wave (which does not affect the state of the gas in front of it) and at the closed end of the pipe. Since the gas acquires a non-zero velocity when the detonation wave passes, the velocity must diminish in the region between the detonation wave and the closed end of the pipe. In order to determine the resulting flow pattern, we notice that in this case there is no length parameter which might characterise the conditions of flow along the pipe (the x-direction). We have seen in §92 that, in such cases, the gas velocity can change either in a shock wave (separating two regions where the velocity is constant) or in a similarity rarefaction wave.

Let us first assume that the detonation wave does not correspond to the Jouguet point on the adiabatic. Then its velocity of propagation relative to the gas behind it is $v_2 < c_2$. It is easy to see that, in this case, neither a shock wave nor a weak discontinuity (the forward front of a rarefaction wave) can follow the detonation wave. For the former would have to move, relative to the gas in front of it, with a velocity exceeding c_2, and the latter with a velocity equal to c_2, and either would overtake the detonation wave. Thus, on the above assumption, the velocity of the gas moving behind the detonation wave cannot decrease, i.e. the boundary condition at $x = 0$ cannot be satisfied.

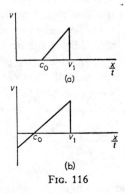

Fig. 116

This condition can be satisfied only for a detonation wave corresponding to the Jouguet point. Then $v_2 = c_2$, and a rarefaction wave can follow the detonation wave. It is formed at $x = 0$ when the detonation begins, and its forward front coincides with the detonation wave.

Thus we reach the important result that a detonation wave propagated in a pipe, with the gas ignited at the closed end, must correspond to the Jouguet point. It moves relative to the gas just behind it with a velocity equal to the local velocity of sound. The detonation wave adjoins a rarefaction wave, in which the gas velocity (relative to the pipe) falls monotonically to zero. The point where the velocity becomes zero is a weak discontinuity. Behind this discontinuity the gas is at rest (Fig. 116a).

Let us now consider a detonation wave propagated from the open end of a pipe. The pressure of the gas in front of the detonation wave must be equal to the original pressure, which clearly equals the external pressure. It is evident that, in this case also, the velocity must decrease somewhere behind the detonation wave. If the gas velocity were constant between the end of the pipe and the detonation wave, it would follow that gas was being sucked into the open end of the pipe from outside; this would be impossible, since the gas pressure in the pipe would be greater than the external pressure on account of the pressure increase in a detonation wave. For the same reasons as in the previous case, the detonation wave must correspond to the Jouguet point. The resulting flow pattern is shown diagrammatically in Fig. 116b. Immediately behind the detonation wave is a similarity rarefaction wave, in which the velocity decreases monotonically towards the end of the pipe, changing sign at some point. This means that, in the end section of the pipe, the gas moves towards the open end and flows out of it; the velocity with which it leaves the pipe equals the local velocity of sound, and its pressure exceeds the external pressure. We have seen in §90 that such a flow is possible.

Let us next consider a spherically symmetrical outgoing detonation wave whose centre is the point where the gas is first ignited (YA. B. ZEL'DOVICH 1942). Since the gas must be at rest both in front of the detonation wave and near the centre, the gas velocity must decrease from the detonation wave towards the centre. As with the flow in a pipe, there are no characteristic parameters having the dimensions of length. The result must therefore be a similarity flow, the co-ordinate x being replaced by the distance r from the centre. Thus all quantities are functions only of the ratio r/t.

For centrally symmetrical flow ($v_r = v(r, t)$, $v_\phi = v_\theta = 0$), the equations of motion are as follows. The equation of continuity is

$$\frac{\partial \rho}{\partial t} + \frac{\partial (v\rho)}{\partial r} + \frac{2v\rho}{r} = 0;$$

Euler's equation is

$$\frac{\partial v}{\partial t} + v\frac{\partial v}{\partial r} = -\frac{1}{\rho}\frac{\partial p}{\partial r},$$

and the equation of conservation of entropy is

$$\frac{\partial s}{\partial t} + v\frac{\partial s}{\partial r} = 0.$$

Introducing the variable $\xi = r/t$ (> 0) and assuming that all quantities are functions of ξ only, we obtain

$$(\xi - v)\rho'/\rho = v' + 2v/\xi, \tag{122.1}$$
$$(\xi - v)v' = p'/\rho, \tag{122.2}$$
$$(\xi - v)s' = 0, \tag{122.3}$$

the prime denoting differentiation with respect to ξ. We cannot have $v = \xi$, since this contradicts the first equation. From the third equation, therefore, $s' = 0$, i.e. $s = $ constant. We can therefore write $p' = (\partial p/\partial \rho)_s\rho' = c^2\rho'$, and equation (122.2) becomes

$$(\xi - v)v' = c^2\rho'/\rho. \tag{122.4}$$

Substituting ρ'/ρ from (122.1), we obtain the relation

$$\left[\frac{(\xi - v)^2}{c^2} - 1\right]v' = \frac{2v}{\xi}. \tag{122.5}$$

Equations (122.4) and (122.5) cannot be integrated analytically, but the properties of their solutions can be investigated.

The region where the gas flow is of the type considered is bounded, as we shall see below, by two spheres, of which the outer is the surface of the detonation wave itself, and the inner is the surface of a weak discontinuity, where the velocity is zero.

Let us first examine the properties of the solution near the point where v is zero. It is easy to see that, where $v = 0$, $\xi = c$ also:

$$v = 0, \qquad \xi = c. \tag{122.6}$$

For, when v tends to zero, $\log v \to -\infty$; hence, when ξ decreases to the value corresponding to the inner boundary of the region in question, the derivative $\mathrm{d} \log v/\mathrm{d}\xi$ must tend to $+\infty$. From (122.5), however, we have for $v = 0$

$$\mathrm{d} \log v/\mathrm{d}\xi = 2c^2/\xi(\xi^2 - c^2).$$

This expression can tend to $+\infty$ only if $\xi \to c$.

At the origin, the radial velocity must vanish, by symmetry. Thus there is a region of gas at rest round the origin; this is the region inside the sphere $\xi = c_0$, where c_0 is the velocity of sound for the point where $v = 0$.

Let us ascertain the properties of the function $v(\xi)$ near the point (122.6). From (122.5) we have

$$v\frac{\mathrm{d}\xi}{\mathrm{d}v} = \tfrac{1}{2}\xi\left[\frac{(\xi - v)^2}{c^2} - 1\right].$$

As far as quantities of the first order (such as v, $\xi - c_0$ and $c - c_0$), we have after a simple calculation $v\mathrm{d}(\xi - c_0)/\mathrm{d}v = (\xi - c_0) - (v + c - c_0)$. According to (95.1) we have $v + c - c_0 = \alpha_0 v$, where α_0 is a positive constant, the value of (95.2) for $v = 0$, and we obtain the following linear first-order differential equation for $\xi - c_0$ as a function of v:

$$v\,\mathrm{d}(\xi - c_0)/\mathrm{d}v - (\xi - c_0) = -\alpha_0 v.$$

The solution of this equation is

$$\xi - c_0 = \alpha_0 v \log(\text{constant}/v). \tag{122.7}$$

This implicitly determines the function $v(\xi)$ near the point where $v = 0$.

We see that the inner boundary is a surface of weak discontinuity: the velocity tends continuously to zero. The curve of $v(\xi)$ has a horizontal tangent at this point ($\mathrm{d}v/\mathrm{d}\xi = 0$). The weak discontinuity involved is very unusual: the first derivative is continuous, but all higher derivatives are infinite (as is easily seen from (122.7)). The ratio r/t for $v = 0$ is clearly just the velocity of motion of the boundary relative to the gas; according to (122.6), it is equal to the local velocity of sound, as it should be for a weak discontinuity.

We have also for small v, by (122.7),

$$\xi - v - c = (\xi - c_0) - (v + c - c_0)$$
$$= \alpha_0 v[\log(\text{constant}/v) - 1].$$

For small v, this quantity is positive: $\xi - v - c > 0$. We shall show that the difference $(\xi - v) - c$ cannot change sign anywhere in the region of the flow considered. Let us consider a point, if there is one, where

$$\xi - v = c, \qquad v \neq 0. \tag{122.8}$$

We see from (122.5) that the derivative v' must be infinite at this point, i.e.

$$d\xi/dv = 0. \tag{122.9}$$

The second derivative $d^2\xi/dv^2$ is shown by a simple calculation (using the conditions (122.8) and (122.9)) to be $d^2\xi/dv^2 = -\alpha_0\xi/c_0 v$, which is not zero. This means that ξ as a function of v has a maximum at the point in question. Thus the function $v(\xi)$ exists only for ξ less than the value corresponding to the conditions (122.8), and this value is the other boundary of the region considered. Since $\xi - v - c$ can vanish only at the boundary of the region, and $\xi - v - c > 0$ for small v, we conclude that

$$\xi - v > c \tag{122.10}$$

everywhere in the region.

It is now easy to see that the outer boundary of the region of the flow considered must in fact be at the point where the conditions (122.8) hold. To see this, we notice that the difference $r/t - v$, where r is the co-ordinate of the boundary, is just the velocity of the boundary relative to the gas behind it. A surface on which $r/t - v > c$, however, cannot be the surface of a detonation wave (where we must have $r/t - v \leqslant c$). We therefore conclude that the outer boundary of the region considered can only be the point where (122.8) holds. On this boundary v falls discontinuously to zero, and the velocity of the boundary relative to the gas just behind it is equal to the local velocity of sound. This means that the detonation wave must correspond to the Jouguet point on the detonation adiabatic.†

We thus have the following flow pattern for spherical propagation of a detonation. The detonation wave, like that in a pipe, must correspond to the Jouguet point. Immediately behind it is a spherical similarity rarefaction wave, in which the gas velocity decreases to zero. The decrease is monotonic, since, by (122.5), the derivative $dv/d\xi$ can vanish only if $v = 0$ also. The gas pressure and density also decrease monotonically, since by (122.4) and (122.10) the derivative p' always has the same sign as v'. The curve giving v as a function of r/t has a vertical tangent at the outer boundary (by (122.9)) and a horizontal tangent at the inner boundary (Fig. 117). The inner boundary is a weak discontinuity, near which the dependence of v on r/t is given

---

† We may notice for completeness that $v =$ constant is not a solution of the equations of centrally symmetrical motion. Hence the detonation wave cannot be followed by a region of constant velocity.

by equation (122.7). The gas within the sphere bounded by the weak discontinuity is at rest. The total mass of gas at rest is, however, very small (cf. the remarks at the end of §99).

Thus, in all the typical cases of spontaneous one-dimensional propagation of detonation which we have considered, the boundary conditions in the region behind the detonation wave give a unique velocity for the latter, which corresponds to the Jouguet point (the whole of the detonation adiabatic below this point being excluded by the arguments of §121). The achievement, in a pipe of constant cross-section, of a detonation corresponding to the part of the adiabatic above the Jouguet point† would require an artificial compression of the combustion products by a piston moving with a supersonic velocity (see Problem 3).

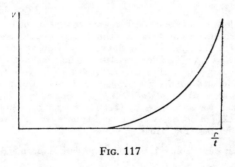

Fig. 117

It should be emphasised, however, that these conclusions are not universally valid, and there are cases of propagation of a detonation where an over-compressed detonation wave occurs spontaneously. In particular, an over-compressed detonation wave is formed when an ordinary detonation wave goes from a wide pipe into a narrow one (B. V. Aïvazov and Ya. B. Zel'dovich 1947). This phenomenon occurs because, when a detonation wave reaches a narrowing of the pipe, it is partly reflected, and the pressure of the combustion products moving from the wide part to the narrow part is considerably increased (cf. Problem 4).

In the foregoing we have entirely neglected the heat losses which may accompany the propagation of a detonation wave. As in the case of slow combustion, these losses may render the propagation of the detonation impossible. In the detonation of gas in a pipe, the source of the losses is primarily the removal of heat through the walls of the pipe and the retardation of the gas by friction. A detonation in a thin rod of explosive is limited mainly by the dispersal of the combustion products: when the rod is too thin

---

† Such detonation waves are sometimes said to be "over-compressed", since the gas is compressed in them to higher pressures than in the "normal" detonation wave corresponding to the Jouguet point.

in comparison with the width of the combustion zone, part of the material is dispersed before the reaction can occur, and the propagation of the detonation is impossible (YU. B. KHARITON 1940).

Under conditions close to the limiting ones for the detonation to be propagated, the curious phenomenon of *spinning detonation* is observed. According to K. I. SHCHELKIN (1945), the shock wave propagated along the pipe in a spinning detonation is no longer of the axially symmetrical (usually almost plane) type, and rotates about the axis of the pipe as it moves along it. The ignition of the gas as it passes through the shock occurs mainly at an eccentrically situated and spirally moving bend in the shock wave front. No quantitative theory of spinning detonation exists.†

PROBLEMS

PROBLEM 1. Determine the gas flow when a detonation wave is propagated from the closed end of a pipe.

SOLUTION. The velocity v_1 of the detonation wave relative to the gas at rest in front of it, and its velocity v_2 relative to the burnt gas just behind it, are given in terms of the temperature T_1 by formulae (121.11), (121.12). v_1 is also the velocity of the wave relative to the pipe, so that its co-ordinate is $x = v_1 t$. The velocity (relative to the pipe) of the combustion products at the detonation wave is $v_1 - v_2$. The velocity v_2 equals the local velocity of sound. Since the velocity of sound is related to the gas velocity v in a similarity rarefaction wave by $c = c_0 + \frac{1}{2}(\gamma - 1)v$, we have $v_2 = c_0 + \frac{1}{2}(\gamma_2 - 1)(v_1 - v_2)$, whence $c_0 = \frac{1}{2}(\gamma_2 + 1)v_2 - \frac{1}{2}(\gamma_2 - 1)v_1$. For a strong detonation wave we have, by (121.14), simply $c_0 = \frac{1}{2}v_1$. The quantity c_0 is the velocity of the backward boundary of the rarefaction wave. The velocity varies linearly between the two boundaries (Fig. 116a).

PROBLEM 2. The same as Problem 1, but for a pipe with an open end.

SOLUTION. The velocities v_1 and v_2 are determined as in the previous case, and so c_0 is the same also. The rarefaction wave, however, now extends, not to the point where $v = 0$, but to the end of the pipe ($x = 0$, Fig. 116b). We see from the formula $x/t = v + c$ (92.5) that the gas leaves the open end of the pipe with a velocity $v = -c$ equal to the local velocity of sound. Putting $-v = c = c_0 + \frac{1}{2}(\gamma_2 - 1)v$, we therefore find the velocity of outflow to be $[-v]_{x=0} = 2c_0/(\gamma_2 + 1)$. For a strong detonation wave this velocity is $v_1/(\gamma_2 + 1)$.

PROBLEM 3. The same as Problem 1, but for a detonation wave propagated in a pipe whose end is closed by a piston which begins to move forward with a constant velocity U.

SOLUTION. If $U < v_1$, the velocity distribution in the gas is of the form shown in Fig. 118a. The gas velocity decreases from $v_1 - v_2$ at $x/t = v_1$ to U at $x/t = c_0 + \frac{1}{2}(\gamma + 1)U$, with the same value of c_0 as before. Then follows a region in which the gas moves with constant velocity U. If $U > v_1$, however, the detonation wave cannot correspond to the Jouguet point (since the piston would overtake it). In this case we have an over-compressed detonation wave, corresponding to a point on the adiabatic above the Jouguet point. It is determined by the fact that the discontinuity of velocity in the detonation wave must equal the velocity of the piston: $v_1 - v_2 = U$. Throughout the region between the detonation wave and the piston, the gas moves with constant velocity U (Fig. 118b).

PROBLEM 4. Determine the pressure at a perfectly rigid wall when a strong plane detonation wave normally incident is reflected from it (K. P. STANYUKOVICH 1946).

SOLUTION. When a detonation wave is incident on a wall, a reflected shock wave is formed and propagated in the opposite direction, through the combustion products. The calculations are entirely similar to those in §93, Problem 1. With the same notation, we obtain the

---

† A qualitative discussion is given by YA. B. ZEL'DOVICH (*Comptes rendus de l'Académie des Sciences de l'URSS* 52, 147, 1946).

three relations

$$p_2(V_1 - V_2) = (p_3 - p_2)(V_2 - V_3), \qquad V_2/V_1 = \gamma_2/(\gamma_2 + 1),$$

$$\frac{V_3}{V_2} = \frac{(\gamma_2 + 1)p_2 + (\gamma_2 - 1)p_3}{(\gamma_2 - 1)p_2 + (\gamma_2 + 1)p_3};$$

here we have neglected p_1 in comparison with p_2, but p_2 and p_3 are of the same order of magnitude. Eliminating the volumes, we obtain a quadratic equation for p_3, and must take the root which is greater than p_2:

$$\frac{p_3}{p_2} = \frac{5\gamma_2 + 1 + \sqrt{(17\gamma_2^2 + 3\gamma_2 + 1)}}{4\gamma_2}.$$

It should be noted that this quantity is almost independent of γ_2, varying from 2·6 to 2·3 as γ_2 varies from 1 to ∞.

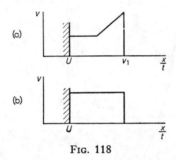

(a)

(b)

FIG. 118

§123. The relation between the different modes of combustion

It has been shown in §121 that detonation corresponds to points on the upper part of the detonation adiabatic for the combustion process concerned. Since the equation of this adiabatic is a consequence only of the conservation laws for mass, momentum and energy (applied to the initial and final states of the burning gas), it is clear that the points representing the state of the reaction products must lie on the same curve for any other mode of combustion in which the combustion zone can be regarded as a surface of discontinuity of some kind. Let us now ascertain the physical significance of the remainder of the curve.

We draw through the point (p_1, V_1) (point 1 in Fig. 119) vertical and horizontal lines $1A$ and $1A'$, and the two tangents $1O$ and $1O'$ to the adiabatic. The points A, A', O, O' where these lines intersect or touch the curve divide the adiabatic into five parts. The part lying above O corresponds to detonation, as we have said. We shall now consider the other parts of the curve.

First of all, it is easy to see that the section AA' has no physical significance. For we have on this section $p_2 > p_1$, $V_2 > V_1$, and so the mass flux $j = \sqrt{[(p_2 - p_1)/(V_1 - V_2)]}$ is imaginary.

At the points of contact O and O', the derivative $d(j^2)/dp_2$ is zero; it has been shown in §84 that at such points we have $v_2/c_2 = 1$ and $d(v_2/c_2)/dp_2 < 0$. Hence it follows that above the points of contact $v_2/c_2 < 1$, and below them $v_2/c_2 > 1$. The relation between v_1 and c_1 is always easily found by considering the slopes of the corresponding chords and tangents, as was done in §121 for the part above O. The result is that the following inequalities hold on the various sections of the adiabatic:

$$\left.\begin{array}{llll} \text{above } O & v_1 > c_1, & v_2 < c_2; \\ \text{on } AO & v_1 > c_1, & v_2 > c_2; \\ \text{on } A'O' & v_1 < c_1, & v_2 < c_2; \\ \text{below } O' & v_1 < c_1, & v_2 > c_2. \end{array}\right\} \quad (123.1)$$

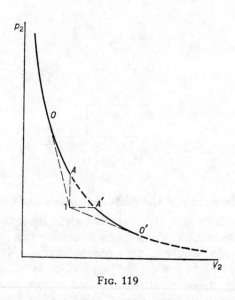

Fig. 119

At O and O', $v_2 = c_2$. As we approach A, the flux j, and therefore the velocities v_1, v_2, tend to infinity. As we approach A', however, j and the velocities v_1, v_2 tend to zero.

In §84 we have investigated the stability of a shock wave with respect to infinitesimal displacements in the direction perpendicular to its plane, and we have seen that the stability depends on the relation between the number of parameters determining the perturbation and the number of boundary conditions which the perturbations must satisfy at the surface of discontinuity.

All these considerations can also be applied to the surfaces of discontinuity here considered. In particular, the calculation made in §84 of the number of parameters of the perturbation for each case (123.1), shown in Fig. 47, remains valid. The only difference is the following. In combustion without detonation, the velocity of propagation is determined only by the properties of the chemical reaction and by the conditions of heat transfer from the combustion zone to the cold gas in front of it. This means that the mass flux j through the combustion zone is a given quantity (more precisely, a given function of the state of the unburnt gas 1), whereas in a shock or detonation wave j can have any value. Hence it follows that, on a discontinuity which is a zone of combustion without detonation, the number of boundary conditions is one more than at a shock wave: the condition that j has a given value is added. Thus there are altogether four conditions, and we now conclude in the same manner as in §84 that the discontinuity is absolutely unstable only in the case $v_1 < c_1$, $v_2 > c_2$, which corresponds to points below O' on the adiabatic. Consequently, this part of the curve does not correspond to any mode of combustion that can be realised in practice.

The section $A'O'$ of the adiabatic, on which both velocities v_1 and v_2 are subsonic, corresponds to the ordinary slow combustion. An increase in the rate of propagation of combustion, i.e. in j, corresponds to a movement from A' (where $j = 0$) towards O'. The formulae (120.5) correspond to the point A' (where $p_1 = p_2$), and are valid if j is sufficiently small, viz. if the velocity of propagation is small compared with that of sound. The point O' corresponds to the "most rapid" combustion of this type. We shall give the formulae pertaining to this limiting case.

The point O', like O, is a point of contact between the curve and the tangent from the point 1. Hence the formulae relating to O' can be obtained immediately from formulae (121.8)–(121.11) for O by appropriately changing the signs (see the footnote to (121.9)). In formulae (121.9) and (121.11) for v_1 and v_2 we change the sign of the second radical, and the sign of the expression (121.12) for $v_1 - v_2$ is therefore changed also. Formulae (121.10) are unchanged if v_1 is taken to have its new value. All these formulae are much simplified if the heat of reaction is large ($q \gg c_{v1}T_1$). We then obtain

$$v_1 = \gamma_2 p_1 V_1/\sqrt{[2(\gamma_2{}^2 - 1)q]}, \qquad v_2 = \sqrt{[2(\gamma_2 - 1)q/(\gamma_2 + 1)]},$$
$$p_2/p_1 = 1/(\gamma_2 + 1), \qquad c_{v2}T_2 = 2q/\gamma_2(\gamma_2 + 1). \tag{123.2}$$

The following remark must be made here. We have seen that, in slow combustion in a closed pipe, a shock wave must be formed in front of the combustion zone. For large velocities of propagation of combustion, this shock wave is strong, and it may considerably affect the state of the gas which enters the combustion zone. It is therefore, strictly speaking, useless to investigate the change in the manner of combustion with increasing velocity, the state p_1, V_1 of the unburnt gas remaining unchanged. In order to reach the point O' we must create conditions of combustion in which no shock wave is formed. This can be done, for instance, in combustion in a pipe open at both ends, with a continuous removal of combustion products at the rear end. The rate of removal must be such that the combustion zone remains at rest, and so no shock wave is formed.

When the transfer of heat is very efficient (for example, transfer by radiation), the value of j may in principle exceed that corresponding to the point O'. The mode of combustion then resulting must correspond to points on the section AO of the adiabatic, since combustion corresponding to a point above O cannot in general occur spontaneously, for the same reasons as in detonation.

Ordinary slow combustion may spontaneously change into detonation. This transition occurs owing to an acceleration of the flame, accompanied by an increase in the intensity of the shock wave preceding it, until the shock becomes strong enough to ignite the gas passing through it. The mechanism of this spontaneous acceleration of the flame is not yet clear; it is possible that turbulence of the flame caused by the walls of the pipe is important (K. I. SHCHELKIN). It is also possible that steady propagation of a flame is unstable when its front is curved by the friction of the gas against the walls of the pipe (YA. B. ZEL'DOVICH).

In conclusion, we may call attention to the following general differences (besides those contained in the inequalities (123.1)) between the modes of combustion corresponding to the upper and lower parts of the adiabatic. Above A we have $p_2 > p_1$, $V_2 < V_1$, $v_2 < v_1$. That is, the reaction products have a pressure and density greater than that of the original gas, and move behind the combustion front with velocity $v_1 - v_2$. In the region below A, however, the inequalities are reversed: $p_2 < p_1$, $V_2 > V_1$, $v_2 > v_1$, and the combustion products are less dense than the original gas.

§124. Condensation discontinuities

There is a formal similarity between detonation waves and what are called *condensation discontinuities*; these occur, for instance, in the flow of a gas containing supersaturated water vapour. The discontinuities are the result of a sudden condensation of vapour occurring very rapidly in a very narrow region, which can be regarded as a surface of discontinuity (condensation discontinuity) separating the original gas from a gas containing condensed vapour (a *fog*). It should be emphasised that condensation discontinuities are a distinct physical phenomenon, and do not result from the compression of gas in an ordinary shock wave; the latter effect cannot lead to condensation, since the increase of pressure in the shock wave has less effect on the degree of supersaturation than the increase of temperature. Condensation discontinuities were first investigated theoretically by S. Z. BELEN'KIĬ(1945).

Like combustion, the condensation of a vapour is an exothermic process. The heat of reaction q is represented by the heat evolved per unit mass of gas by the condensation of the vapour.† The condensation adiabatic which gives p_2 as a function of V_2 for a given state p_1, V_1 of the original uncondensed gas

---

† The heat q is not, strictly speaking, the usual latent heat of condensation, since the process occurring in the condensation zone includes not only the isothermal condensation of the vapour, but also a general change in the gas temperature. However, if the degree of supersaturation is not too small (a condition usually satisfied), the difference is unimportant.

is of the same form as the combustion adiabatic shown in Fig. 119. The relations between the velocities of propagation of the discontinuity v_1, v_2 and the velocities of sound c_1, c_2 for the various parts of the condensation adiabatic are given by the inequalities (123.1). However, not all the four cases enumerated in (123.1) can actually occur.

First of all, the question arises whether condensation discontinuities are stable with respect to small perturbations in a direction perpendicular to the surface. In this respect their properties are entirely similar to those of combustion zones. We have seen (§123) that the difference in stability between combustion zones and ordinary shock waves is due to the existence of a further condition (that the flux j has a given value) which must be satisfied at the surface. In the case of condensation discontinuities there is again a further condition: the thermodynamic state of the gas 1 in front of the discontinuity must be one for which rapid condensation of the vapour begins.† We therefore conclude immediately that the whole of the adiabatic below O', for which $v_1 < c_1$, $v_2 > c_2$, is excluded, since it does not correspond to stable discontinuities.

It is easy to see that discontinuities corresponding to the part above O, for which $v_1 > c_1$, $v_2 < c_2$, also cannot occur in practice. Such a discontinuity would move with supersonic velocity relative to the gas in front of it, and so its presence would have no effect on the state of that gas. Consequently, the discontinuity would have to be formed along a surface determined by the conditions of flow, namely the surface on which the necessary conditions for the onset of rapid condensation would be fulfilled in continuous flow. The velocity of the discontinuity relative to the gas behind it, on the other hand, would be subsonic in this case. The equations of subsonic flow, however, in general have no solution for which all quantities take prescribed values on a given surface.‡

Thus only two types of condensation discontinuity are possible: (1) supersonic discontinuities (the section AO of the adiabatic) for which

$$v_1 > c_1, \qquad v_2 > c_2, \qquad p_2 > p_1, \qquad V_2 < V_1 \qquad (124.1)$$

and the condensation involves a compression, (2) subsonic discontinuities (the section $A'O'$ of the adiabatic), for which

$$v_1 < c_1, \qquad v_2 < c_2, \qquad p_2 < p_1, \qquad V_2 > V_1 \qquad (124.2)$$

and the condensation involves a rarefaction.

The value of the flux j increases monotonically along the section $A'O'$ from A' (where $j = 0$) to O', and decreases monotonically along AO from A

---

† This condition gives a relation between the pressure and temperature of gas 1.

‡ Similar arguments hold in the case where the total velocity $\mathbf{v}_2$ (of which $v_2 < c_2$ is the component normal to the discontinuity) is supersonic.

To avoid misunderstanding, it should be mentioned that a condensation discontinuity with $v_1 > c_1$, $v_2 < c_2$ may actually (for certain conditions of vapour content and shape of the surface past which the flow occurs) be simulated by a true condensation discontinuity with $v_1 > c_1$, $v_2 > c_2$, closely followed by a shock wave which renders the flow subsonic.

(where $j = \infty$) to O. The range of values of j (and therefore the range of values of the velocity $v_1 = jV_1$) between those corresponding to O and O' is "forbidden", and cannot occur in a condensation discontinuity. The total mass of condensed vapour is usually very small compared with the mass of the original gas. We can therefore regard both gases 1 and 2 as perfect gases; for the same reason, the specific heats of the two gases may be supposed equal. Then the value of v_1 at the point O is determined by formula (121.9), and its value at O' by the same formula with the sign of the second radical changed; putting $\gamma_1 = \gamma_2 \equiv \gamma$ and $c_1^2 = \gamma(\gamma-1)c_v T_1$, we find the forbidden range of values of v_1 to be

$$\sqrt{[c_1^2 + \tfrac{1}{2}(\gamma^2 - 1)q]} - \sqrt{[\tfrac{1}{2}(\gamma^2 - 1)q]} < v_1$$
$$< \sqrt{[c_1^2 + \tfrac{1}{2}(\gamma^2 - 1)q]} + \sqrt{[\tfrac{1}{2}(\gamma^2 - 1)q]}. \qquad (124.3)$$

PROBLEM

Determine the limiting values of the ratio of pressures p_2/p_1 in a condensation discontinuity, assuming that $q/c_1^2 \ll 1$.

SOLUTION. On the section $A'O'$ of the condensation adiabatic (Fig. 119), the ratio p_2/p_1 increases monotonically from O' to A', taking values in the range

$$1 - \gamma\sqrt{[2(\gamma-1)q/(\gamma+1)c_1^2]} \leqslant p_2/p_1 \leqslant 1.$$

On the section AO, this ratio increases from A to O, taking values in the range

$$1 + \gamma(\gamma-1)q/c_1^2 \leqslant p_2/p_1 \leqslant 1 + \gamma\sqrt{[2(\gamma-1)q/(\gamma+1)c_1^2]}.$$

CHAPTER XV

RELATIVISTIC FLUID DYNAMICS

§125. The energy-momentum tensor

THE establishment of the relativistic equations of fluid motion is of fundamental importance. The necessity of allowing for relativistic effects may
be due not only to a large velocity of the macroscopic motion (comparable
with that of light), but also, as we shall see, to a large velocity of the microscopic motion of the fluid particles.

We must first of all determine the form of the energy-momentum 4-tensor
T_{ik} for a fluid in motion.† The momentum flux through an element df of the
surface of a body is just the force on that element. Hence $T_{\alpha\beta}\, df^\beta$ is the
α-component of the force on a surface element.‡ Let us consider some
element of volume in the fluid, and use a frame of reference in which this
element is at rest (the "proper" frame). In such a frame Pascal's law holds:
the pressure exerted by a given portion of fluid is the same in all directions and
perpendicular to the area on which it acts. We can therefore write $T_{\alpha\beta}\, df^\beta$
$= p df_\alpha$, whence†† $T_{\alpha\beta} = p g_{\alpha\beta}$.

The components $T_{0\alpha}$ which give the momentum density are zero for a given
volume element in its proper frame. The component T_{00} is the proper
internal energy density of the fluid, which we shall denote in this chapter by e.

Thus the energy-momentum tensor for a given portion of fluid is, in the
proper frame,

$$T_{ik} = \begin{bmatrix} p & 0 & 0 & 0 \\ 0 & p & 0 & 0 \\ 0 & 0 & p & 0 \\ 0 & 0 & 0 & e \end{bmatrix}. \tag{125.1}$$

It is now easy to find the expression for the energy-momentum tensor in
any frame. To do so, we introduce the fluid 4-velocity u^i. In the proper frame
of the element concerned, the 4-velocity components are $u^\alpha = 0$, $u^0 = 1$.

---

† The notation in this chapter corresponds to that used in chapters 10 and 11 of *The Classical
Theory of Fields*, Addison-Wesley Press, Cambridge (Mass.) 1951. The Latin affixes $i, k, l, \ldots$ take
the values 0, 1, 2, 3, $x^0 = ct$ being the real time co-ordinate. The Greek affixes $\alpha, \beta, \gamma, \ldots$ take the
values 1, 2, 3. The metric tensor is given by the expression $-ds^2 = g_{ik}dx^i dx^k$ for the interval, the
Galilean values of the g_{ik} being $g_{11} = g_{22} = g_{33} = 1$, $g_{00} = -1$.

‡ We may recall that, in Galilean co-ordinates, $T_{00} = T^{00} = -T_0{}^0$ is the energy density, and
$T_\alpha{}^0/c = -T_{0\alpha}/c = T^{0\alpha}/c$ is the momentum component density; the quantities $T_{\alpha\beta} = T^{\alpha\beta} = T_\alpha{}^\beta$
form the momentum flux density tensor.

†† We write all expressions in the covariant form, since we shall use them when gravitational fields
are present, i.e. in the general theory of relativity.

The expression for T_{ik} which becomes (125.1) for these values of u^i is

$$T_{ik} = w u_i u_k + p g_{ik}, \tag{125.2}$$

where $w = e + p$ is the heat function per unit volume.† This is the required expression for the energy-momentum tensor.

The components T_{ik} in three-dimensional form are

$$T_{\alpha\beta} = \frac{w v_\alpha v_\beta}{c^2(1 - v^2/c^2)} + p\delta_{\alpha\beta}, \tag{125.3}$$

$$T_{\alpha 0} = -\frac{w v_\alpha}{c(1 - v^2/c^2)}, \qquad T_{00} = \frac{w}{1 - v^2/c^2} - p = \frac{e + p v^2/c^2}{1 - v^2/c^2}.$$

The non-relativistic case is that of small velocities ($v \ll c$) and small velocities of the internal (microscopic) motion of the fluid particles. In passing to the limit it must be borne in mind that the relativistic internal energy e includes the rest energy nmc^2 of the fluid particles (m being the rest mass of one particle). It must also be remembered that the particle density n is referred to unit proper volume; in non-relativistic expressions, however, the energy density is referred to unit volume in the laboratory frame, in which the fluid element concerned is in motion. We must therefore put, in the limit, $mn \to \rho\sqrt{(1 - v^2/c^2)} \approx \rho - \rho v^2/2c^2$, where ρ is the ordinary non-relativistic mass density. Both the non-relativistic energy density $\rho\epsilon$ and the pressure are small compared with ρc^2.

We thus find that the limiting value of T_{00} is $\rho c^2 + \rho\epsilon + \frac{1}{2}\rho v^2$, i.e. it is ρc^2 together with the non-relativistic energy density. The corresponding limiting form of the tensor $T_{\alpha\beta}$ is $\rho v_\alpha v_\beta + p\delta_{\alpha\beta}$, i.e. it coincides, as it should, with the usual expression for the momentum flux density, which we have denoted in §7 by $\Pi_{\alpha\beta}$.

The relativistic momentum density $-T_{\alpha 0}/c$ is at the same time the energy flux density (divided by c^2). This simple relation no longer holds, however, in the non-relativistic limit, because the non-relativistic energy does not include the rest energy. We have $-T_{\alpha 0}/c \cong \rho v_\alpha + (v_\alpha/c^2)(\rho\epsilon + p + \frac{1}{2}\rho v^2)$. Hence we see that the limiting value of the momentum density is just ρv_α, as it should be; for the energy flux density $-cT_{\alpha 0}$ we have, omitting the term $\rho c^2 v_\alpha$, the expression $(\rho\epsilon + p + \frac{1}{2}\rho v^2)v_\alpha$, in agreement with the result obtained in §6.

§126. The equations of relativistic fluid dynamics

The equations of motion are contained in

$$\partial T_i{}^k / \partial x^k = 0, \tag{126.1}$$

---

† In all the formulae in this chapter the thermodynamic quantities are taken to have their values in the proper frame of the fluid element concerned. Such quantities as the internal energy, the heat function, and the entropy are referred to unit proper volume and are denoted by e, w, σ respectively.

which expresses the laws of conservation of energy and momentum for the physical system to which the tensor T_{ik} pertains. Using the expression (125.2) for T_{ik}, we obtain the equations of fluid motion; it is necessary, however, to use also the law of conservation of numbers of particles, which is not contained in (126.1).

Let us derive the equation of continuity, which expresses the fact that the number of fluid particles is conserved. To do so, we use the particle flux 4-vector n^i. Its time component is the number density of particles, and the three space components form the three-dimensional particle flux vector. It is evident that the vector n^i must be proportional to the 4-velocity u^i, so that

$$n^i = nu^i, \qquad (126.2)$$

where n is a scalar; it is clear from the definition of n that this scalar is just the number density of particles in the frame in which the fluid volume element concerned is at rest.† The equation of continuity is obtained by simply equating to zero the 4-divergence of the flux vector:‡

$$\partial(nu^i)/\partial x^i = 0. \qquad (126.3)$$

Let us now return to equation (126.1). Differentiating the expression (125.2) for the energy-momentum tensor, we obtain

$$\frac{\partial T_i{}^k}{\partial x^k} = u_i \frac{\partial(wu^k)}{\partial x^k} + wu^k \frac{\partial u_i}{\partial x^k} + \frac{\partial p}{\partial x^i} = 0. \qquad (126.4)$$

We multiply this equation by u^i, i.e. project it on the direction of the 4-velocity. Since $u_i u^i = -1$, $u_i \partial u^i/\partial x^k = 0$, we find

$$-\frac{\partial(wu^k)}{\partial x^k} + u^k \frac{\partial p}{\partial x^k} = 0.$$

We can rewrite this equation as

$$\frac{\partial}{\partial x^k}\left(\frac{w}{n}nu^k\right) - \frac{1}{n}\frac{\partial p}{\partial x^k}nu^k = 0$$

and, by virtue of the equation of continuity (126.3), obtain

$$nu^k\left[\frac{\partial}{\partial x^k}\left(\frac{w}{n}\right) - \frac{1}{n}\frac{\partial p}{\partial x^k}\right] = 0.$$

But $1/n$ is just the molecular volume of the substance, and w/n is its heat function per particle. By the thermodynamic identity $d(w/n) - dp/n = Td(\sigma/n)$ (where T is the temperature and σ the entropy per unit proper volume),

---

† At very high temperatures, new particles (for instance, electron pairs) may be formed in the substance, so that the total number of particles of all kinds is changed. In such cases n must be taken as (e.g.) the number of electrons which would remain if all pairs were annihilated.

‡ Cf. the equation of continuity in electrodynamics (*The Classical Theory of Fields*, §4–4).

we therefore have $nTu^k\partial(\sigma/n)/\partial x^k = 0$, or

$$d(\sigma/n)/ds = 0, \tag{126.5}$$

where the derivative is taken along the world line of the fluid element concerned.

By the equation of continuity (126.3), equation (126.5) can also be written as the vanishing of the 4-divergence of the entropy flux σu^i:

$$\partial(\sigma u^i)/\partial x^i = 0. \tag{126.6}$$

Both these equations show that the flow is adiabatic; the energy-momentum tensor (125.2) does not take account of processes of internal friction and thermal conduction, i.e. we consider an ideal fluid.

We now project equation (126.1) on a direction perpendicular to u^i. Such a projection of the vector $\partial T_i{}^k/\partial x^k$ is evidently $\partial T_i{}^k/\partial x^k + u_i u^k \, \partial T_k{}^l/\partial x^l$; it gives zero on scalar multiplication by u^i. A simple calculation leads to the equation

$$wu^k\frac{\partial u_i}{\partial x^k} = -\frac{\partial p}{\partial x^i} - u_i u^k\frac{\partial p}{\partial x^k}. \tag{126.7}$$

The three space components of this equation are the relativistic generalisation of Euler's equation (the time component is a consequence of the other three).

As an interesting application, let us consider the propagation of sound in a substance having a relativistic equation of state (i.e. one in which the pressure is comparable with the internal energy density, including the rest energy). The equations of fluid dynamics for the sound waves can be linearised; it is convenient to start from the equations of motion in the original form (126.1), and not the equivalent form (126.6), (126.7). Substituting the expressions (125.3) for the components of the energy-momentum tensor and retaining only quantities of the same order of smallness as the wave amplitude, we obtain the equations

$$\partial e'/\partial t = -w \operatorname{div}\mathbf{v}, \qquad (w/c^2)\,\partial\mathbf{v}/\partial t = -\mathbf{grad}\,p', \tag{126.8}$$

where the prime denotes the variable parts of quantities. Eliminating $\mathbf{v}$, we find $\partial^2 e'/\partial t^2 = c^2\triangle p'$. Finally, putting $e' = (\partial e/\partial p)_{\mathrm{ad}}p'$, we obtain the wave equation for p', with the velocity of sound

$$u = c\sqrt{(\partial p/\partial e)_{\mathrm{ad}}}; \tag{126.9}$$

the suffix ad signifies that the derivative is taken for an adiabatic process, i.e. for constant σ/n. This formula differs from the corresponding non-relativistic expression in that the mass density is replaced by e/c^2.

In the ultra-relativistic case, the equation of state for any substance is $p = \frac{1}{3}e$. The velocity of sound is then $u = c/\sqrt{3}$, which is less than the velocity of light by a factor $\sqrt{3}$.

Finally, let us discuss briefly the equations of fluid dynamics in the presence of gravitational fields. They are obtained from equations (126.6) and (126.7) by simply replacing the ordinary derivatives by the covariant ones:

$$wu^k u_{i;k} = -\partial p/\partial x^i - u_i u^k \, \partial p/\partial x^k, \qquad (\sigma u^i)_{;i} = 0. \qquad (126.10)$$

From these equations we can derive the condition of mechanical equilibrium in a gravitational field. In equilibrium, the field is static; we can take a frame of reference in which the substance is at rest ($u^\alpha = 0$, $u^0 = 1/\sqrt{(-g_{00})}$), all quantities are independent of time, and the mixed components of the metric tensor are zero ($g_{0\alpha} = 0$). The space components of equation (126.10) then give

$$-w\Gamma_{\alpha0}^{0} u^0 u_0 = \tfrac{1}{2}(w/g_{00}) \, \partial g_{00}/\partial x^\alpha = -\partial p/\partial x^\alpha,$$

or

$$\frac{1}{w}\frac{\partial p}{\partial x^\alpha} = -\frac{\partial}{\partial x^\alpha}\log\sqrt{(-g_{00})}. \qquad (126.11)$$

This is the required equation of equilibrium. In the non-relativistic limit $w \cong \rho c^2$, $-g_{00} = 1 + 2\phi/c^2$ (ϕ being the Newtonian gravitational potential), and equation (126.11) becomes $\mathbf{grad}\, p = -\rho\,\mathbf{grad}\,\phi$, i.e. the usual equation of hydrostatics.

PROBLEMS

PROBLEM 1. Find the solution of the equations of relativistic fluid dynamics which describes a one-dimensional non-steady simple wave.

SOLUTION. In a simple wave, all quantities can be expressed as functions of any one of them (see §94). Writing the equations of motion in the form

$$\frac{\partial T_{00}}{c\,\partial t} + \frac{\partial T_{01}}{\partial x} = 0, \qquad \frac{\partial T_{01}}{c\,\partial t} + \frac{\partial T_{11}}{\partial x} = 0, \qquad (1)$$

and supposing T_{00}, T_{01}, T_{11} to be functions of one another, we obtain $dT_{00}dT_{11} = (dT_{01})^2$. Here we must substitute $T_{00} = eu_0^2 + pu_1^2$, $T_{01} = wu_0u_1$, $T_{11} = eu_1^2 + pu_0^2$, using the fact that $u_1^2 - u_0^2 = -1$; it is convenient to introduce a parameter η such that $u_0 = \cosh\eta$, $u_1 = \sinh\eta$. The result is

$$\tanh^{-1}(v/c) = \pm(1/c)\int (u/w)\,de, \qquad (2)$$

where u is the velocity of sound. Next, from (1) we find $\partial x/\partial t = cdT_{01}/dT_{00}$, and a calculation of the derivative gives

$$x = \frac{t(v \pm u)}{1 \pm uv/c^2} + f(v). \qquad (3)$$

Formulae (2) and (3) give the required solution.

PROBLEM 2. Find the relativistic generalisation of Bernoulli's equation.

SOLUTION. In steady flow, all quantities are independent of time. The space component of equation (126.7) give

$$\frac{w}{\sqrt{(1-v^2/c^2)}}(\mathbf{v}\cdot\mathbf{grad})\frac{\mathbf{v}}{\sqrt{(1-v^2/c^2)}} = -c^2\,\mathbf{grad}\,p - \frac{\mathbf{v}}{1-v^2/c^2}\mathbf{v}\cdot\mathbf{grad}\,p.$$

Multiplying this equation scalarly by $\mathbf{v}/n$ and using the fact that, for constant entropy σ/n, we have $d(e/n) = -pd(1/n)$, we finally obtain

$$\mathbf{v}\cdot\mathbf{grad}\frac{w/n}{\sqrt{(1-v^2/c^2)}} = 0,$$

whence it follows that the quantity

$$\frac{w/n}{\sqrt{(1-v^2/c^2)}}$$

is constant along any streamline. For $v \ll c$ this becomes the usual Bernoulli's equation $p/\rho + \frac{1}{2}v^2 = $ constant.

PROBLEM 3. Find the relativistic generalisation of potential flow (I. M. KHALATNIKOV 1954).

SOLUTION. For isentropic flow we have $\sigma/n = $ constant, and therefore

$$\frac{\partial}{\partial x^k}\left(\frac{\sigma}{n}\right) = \frac{\partial}{\partial x^k}\left(\frac{w}{n}\right) - \frac{1}{n}\frac{\partial p}{\partial x^k} = 0.$$

The equation of motion (126.7) then becomes

$$\frac{w}{n}u^k\frac{\partial u_i}{\partial x^k} + u_iu^k\frac{\partial}{\partial x^k}\left(\frac{w}{n}\right)$$

$$\equiv u^k\frac{\partial}{\partial x^k}\left(\frac{w}{n}u_i\right) = -\frac{\partial}{\partial x^i}\left(\frac{w}{n}\right)$$

or $u^k\omega_{ik} = 0$, where

$$\omega_{ik} = \frac{\partial}{\partial x^k}\left(\frac{w}{n}u_i\right) - \frac{\partial}{\partial x^i}\left(\frac{w}{n}u_k\right).$$

The solutions of this equation which correspond to potential flow are those for which $\omega_{ik} = 0$, i.e.

$$\frac{w}{n}u_i = \frac{\partial\phi}{\partial x^i}. \tag{1}$$

In the non-relativistic limit this gives the usual condition $v_i = \partial\phi/\partial x^i$.

For a steady flow we obtain from (1) Bernoulli's equation: wu_0/n is constant everywhere in the fluid.

PROBLEM 4. Obtain the equation of the shock adiabatic and the formulae for the gas velocities in a shock wave in relativistic fluid dynamics (A. H. TAUB 1948).

SOLUTION. Let us consider the discontinuity in a frame in which it is at rest, and let the x^1-axis (the x-axis) be perpendicular to its plane, i.e. in the direction of the gas velocity. The conditions of continuity for the energy and momentum flux densities are

$$-[cT_{0x}] = -c[wu_0u_x] = [wv/(1-v^2/c^2)] = 0, \tag{1}$$

$$[T_{xx}] = [wu_x^2+p] = [wv^2/(c^2-v^2)+p] = 0. \tag{2}$$

From these conditions we easily obtain (putting $v/c = \tanh\phi$, $u_x = \sinh\phi$, $u_0 = -\cosh\phi$) the following expressions for the gas velocities on the two sides of the shock wave:

$$\frac{v_1}{c} = \sqrt{\frac{(p_2-p_1)(e_2+p_1)}{(e_2-e_1)(e_1+p_2)}}, \qquad \frac{v_2}{c} = \sqrt{\frac{(p_2-p_1)(e_1+p_2)}{(e_2-e_1)(e_2+p_1)}}, \tag{3}$$

where the suffixes 1 and 2 denote quantities pertaining to the two sides of the discontinuity.

The relative velocity of the gas on the two sides is given by the relativistic law of addition of velocities:

$$v_{12} = \frac{v_1 - v_2}{1 - v_1 v_2/c^2} = c\sqrt{\frac{(p_2 - p_1)(e_2 - e_1)}{(e_1 + p_2)(e_2 + p_1)}}. \tag{4}$$

In the non-relativistic limit we put $e \cong mc^2 n = c^2/V$ (V being the specific volume) and neglect p in comparison with e. Formulae (3) and (4) then become (82.6) and (82.7). In the ultra-relativistic case, $p = \frac{1}{3}e$, and we have from (3)

$$\frac{v_1}{c} = \sqrt{\frac{3e_2 + e_1}{3(3e_1 + e_2)}}, \qquad \frac{v_2}{c} = \sqrt{\frac{3e_1 + e_2}{3(3e_2 + e_1)}}.$$

As the shock wave increases in strength ($e_2 \to \infty$), v_1 tends to the velocity of light c, and v_2 to $c/\sqrt{3}$.

To obtain the equation of the shock adiabatic, (1) and (2) must be supplemented by the condition of continuity of the particle flux density:

$$[nu_x] = \left[\frac{nv}{c\sqrt{(1 - v^2/c^2)}}\right] = 0. \tag{5}$$

Eliminating the velocities from (1), (2) and (5), we obtain the required equation:

$$\frac{w_1^2}{n_1^2} - \frac{w_2^2}{n_2^2} + (p_2 - p_1)\left(\frac{w_1}{n_1^2} + \frac{w_2}{n_2^2}\right) = 0. \tag{6}$$

In the non-relativistic limit, this formula becomes (82.9).

§127. Relativistic equations for dissipative processes

The finding of the relativistic equations of fluid dynamics in the presence of dissipative processes (viscosity and thermal conduction) amounts to determining the form of the additional terms in the energy-momentum tensor and in the particle flux density vector. Denoting these terms by τ_{ik} and ν_i respectively, we write

$$T_{ik} = pg_{ik} + wu_i u_k + \tau_{ik}, \tag{127.1}$$

$$n_i = nu_i + \nu_i. \tag{127.2}$$

The equations of motion are again contained in $\partial T_i{}^k/\partial x^k = 0$, $\partial n^i/\partial x^i = 0$.

First of all, however, we must discuss more closely the concept of the velocity u^i itself. In relativistic mechanics, an energy flux necessarily involves a mass flux. Hence, when there is (e.g.) a heat flux, the definition of the velocity in terms of the mass flux density (as in non-relativistic fluid dynamics) has no direct meaning. We now define the velocity by the condition that, in the proper frame of any given fluid element, the momentum of the element is zero and its energy is expressible in terms of the other thermodynamic quantities by the same formulae as when dissipative processes are absent. This means that, in the proper frame, the components τ_{00} and $\tau_{0\alpha}$ of the tensor τ_{ik} are zero; since, in this frame, $u^\alpha = 0$ also, we have (in any frame) the tensor equation

$$\tau_{ik} u^k = 0. \tag{127.3}$$

A similar relation,

$$\nu_i u^i = 0, \tag{127.4}$$

must hold for the vector v_i, since the component n^0 of the particle flux 4-vector n^i in the proper frame must, by definition, equal the particle number density n.

The required form of the tensor τ_{ik} and the vector v_i can be established from the requirements of the law of increase of entropy. This law must be contained in the equations of motion (in the same way as the condition of constant entropy for an ideal fluid was obtained in §2 from these equations). By simple transformations, using the equation of continuity, we easily obtain the equation

$$u^i \frac{\partial T_i{}^k}{\partial x^k} = -T\frac{\partial}{\partial x^i}(\sigma u^i) + \mu\frac{\partial v^i}{\partial x^i} + u^i\frac{\partial \tau_i{}^k}{\partial x^k},$$

where $\mu = (w - T\sigma)/n$ is the relativistic chemical potential. Finally, using the relation (127.3), we can rewrite this equation as

$$\frac{\partial}{\partial x^i}\left(\sigma u^i - \frac{\mu}{T}v^i\right) = -v^i\frac{\partial}{\partial x^i}\left(\frac{\mu}{T}\right) - \frac{\tau_i{}^k}{T}\frac{\partial u^i}{\partial x^k}. \tag{127.5}$$

The expression on the left must be the 4-divergence of the entropy flux, and that on the right the increase in entropy owing to dissipative processes. Thus the entropy flux density 4-vector is

$$\sigma^i = \sigma u^i - (\mu/T)v^i, \tag{127.6}$$

and τ_{ik} and v^i must be linear functions of the gradients of velocity and thermodynamic quantities, such as to make the right-hand side of equation (127.5) necessarily positive. This condition, together with (127.3) and (127.4), uniquely determines the form of the 4-tensor τ_{ik} and the 4-vector v_i:

$$\tau_{ik} = -\eta\left(\frac{\partial u_i}{\partial x^k} + \frac{\partial u_k}{\partial x^i} + u_k u^l\frac{\partial u_i}{\partial x^l} + u_i u^l\frac{\partial u_k}{\partial x^l}\right) - (\zeta - \tfrac{2}{3}\eta)\frac{\partial u_l}{\partial x}(g_{ik} + u_i u_k), \tag{127.7}$$

$$v_i = -\frac{\kappa}{c}\left(\frac{nT}{w}\right)^2\left[\frac{\partial}{\partial x^i}\left(\frac{\mu}{T}\right) + u_i u^k\frac{\partial}{\partial x^k}\left(\frac{\mu}{T}\right)\right]. \tag{127.8}$$

Here η and ζ are the two viscosity coefficients, and κ the thermal conductivity, taken in accordance with their non-relativistic definitions.

In particular, pure thermal conduction corresponds to an energy flux with no particle flux. The condition of zero particle flux is $nu^\alpha + v^\alpha = 0$; the energy flux density is then, as far as terms of the first order in the gradients,

$$cT_\alpha{}^0 = cwu^0 u_\alpha = -\frac{cw}{n}v_\alpha = \frac{\kappa nT^2}{w}\frac{\partial}{\partial x^\alpha}\left(\frac{\mu}{T}\right).$$

Using the thermodynamic identity

$$d(\mu/T) = -(w/nT^2)dT + dp/nT,$$

we find the energy flux $-\kappa[\mathbf{grad}\, T - (T/w)\mathbf{grad}\, p]$. We see that, in the relativistic case of thermal conduction, the heat flux is proportional, not to the temperature gradient simply, but to a certain combination of the temperature and pressure gradients.

CHAPTER XVI

DYNAMICS OF SUPERFLUIDS

§128. Principal properties of superfluids

AT temperatures close to absolute zero, quantum effects begin to be of importance in the properties of fluids. There is in Nature only one substance which remains fluid at absolute zero, namely helium; all other fluids solidify long before quantum effects become noticeable. At a temperature of 2·19°K, liquid helium has a λ-*point* (a second-order phase transition); at temperatures below this point liquid helium (helium II) has a number of remarkable properties, the most important of which is the *superfluidity* discovered by P. L. KAPITZA in 1938. This is the property of being able to flow without viscosity in narrow capillaries or gaps.†

The theory of superfluids was developed by L. LANDAU (1941). We shall discuss here only the part of the theory which gives a macroscopic description of the dynamical properties of superfluids.

The basis of the dynamics of helium II is the following fundamental result of the microscopic theory.‡ At temperatures other than zero, helium II behaves as if it were a mixture of two different liquids. One of these is a superfluid, and moves with zero viscosity along a solid surface. The other is a normal viscous fluid. It is of great importance that no friction occurs between these two parts of the liquid in their relative motion, i.e. no momentum is transferred from one to the other.

It should, however, be most decidedly emphasised that regarding the liquid as a mixture of normal and superfluid parts is no more than a convenient description of the phenomena which occur in a fluid where quantum effects are important. Like any description of quantum phenomena in classical terms, it falls short of adequacy. In reality, we ought to say that a quantum fluid, such as helium II, can execute two motions at once, each of which involves its own "effective mass" (the sum of the two effective masses being equal to the total mass of the fluid). One of these motions is normal, i.e. has the same properties as the motion of an ordinary viscous fluid, but the other is the motion of a superfluid. The two motions occur without any transfer of momentum from one to the other. We can, in a certain sense, speak of the superfluid and normal parts of the fluid, but this does not mean that the fluid can actually be separated into two such parts.

With careful note taken of these reservations concerning the true nature of the phenomena in helium II, we can use the terms *superfluid part* and *normal*

---

† Only one isotope of helium, He⁴, is a superfluid. The liquid isotope He³ does not become superfluid.

‡ See *Statistical Physics*, §§66, 67, Pergamon Press, London 1958.

part of the fluid to give a convenient concise description of these phenomena. We shall, however, prefer to use the more exact terms *superfluid flow* and *normal flow*, without associating them with the components of a "mixture of two parts" of the fluid.

The concept of two kinds of flow enables us to give a simple explanation of the main observed dynamical properties of helium II. The absence of viscosity when helium II flows in a narrow passage is the result of frictionless superfluid flow in the passage; we can say that the normal part remains in the vessel, flowing much more slowly through the passage at a velocity in accordance with its viscosity and the passage width. The measurement of the viscosity of helium II from the damping of torsional oscillations of a disk immersed in it, on the other hand, gives non-zero values; the rotation of the disk causes a normal flow near it, which brings the disk to rest by virtue of the viscosity pertaining to that flow. Thus, in experiments on flow through a capillary, the superfluid flow is observed, whereas in experiments on the rotation of a disk in helium II the normal flow is observed. The existence of these two flows is seen especially clearly when a cylindrical vessel filled with helium II rotates about its axis. The walls of the rotating cylinder cause a normal flow and carry with them only part of the fluid, the superfluid part remaining at rest. Consequently, the total moment of inertia I of the rotating vessel is less than the moment of inertia I_0 calculated on the assumption that the whole fluid rotates with the vessel, and a measurement of the ratio I/I_0 enables us to find at once what parts of the fluid are normal and superfluid.

Besides the absence of viscosity, the superfluid flow has two other important properties: it does not involve heat transfer, and it is always potential flow. Both these properties also follow from the microscopic theory, according to which the normal flow is actually the flow of an "excitation gas" (we may recall that the collective thermal motion of the atoms in a quantum fluid can be regarded as a system of excitations, which behave like quasi-particles moving in the volume occupied by the fluid and have definite momenta and energies).

The entropy of helium II is determined by the statistical distribution of the elementary excitations. In any flow, therefore, in which the excitation gas is at rest, there is no macroscopic transfer of entropy. This means that the superfluid flow involves no entropy transfer, and therefore no heat transfer. Hence it follows that a superfluid flow of helium II is thermodynamically reversible, a result actually found.

The transfer of heat by the normal flow is the only mechanism of heat transfer in helium II. It is therefore of the nature of convection, and is fundamentally different from ordinary thermal conduction. Any difference of temperature in helium II causes internal flow, both normal and superfluid; the two flows may balance as regards mass transfer, so that no macroscopic mass transfer occurs in the fluid.

In what follows we shall denote by $\mathbf{v}_s$ and $\mathbf{v}_n$ the velocities of the superfluid and normal flow respectively. The heat-transfer mechanism described above means that the entropy flux density is the product $\mathbf{v}_n\rho s$ of the velocity $\mathbf{v}_n$ and the entropy per unit volume (s being the entropy per unit mass). The heat flux density is obtained by multiplying the entropy flux density by T, i.e. it is

$$\mathbf{q} = \rho T s \mathbf{v}_n. \tag{128.1}$$

The potential flow of the superfluid part corresponds to the equation

$$\mathbf{curl}\,\mathbf{v}_s = 0, \tag{128.2}$$

which must hold at any instant throughout the volume of the fluid. This property is the macroscopic expression of a property of the helium II energy spectrum which underlies the microscopic theory of superfluidity: the elementary excitations which have long wavelengths (i.e. small momenta and energies) are sound quanta or *phonons*. Hence the macroscopic superfluid dynamics can include only sound vibrations, a result which follows from the condition (128.2).

Since it is potential flow, a steady superfluid flow exerts no force on a solid body (d'Alembert's paradox; see §11). The normal flow, on the other hand, exerts a drag force. If the flow is such that the superfluid and normal mass transfers balance, we have a very unusual flow: a force acts on a body immersed in helium II, but there is no net mass transfer.

§129. The thermo-mechanical effect

The *thermo-mechanical effect* in helium II is as follows: when helium flows out of a vessel through a narrow capillary, a rise in temperature occurs in the vessel, and a cooling where the helium flows out of the capillary into another vessel.† This phenomenon has the natural explanation that the flow into a capillary is mainly superfluid, and therefore transfers no heat, so that the heat remaining in the vessel is distributed over a smaller quantity of helium II. In flow out of a capillary the opposite effect is seen.

It is easy to find the quantity of heat Q absorbed when unit mass of helium enters a vessel through a capillary. The incoming fluid transfers no entropy. If the helium in the vessel were to remain at its initial temperature T, an amount of heat Ts would be needed, to compensate the decrease in entropy per unit mass due to the addition of unit mass of helium of zero entropy. This means that, when unit mass of helium enters a vessel containing helium at temperature T, an amount of heat

$$Q = Ts \tag{129.1}$$

is absorbed. Conversely when unit mass of helium leaves a vessel containing helium at temperature T, an amount of heat Ts is evolved.

---

† A very slight thermo-mechanical effect must, strictly speaking, occur for any fluid; the anomaly in helium II is the magnitude of the effect. The effect in ordinary fluids is an irreversible phenomenon similar to the thermo-electric Peltier effect (and is actually observed in rarefied gases). Such an effect occurs in helium II also, but is masked by another considerably larger effect described below, which occurs only in helium II and is not an irreversible phenomenon like the Peltier effect.

Let us now consider two vessels containing helium II at temperatures T_1 and T_2, connected by a narrow capillary. Since the superfluid can flow freely along the capillary, mechanical equilibrium is rapidly established. The superfluid, however, does not transfer heat, and so thermal equilibrium (in which the temperature of the helium in the two vessels is the same) is established considerably more slowly.

The condition of mechanical equilibrium is easily written down by using the fact that this equilibrium is established for constant entropies s_1, s_2 of the helium in the two vessels. If ϵ_1, ϵ_2 are the internal energies per unit mass of helium at temperatures T_1, T_2, the condition of mechanical equilibrium (minimum energy) effected by superfluid flow is $(\partial \epsilon_1/\partial N)_{s_1} = (\partial \epsilon_2/\partial N)_{s_2}$, where N is the number of atoms in unit mass of helium. The derivative $(\partial \epsilon/\partial N)_s$ is the chemical potential μ. We therefore obtain the equilibrium condition

$$\mu(p_1, T_1) = \mu(p_2, T_2), \tag{129.2}$$

where p_1 and p_2 are the pressures in the two vessels.

In what follows we shall understand by the chemical potential μ not the usual thermodynamic potential per particle (atom), but the thermodynamic potential per unit mass of helium. These differ only by a constant factor, the mass of a helium atom.

If the pressures p_1, p_2 are small, then, expanding in powers of the pressures and recalling that $(\partial \mu/\partial p)_T$ is the specific volume (which depends only slightly on the temperature), we obtain

$$\frac{\Delta p}{\rho} = \mu(0, T_1) - \mu(0, T_2) = \int_{T_1}^{T_2} s \, dT,$$

where $\Delta p = p_2 - p_1$. If the temperature difference $\Delta T = T_2 - T_1$ is also small, then, expanding in powers of ΔT and recalling that $(\partial \mu/\partial T)_p = -s$, we obtain

$$\Delta p/\Delta T = \rho s. \tag{129.3}$$

Since $s > 0$, $\Delta p/\Delta T > 0$. The relation (129.3) was first derived by H. London (1939).

§130. The equations of superfluid dynamics

We shall now derive a complete system of equations describing macroscopically (phenomenologically) the flow of helium II. From the above discussion, we are concerned with equations of motion which involve at every point two velocities $\mathbf{v}_s$ and $\mathbf{v}_n$, and not one as in ordinary fluid dynamics. It is found that the required system of equations can be uniquely determined simply from the requirements imposed by Galileo's relativity principle and by the necessary conservation laws (using also the properties of the motion expressed by equations (128.1) and (128.2)).

It should be borne in mind that helium II actually ceases to be superfluid at high velocities. We shall not discuss the nature of this phenomenon of *critical velocities*, but merely note that its existence means that the equations of superfluid dynamics for helium II are physically significant only when the velocities v_s and v_n are not too large. Nevertheless, we shall first derive these equations without making any assumptions concerning the velocities $\mathbf{v}_s$ and $\mathbf{v}_n$, since, if higher powers of the velocities are neglected, the equations cannot be consistently derived from the conservation laws. The transition to the physically significant case of small velocities will be made in the final equations.

We denote by $\mathbf{j}$ the mass flux density; this quantity is also the momentum of unit volume (cf. the footnote to §49). We write

$$\mathbf{j} = \rho_s \mathbf{v}_s + \rho_n \mathbf{v}_n \tag{130.1}$$

as a sum of the fluxes pertaining to the superfluid and normal flows. The coefficients ρ_s and ρ_n may be called the superfluid and normal densities. Their sum is the actual density ρ of helium II:

$$\rho = \rho_s + \rho_n. \tag{130.2}$$

The quantities ρ_s and ρ_n are, of course, functions of the temperature; ρ_n vanishes at absolute zero, where helium II becomes wholly superfluid,[†] while ρ_s vanishes at the λ-point, where the liquid becomes wholly normal. It should also be noted that ρ_n and ρ_s in general depend on the velocities themselves;[‡] only at small velocities can this dependence be neglected, and ρ_n, ρ_s regarded as functions of the temperature (and pressure) only.

The density ρ and the flux $\mathbf{j}$ must satisfy the equation of continuity

$$\partial\rho/\partial t + \operatorname{div}\mathbf{j} = 0, \tag{130.3}$$

which expresses the law of conservation of mass. The law of conservation of momentum gives an equation

$$\frac{\partial j_i}{\partial t} + \frac{\partial \Pi_{ik}}{\partial x_k} = 0, \tag{130.4}$$

where Π_{ik} is the momentum flux density tensor.

We shall not at present consider dissipative processes. Then the flow is reversible, and the entropy of the fluid is also conserved. Since the entropy flux is $\rho s \mathbf{v}_n$, we can write the law of conservation of entropy as

$$\partial(\rho s)/\partial t + \operatorname{div}(\rho s \mathbf{v}_n) = 0. \tag{130.5}$$

Equations (130.3)–(130.5) must be supplemented by an equation which gives the time derivative of the velocity $\mathbf{v}_s$. This equation must be such that

---

† If the helium II contains an admixture (of the isotope He³), then ρ_n is not zero even at 0°K.
‡ More precisely, on the velocity difference $\mathbf{v}_n - \mathbf{v}_s$, since the movement of the fluid as a whole with velocity $\mathbf{v}_s = \mathbf{v}_n$ cannot, of course, affect its thermodynamic properties.

we have potential flow at all times; this means that the derivative of $\mathbf{v}_s$ must be the gradient of a scalar. We can write the equation as

$$\frac{\partial \mathbf{v}_s}{\partial t} + \mathbf{grad}(\tfrac{1}{2}v_s^2 + \mu) = 0, \tag{130.6}$$

where μ is some scalar.

Equations (130.4) and (130.6) become significant, of course, only when we obtain values for the still undefined quantities Π_{ik} and μ. To do so, we must use the law of conservation of energy and arguments based on Galileo's relativity principle. The equations (130.3)–(130.6) must imply the law of conservation of energy, which is expressed by an equation of the form

$$\partial E/\partial t + \operatorname{div}\mathbf{Q} = 0, \tag{130.7}$$

where E is the energy in unit volume of the fluid and $\mathbf{Q}$ the energy flux density. Galileo's relativity principle enables us to determine all quantities as functions of one velocity ($\mathbf{v}_s$) and the given relative velocity $\mathbf{v}_n - \mathbf{v}_s$ of the two simultaneous motions.

We use both the original co-ordinate system K and a system K_0 in which the velocity of the superfluid flow of a given fluid element is zero. The system K_0 moves relative to the system K with a velocity equal to the superfluid velocity in the original system. The values of all quantities in the system K are related to their values in K_0 (which we distinguish by the suffix 0) by the following transformation formulae of mechanics:[†]

$$\mathbf{j} = \rho\mathbf{v}_s + \mathbf{j}_0,$$
$$E = \tfrac{1}{2}\rho v_s^2 + \mathbf{j}_0\cdot\mathbf{v}_s + E_0,$$
$$\mathbf{Q} = (\tfrac{1}{2}\rho v_s^2 + \mathbf{j}_0\cdot\mathbf{v}_s + E_0)\mathbf{v}_s + \tfrac{1}{2}v_s^2\mathbf{j}_0 + \Pi_0\cdot\mathbf{v}_s + \mathbf{Q}_0, \tag{130.8}$$
$$\Pi_{ik} = \rho v_{si}v_{sk} + v_{si}j_{0k} + v_{sk}j_{0i} + \Pi_{0ik}.$$

Here $\Pi_0\cdot\mathbf{v}_s$ denotes the vector whose components are $\Pi_{0ik}v_{sk}$.

In the system K_0, the fluid element considered executes only one motion, a normal flow with velocity $\mathbf{v}_n - \mathbf{v}_s$. Hence the quantities $\mathbf{j}_0$, E_0, $\mathbf{Q}_0$ and Π_{0ik} can depend only on the difference $\mathbf{v}_n - \mathbf{v}_s$, and not on $\mathbf{v}_n$ and $\mathbf{v}_s$ separately; in particular, the vectors $\mathbf{j}_0$ and $\mathbf{Q}_0$ must be parallel to the vector $\mathbf{v}_n - \mathbf{v}_s$ (the mass flux $\mathbf{j}_0$ is simply $\rho_n(\mathbf{v}_n - \mathbf{v}_s)$). Thus formulae (130.8) give the dependence of the quantities concerned on $\mathbf{v}_s$ for given $\mathbf{v}_n - \mathbf{v}_s$.

[†] These formulae are a direct consequence of Galileo's relativity principle, and therefore hold for any particular system. They can be derived by considering, for instance, an ordinary fluid. The momentum flux density tensor in ordinary fluid dynamics is $\Pi_{ik} = \rho v_i v_k + p\delta_{ik}$. The fluid velocity $\mathbf{v}$ in the system K is related to the velocity $\mathbf{v}_0$ in K_0 by $\mathbf{v} = \mathbf{v}_0 + \mathbf{u}$, where $\mathbf{u}$ is the relative velocity of the two systems. Substituting in Π_{ik}, we have

$$\Pi_{ik} = p\delta_{ik} + \rho v_{0i}v_{0k} + \rho v_{0i}u_k + \rho u_i v_{0k} + \rho u_i u_k.$$

Putting $\Pi_{0ik} = p\delta_{ik} + \rho v_{0i}v_{0k}$ and $\mathbf{j}_0 = \rho\mathbf{v}_0$, we obtain the transformation formula for the tensor Π_{ik} given in (130.8). The remaining formulae are obtained similarly.

The energy E_0, as a function of ρ, s and the momentum $\mathbf{j}_0$ per unit volume, satisfies the thermodynamic identity

$$dE_0 = \mu \, d\rho + T d(\rho s) + (\mathbf{v}_n - \mathbf{v}_s) \cdot d\mathbf{j}_0, \qquad (130.9)$$

where μ is the (thermodynamic) chemical potential per unit mass. The first two terms correspond to the usual thermodynamic identity for a fluid at rest with constant volume (in this case unity), and the last term shows that the derivative of the energy with respect to the momentum is the velocity.

We shall not give here the subsequent calculations, which are fairly laborious, but give only their general outline. In the equation of conservation of energy (130.7) we substitute E and $\mathbf{Q}$ from (130.8), calculating the derivative $\partial E_0/\partial t$ by means of the identity (130.9). We then eliminate all the time derivatives ($\dot\rho$, $\dot{\mathbf{v}}_s$, etc.) by means of the equations (130.3)–(130.6); the equation of conservation of energy must then be satisfied identically. If we take into account the fact that the fluxes $\mathbf{Q}_0$, Π_{0ik} and the scalar μ in equation (130.6) can depend only on the thermodynamic variables and the velocity $\mathbf{v}_n - \mathbf{v}_s$, and not on their gradients (since we neglect dissipative processes), we find that this identity can be achieved only if the quantities mentioned above are chosen in a uniquely defined way.

It is found also that the scalar μ is the chemical potential (for which reason we have denoted it by the same letter), and the final expressions for the energy flux density and the momentum flux density tensor are

$$\mathbf{Q} = (\mu + \tfrac{1}{2}v_s^2)\mathbf{j} + T\rho s\mathbf{v}_n + \rho_n \mathbf{v}_n[\mathbf{v}_n \cdot (\mathbf{v}_n - \mathbf{v}_s)], \qquad (130.10)$$

$$\Pi_{ik} = \rho_n v_{ni} v_{nk} + \rho_s v_{si} v_{sk} + p\delta_{ik}, \qquad (130.11)$$

where

$$p = -E_0 + T\rho s + \mu\rho + \rho_n(\mathbf{v}_n - \mathbf{v}_s)^2. \qquad (130.12)$$

The expression for Π_{ik} is the natural generalisation of the formula $\Pi_{ik} = \rho v_i v_k + p\delta_{ik}$ of ordinary fluid dynamics. The quantity p defined by formula (130.12) can be naturally regarded as the fluid pressure.†

Equations (130.3)–(130.6), with $\mathbf{j}$ and Π_{ik} defined by (130.10) and (130.11), form the required complete system of equations of superfluid dynamics.‡ These are very complex, largely because the quantities ρ_s, ρ_n, μ, etc. which appear in the equations are functions of the velocities (more precisely, of the difference $\mathbf{v}_n - \mathbf{v}_s$). The form of these functions can in principle be determined only from the microscopic theory.

---

† The usual thermodynamic definition of the pressure as the mean force acting on unit area relates to a medium at rest. In ordinary fluid dynamics, however, there is no ambiguity in the definition of pressure (dissipative processes being neglected), since we can always take a co-ordinate system in which the fluid volume element considered is at rest. In superfluid dynamics, however, we can eliminate only one of the two simultaneous motions by a suitable choice of the co-ordinate system, and so the usual definition of pressure cannot be applied.

In a fluid entirely at rest, the definition (130.12) is of course the same as the ordinary definition, since in that case $\mu\rho + T\rho s - E_0 = p$ by the definition of the chemical potential.

‡ The system of equations can also be established in a general form for a mixture of helium II with other substances (in practice, the isotope He^3), for any concentrations. This is due to I. M. KHALATNIKOV (*Zhurnal éksperimental'noi i teoreticheskoi fiziki* 23, 169, 1952).

The equations are much simplified, however, in the physically interesting case of small velocities.† In this case we can first of all neglect, as already mentioned, the velocity dependence of ρ_n and ρ_s; then the expression (130.1) for the flux **j** gives essentially the first terms in an expansion of **j** in powers of $\mathbf{v}_s$ and $\mathbf{v}_n$. The expansion in powers of the velocities must also be made for the other thermodynamic quantities appearing in the equations.

We take the pressure and temperature as independent thermodynamic variables. The thermodynamic identity for the chemical potential is

$$d\mu = -s\,dT + (1/\rho)dp - (\rho_n/\rho)(\mathbf{v}_n - \mathbf{v}_s)\cdot d(\mathbf{v}_n - \mathbf{v}_s);$$

this can be obtained by differentiating the expression (130.12) and using the identity (130.9). Hence we see that the first two terms in the expansion of μ in powers of the velocity difference are

$$\mu(p, T, \mathbf{v}_n - \mathbf{v}_s) \approx \mu(p, T) - \tfrac{1}{2}(\rho_n/\rho)(\mathbf{v}_n - \mathbf{v}_s)^2, \qquad (130.13)$$

where the right-hand side contains the ordinary chemical potential $\mu(p, T)$ and density $\rho(p, T)$ of the fluid at rest. Differentiating with respect to temperature and pressure, we find the corresponding expansions for entropy and density:

$$s(p, T, \mathbf{v}_n - \mathbf{v}_s) \approx s(p, T) + \tfrac{1}{2}(\mathbf{v}_n - \mathbf{v}_s)^2 \frac{\partial}{\partial T}\left(\frac{\rho_n}{\rho}\right), \qquad (130.14)$$

$$\rho(p, T, \mathbf{v}_n - \mathbf{v}_s) \approx \rho(p, T) + \tfrac{1}{2}\rho^2(\mathbf{v}_n - \mathbf{v}_s)^2 \frac{\partial}{\partial p}\left(\frac{\rho_n}{\rho}\right). \qquad (130.15)$$

These expressions are to be substituted in the dynamical equations, which are then valid as far as second-order terms in the velocities.

Let us briefly consider the subject of the dissipative terms in the equations of superfluid dynamics. The form of these terms is restricted only by the conditions imposed by the law of increase of entropy and by the symmetry of the kinetic coefficients. A detailed analysis, due to I. M. KHALATNIKOV,‡ shows that there are five independent dissipation coefficients (instead of the three coefficients η, ζ, κ for an ordinary fluid). Of these, one is the first viscosity η, due to the normal flow and entirely analogous to the viscosity of an ordinary fluid. The momentum flux tensor Π_{ik} and the quantity whose gradient appears in (130.6) involve further terms proportional to div $\mathbf{v}_n$ and div $[\rho_s(\mathbf{v}_n - \mathbf{v}_s)]$; of the four proportionality coefficients, two are equal on account of the symmetry of the kinetic coefficients, so that there are three "second viscosities" ζ_1, ζ_2, ζ_3. Finally, the right-hand side of the entropy equation (130.5) involves a term of the form $(1/T)$ div $(\kappa\,\mathbf{grad}\,T)$, with a coefficient κ which is formally analogous to the thermal conductivity of an ordinary fluid, and also terms quadratic in the velocity gradients, which result from the viscosity effect in heat transfer (cf. equation (49.5)).

---

† That is, when the ratio of the velocities to the velocity of propagation of second sound (see §131) is a small quantity.

‡ See *Zhurnal éksperimental'noǐ i teoreticheskoǐ fiziki* 23, 265, 1952.

The boundary conditions on the equations of superfluid dynamics are as follows. Firstly, the perpendicular component of the mass flux **j** must vanish at any solid surface at rest. To determine the conditions on $\mathbf{v}_n$, we must recall that the normal flow is actually a flow of a thermal excitation gas. In flow along a solid surface, the excitation quanta interact with the surface, and this must be described macroscopically as the "adhesion" of the normal fluid to the surface, as in ordinary viscous fluids. In other words, the tangential component of the velocity $\mathbf{v}_n$ must be zero at a solid surface.

The component of $\mathbf{v}_n$ perpendicular to the surface need not vanish, since the excitation quanta can be absorbed or emitted by the surface, corresponding simply to heat transfer between the fluid and the surface. The boundary condition requires only that the heat flux perpendicular to the surface is continuous. The temperature itself has a discontinuity at the boundary which is proportional to the heat flux: $\Delta T = Kq$, with a proportionality coefficient which depends on the properties of both the fluid and the solid. The occurrence of this discontinuity is due to the peculiar nature of heat transfer in helium II. All the resistance to heat transfer between the solid and the fluid is in the fluid adjoining the surface, since the convective propagation of heat in the fluid meets with almost no resistance. Consequently, the whole of the temperature drop which causes the heat transfer occurs at the surface itself.

An interesting property of these boundary conditions is that the heat exchange between the solid surface and the moving fluid results in tangential forces on the surface. If the x-axis is perpendicular to the surface, and the y-axis tangential, the tangential force per unit area is equal to the component Π_{xy} of the momentum flux tensor. Since we must have $j_x = \rho_n v_{nx} + \rho_s v_{sx} = 0$ on the surface, we find for this force the non-zero expression $\Pi_{xy} = \rho_s v_{sx} v_{sy} + \rho_n v_{nx} v_{ny} = \rho_n v_{nx}(v_{ny} - v_{sy})$. We can write this in terms of the heat flux $\mathbf{q} = \rho s T \mathbf{v}_n$ as $\Pi_{xy} = (\rho_n/\rho s T) q_x (v_{ny} - v_{sy})$, where q_x is the heat flux from the solid surface to the fluid, which is continuous at the surface.

In the absence of heat transfer between the solid surface and the fluid, the component of $\mathbf{v}_n$ perpendicular to the surface is also zero. The boundary conditions $j_x = 0$ and $\mathbf{v}_n = 0$ (with the x-axis perpendicular to the surface) are equivalent to $v_{sx} = 0$ and $\mathbf{v}_n = 0$. In this case, therefore, we obtain the usual boundary conditions for an ideal fluid for $\mathbf{v}_s$, and those for a viscous fluid for $\mathbf{v}_n$.

Finally, let us consider motions of helium II for which it may be regarded as incompressible, as usually happens in flow past bodies. We shall also take into account the viscosity of the normal flow. To do so, we must add to the tensor Π_{ik} a term which involves, as usual, the viscosity coefficient η and the spatial derivatives of the velocity $\mathbf{v}_n$:

$$\Pi_{ik} = p\delta_{ik} + \rho_s v_{si} v_{sk} + \rho_n v_{ni} v_{nk} - \eta\left(\frac{\partial v_{ni}}{\partial x_k} + \frac{\partial v_{nk}}{\partial x_i}\right). \qquad (130.16)$$

The second viscosity coefficients do not appear for an incompressible fluid.

The dissipative terms in the entropy equation are, in the case considered, small quantities of higher order, and can be neglected. Assuming the densities ρ_n, ρ_s and the entropy s to be constants, we obtain from equation (130.5) div $\mathbf{v}_n = 0$ and from (130.3) div $\mathbf{j} = 0$, so that div $\mathbf{v}_s =$ div $\mathbf{v}_n = 0$. Using these equations and substituting (130.16) in (130.4), we obtain the equation

$$\rho_s \frac{\partial \mathbf{v}_s}{\partial t} + \rho_n \frac{\partial \mathbf{v}_n}{\partial t} + \rho_s(\mathbf{v}_s \cdot \mathbf{grad})\mathbf{v}_s + \rho_n(\mathbf{v}_n \cdot \mathbf{grad})\mathbf{v}_n$$

$$= -\mathbf{grad}\, p + \eta \triangle \mathbf{v}_n. \tag{130.17}$$

Equation (130.6) remains unchanged.

Since the superfluid flow is potential flow, we can introduce the velocity potential by $\mathbf{v}_s = \mathbf{grad}\,\phi_s$ and, since div $\mathbf{v}_s = 0$, the potential will satisfy Laplace's equation

$$\triangle \phi_s = 0. \tag{130.18}$$

Introducing ϕ_s in equation (130.17) and putting $(\mathbf{v}_s \cdot \mathbf{grad})\mathbf{v}_s = \mathbf{grad}\,\tfrac{1}{2}v_s^2$, we obtain

$$\rho_n \frac{\partial \mathbf{v}_n}{\partial t} + \rho_n(\mathbf{v}_n \cdot \mathbf{grad})\mathbf{v}_n + \rho_s\,\mathbf{grad}\,\tfrac{1}{2}v_s^2 + \rho_s\,\mathbf{grad}(\partial\phi_s/\partial t)$$

$$= -\mathbf{grad}\, p + \eta \triangle \mathbf{v}_n.$$

We use as auxiliary quantities the "pressures" p_n, p_s of the normal and superfluid flows:

$$p = p_0 + p_n + p_s, \tag{130.19}$$

where p_0 is the pressure at infinity, and p_s is defined by the usual formula for an ideal fluid,

$$p_s = -\rho_s\,\partial\phi_s/\partial t - \tfrac{1}{2}\rho_s v_s^2. \tag{130.20}$$

The equation for the velocity $\mathbf{v}_n$ then becomes

$$\frac{\partial \mathbf{v}_n}{\partial t} + (\mathbf{v}_n \cdot \mathbf{grad})\mathbf{v}_n = -\frac{1}{\rho_n}\mathbf{grad}\,p_n + \frac{\eta}{\rho_n}\triangle \mathbf{v}_n. \tag{130.21}$$

This equation is formally identical with the Navier–Stokes equation for a fluid of density ρ_n and viscosity η (and therefore kinematic viscosity η/ρ_n).

Thus the problem of the flow of incompressible helium II reduces to two problems of ordinary fluid dynamics, one for an ideal fluid and the other for a viscous fluid. The superfluid flow is determined by Laplace's equation (130.18) with a boundary condition on the normal derivative $\partial\phi_s/\partial n$, as in the ordinary problem of potential flow of an ideal fluid past a body. The normal flow is determined by the Navier–Stokes equation (130.21), with the same boundary conditions on $\mathbf{v}_n$ (in the absence of heat exchange between the surface and the fluid) as in ordinary flow of a viscous fluid. The pressure distribution is then determined by formula (130.19).

PROBLEMS

PROBLEM 1. A small temperature difference ΔT is maintained between the ends of a capillary containing helium II. Determine the heat flux along the capillary.

SOLUTION. According to formula (129.3), the pressure drop between the two ends of the capillary is $\Delta p = \rho s \Delta T$. This causes a normal flow whose mean (over the cross-section of the capillary) velocity is $\bar{v}_n = R^2 \Delta p / 8 \eta l$ (R being the radius and l the length of the capillary, and η the viscosity of normal flow; cf. formula (17.10)). The total heat flux is $\rho s \bar{v}_n \pi R^2 = \pi R^4 \rho^2 s^2 \Delta T / 8 \eta l$. A superfluid flow occurs in the opposite direction, its velocity being given by the condition of zero total mass transfer: $\bar{v}_s = -\rho_n \bar{v}_n / \rho_s$.

PROBLEM 2. Derive the formula for the temperature distribution in helium II in incompressible flow.

SOLUTION. Writing in equation (130.6) (with μ given by (130.13)) $\mathbf{v}_s = \mathbf{grad}\, \phi_s$ and integrating, we obtain $\mu(p, T) + \frac{1}{2}v_s^2 - \frac{1}{2}(\rho_n/\rho)(\mathbf{v}_n - \mathbf{v}_s)^2 + \partial \phi_s / \partial t = $ constant. The changes of temperature and pressure in an incompressible fluid are small, and we have as far as terms of the first order $\mu - \mu_0 = -s(T - T_0) + (p - p_0)/\rho$, where T_0 and p_0 are the temperature and pressure at infinity. Substituting this expression in the above integral, and using p_n and p_s, we obtain

$$T - T_0 = \frac{\rho_n}{\rho s} \left[\frac{p_n}{\rho_n} - \frac{p_s}{\rho_s} - \frac{1}{2}(\mathbf{v}_n - \mathbf{v}_s)^2 \right].$$

§131. The propagation of sound in a superfluid

Let us apply the equations of fluid dynamics for helium II to the propagation of sound in it. As usual, the velocities in the sound wave are supposed small, and the density, pressure and entropy almost equal to their constant equilibrium values. Then we can linearise the equations, neglecting the terms quadratic in the velocity in (130.11), (130.13) and (130.14), and regard the entropy ρs as constant in the term $\mathrm{div}(\rho s \mathbf{v}_n)$ in (130.5) (since the term already contains the small quantity $\mathbf{v}_n$). Thus the equations of fluid dynamics become

$$\partial \rho / \partial t + \mathrm{div}\, \mathbf{j} = 0, \tag{131.1}$$

$$\partial(\rho s)/\partial t + \rho s\, \mathrm{div}\, \mathbf{v}_n = 0, \tag{131.2}$$

$$\partial \mathbf{j} / \partial t + \mathbf{grad}\, p = 0, \tag{131.3}$$

$$\partial \mathbf{v}_s / \partial t + \mathbf{grad}\, \mu = 0. \tag{131.4}$$

Differentiating (131.1) with respect to time and substituting (131.3), we obtain

$$\partial^2 \rho / \partial t^2 = \triangle p. \tag{131.5}$$

By the thermodynamic identity $\mathrm{d}\mu = -s\mathrm{d}T + \mathrm{d}p/\rho$, we have $\mathbf{grad}\, p = \rho s\, \mathbf{grad}\, T + \rho\, \mathbf{grad}\, \mu$. Substituting $\mathbf{grad}\, p$ from (131.3) and $\mathbf{grad}\, \mu$ from (131.4), we obtain

$$\rho n \frac{\partial}{\partial t}(\mathbf{v}_n - \mathbf{v}_s) + \rho s\, \mathbf{grad}\, T = 0.$$

We take the divergence of this equation, substituting for $\operatorname{div}(\mathbf{v}_n - \mathbf{v}_s)$ the expression $(\rho/s\rho_s)\,\partial s/\partial t$, which follows from the equation

$$\frac{\partial s}{\partial t} = \frac{1}{\rho}\frac{\partial(\rho s)}{\partial t} - \frac{s}{\rho}\frac{\partial\rho}{\partial t}$$

$$= -s\operatorname{div}\mathbf{v}_n + (s/\rho)\operatorname{div}\mathbf{j}$$

$$= (s\rho_s/\rho)\operatorname{div}(\mathbf{v}_s - \mathbf{v}_n).$$

The result is

$$\partial^2 s/\partial t^2 = (\rho_s s^2/\rho_n)\triangle T. \tag{131.6}$$

Equations (131.5) and (131.6) determine the propagation of sound in a superfluid. Since there are two equations, we see that there are two velocities of propagation of sound.

We write s, p, ρ and T as $s = s_0 + s'$, $p = p_0 + p'$, etc., where the primed letters are the small changes in the corresponding quantities in the sound wave, and those with the suffix zero (which we omit, for brevity) their constant equilibrium values. Then we can write

$$\rho' = \frac{\partial\rho}{\partial p}p' + \frac{\partial\rho}{\partial T}T', \qquad s' = \frac{\partial s}{\partial p}p' + \frac{\partial s}{\partial T}T',$$

and the equations (131.5) and (131.6) become

$$\frac{\partial\rho}{\partial p}\frac{\partial^2 p'}{\partial t^2} - \triangle p' + \frac{\partial\rho}{\partial T}\frac{\partial^2 T'}{\partial t^2} = 0,$$

$$\frac{\partial s}{\partial p}\frac{\partial^2 p'}{\partial t^2} + \frac{\partial s}{\partial T}\frac{\partial^2 T'}{\partial t^2} - \frac{\rho_s s^2}{\rho_n}\triangle T' = 0.$$

We seek a solution of these equations in the form of a plane wave, in which p' and T' are proportional to a factor $e^{-i\omega(t-x/u)}$ (the velocity of sound being here denoted by u). The condition of compatibility of the two equations is

$$u^4\frac{\partial(s,\rho)}{\partial(T,p)} - u^2\left(\frac{\partial s}{\partial T} + \frac{\rho_s s^2}{\rho_n}\frac{\partial\rho}{\partial p}\right) + \frac{\rho_s s^2}{\rho_n} = 0,$$

where $\partial(s,\rho)/\partial(T,p)$ denotes the Jacobian of the transformation from s, ρ to T, p. By a simple transformation, using the thermodynamic relations, this equation can be reduced to

$$u^4 - u^2\left[\left(\frac{\partial p}{\partial\rho}\right)_s + \frac{\rho_s T s^2}{\rho_n c_v}\right] + \frac{\rho_s T s^2}{\rho_n c_v}\left(\frac{\partial p}{\partial\rho}\right)_T = 0, \tag{131.7}$$

c_v being the specific heat per unit mass. This quadratic equation in u^2 gives the two velocities of propagation of sound in helium II. For $\rho_s = 0$, one root is zero, and we obtain, as we should expect, only the ordinary velocity of sound $u = \sqrt{(\partial p/\partial\rho)_s}$.

The specific heats c_p and c_v of helium II are actually very nearly the same at all temperatures (since the coefficient of thermal expansion is small). By a well-known thermodynamic formula, the isothermal and adiabatic compressibilities are then very nearly the same also: $(\partial p/\partial \rho)_T \approx (\partial p/\partial \rho)_s$. Denoting the common value of c_p and c_v by c, and that of $(\partial p/\partial \rho)_T$ and $(\partial p/\partial \rho)_s$ by $\partial p/\partial \rho$, we obtain from equation (131.7) the following expressions for the velocities of sound:

$$u_1 = \sqrt{(\partial p/\partial \rho)}, \qquad u_2 = \sqrt{(Ts^2 \rho_s/c\rho_n)}. \tag{131.8}$$

One of these, u_1, is almost constant, while the other, u_2, depends markedly on temperature, vanishing with ρ_s at the λ-point.[†]

At very low temperatures, where nearly all the elementary excitations in the fluid are phonons, the quantities ρ_n, c and s are related by[‡] $c = 3s$, $\rho_n = cT\rho/3u_1^2$, and $\rho_s \cong \rho$. Substituting these expressions in formula (131.8) for u_2, we find $u_2 = u_1/\sqrt{3}$. Thus, as the temperature tends to zero, the velocities u_1 and u_2 tend to finite limits, and their ratio tends to $\sqrt{3}$.

In order to elucidate more clearly the physical nature of the two kinds of sound wave in helium II, let us consider a plane sound wave (E. LIFSHITZ 1944). In such a wave, the velocities $\mathbf{v}_n$, $\mathbf{v}_s$ and the variable parts T', p' of the temperature and pressure are proportional to one another. We introduce proportionality coefficients by

$$\mathbf{v}_n = a\mathbf{v}_s, \qquad p' = bv_s, \qquad T' = cv_s. \tag{131.9}$$

A simple calculation, using equations (131.1)–(131.6), and working to the necessary accuracy, gives for *first sound*

$$a_1 = 1 + \frac{\beta\rho}{\rho_s s}\frac{u_1^2 u_2^2}{(u_1^2 - u_2^2)}, \qquad b_1 = \rho u_1, \qquad c_1 = \frac{\beta T u_1^3}{c(u_1^2 - u_2^2)}, \tag{131.10}$$

and for *second sound*

$$a_2 = -\frac{\rho_s}{\rho_n} + \frac{\beta\rho}{\rho_n s}\frac{u_1^2 u_2^2}{(u_1^2 - u_2^2)}, \qquad b_2 = \frac{\beta\rho u_1^2 u_2^3}{s(u_1^2 - u_2^2)}, \qquad c_2 = -u_2/s. \tag{131.11}$$

Here $\beta = -(1/\rho)\partial\rho/\partial T$ is the coefficient of thermal expansion; since it is small, the quantities which involve β are small in comparison with those which do not.

We see that, in a sound wave of the first type, $\mathbf{v}_n \approx \mathbf{v}_s$, i.e. to a first approximation the fluid in any given volume element oscillates as a whole in such a wave, the normal and superfluid parts moving together. This type of wave clearly corresponds to an ordinary sound wave in an ordinary fluid.

---

† The problem of sound propagation in solutions of admixtures in helium II is discussed by I. YA. POMERANCHUK, *Zhurnal éksperimental'noĭ i teoretícheskoĭ fiziki* 19, 42, 1949, for the case of small concentrations, and by I. M. KHALATNIKOV, *ibid.* 23, 265, 1952, for arbitrary concentrations.

‡ See *Statistical Physics*, §§66, 67.

In a wave of the second type, however, we have $\mathbf{v}_n \approx -\rho_s\mathbf{v}_s/\rho_n$, i.e. the total flux density $\mathbf{j} = \rho_s\mathbf{v}_s + \rho_n\mathbf{v}_n \approx 0$. Thus, in a second-sound wave the superfluid and normal parts move in opposition, the centre of mass of any given volume element remaining at rest to a first approximation, and the total mass flux being zero. Such a wave is evidently peculiar to superfluids.

There is another important difference between the two types of wave, which is seen from formulae (131.10) and (131.11). In a sound wave of the ordinary type, the amplitude of the pressure oscillations is relatively large, while that of the temperature oscillations is small. In a second-sound wave, however, the relative amplitude of the temperature oscillations is large compared with that of the pressure oscillations. In this sense we can say that second-sound waves are undamped temperature waves.†

In an approximation in which the thermal expansion is neglected, second-sound waves are purely temperature oscillations (with $\mathbf{j} = 0$), while first-sound waves are pressure oscillations (with $\mathbf{v}_s = \mathbf{v}_n$). Accordingly, their equations of motion are completely separable: in equation (131.6), we write $s' = cT'/T$, obtaining $\partial^2 T'/\partial t^2 = u_2{}^2 \triangle T'$, and in equation (131.5) we write $\rho' = p'\partial\rho/\partial p$, obtaining $\partial^2 p'/\partial t^2 = u_1{}^2 \triangle p'$.

The subject of the various methods of exciting sound waves in helium II has been discussed by E. M. LIFSHITZ (1944). It is found (see the Problems) that the usual mechanical means of exciting sound waves (oscillation of solid bodies) is very unsuitable for creating second sound, the intensity of the second sound emitted being negligible compared with that of the ordinary sound. Other methods of exciting sound waves are possible in helium II, however. Such is the emission of sound by solid bodies whose temperature varies periodically; the intensity of the second sound emitted is then large compared with that of the first sound, as we should expect in view of the above-mentioned difference in the nature of the temperature oscillations.

PROBLEMS

PROBLEM 1. Determine the ratio of intensities of the first and second sound emitted by a plane oscillating in a direction perpendicular to itself.

SOLUTION. We seek the velocities v_s (along the x-axis, which is perpendicular to the plane) in the first and second sound waves in the forms

$$v_{s1} = A_1 \cos \omega(t - x/u_1), \qquad v_{s2} = A_2 \cos \omega(t - x/u_2)$$

respectively. At the surface of the oscillating plane, the velocities v_s and v_n must be equal to the velocity of the plane, which we denote by $v_0 \cos \omega t$. This gives the equations $A_1 + A_2 = v_0$, $a_1A_1 + a_2A_2 = v_0$, where the coefficients a_1 and a_2 are given by (131.10) and (131.11). The (time) average energy density in a sound wave in helium II is $\rho_s v_s{}^2 + \rho_n v_n{}^2 = \frac{1}{2}A^2(\rho_s + \rho_n a^2)$; the energy flux (intensity) is obtained by multiplying by the corresponding velocity of sound u. The ratio of the intensities of the second and first sound waves is found to be

$$\frac{I_2}{I_1} = \frac{A_2{}^2(\rho_s + \rho_n a_2{}^2)u_2}{A_1{}^2(\rho_s + \rho_n a_1{}^2)u_1} \approx \frac{\beta^2 T u_2{}^3}{c u_1}.$$

---

† They have, of course, no connection with the damped temperature waves in an ordinary thermally conducting medium (§52).

Here we have assumed that $u_2 \ll u_1$, which is valid down to very low temperatures. The ratio is always small.

PROBLEM 2. The same as Problem 1, but for a surface whose temperature varies periodically.

SOLUTION. It is sufficient to use the boundary condition $j = 0$, which must hold at a fixed surface. This gives $\rho_s(A_1 + A_2) + \rho_n(a_1 A_1 + a_2 A_2) = 0$, whence

$$|A_2/A_1| = (\rho_n a_1 + \rho_s)/(\rho_n a_2 + \rho_s) \approx s/\beta u_2^2.$$

The ratio of intensities is found to be $I_2/I_1 = c/T\beta^2 u_1 u_2$. This is very large.

PROBLEM 3. Determine the velocity of points on the profile of a one-dimensional travelling second-sound wave of large amplitude, and the velocity of propagation of the discontinuities which occur in the wave as a result of the deformation of the profile (I. M. KHALATNIKOV 1952).

SOLUTION. In a one-dimensional travelling wave, all quantities (ρ, p, T, v_s, v_n) can be expressed as functions of one parameter, which may be one of these quantities (cf. §94). The velocity U of points on the wave profile is equal to the derivative dx/dt taken for a given value of the parameter. The space and time derivatives of each quantity are related by $\partial/\partial t = -U \partial/\partial x$. The derivatives of quantities with respect to the parameter will be denoted by adding primes.

Instead of the velocities v_s and v_n, it is convenient to use $v = j/\rho$ and $w = v_n - v_s$, and we take a co-ordinate system in which the velocity v is zero at the point considered. The equations (130.3)–(130.6), with Π_{ik}, μ, ρ, s given by (130.11), (130.13)–(130.15), lead to the equations

$$-U\frac{\partial\rho}{\partial p}p' - U\rho^2\frac{\partial}{\partial p}\left(\frac{\rho_n}{\rho}\right)ww' + \rho v' = 0, \tag{1}$$

$$p' + 2\rho_s\rho_n\, ww'/\rho - U\rho v' = 0, \tag{2}$$

$$\left[-\rho U\frac{\partial s}{\partial T} + w\frac{\partial}{\partial T}(s\rho_s)\right]T' + sw\frac{\partial\rho_s}{\partial p}p' + \left[\rho_s s - Uw\frac{\partial\rho_n}{\partial T}\right]w' = 0, \tag{3}$$

$$\left[-\rho s + Uw\frac{\partial\rho_n}{\partial T}\right]T' + \left[1 + Uwp\frac{\partial}{\partial p}\left(\frac{\rho_n}{\rho}\right)\right]p' +$$

$$+ \left[\rho_n U - \frac{\rho_n\rho_s}{\rho}w\right]w' - [U\rho + w\rho_n]v' = 0. \tag{4}$$

Here all terms above the second order of smallness have been omitted, and so have all terms containing the thermal-expansion coefficient.

In a second-sound wave, the relative amplitude of the oscillations of p and v is small compared with that of T and w; we can therefore also omit terms in wp' and wv'. To determine U, it is sufficient to take equation (3) and the difference of equations (2) and (4); the condition of compatibility of the two linear equations obtained for T' and w' gives the quadratic equation

$$\rho_n U^2\frac{\partial s}{\partial T} - Uw\left[\frac{4\rho_s\rho_n}{\rho}\frac{\partial s}{\partial T} - 2s\frac{\partial\rho_n}{\partial T}\right] - \rho_s s^2 = 0$$

whence

$$U = u_2 + w\left(\frac{2\rho_s}{\rho} - \frac{sT}{\rho_n c}\frac{\partial\rho_n}{\partial T}\right).$$

Here u_2 is the local velocity of second sound, which varies over the wave profile together with the deviation δT of the temperature from its equilibrium value. Expanding u_2 in powers of δT, we obtain

$$u_2 = u_{20} + \frac{\partial u_2}{\partial T}\delta T = u_{20} + \frac{\partial u_2}{\partial T}\frac{\rho_n u_2}{\rho s}w,$$

where u_{20} is the equilibrium value of u_2. We have finally

$$U = u_{20} + w\frac{\rho_s s T}{\rho c}\frac{\partial}{\partial T}\log\frac{u_{20}^3 c}{T}. \tag{5}$$

When the wave profile changes sufficiently, discontinuities (in this case, discontinuities of temperature) occur in it; cf. §§94, 95. The velocity of propagation of the discontinuity is half the sum of the velocities U on the two sides of the discontinuity, i.e. it is

$$c_{20} + \frac{w_1 + w_2}{2}\frac{\rho_s s T}{\rho c}\frac{\partial}{\partial T}\log\frac{u_{20}^3 c}{T},$$

where, w_1, w_2 are the values of w on the two sides of the discontinuity.

The coefficient of w in (5) may be either positive or negative. Correspondingly, the points with large values of w may be either ahead of or behind those with small w, and the discontinuity may be formed at either the forward or backward front of the wave (whereas, an ordinary sound wave, the shock wave always appears at the forward front).

CHAPTER XVII

FLUCTUATIONS IN FLUID DYNAMICS

§132. The general theory of fluctuations in fluid dynamics

THE calculation of the root-mean-square fluctuations of density, temperature, velocity etc. at each point in a fluid at rest requires no special discussion: these fluctuations (in the classical, i.e. non-quantum, case) are given by the usual formulae of thermodynamics, which are valid for fluctuations in any medium in thermal equilibrium.†

A problem peculiar to fluid dynamics, however, is that of the time correlations in the fluctuations of these quantities, and so is that of fluctuations in a fluid in motion. The solution of these problems must include an allowance for dissipative processes (viscosity and thermal conduction) in the fluid.

The construction of the general theory of fluctuations in fluid dynamics amounts to setting up the "equations of motion" for the fluctuating quantities. This can be done by introducing the appropriate additional terms in the general equations of fluid dynamics.

The equations of fluid dynamics in the form

$$\partial \rho / \partial t + \operatorname{div}(\rho \mathbf{v}) = 0, \tag{132.1}$$

$$\rho \frac{\partial v_i}{\partial t} = -\frac{\partial p}{\partial x_i} + \frac{\partial \sigma'_{ik}}{\partial x_k}, \tag{132.2}$$

$$\rho T \left(\frac{\partial s}{\partial t} + \mathbf{v} \cdot \mathbf{grad}\, s \right) = \tfrac{1}{2} \sigma'_{ik} \left(\frac{\partial v_i}{\partial x_k} + \frac{\partial v_k}{\partial x_i} \right) - \operatorname{div} \mathbf{q}, \tag{132.3}$$

without any specific form of the stress tensor σ'_{ik} or the heat flux vector $\mathbf{q}$, simply express the conservation of mass, momentum and energy in the moving fluid. They are therefore valid in this form for any motion, including fluctuating changes in the state of the fluid. In that case ρ, p, $\mathbf{v}$, etc. must be understood as the sums of the values of the corresponding quantities in the main motion of the fluid and their fluctuations; the equations can, of course, always be linearised with respect to the latter.

The general expressions (15.3) for the stress tensor and (49.1) for the heat flux relate these quantities to the velocity and temperature gradients respectively. In the presence of fluctuations, however, there are also spontaneous local stresses and heat fluxes in the fluid, which are not related to the velocity and temperature gradients; we denote these by s_{ik} and $\mathbf{g}$, and call them

---

† See *Statistical Physics*, §111, Pergamon Press, London 1958. Sections and formulae in this book will be referred to by means of the prefix SP.

random quantities.† Thus we can write‡

$$\sigma'_{ik} = \eta\left(\frac{\partial v_i}{\partial x_k} + \frac{\partial v_k}{\partial x_i} - \tfrac{2}{3}\delta_{ik}\frac{\partial v_l}{\partial x_l}\right) + \zeta\frac{\partial v_l}{\partial x_l}\delta_{ik} + s_{ik},$$

$$\tag{132.4}$$

$$\mathbf{q} = -\kappa\,\mathbf{grad}\,T + \mathbf{g}.$$

The problem is now to establish certain properties of s_{ik} and $\mathbf{g}$, viz. their mean squares and the correlation between their values at various points in the fluid at various instants. This can be done by means of the general formulae of fluctuation theory. For simplicity, we shall give the argument for the case of non-quantised fluctuations which usually occurs in fluid mechanics,†† and suppose that the coefficients of viscosity and thermal conduction are independent of the frequency of the fluctuations, i.e. do not exhibit dispersion.

In the general theory of fluctuations (SP §§110, 121) we have discussed a discrete series of fluctuating quantities x_1, x_2, ..., whereas here we have a continuous series (the values of ρ, p, $\mathbf{v}$, ... at every point in the fluid). We shall evade this unimportant difficulty in a purely formal manner, by dividing the volume of the fluid into small but finite portions ΔV and taking some mean values of the quantities in each portion; the passage to infinitesimal portions will be made in the final formulae.

We shall take formulae (132.4) as the equations

$$\dot{x}_a = -\sum_b \gamma_{ab}X_b + y_a \tag{132.5}$$

of the general theory (see SP (121.9)), the quantities $\dot{x}_a$ being the components of the tensor σ'_{ik} and of the vector $\mathbf{q}$ in each portion ΔV:

$$\dot{x}_a \to \sigma'_{ik}, \quad q_i. \tag{132.6}$$

The random quantities s_{ik} and $\mathbf{g}$ are then the corresponding quantities y_a:

$$y_a \to s_{ik}, \quad g_i. \tag{132.7}$$

The meaning of X_a is found, according to the general rules, by using the formula for the rate of change of the total entropy S of the fluid. As in §49, we find from equation (132.3) that

$$\dot{S} = \int\left\{\frac{\sigma'_{ik}}{2T}\left(\frac{\partial v_i}{\partial x_k} + \frac{\partial v_k}{\partial x_i}\right) - \frac{\mathbf{q}\cdot\mathbf{grad}\,T}{T^2}\right\}dV$$

(for $s_{ik} = 0$, $\mathbf{g} = 0$ this gives (49.6)) or, replacing the integral by a sum over

---

† In correspondence with the term *random forces* in the general theory of fluctuations.

‡ In this chapter we shall understand by T the temperature measured in energy units, i.e. Boltzmann's constant will be omitted.

†† This means that the frequencies ω occurring in the fluctuations are assumed such that $\hbar\omega \ll T$; see SP (109.2).

the portions ΔV,

$$\dot{S} = \sum \left\{ \frac{\sigma'_{ik}}{2T} \left(\frac{\partial v_i}{\partial x_k} + \frac{\partial v_k}{\partial x_i} \right) - \frac{\mathbf{q} \cdot \mathbf{grad}\, T}{T^2} \right\} \Delta V. \tag{132.8}$$

We must also have $\dot{S} = -\Sigma X_a \dot{x}_a$; see (58.4). Substituting (132.6) and comparing with (132.8), we find that the corresponding quantities X_a are

$$X_a \to -\frac{1}{2T} \left(\frac{\partial v_i}{\partial x_k} + \frac{\partial v_k}{\partial x_i} \right) \Delta V, \qquad \frac{1}{T^2} \frac{\partial T}{\partial x_i} \Delta V. \tag{132.9}$$

It is now easy to find the coefficients γ_{ab}, which give at once the required correlations by

$$\overline{y_a(t_1) y_b(t_2)} = (\gamma_{ab} + \gamma_{ba}) \delta(t_1 - t_2); \tag{132.10}$$

see SP (121.10a). The averaging is here taken in the usual statistical sense, i.e. we average with respect to the probabilities of all values which the quantities can have at the instants t_1 and t_2; it can also be written as an average with respect to t_1 or t_2 for a given difference $t_1 - t_2$.

We note first of all that formulae (132.4) contain no terms which would relate σ'_{ik} to the temperature gradient or $\mathbf{q}$ to the velocity gradient. This means that the corresponding coefficients γ_{ab} are zero, and by (132.10) we have

$$\overline{s_{ik}(\mathbf{r}_1, t_1) g_l(\mathbf{r}_2, t_2)} = 0, \tag{132.11}$$

i.e. the values of s_{ik} and g_l are entirely uncorrelated ($\mathbf{r}_1$ and $\mathbf{r}_2$ being the co-ordinates of two points in the fluid).

Next, the coefficients relating q_i to the values of $(1/T^2)(\partial T/\partial x_k)\Delta V$ are zero if these two quantities are taken in different volumes ΔV, and $\kappa T^2 \delta_{ik}/\Delta V$ if the same volume is involved. Hence we have

$$\overline{g_i(\mathbf{r}_1, t_1) g_k(\mathbf{r}_2, t_2)} = 0 \qquad \text{if} \quad \mathbf{r}_1 \neq \mathbf{r}_2,$$

$$\overline{g_i(\mathbf{r}, t_1) g_k(\mathbf{r}, t_2)} = \frac{2\kappa T^2}{\Delta V} \delta_{ik}\, \delta(t_1 - t_2).$$

Passing now to the limit $\Delta V \to 0$, we can evidently write both these formulae together as

$$\overline{g_i(\mathbf{r}_1, t_1) g_k(\mathbf{r}_2, t_2)} = 2\kappa T^2 \delta_{ik}\, \delta(t_1 - t_2) \delta(\mathbf{r}_1 - \mathbf{r}_2). \tag{132.12}$$

Finally, we can similarly obtain formulae for the correlation between the components of the random stress tensor:

$$\overline{s_{ik}(\mathbf{r}_1, t_1) s_{lm}(\mathbf{r}_2, t_2)}$$
$$= 2T[\eta(\delta_{il}\delta_{km} + \delta_{im}\delta_{kl}) + (\zeta - \tfrac{2}{3}\eta)\delta_{ik}\delta_{lm}]\delta(\mathbf{r}_1 - \mathbf{r}_2)\delta(t_1 - t_2). \tag{132.13}$$

Formulae (132.11)–(132.13) give the solution of the problem (L. D. LANDAU and E. M. LIFSHITZ 1957).†

These formulae can be rewritten in terms of the Fourier time components of the quantities concerned. The Fourier component of the fluctuating quantity $x_a(t)$ is defined as

$$x_{a\omega} = (1/2\pi) \int_{-\infty}^{\infty} x_a(t)e^{i\omega t}\,dt, \qquad x_a(t) = \int_{-\infty}^{\infty} x_{a\omega}e^{-i\omega t}\,d\omega; \qquad (132.14)$$

see SP §118. For the components $y_{a\omega}$ and $y_{b\omega}$ thus defined we have

$$\overline{y_{a\omega}y_{b\omega'}} = (1/2\pi)(\gamma_{ab}+\gamma_{ba})\delta(\omega+\omega');$$

see SP (121.10). Thus, replacing $\delta(t_1-t_2)$ by $(1/2\pi)\delta(\omega+\omega')$, we obtain instead of (132.11)–(132.13)

$$\overline{s_{ik\omega}(\mathbf{r}_1)g_{l\omega'}(\mathbf{r}_2)} = 0, \qquad (132.15)$$

$$\overline{g_{i\omega}(\mathbf{r}_1)g_{k\omega'}(\mathbf{r}_2)} = \frac{\kappa T^2}{\pi}\delta_{ik}\delta(\mathbf{r}_1-\mathbf{r}_2)\delta(\omega+\omega'), \qquad (132.16)$$

$$\overline{s_{ik\omega}(\mathbf{r}_1)s_{lm\omega'}(\mathbf{r}_2)}$$
$$= (T/\pi)[\eta(\delta_{il}\delta_{km}+\delta_{im}\delta_{kl})+(\zeta-\tfrac{2}{3}\eta)\delta_{ik}\delta_{lm}]\delta(\mathbf{r}_1-\mathbf{r}_2)\delta(\omega+\omega').$$
$$(132.17)$$

The generalisation of these formulae to the case of quantised fluctuations is made by simply including a factor $(\hbar\omega/2T)\coth(\hbar\omega/2T)$ on the right-hand sides of formulae (132.15)–(132.17); see SP (124.21). When the viscosity and thermal conductivity exhibit dispersion, η, ζ, κ are complex functions of the frequency. The corresponding generalisation of formulae (132.15)–(132.17) is made by replacing η, ζ, κ by the real parts of the functions $\eta(\omega)$, $\zeta(\omega)$, $\kappa(\omega)$, as is easily demonstrated.

§133. Fluctuations in an infinite medium

The formulae obtained above give in principle the fluctuations in any particular case. The problem is solved as follows. Regarding s_{ik} and $\mathbf{g}$ as known functions of co-ordinates and time, we formally solve equations (132.1)–(132.3) for $\mathbf{v}$, ρ, ..., taking into account the appropriate boundary conditions of fluid mechanics. We thus obtain $\mathbf{v}$, ρ, ... as linear functionals of s_{ik} and $\mathbf{g}$. Accordingly, any quantity quadratic (or bilinear) in $\mathbf{v}$, ρ,... can be expressed in terms of quadratic functionals of s_{ik} and $\mathbf{g}$, and the

---

† To convert to the ordinary units of temperature measurement (degrees) we must replace T by kT and κ by κ/k, where k is Boltzmann's constant.

mean values are calculated from formulae (132.11)–(132.13); the auxiliary quantities s_{ik} and **g** do not appear in the final result.

As an illustration of the method just described, let us consider the fluctuations of pressure in an infinite medium at rest, having a large second viscosity $\zeta(\omega)$ which exhibits dispersion; the effects of ordinary viscosity and thermal conduction are assumed negligible in comparison (as may happen in the conditions described in §78).

The solution of equations (132.1)–(132.3) can be effected by expanding all quantities (already Fourier-expanded with respect to time) as space Fourier integrals; for any quantity $f(\mathbf{r})$ we put

$$f(\mathbf{r}) = \int_{-\infty}^{\infty} f_{\mathbf{k}} \exp(i\mathbf{k}\cdot\mathbf{r})\, d\mathbf{k}, \qquad d\mathbf{k} = dk_x dk_y dk_z, \tag{133.1}$$

where

$$f_{\mathbf{k}} = \frac{1}{(2\pi)^3} \int f(\mathbf{r}) \exp(-i\mathbf{k}\cdot\mathbf{r})\, dV. \tag{133.2}$$

The correlation functions for the Fourier components are found at once from those for the quantities themselves. For instance, if

$$\overline{f(\mathbf{r}_1)f(\mathbf{r}_2)} = A\delta(\mathbf{r}_1 - \mathbf{r}_2), \tag{133.3}$$

then

$$\overline{f_{\mathbf{k}}f_{\mathbf{k}'}} = \frac{1}{(2\pi)^6} \int\!\!\int \overline{f(\mathbf{r}_1)f(\mathbf{r}_2)} \exp[-i(\mathbf{k}\cdot\mathbf{r}_1 + \mathbf{k}'\cdot\mathbf{r}_2)]\, dV_1\, dV_2$$

$$= \frac{A}{(2\pi)^6} \int\!\!\int \delta(\mathbf{r}_1 - \mathbf{r}_2) \exp[-i(\mathbf{k}\cdot\mathbf{r}_1 + \mathbf{k}'\cdot\mathbf{r}_2)]\, dV_1\, dV_2$$

$$= \frac{A}{(2\pi)^6} \int \exp[-i(\mathbf{k}+\mathbf{k}')\cdot\mathbf{r}]\, dV,$$

or, finally,

$$\overline{f_{\mathbf{k}}f_{\mathbf{k}'}} = \frac{A}{(2\pi)^3}\delta(\mathbf{k}+\mathbf{k}'). \tag{133.4}$$

In particular, instead of formula (132.17), in which we retain (in this case) only the term in ζ, we have

$$\overline{(s_{ik})_{\omega\mathbf{k}}(s_{lm})_{\omega'\mathbf{k}'}} = \frac{T}{8\pi^4}\, \mathrm{re}\, \zeta(\omega)\,.\,\delta(\omega+\omega')\delta(\mathbf{k}+\mathbf{k}'). \tag{133.5}$$

Under the above conditions (small thermal conductivity) the effect of the change in entropy on the change in pressure is relatively small; the pressure fluctuations may therefore be regarded as adiabatic. Accordingly, we write the linearised equation (132.1) as

$$\frac{1}{c_0{}^2}\frac{\partial \delta p}{\partial t} + \rho \operatorname{div}\delta \mathbf{v} = 0,$$

where δp, $\delta \mathbf{v}$ are the fluctuations in pressure and velocity, ρ is the equilibrium density, and c_0 the velocity of sound calculated from the equilibrium equation of state (§78). Taking the time and space Fourier components, we have

$$-\frac{\omega}{c_0{}^2}\delta p_{\omega\mathbf{k}} + \rho \mathbf{k}\cdot \delta \mathbf{v}_{\omega\mathbf{k}} = 0. \tag{133.6}$$

Similarly, equation (132.2) becomes

$$-\omega\rho(\delta v_i)_{\omega\mathbf{k}} = -k_i\delta p_{\omega\mathbf{k}} + i\zeta(\omega)k_i\mathbf{k}\cdot \delta \mathbf{v}_{\omega\mathbf{k}} + k_k s_{ik}, \tag{133.7}$$

where $\zeta(\omega)$ may, for instance, be defined by formula (78.6).

To find $\delta p_{\omega\mathbf{k}}$, we multiply equation (133.7) scalarly by $\mathbf{k}$:

$$-\omega\rho\,\mathbf{k}\cdot \delta \mathbf{v}_{\omega\mathbf{k}} = -k^2\delta p_{\omega\mathbf{k}} + i\zeta k^2(\mathbf{k}\cdot \delta \mathbf{v}_{\omega\mathbf{k}}) + k_i k_k s_{ik},$$

and eliminate $\mathbf{k}\cdot \delta \mathbf{v}_{\omega\mathbf{k}}$ by means of (133.6). The result is

$$\delta p_{\omega\mathbf{k}} = k_i k_k s_{ik}\Big/\left(k^2 - \frac{\omega^2}{c_0{}^2} - i\zeta k^2\frac{\omega}{\rho c_0{}^2}\right).$$

Finally, taking the product $\delta p_{\omega\mathbf{k}}\delta p_{\omega'\mathbf{k}'}$ and averaging with the aid of formula (133.6), we have

$$\overline{\delta p_{\omega\mathbf{k}}\delta p_{\omega'\mathbf{k}'}} = \frac{Tc_0{}^4}{8\pi^4}\frac{\operatorname{re}\zeta(\omega)\,.\,\delta(\omega+\omega')\delta(\mathbf{k}+\mathbf{k}')}{[c_0{}^2-(\omega^2/k^2)-i\omega(\zeta/\rho)][c_0{}^2-(\omega^2/k^2)+i\omega(\zeta^*/\rho)]}. \tag{133.8}$$

For example, if we substitute $\zeta(\omega)$ from (78.6), we obtain

$$\overline{\delta p_{\omega\mathbf{k}}\delta p_{\omega'\mathbf{k}'}} = \frac{T}{8\pi^4}\frac{\tau\rho c_0{}^4(c_\infty{}^2-c_0{}^2)\delta(\omega+\omega')\delta(\mathbf{k}+\mathbf{k}')}{[c_0{}^2-(\omega^2/k^2)]^2+\omega^2\tau^2[c_\infty{}^2-(\omega^2/k^2)]^2}. \tag{133.9}$$

To conclude, we shall show how the usual (classical) formula for the mean square pressure fluctuation at a given point in the fluid can be derived from (133.8). To simplify the calculations we shall assume that ζ is independent of ω.†

Returning by formulae (132.14) and (133.2) from the Fourier expansions to the functions $\delta p(\mathbf{r}_1, t_1)$, $\delta p(\mathbf{r}_2, t_2)$ themselves, and taking $t_1 = t_2 \equiv t$, we

---

† The (classical) result is actually independent of the form of the function $\zeta(\omega)$; the only properties of this function which are used in the derivation are ones which it must have in any case (see SP §122).

can write

$$\overline{\delta p(\mathbf{r}_1, t)\delta p(\mathbf{r}_2, t)}$$

$$= \int_{-\infty}^{\infty}\int_{-\infty}^{\infty}\int_{-\infty}^{\infty}\int_{-\infty}^{\infty} \overline{\delta p_{\omega\mathbf{k}}\delta p_{\omega'\mathbf{k}'}}\exp[-i(\omega+\omega')t]\,\exp[i(\mathbf{k}\cdot\mathbf{r}_1+\mathbf{k}'\cdot\mathbf{r}_2)] \times$$

$$\mathrm{d}\omega\,\mathrm{d}\omega'\,\mathrm{d}\mathbf{k}\,\mathrm{d}\mathbf{k}',$$

where we must substitute (133.8). One integration over frequency results in replacing $\omega+\omega'$ by zero in the integrand, owing to the factor $\delta(\omega+\omega')$. A second integration gives†

$$\int_{-\infty}^{\infty} \frac{(c_0{}^2\zeta/\rho)\mathrm{d}\omega}{[c_0{}^2-(\omega^2/k^2)-(i\zeta\omega/\rho)][c_0{}^2-(\omega^2/k^2)+(i\zeta\omega/\rho)]} = \pi.$$

Thus

$$\overline{\delta p(\mathbf{r}_1, t)\delta p(\mathbf{r}_2, t)} = \frac{T\rho c_0{}^2}{(2\pi)^3}\int\int \delta(\mathbf{k}+\mathbf{k}')\exp[i(\mathbf{k}\cdot\mathbf{r}_1+\mathbf{k}'\cdot\mathbf{r}_2)]\,\mathrm{d}\mathbf{k}\,\mathrm{d}\mathbf{k}'$$

$$= \frac{T\rho c_0{}^2}{(2\pi)^3}\int \exp[i\mathbf{k}\cdot(\mathbf{r}_1-\mathbf{r}_2)]\,\mathrm{d}\mathbf{k}$$

$$= \rho T c_0{}^2\delta(\mathbf{r}_2-\mathbf{r}_1). \tag{133.10}$$

This is the required result: averaging (133.10) over a small volume ΔV, we obtain

$$\overline{(\delta p)^2} = \rho T c_0{}^2/\Delta V,$$

in agreement with a formula of thermodynamics, SP (111.11).

---

† The integration is easily effected, fori nstance, by means of the calculus of residues, taking the path of integration as a large semicircle in the plane of the complex variable ω.

INDEX

Activation energy, 474n.
Adiabatic flow, 4
Adsorbed films, effect of, on motion of liquid, 241ff.
Amplitude, complex, 248
Angle
  ideal potential flow near, 27
  supersonic flow round, 413ff.
  turbulent flow near, 130ff., 156
Antinodes, 264
Attack, angle of, 173

Barodiffusion coefficient, 225ff.
Bernoulli's equation
  for ideal fluid, 10
  for incompressible flow, 20f.
  for potential flow, 19f.
  relativistic, 503f.
Boundary layers, 18, 130, (IV) 145ff.
  heat transfer in, 205ff.
  laminar, 145ff.
    in compressible fluid, 150n.
    in converging channel, 150
    in a pipe, 150
    on a plate, 147ff.
    stability of flow in, 156ff.
  logarithmic velocity profile in, 159ff.
  PRANDTL's theory of, 79n., 145ff.
  separation of, 151
  in intersection of shock wave with surface, 410
  turbulent, 166ff.
    on a plate, 166f.
Bow wave, 457
Brownian motion, 227ff.

Capillary constant, 232
Chaplygin's equation, 432
  for perfect gas, 433
  for transonic flow, 437
Characteristic lines, 312
Characteristics, 312, 378ff.
  leaving a point, 312
  reaching a point, 312
  in steady two-dimensional flow, 312f., 433ff., 442ff.
Characteristic surface, 311
Circulation, velocity, 14
  law of conservation of, 15

Combustion
  detonation, 480ff.
  fluid dynamics of, (XIV) 474ff.
  on liquid, 479
  slow, 474ff.
  turbulent, 478, 496
  zone, 475
Compressible flow, 245ff.; see Gas flow
Condensation adiabatic, 496f.
Condensation discontinuities, 496ff.
Conduction, thermal
  equation of, 189
    boundary conditions on, 190
  in finite media, 196ff.
  in fluids, (V) 183ff.
  in incompressible fluids, 188ff.
  in infinite media, 192ff.
  pure, 224
  relaxation time for, 194
  turbulent, 205
Conductivity
  thermal, 184
  thermometric, 189
  turbulent, 205
Cone, gas flow past, 418ff.
Conical flows, 421
Contact, angles of, 233
Continuity, equation of, 2; see Equation of continuity
Convection, 8, 202
  in pipe, 214f., 217f.
  between parallel planes, 214, 216f.
  on vertical plate, 215
  condition for absence of, 9
  free, 212ff.
Converging channel, boundary layer in, 150
  flow in, 81ff.
Critical velocity, 315
  in superfluids, 511
Cross-section, effective, 296
  for absorption of sound by sphere, 304
  for scattering of sound
    by drop, 297
    by sphere, 297f.
  total, 296
Cylinders, flow between rotating, 60f.
  stability of, 107ff.

d'Alembert's paradox, 34, 322
Damping
  coefficient, 99, 243f.
  of gravity waves, 98ff.

530